Single Variable
CALCULUS

Mansik Min · Sangmin Hwang · Jungbin Yoon · Sangwoo Jeon

자유아카데미

Introduction

We wrote this calculus textbook for students seriously considering a career in STEM: Science, Technology, Engineering, and Mathematics. Having taught calculus for over thirty years at universities, we have gained invaluable insights into making the material more accessible and less daunting through the use of simple, direct language. Thus, even for students with weak mathematical foundations or for those who are approaching calculus for the very first time, we believe that this textbook can facilitate their comprehension. Numerous examples and problem sets were designed so that the major concepts of calculus could be understood, then solved, and finally applied.

The contents of this book are as follows. Chapter 1 introduces the definitions of functions and deals with inverse and composite functions. Specifically, exponential functions, logarithmic functions, and trigonometric functions—all crucial parts in science and engineering—are thoroughly examined. Chapter 2 deals with the limits of functions and the concept of continuity. Special attention is paid to the limits of transcendental functions and the limits at infinity, as well as the mean value theorem through continuity. Chapter 3 introduces derivatives and deals with tangent and normal line equations, as well as various differential calculations using slope. Chapter 4 examines the close relationship between mathematics and its practical applications in real life via local maxima, local minima, and optimization problems. Chapter 5 introduces indefinite integrals and definite integrals, focusing on the relationship between the derivative and the integral through the fundamental theorem of calculus. Chapter 6 introduces the application of definite integrals. Mainly examined are the areas between curves, the volume of revolution, the motion of curves, the length of arcs on a plane, and improper integrals. This chapter expands the concepts of derivatives or integrals from single-variable calculus to solve problems related to maximum or minimum values of multivariate functions or volumes of solids, helping students understand physical phenomena using vector functions. Chapter 7 deals with differential equations.

We received a lot of help from Jungbin Yoon in writing this book. In reviewing the text, examples, and exercises, they offered creative solutions and multiple-choice questions, providing student perspectives that allowed us to anticipate potential difficulties in the material. Among the many examples that come to mind, while calculating the limit of a trigonometric function, instead of obtaining the derivative first, Jungbin Yoon suggested an easier method of obtaining the graph by transforming the given trigonometric function. It was through their diligence that, in the process of substitution integration, they organized the solution themselves to solve a first-order linear differential equation, thereby greatly improving the quality of this book.

In addition, Jungbin Yoon's careful review of the text and scope of this book, correcting formulas as well as English grammar, allowed us to devote our energies to the mathematical concepts and the overarching whole of the project. Finally, we would like to extend our special gratitude to our publisher, Freedom Academy. Any corrections or updates identified after publication will be made available in the Resources section of the Freedom Academy website (www.freeaca.com) for reference. We hope that this textbook can help students develop a better understanding and appreciation of calculus and the benefits it will have in their professional lives.

January 2026

Authors

Contents

01 Functions

02 Limits and Continuity

03 Derivatives

01

Functions

1.1 Functions

1.1.1 Functions and Relations

A relation is a set of ordered pairs (x,y). A function is a relation such that for each value of x there is exactly one value of y. Domain is the set of x value of the ordered pairs. Range is the set of y value of the ordered pairs.

Vertical Line Test : If a vertical line passes through a graph at only one points, then the graph is a function.

Example 1 Determine whether each of the following relations is a function.

(1) $\{(1,3),(2,4),(3,5),(4,6)\}$
(2) $\{(1,3),(3,2),(1.7),(-1,4)\}$
(3) $\{(2,-1),(2,0),(2,3),(2,11)\}$
(4) $\{(7,6),(5,6),(3,6),(-4,6)\}$
(5) $\{(0,0),(1,0),(3,0),(5,0)\}$
(6) $\{(0,0),(0,-2),(0,2),(0,4)\}$

Solution

(1), (4), (5)

Example 2 Determine whether each relation is a function?

(1)
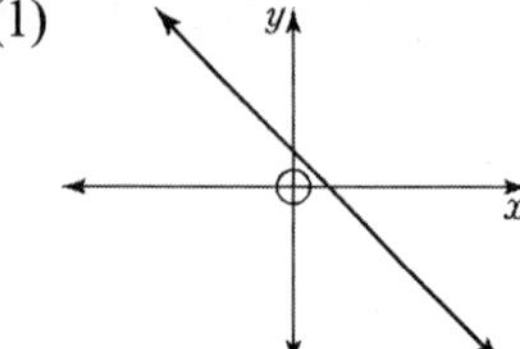

(2)
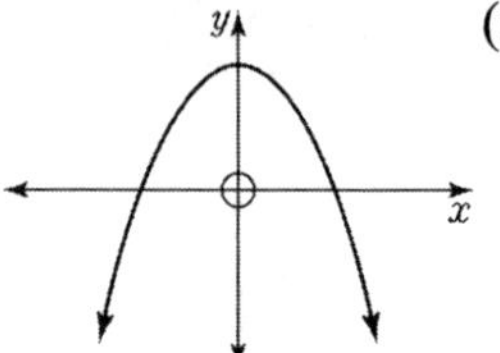

(3)
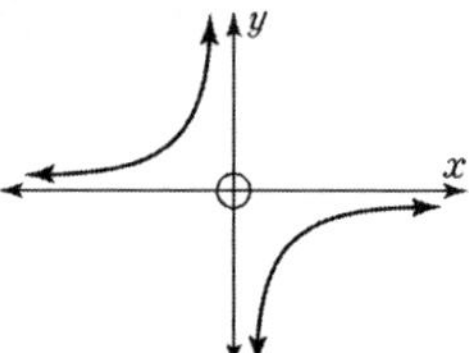

(4)
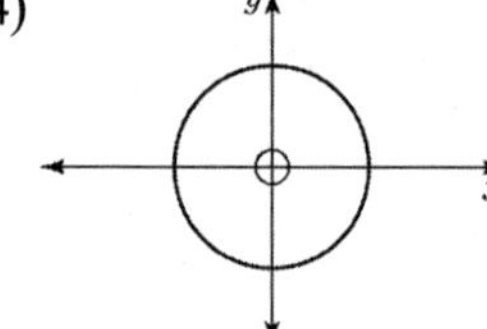

(5)
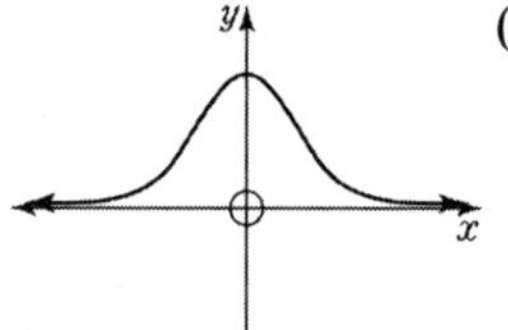

(6)
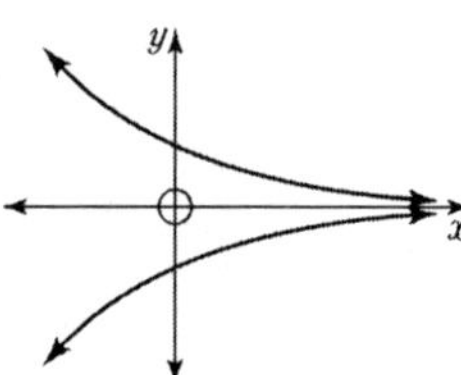

(7)
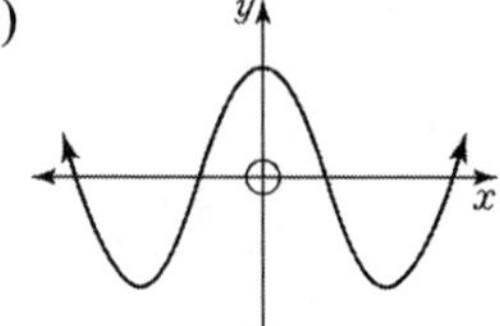

(8)
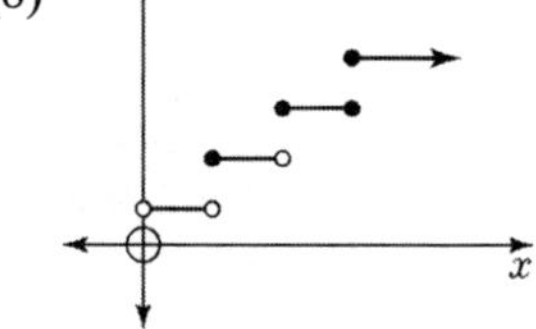

(9)
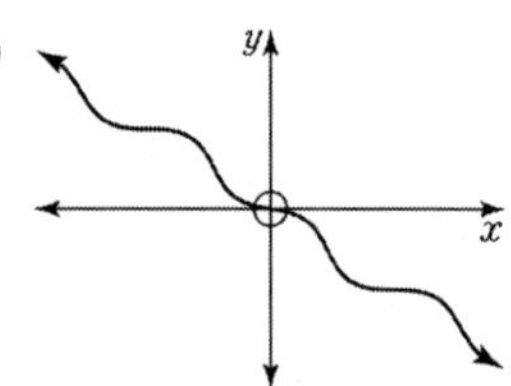

Solution

(1), (2), (3), (5), (7), (9)

Example 3 For each of the following graphs, state the domain and range.

(1)

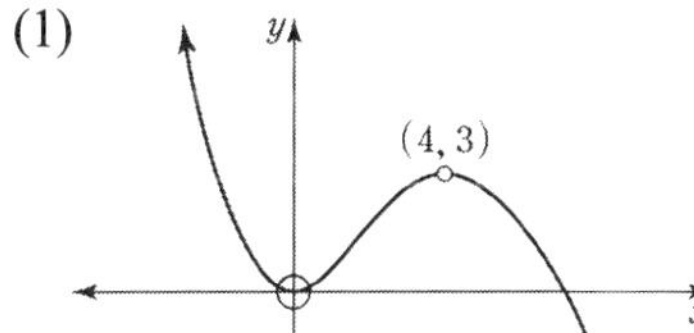

(2)

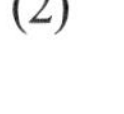

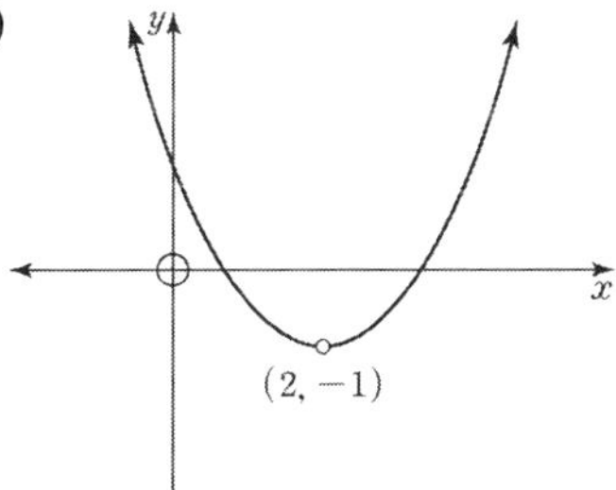

Solution

(1) Domain: $x \leq 8$

Range: $y \geq -2$

(2) Domain: $x \in R$

Range: $y \geq -1$

EX 1 For each of the following graphs, state tne domain and range.

(1)

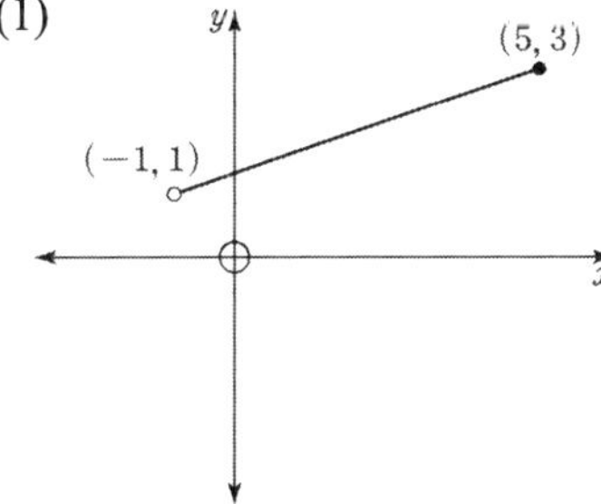

(2)

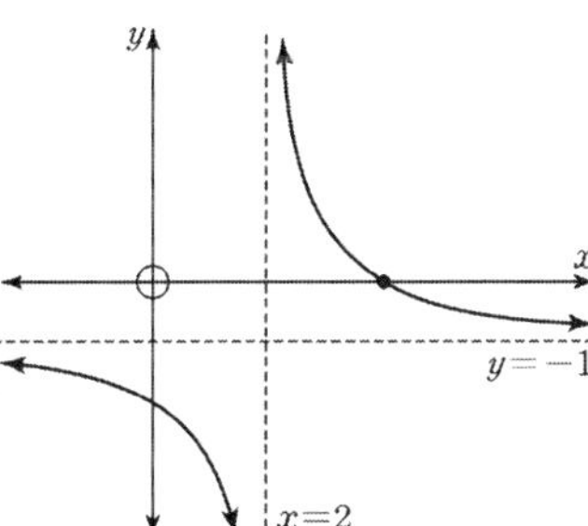

(3)

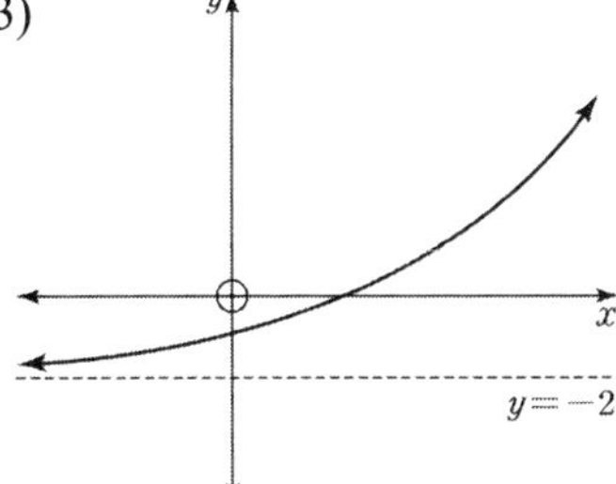

(4)

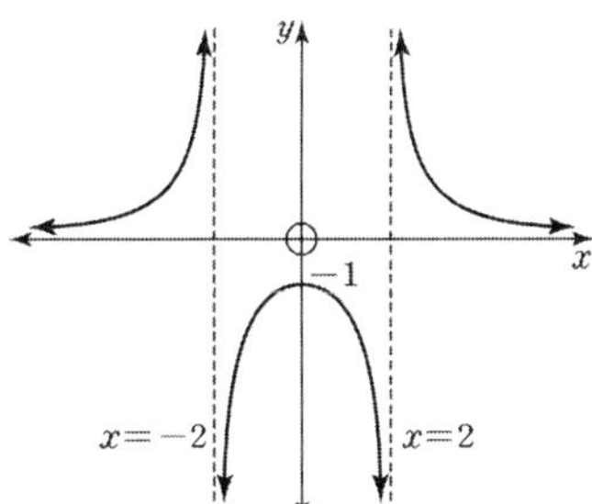

Example 4 State the domain and range of the following functions.

(1) $f(x)=\sqrt{x-5}$ (2) $f(x)=\dfrac{1}{x-5}$ (3) $f(x)=\dfrac{1}{\sqrt{x-5}}$

Solution

(1) Domain: $x-5\geq 5, x\geq 5$ (2) Domain: $x\neq 5$ (3) Domain: $x>5$

Range: $y\geq 0$ Range: $y\neq 0$ Range: $y>0$

EX 2 State the domain and range of the following functions.

(1) $f(x)=\sqrt{x-2}$ (2) $f: x\rightarrow\sqrt{4-x}$

(3) $f: x\rightarrow 5x-3x^2$ (4) $y=\dfrac{x+4}{x-2}$

Example 5 Given $f(x)=2x^2-4$ find each of the following.

(1) $f(-\sqrt{3})$ (2) $2f(x+1)$

(3) $t^2\cdot f\left(\dfrac{1}{t}\right)$ (4) $\dfrac{f(x+h)-f(x)}{h}$

Solution

(1)
$$f(-\sqrt{3})=2(-\sqrt{3})^2-4$$
$$=6-4$$
$$=2$$

(2)
$$2f(x+1)=2[2(x+1)^2-4]$$
$$=2[2(x^2+2x+1)-4]$$
$$=2[2x^2+4x-2]$$
$$=4x^2+8x-4$$

(3)
$$t^2\cdot f\left(\frac{1}{t}\right)=t^2\cdot\left[2\left(\frac{1}{t}\right)^2-4\right]$$
$$=t^2\cdot\left[\frac{2}{t^2}-4\right]$$
$$=2-4t^2$$

(4)
$$\frac{f(x+h)-f(x)}{h}=\frac{2(x+h)^2-4-[2x^2-4]}{h}$$
$$=\frac{2(x^2+2xh+h^2)-4-2x^2+4}{h}$$
$$=\frac{2x^2+4xh+2h^2-4-2x^2+4}{h}$$
$$=\frac{4xh+2h^2}{h}$$
$$=4x+2h$$

EX 3 Given $f(x) = x^2 - 3x + 4,\ \ g(x) = \sqrt{x-2}$, find each of the following.

(1) $f(3)+1$

(2) $f(2x)$

(3) $f(x-2)$

(4) $\dfrac{f(x+2)-f(x)}{2}$

Example 6 Given $f(x+2) = \frac{1}{2}x + 2$, find $f(4)$ and $f(x)$.

Solution

$x=2,$

$f(2+2) = \frac{1}{2}(2)+2$

$\therefore f(4) = 3$

$x+2=u,$

$x=u-2$

$f(x+2) = \frac{1}{2}x+2 \Rightarrow f(u) = \frac{1}{2}(u-2)+2$

$f(u) = \frac{1}{2}u-1+2$

$f(u) = \frac{1}{2}u+1 \quad \therefore f(x) = \frac{1}{2}x+1$

1.1.2 Composition of Functions

Given two functions f and g, the composite function $f \circ g$ is defined by

$$(f \circ g)(x) = f(g(x)) \text{ or } f \circ g : x \rightarrow f(g(x))$$

Where the domain of f include the range of g.

Generally, $f \circ g \neq g \circ f$.

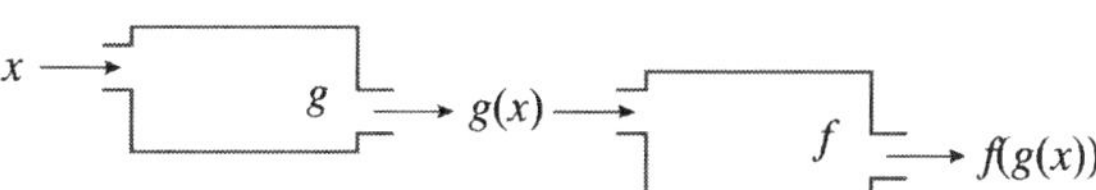

Example 7 Given

$f = \{(0,2),(1,5),(2,7),(3,9)\}$ and

$g = \{(2,2),(5,0),(7,1),(9,3)\}$, Find the set of ordered pairs of $f \circ g$ and $g \circ f$.

(1) $f \circ g$

(2) $g \circ f$

Solution

(1) $f \circ g = \{(2,7),(5,2),(7,5),(9,9)\}$

(2) $g \circ f = \{(0,2),(1,0),(2,1),(3,3)\}$

EX 4 Given $f=\{(0,3),(1,0),(2,1),(3,2)\}$, $g=\{(0,1),(1,2),(2,3),(9,3)\}$, find $f\circ g$.

Example 8 Given $f(x)=x^2+2$ and $g(x)=2x+1$, find each of the following.

(1) $f\circ g$ (2) $g\circ f$

(3) $f\circ g(3)$ (4) $g\circ g(-2)$

Solution

(1) $(f\circ g)(x)=f(g(x))$
$=f(2x+1)$
$=(2x+1)^2+2$
$=4x^2+4x+3$

(2) $(g\circ f)(x)=g(f(x))$
$=g(x^2+2)$
$=2(x^2+2)+1$
$=2x^2+5$

(3) $(f\circ g)(3)=f(g(3))$
$=f(2(3)+1)$
$=f(7)$
$=7^2+2=51$

(4) $(g\circ g)(-2)=g(g(-2))$
$=g(2(-2)+1)$
$=g(-3)$
$=2(-3)+1=-5$

EX 5 Given $f(x)=\dfrac{x}{1-x}$, $g(x)=\dfrac{1}{x}$ find each of the following.

(1) $(g\circ f)(4)$ (2) $(f\circ f)(-2)$

(3) $(f\circ g)(x)$ (4) $(g\circ f)(x)$

1.1.3 Inverse Functions : $y=f^{-1}(x)$

- **The inverse of a function is the set of ordered**

Pairs obtaines by interchanging the x and y value of each pair in the original function.

(1) The graph of $y=f^{-1}(x)$ is the reflection of the graph of $y=f(x)$ over the line $y=x$.

(2) If $y=f(x)$ is one-to-one, $y=f^{-1}(x)$ is also a fucntion.

(3) $f(f^{-1}(x))=f^{-1}(f(x))=x$

(4) $f^{-1}\neq\dfrac{1}{f}$

(5) The domain of f^{-1} is the range of f

(6) The range of f^{-1} is the domain of f

Example 9 Evaluate

(1) Given $f=\{(-5,-1),(-1,0),(1,3),(3,6)\}$, find the set of orderd pairs for f^{-1}.

(2) Draw the arrow diagrams of f and f^{-1}.

(3) Sketch the graph of $y=f(x)$ and $y=f^{-1}(x)$ on the same grid.

Solution

(1) $f^{-1}=\{(-1,\ -5),(0,-1),(3,1),(6,3)\}$

(2)

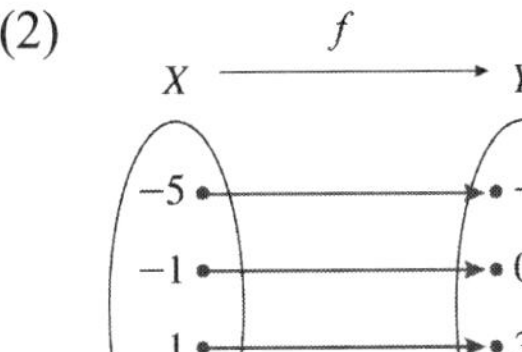

Domain
Range

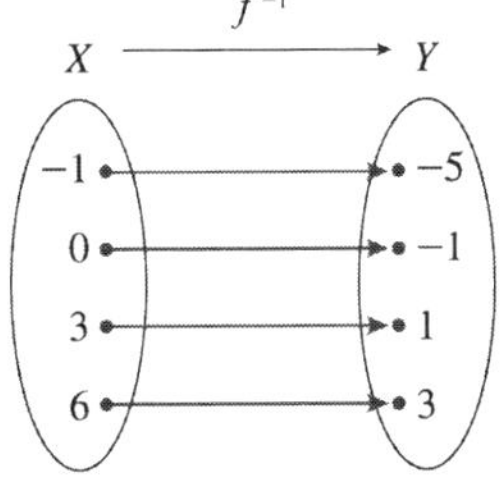

Domain
Range

(3)

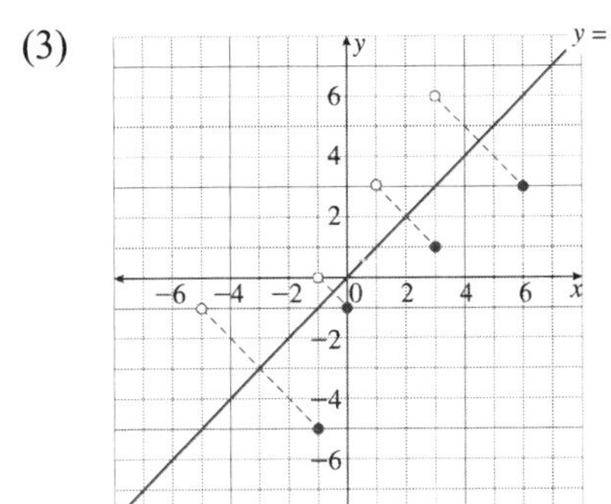

- **Finding the inverse of a function**

Step 1. Replace $f(x)$ with y.

Step 2. Interchange x and y.

Step 3. Solve for y.

Step 4. Replace y with $f^{-1}(x)$.

Example 10 (1) Find the inverse of each of the following functions in the form $f^{-1}(x)$.

(2) State the domain of the inverse, $f^{-1}(x)$

(1) $y=x^2+1\ \ (x\geq 0)$

(2) $f(x)=\sqrt{x-1}+2$

Solution

D : Domain, R: Range.

(1) $D=\{x|x\geq 0\}$, $R=\{y|y\geq 1\}$, $D\rightarrow R$; one-to-one.

$y=x^2+1(x\geq 0,\ y\geq 1)$

$x^2=y-1$

$\therefore\ x=\sqrt{y-1}(y\geq 1,\ x\geq 0)$

$y=\sqrt{x-1}(x\geq 1,\ y\geq 0)$

$f^{-1}(x)=\sqrt{x-1}(x\geq 1)$

(2) $y=\sqrt{x-1}+2$

$D=\{x|x\geq 1\}$, $R=\{y|y\geq 2\}$, $D\rightarrow R$; one-to-one.

$y=\sqrt{x-1}+2(x\geq 1,\ y\geq 2)$

$\sqrt{x-1}=y-2,\ x-1=(y-2)^2$

$\therefore\ x=(y-2)^2+1(y\geq 2,\ x\geq 1)$

$y=(x-2)^2+1(x\geq 2,\ y\geq 1)$

$f^{-1}(x)=(x-2)^2+1(x\geq 2)$

EX 6 Find the inverse of each of the following functions in the form, $f^{-1}(x)$ form.

(1) $f(x)=3x-2$

(2) $y=x^3-5$

(3) $f(x)=2(x-4)^2+1\ (x\geq 4)$

(4) $g(x)=\dfrac{x-2}{2x-3}$

- **Inverse Function Property**

(1) $f(f^{-1}(x))=x$

(2) $f^{-1}(f(x))=x$

(3) $y=f(x)\Leftrightarrow x=f^{-1}(y)$

(4) $f^{-1}(f(a))=a$

Proof

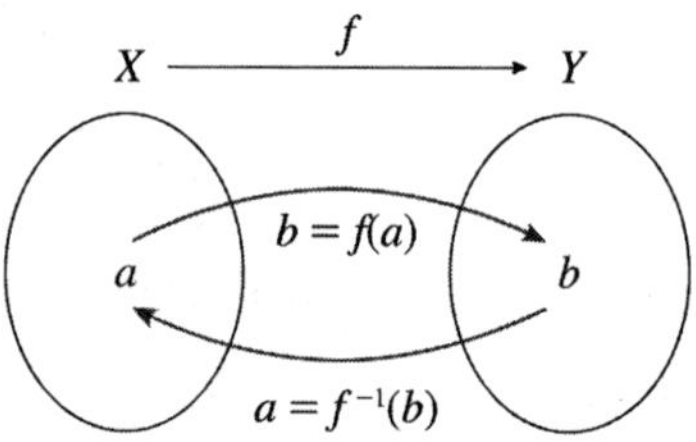

$f(f^{-1}(b))=f(a)=b$

$\therefore f(f^{-1}(b))=b$

$f^{-1}(f(a))=f^{-1}(b)=a$

$\therefore f^{-1}(f(a))=a$

Example 11 Given $f(x) = x^3 + 15$, find the values of $f^{-1}(7)$.

Solution

$f^{-1}(7) = x \Rightarrow 7 = f(x)$

$7 = x^3 + 15$

$x^3 = -8 \qquad \therefore x = -2$

EX 7 Evaluate.

(1) $f(x) = 2x^3 - 11,\ f^{-1}(5)$

(2) $f(x) = \sqrt[3]{x-5}+4,\ f^{-1}(2)$

(3) $f(x) = \dfrac{6}{x+1},\ f(g(x)) = x,\ g(x)$

(4) $f(x) = (x+3)^3 - 2,\ f(g(x)) = x,\ g(x)$

1.1.4 Definition of Slope

The slope m of the line passing through points (x_1, y_1) and (x_2, y_2) is given by

$$m = \frac{y_2 - y_1}{x_2 - x_1},$$

where $x_1 \neq x_2$.

Example 12 If a line passing through the points $(t, 3)$ and (1,8) has a slope of $\frac{5}{3}$, what is the value of t?

Solution

$$m = \frac{8-3}{1-t} = \frac{5}{3} \Rightarrow \frac{5}{1-t} = \frac{5}{3}$$

$\Rightarrow 1 - t = 3$

$\Rightarrow t = -2$

1.1.5 Linear Functions

A function f is a linear function if it can be written in the form $f(x) = \mathrm{mx} + \mathrm{b}$, where m and b are constants.

The Graph of a linear function is a nonvertical straight line with slope m and y-intercept b.

Example 13 The function f is linear. If $f(0)=3$ and $f(-2)=7$, what is the value of $f(3)$?

Solution

$f(x)=mx+b$

$f(0)=m(0)+b=3 \Rightarrow b=3$

$\Rightarrow f(x)=mx+3$

$f(-2)=m(-2)+3=7$

$\Rightarrow -2m=4 \Rightarrow m=-2$

$f(x)=-2x+3$

$f(3)=-2(3)+3=-3$

1.1.6 Equations of Lines

(1) Slope-Intercept Form: $y=mx+b$, where m is the slope and b is the y-intercept

(2) Point-Slope Form: $y-y_1=m(x-x_1)$, where m is the slope and (x_1,y_1) is a point on the line.

(3) Standard Form: $Ax+By=C$, where A, B, and C are integers.

(4) Vertical Line: $x=a$, where a is the x-intercept. the slope is undefined.

(5) Horizontal Line: $y=b$, where b is the y-intercept. The slope is zero.

Example 14 Find an equation of a line if the x-intercept is 2 and the y-intercept is 5.

Solution

If the x-intercept is 2, the line passes through $(2,0)$.

If the y-intercept is 5, the line passes through (0,5).

$$m=\frac{5-0}{0-2}=-\frac{5}{2}$$

$$y=-\frac{5}{2}x+5$$

1.1.7 Parallel and Perpendicular Lines

Two different lines with equations $y=m_1x+b_1$ and $y=m_2x+b_2$ are parallel if $m_1=m_2$(the slopes are equal). They are perpendicular if $m_1 \cdot m_2=-1$(the slopes are negative reciprocals).

Example 15

a. Find an equation of the line which passes through the point $(3,-1)$ and is parallel to $2x-y=10$

b. Find an equation of the line which passes through the point $(2,5)$ and is perpendicular to $x-2y=-6$

Solution

a. First solve the equation of the given line for y.

$$y=2x-10$$

Then the slope of the line parallel to this line is 2.

$$y=2x+b$$
$$-1=2(3)+b$$
$$b=-7$$
$$y=2x-7$$

b. First solve the equation of the given line for y.

$$y=\frac{1}{2}x+3$$

Then the slope of the line perpendicular to this line is -2.

$$y=-2x+b$$
$$5=-2(2)+b$$
$$b=9$$
$$y=-2x+9$$

1.1.8 Quadratic Functions

A quadratic function is function that can be written in the form $f(x)=ax^2+bx+c$ $(a\neq 0)$.

The graph of a quadratic function is a parabola .

The line of symmetry for a quadratic function is $x=-\frac{b}{2a}$, and the vertex is $(-\frac{b}{2a},f(-\frac{b}{2a}))$.

If $a<0$, f has a maximum value at the vertex, that is when $x=-\frac{b}{2a}$.

If $a>0$, f has a minimum value at the vertex, that is when $x=-\frac{b}{2a}$.

• Solutions, Roots, and Zeros of Functions

The solution of a function is the value(s) of x for which $f(x)=0$. Solutions of functions are also called roots or zeros On a graph, the solution of the function is the x-intercept(s).

• Discriminant and Roots of a Quadratic Function

The solutions of the quadratic equation $ax^2+bx+c=0$ are given by the quadratic formula: $x=\dfrac{-b\pm\sqrt{b^2-4ac}}{2a}$. The expression b^2-4ac is called the discriminant.

(1) If the discriminant $b^2-4ac>0$, then there are two real roots, and the graph crosses the x-axis twice.

(2) If the discriminant $b^2-4ac=0$, then there is one real root, and the graph is tangent to the x-axis.

(3) If $b^2-4ac>0$, then there are no real roots, and the graph does not cross the x-axis.

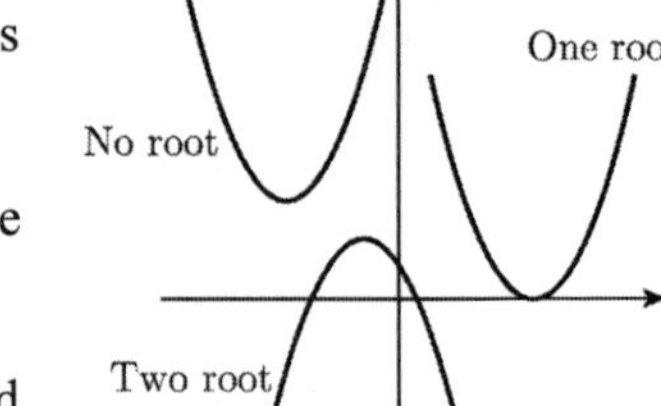

Example 16

a. Find the line of symmetry, minimum value, and the vertex of the graph

$$f(x)=2x^2-4x+5.$$

b. If k is a positive integer and $x^2+4x+k=0$ has two real roots, what is one possible value of k?

Solution

a. The line of symmetry is

$$x=-\frac{b}{2a}=-\frac{-4}{2(2)}=\frac{4}{4}=1$$

The minimum occurs when $x=1$.

$$f(-\frac{b}{2a})=f(1)=2(1)^2-4(1)+5=3$$

The minimum value of f is 3.

The vertex is $(1,3)$.

b. If a quadratic function has two real roots, $b^2-4ac>0$.

$$4^2-4(1)k>0\Rightarrow 16-4k>0\Rightarrow 16>4k\Rightarrow 4>k$$

Since k is a positive integer, the possible values of k are 1, 2 and 3.

1.2 Exponents

1.2.1 Exponents Laws($a \neq 0,\ b \neq 0$)

$a^m \cdot a^n = a^{m+n}$ $\quad \dfrac{a^m}{a^n} = a^{m-n}$ $\quad (a^m)^n = a^{mn}$

$(ab)^n = a^n b^n$ $\quad \left(\dfrac{a}{b}\right)^n = \dfrac{a^n}{b^n}$ $\quad a^0 = 1$

$a^{-n} = \dfrac{1}{a^n}$ $\quad \left(\dfrac{a}{b}\right)^{-n} = \left(\dfrac{b}{a}\right)^n$ $\quad a^{\frac{m}{n}} = (\sqrt[n]{a})^m$

Example 1 Evaluate.

(1) $(-8)^{-\frac{2}{3}}$ (2) $\left(-\dfrac{8}{27}\right)^{\frac{2}{3}}$ (3) $\left(\dfrac{25}{4}\right)^{-1.5}$

Solution

(1) $(-8)^{-\frac{2}{3}} = \dfrac{1}{(-8)^{\frac{2}{3}}} = \dfrac{1}{(\sqrt[3]{-8})^2} = \dfrac{1}{4}$

(2) $(-\dfrac{8}{27})^{\frac{2}{3}} = (\sqrt[3]{-\dfrac{8}{27}})^2 = (-\dfrac{2}{3})^2 = \dfrac{4}{9}$

(3) $(\dfrac{25}{4})^{-1.5} = (\dfrac{25}{4})^{-\frac{3}{2}} = (\dfrac{4}{25})^{\frac{3}{2}} = (\sqrt{\dfrac{4}{25}})^3 = \dfrac{8}{125}$

EX 1 Evaluate.

(1) $64^{\frac{2}{3}}$ (2) $-27^{-\frac{2}{3}}$

(3) $\left(\dfrac{81}{16}\right)^{-\frac{3}{4}}$ (4) $-32^{-0.8}$

Example 2 Simplify using positive exponents.

(1) $(-3a^2b^3)^2(-2a^{-3}b^2)^3$ (2) $\dfrac{-4\sqrt[3]{x^2} \cdot 12\sqrt{x^3}}{32\sqrt[4]{x^3}}$

Solution

(1) $(-3a^2b^3)^2(-2a^{-3}b^2)^3 = 9a^4b^6(-8a^{-9}b^6)$

$= -72a^{-5}b^{12}$

$= -\dfrac{72b^{12}}{a^5}$

(2) $\dfrac{-4\sqrt[3]{x^2} \cdot 12\sqrt{x^3}}{32\sqrt[4]{x^3}} = \dfrac{-4x^{\frac{2}{3}} \cdot 12x^{\frac{3}{2}}}{32x^{\frac{3}{4}}}$

$= -\dfrac{3}{2}x^{\frac{2}{3}+\frac{3}{2}-\frac{3}{4}}$

$= -\dfrac{3}{2}x^{\frac{8+18-9}{12}} = -\dfrac{3}{2}x^{\frac{17}{12}}$

EX 2 Simplify.

(1) $\dfrac{a^3b}{ab^3}$

(2) $\dfrac{y^{-5}}{4x^{-3}}$

(3) $\dfrac{(16a^3b^{-2})(-9a^{-7}b^5)}{(6a^{-2}b^3)^2}$

(4) $(3m^3n^2)^{-2}\left(\dfrac{9m^3n}{mn^3}\right)^2$

1.2.2 Expand and factor of exponents

(1) $a(b+c) = ab + ac$

(2) $(a+b)(c+d) = ac + ad + bc + bd$

(3) $(a+b)(a-b) = a^2 - b^2$

(4) $(a+b)^2 = a^2 + 2ab + b^2$

(5) $(a-b)^2 = a^2 - 2ab + b^2$

Example 3 Simplify using positive exponents.

(1) $x^{-\frac{1}{2}}(x^{\frac{3}{2}} + 2x^{\frac{1}{2}} - 3x^{-\frac{1}{2}})$

(2) $(2^x + 3)(2^x + 1)$

(3) $(7^x + 7^{-x})^2$

Solution

(1) $x^{-\frac{1}{2}}(x^{\frac{3}{2}}+2x^{\frac{1}{2}}-3x^{-\frac{1}{2}})$

$= x^{-\frac{1}{2}}\times x^{\frac{3}{2}}+x^{-\frac{1}{2}}\times 2x^{\frac{1}{2}}-x^{-\frac{1}{2}}\times 3x^{-\frac{1}{2}}$

$= x^{1}+2x^{0}-3x^{-1}$

$= x+2-\dfrac{3}{x}$

(2) $(2^{x}+3)(2^{x}+1)$

$= 2^{x}\times 2^{x}+2^{x}+3\times 2^{x}+3$

$= 2^{2x}+4\times 2^{x}+3$

$= 4^{x}+2^{2+x}+3$

(3) $(7^{x}+7^{-x})^{2}$

$= (7^{x})^{2}+2\times 7^{x}\times 7^{-x}+(7^{-x})^{2}$

$= 7^{2x}+2\times 7^{0}+7^{-2x}$

$= 7^{2x}+2+7^{-2x}$

Example 4 Factor and simplify.

(1) $2^{n+3}+8$

(2) $2^{3n}+2^{2n}$

(3) $4^{x}-9$

(4) $9^{x}+4(3^{x})+4$

Solution

(1) $2^{n+3}+8$

$= 2^{n}2^{3}+8$

$= 8(2^{n})+8$

$= 8(2^{n}+1)$

(2) $2^{3n}+2^{2n}$

$= 2^{2n}2^{n}+2^{2n}$

$= 2^{2n}(2^{n}+1)$

(3) $4^{x}-9$

$= (2^{x})^{2}-3^{2}$

$= (2^{x}+3)(2^{x}-3)$

(4) $9^{x}+4(3^{x})+4$

$= (3^{x})^{2}+4(3^{x})+4$

$= (3^{x}+2)^{2}$

1.2.3 Exponential Equations

Exponential Equations can be solved without using logarithms when the bases are equal.

Step 1. Simplify each side of the equation into a single power of the same base.

$$a^x = a^y$$

Step 2. Since the bases are the same, the exponents must eqal one another.

Step 3. Equate Equations and solve the equation.

$$x = y$$

Example 5 Solve for x.

(1) $3(27)^{2-x} = \dfrac{\sqrt{3}}{9^{3x}}$

(2) $5^{4x+1} + 3 = 128$

Solution

(1) $3 \cdot 3^{3(2-x)} = \dfrac{3^{1/2}}{(3^2)^{3x}}$

$3^{1+3(2-x)} = 3^{\frac{1}{2}-6x}$

$1+6-3x = \dfrac{1}{2} - 6x$

$3x = \dfrac{1}{2} - 7$

$3x = -\dfrac{13}{2}$, $\quad \therefore x = -\dfrac{13}{6}$

(2) $5^{4x+1} = 128 - 3$

$5^{4x+1} = 125$

$5^{4x+1} = 5^3$

$4x + 1 = 3$

$4x = 2$, $\quad \therefore x = \dfrac{1}{2}$

EX3 Solve for x.

(1) $9(3)^{-x} = 27^{1+x}$

(2) $2(8)^{x-1} = \dfrac{\sqrt{2}}{4^x}$

(3) $\left(\dfrac{64}{125}\right)^{-x} = \left(\dfrac{5}{4}\right)^{2x-1}$

(4) $\dfrac{1}{16}(8)^{x+2} = \left(\sqrt[3]{4}\right)^x$

Example 6 Solve for x.

$$10(10)^{2x} - 101(10^x) + 10 = 0$$

Solution

$10^x = t$

$10t^2 - 101t + 10 = 0,\ (10t-1)(t-10) = 0$

$t = \frac{1}{10}, 10$

$10^x = \frac{1}{10}, 10 \quad \therefore x = -1, 1$

EX 4 Solve for x.

(1) $5^{2x} - 26(5)^x + 25 = 0$

(2) $2(2)^{2x} - 9(2)^x + 4 = 0$

(3) $2^{2x} + 3(2)^{x+1} - 40 = 0$

(4) $3^{x+1} = \frac{9}{3^x} + 26$

1.2.4 Graphs of Exponential Functions

(1) $y = a^x$ (a : constant, $a > 0$, $a \neq 1$)

- The graphs of $y = a^x$ passes through point (0,1).
- If $a > 1$, the graph of $y = a^x$ increases ∞ x increases. (Increasing function)

$y = 2^x$

x	y
$-\infty$	0^+
⋮	⋮
−3	1/8
−2	1/4
−1	1/2
0	1
1	2
2	4
3	8
⋮	⋮
$+\infty$	$+\infty$

$y = 3^x$

x	y
$-\infty$	0^+
⋮	⋮
−3	1/27
−2	1/9
−1	1/3
0	1
1	3
2	9
3	27
⋮	⋮
$+\infty$	$+\infty$

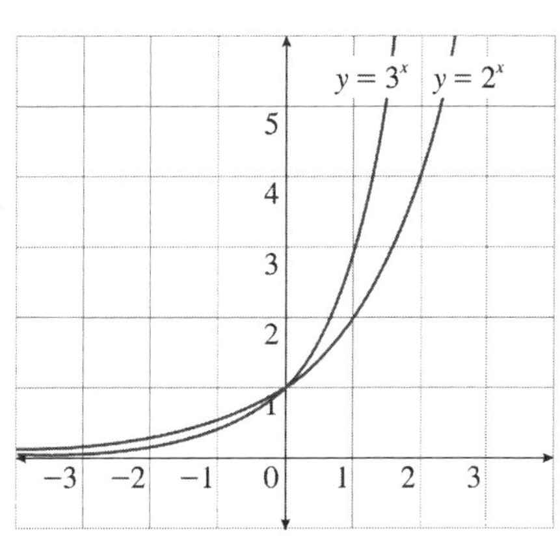

- If $0 < a < 1$, the graphs of $y = a^x$ decreases ∞ x increases. (Decreasing function)

$y=\left(\frac{1}{2}\right)^x$

x	y
$-\infty$	$+\infty$
$\vdots$	$\vdots$
−3	8
−2	4
−1	2
0	1
1	1/2
2	1/4
3	1/8
$\vdots$	$\vdots$
$+\infty$	0^+

$y=\left(\frac{1}{3}\right)^x$

x	y
$-\infty$	$+\infty$
$\vdots$	$\vdots$
−3	27
−2	9
−1	3
0	1
1	1/3
2	1/9
3	1/27
$\vdots$	$\vdots$
$+\infty$	0^+

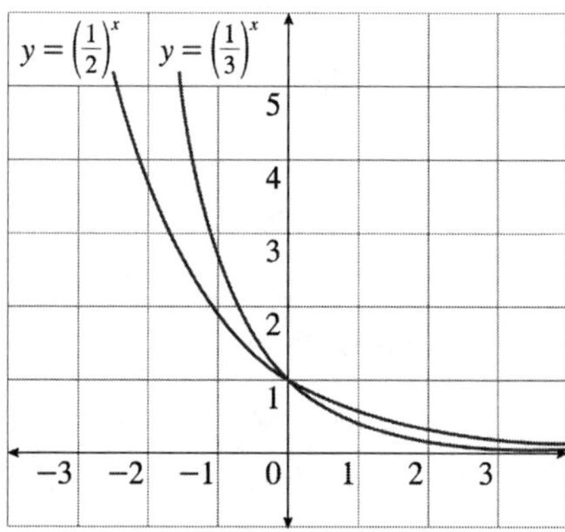

- The graph of $y=a^x$ has a horitontal asymptote of $y=0$.
- Domain : $\{x|x\in R\}$
- Range : $\{y|y>0, y\in R\}$

(2) Exponential e

$$e=2.718281828459045235....$$

x	$\left(1+\frac{1}{x}\right)^x$	$\left(1+\frac{1}{x}\right)^x$
10	$(1.1)^{10}$	2.593742460···
100	$(1.01)^{100}$	2.704813829···
1000	$(1.001)^{1000}$	2.716923932···
10000	$(1.0001)^{10000}$	2.718145927···
100000	$(1.00001)^{100000}$	2.718268237···
1000000	$(1.000001)^{1000000}$	2.718280469···
$\vdots$	$\vdots$	$\vdots$
$+\infty$		e

As x increases, $\left(1+\frac{1}{x}\right)^x$ approaches $e(=2.718....)$.

(3) The Graph of Natural Exponential Function

$y = e^x$

x	y
$-\infty$	0^+
$\vdots$	$\vdots$
-1	1/e
0	1
1	e
2	$e^2 \simeq 7.4$
$\vdots$	$\vdots$
$+\infty$	$+\infty$

$y = e^{-x}$

x	y
$-\infty$	$+\infty$
$\vdots$	$\vdots$
-2	$e^2 \simeq 7.4$
-1	$e \simeq 2.7$
0	1
1	$1/e$
$\vdots$	$\vdots$
$+\infty$	0^+

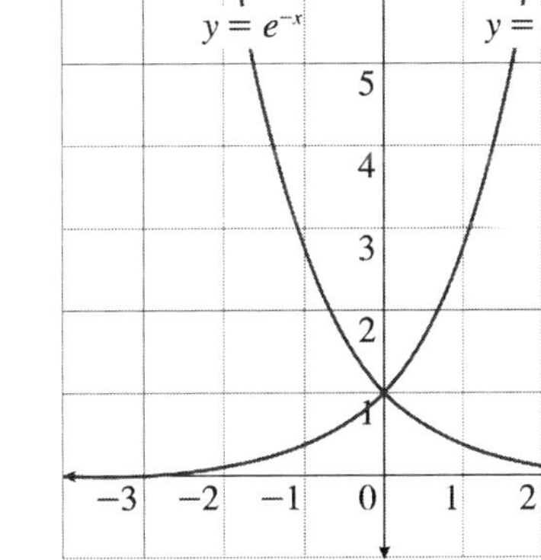

– The graph of $y = e^x$ and $y = e^{-x}$ have horizontal asymptote of $y = 0$.

Example 7 Sketch the graph of $y = 1 - e^{x-1}$.

Solution

$y = e^x$ ⟶ $y = -e^x$ ⟶ $y = -e^{x-1} + 1$

Reflection in the $x-axis$ $\qquad$ $x \to 1$, $y \to 1$

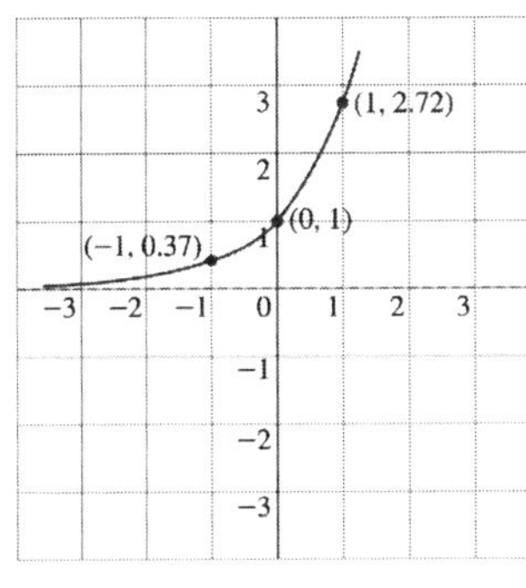

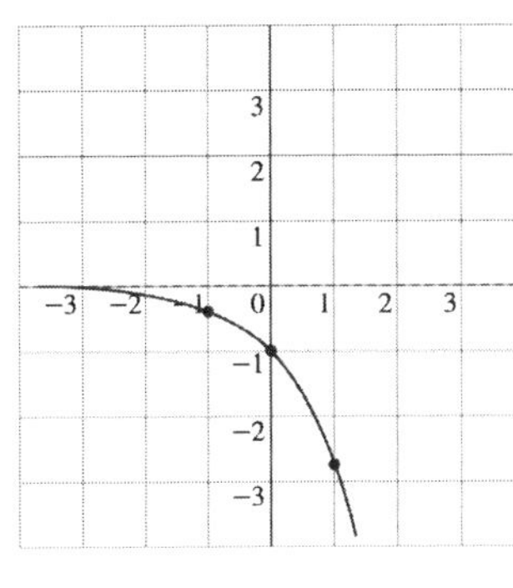

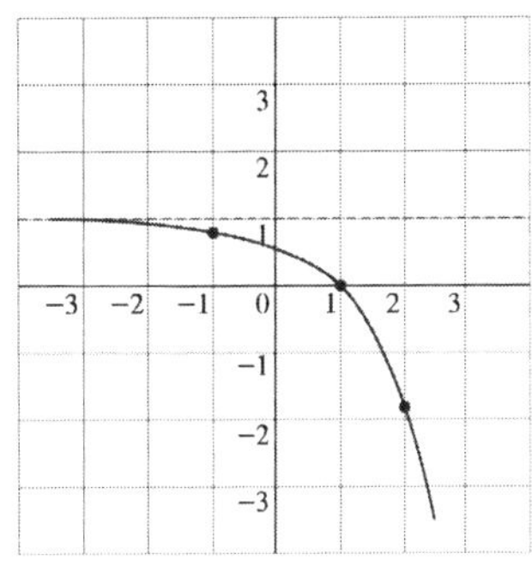

EX 5 Sketch each of the following .

(1) $y = 2^x$

(2) $y = e^x - 2$

(3) $y = -3^x$

(4) $y = e^{-x} + 2$

(5) $y = 2(3)^{x-1}$

(6) $y = -e^{x-3} + 2$

(7) $y = 3(2)^{x-1} - 2$

(8) $y = 1 - (3)^{-x+2}$

1.3 Logarithmic Functions

1.3.1 Definition of log

$$a^x = b \Leftrightarrow x = \log_a b \text{ where } a > 0, a \neq 1, b > 0$$

(base $a > 0$, base $\neq 1$ and argument > 0)

Example 1 Express each of the following in logarithmic form.

(1) $2^{-3} = \frac{1}{8}$

(2) $x^5 = 32$

Solution

(1) $2^{-3} = \frac{1}{8}$

$\Rightarrow -3 = \log_2 \frac{1}{8}$

(2) $x^5 = 32$

$\Rightarrow 5 = \log_x 32$

Example 2 Express each of the following in exponential form.

(1) $\log_{10} 0.001 = -3$

(2) $\log_M X = T$

(3) $\log_7 \frac{1}{343} = -3$

Solution

(1) $\log_{10} 0.001 = -3$

$\Rightarrow 0.001 = 10^{-3}$

(2) $\log_M X = T$

$\Rightarrow X = M^T$

(3) $\log_7 \frac{1}{343} = -3$

$\Rightarrow \frac{1}{343} = 7^{-3}$

- **Logarithmic Notation**
  - Common Logarithm $\quad \log x = \log_{10} x$
  - Natural Logarithm $\quad \ln x = \log_e x$

Example 3 Solve for x.

(1) $\log \sqrt[3]{10} = x$

(2) $\log x = -\frac{1}{2}$

(3) $\ln e = x$

(4) $\ln x = -2$

Solution

(1) $\log \sqrt[3]{10} = x$

$\Rightarrow \sqrt[3]{10} = 10^x$

$x = \frac{1}{3}$

(2) $\log x = -\frac{1}{2}$

$\Rightarrow x = 10^{-\frac{1}{2}}$

$= \frac{1}{\sqrt{10}}$

(3) $\ln e = x$

$\Rightarrow e = e^x$

$x = 1$

(4) $\ln x = -2$

$\Rightarrow x = e^{-2}$

Solve for x.

(1) $\log 1000 = x$

(2) $\log \sqrt[3]{10} = x$

(3) $\ln e = x$

(4) $\ln x = 3$

1.3.2 Logarithmic Functions

$y = \log_a x$ (a : constant, $a > 0$, $a \neq 1$)

Exponential Function		Logarithmic Function
$y = a^x$	$\Leftrightarrow$	$x = a^y$
$(a > 0, a \neq 1)$	Inverse	$y = \log_a x$ $(a > 0, a \neq 1)$

- The graph of $y = \log_a x$ pass through $(1, 0)$.
- If $a > 1$, the graph of $y = \log_a x$ increases ∞ x increases. (Increasing function)

$y = \log_2 x$

x	y
0^+	$-\infty$
⋮	⋮
1/8	−3
1/4	−2
1/2	−1
1	0
2	1
4	2
8	3
⋮	⋮
$+\infty$	$+\infty$

$y = \log_3 x$

x	y
0^+	$-\infty$
⋮	⋮
1/27	−3
1/9	−2
1/3	−1
1	0
3	1
9	2
27	3
⋮	⋮
$+\infty$	$+\infty$

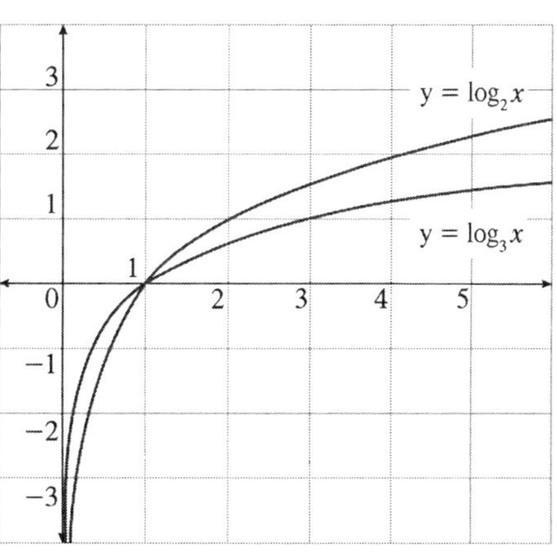

- If $0 < a < 1$, the graph of $y = \log_a x$ decreases ∞ x increases. (Decreasing function)

$y=\log_{\frac{1}{2}} x$ $\quad$ $y=\log_{\frac{1}{3}} x$

x	y	x	y
0^+	$+\infty$	0^+	$+\infty$
⋮	⋮	⋮	⋮
1/8	3	1/27	3
1/4	2	1/9	2
1/2	1	1/3	1
1	0	1	0
2	−1	3	−1
4	−2	9	−2
8	−3	27	−3
⋮	⋮	⋮	⋮
$+\infty$	$-\infty$	$+\infty$	$-\infty$

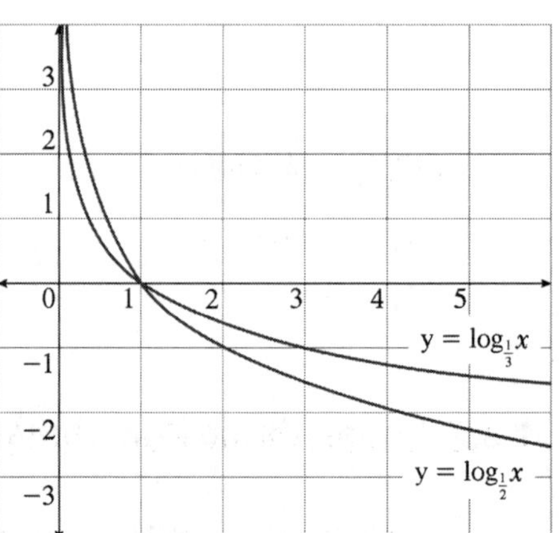

– The graph of $y=\log_a x$ has a vertical asymptote of $x=0$.

Domain : $\{x \mid x>0, x \in R\}$

Range : $\{y \mid y \in R\}$

Example 4 Sketch the graph of $f(x)=\ln(2-x)+1$.

Solution

$f(x)=\ln(-(x-2))+1$:

$y=\ln x \quad \rightarrow \quad y=\ln(-x) \quad \rightarrow \quad y=\ln(-(x-2))+1$

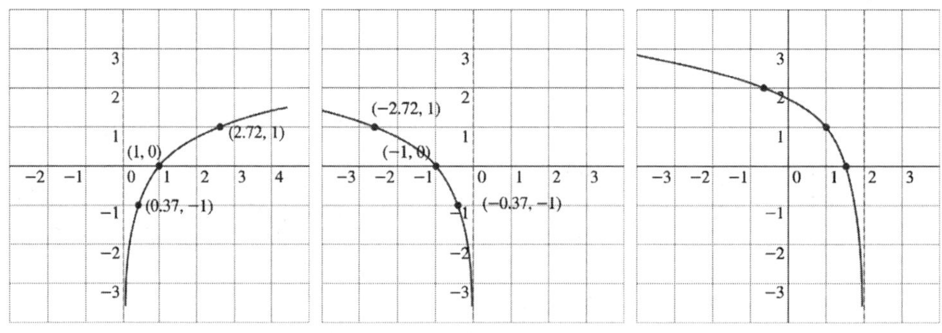

EX 2 Sketch each of the following functions.

(1) $y=\ln x$

(2) $y=\ln(-x)+1$

(3) $y=-\log_2(x-1)+2$

(4) $y=2\log_2(x+4)-2$

Example 5 Find the inverse of each of the following functions in the form, $f^{-1}(x)$.

(1) $f(x)=2(7)^x$ (2) $f(x)=-\log(x-1)$

Solution

(1) $y=2(7)^x \rightarrow x=2(7)^y$

$\frac{x}{2}=7^y$

$y=\log_7\left(\frac{x}{2}\right)$

$\therefore f^{-1}(x)=\log_7\left(\frac{x}{2}\right)$

(2) $y=-\log(x-1) \rightarrow x=-\log(y-1)$

$-x=\log(y-1)$

$y-1=10^{-x}$

$\therefore f^{-1}(x)=10^{-x}+1$

EX 3 Find the inverse of each of the following functions in the form,

(1) $f(x)=3(5)^x$ (2) $y=e^{-x}$

(3) $f(x)=2(3)^{x-1}+4$ (4) $f(x)=-e^{x+3}+1$

1.3.3 Laws of Logarithms and Logarithms Equations

$a \neq 1, a>0, b>0, x>0, y>0$

(1) $\log_a a=1$, $\log_a 1=0$ (2) $\log_a xy=\log_a x+\log_a y$

(3) $\log_a \frac{x}{y}=\log_a x-\log_a y$ (4) $\log_a x^n=n\log_a x$ (n은 실수)

(5) $\log_a b=\frac{\log_c b}{\log_c a}(c \neq 1, c>0)$ (6) $\log_a b=\frac{1}{\log_b a}(b \neq 1)$

Example 6 Express a single $\log$arithm and simplify.

(1) $\log_2 24+\log_2 6-\log_2 9$ (2) $\frac{1}{2}\log 15+\log\left(\frac{\sqrt{10}}{3}\right)+\frac{1}{2}\log 6$

(3) $2-\ln 5-\ln 2$ (4) $\frac{2}{3}(\ln A+3\ln B)-\frac{1}{2}(\ln\sqrt[3]{A}+2\ln B)$

Solution

(1) $\log_2\left(\frac{24 \cdot 6}{9}\right)=\log_2 16=4$

(2) $\log\sqrt{15}+\log\frac{\sqrt{10}}{3}+\log\sqrt{6}=\log\left(\frac{\sqrt{900}}{3}\right)=\log 10=1$

(3) $2-\ln 5-\ln 2=\ln\left(\frac{e^2}{5 \cdot 2}\right)=\ln\left(\frac{e^2}{10}\right)$

(4) $\ln A^{\frac{2}{3}} + \ln B^2 - \ln A^{\frac{1}{6}} - \ln B$

$$= \ln\left(\frac{A^{\frac{2}{3}} \cdot B^2}{A^{\frac{1}{6}} \cdot B}\right) = \ln(A^{\frac{1}{2}} B)$$

EX 4 Simplify.

(1) $\log_6 4 + \log_6 9$

(2) $\log 50 + \log 4 - \log 2$

(3) $\log_7 35 + \log_7\left(\frac{14}{5}\right) - \log_7 2$

(4) $\frac{1}{3}\log 8 + \frac{1}{2}\log 4 - \log 40$

EX 5 Expand the following ∞ the sum and diffence of individual logarithms.

(1) $\log\left(\frac{4a^2}{9b^3}\right)$

(2) $\log_b \frac{b^2\sqrt{x+1}}{(x-1)^3}$

(3) $\ln \frac{\sqrt[4]{2x-1}\,\sqrt{x-1}}{\sqrt{x+1}\,\sqrt{5-x}}$

(4) $\ln \frac{e^x \cos x}{\sqrt{x^2+x}}$

Example 7 Simplify.

(1) $5^{\log_3 36} \div 5^{\log_3 4}$

(2) $\log_2 3 \cdot \log_3 5 \cdot \log_5 2$

(3) $\frac{\log_5 32}{\log_5 4} \cdot \frac{\log 9}{\log 27}$

(4) $e^{\ln 2} \cdot 4^{\log_4 3}$

Solution

(1) $5^{\log_3 36} \div 5^{\log_3 4} = 5^{\log_3 36 - \log_3 4}4$

$= 5^{\log_3 9}$

$= 5^2 = 25$

(2) $\log_2 3 \cdot \log_3 5 \cdot \log_5 2 = \frac{\log 3}{\log 2} \cdot \frac{\log 5}{\log 3} \cdot \frac{\log 2}{\log 5}$

$= 1$

(3) $\frac{\log_5 32}{\log_5 4} \cdot \frac{\log 9}{\log 27} = \frac{5\log_5 2}{2\log_5 2} \cdot \frac{2\log 3}{3\log 3}$

$= \frac{5}{2} \cdot \frac{2}{3} = \frac{5}{3}$

(4) $e^{\ln 2} \cdot 4^{\log_4 3} = 2 \cdot 3 = 6$

• Solving Exponential and Logarithmic Equations

Solving $a^x = b$

(Method 1)

Step 1. Isolate the term involving x to one side of the equation.

Step 2. Take logarithms of both sides and bring down the exponents.

Step 3. Solve for x.

(Method 2)

Step 1. Isolate the term involving x to one side of the equation.

Step 2. Change the equation to the logarithmic form.

Step 3. Solve for x.

Example 8 Solve for x.

(1) $3^{x+1} = 2 \cdot 5^x$

(2) $3^{2x} + 5 = 9$

(3) $\log_6 x = 1 - \log_6(x-1)$

(4) $(\log x)^2 - \log x^2 - 3 = 0$

(5) $2\log(-x) = \log 2 + \log 8$

Solution

(1) $\log 3^{x+1} = \log(2 \cdot 5^x)$, $\log 3^{x+1} = \log 2 + \log 5^x$

$(x+1)\log 3 = \log 2 + x\log 5$, $x\log 3 + \log 3 = \log 2 + x\log 5$

$x\log 3 - x\log 5 = \log 2 - \log 3$, $x(\log 3 - \log 5) = \log 2 - \log 3$

$\therefore x = \dfrac{\log 2 - \log 3}{\log 3 - \log 5} \quad \left(\text{or } \dfrac{\log(2/3)}{\log(3/5)}\right)$

(2) $3^{2x} = 4$, $\log 3^{2x} = \log 4$ or $\ln 3^{2x} = \ln 4$

$2x\log 3 = \log 4$, $2x\ln 3 = \ln 4$

$x = \dfrac{\log 4}{2\log 3}$ or $x = \dfrac{\ln 4}{2\ln 3}$

(3) $\log_6 x + \log_6(x-1) = 1$, $\log_6 x(x-1) = 1$

$x(x-1) = 6^1$, $x^2 - x - 6 = 0$

$(x-3)(x+2) = 0$, $x = 3$ $(\because x > 1)$

$\therefore x = 3$

(4) $(\log x)^2 - 2\log x - 3 = 0$, $\log x = t$

$t^2 - 2t - 3 = 0$, $(t-3)(t+1) = 0$

$t = 3, -1 \Rightarrow \log x = 3, -1 \Rightarrow x = 10^3, 10^{-1}$

$\therefore x = 1000, \dfrac{1}{10}$

(5) $\log(-x)^2 = \log16$, $(-x)^2 = 16$

$x^2 = 16$, $x = -4, 4$ $\quad \therefore x = -4$

EX6 Solve for x.

(1) $5^{2x} = 8^{x+1}$

(2) $e^x = 4^{2-x}$

(3) $5(4)^x = 3^{2x+1}$

(4) $4(3)^{x-2} = 5^{2x}$

EX7 Solve for x.

(1) $\log_2(2x+4) = \log_2(x-1) + 3$

(2) $\ln x - \ln(x+1) = 1$

(3) $(\log_3 x)^2 = 2\log_3 x$

(4) $2(\ln x)^2 - \ln x^5 + 2 = 0$

(5) $\log(3x-2) = \log 2 + \log(x+2)$

(6) $\ln(3-x) - \ln(x+5) = \ln 3$

1.4 Trigonometric Equations

1.4.1 Angular Measure and Radians

An angle is in standard position if its vertex is at the origin of a coordinate plane and its initial arm lies along the positive x-axis.

– If the direction of rotation is counterclockwise, the angle is positive.
– If the direction of rotation is clockwise, the angle is negative.

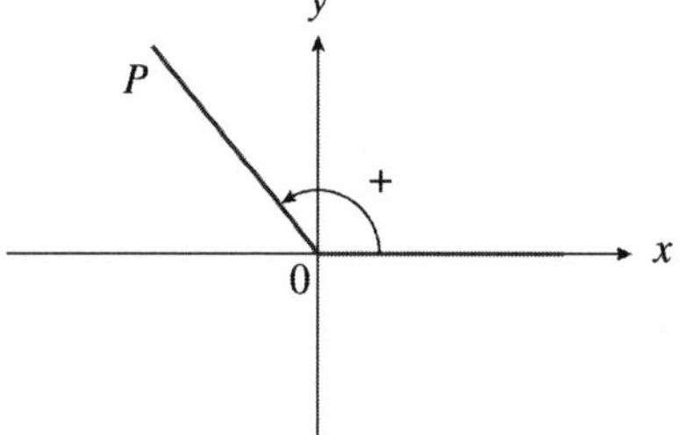

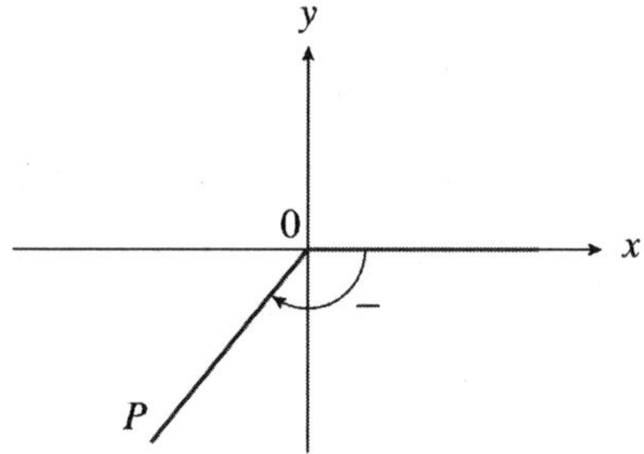

Example 1 Sketch each angle in standard position.

(1) $120°$ (2) $-70°$

(3) $560°$ (4) $-450°$

Solution

(1)

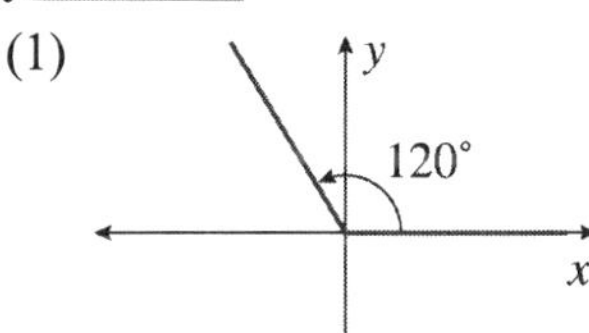

(2)

y
−70°
x

(3)

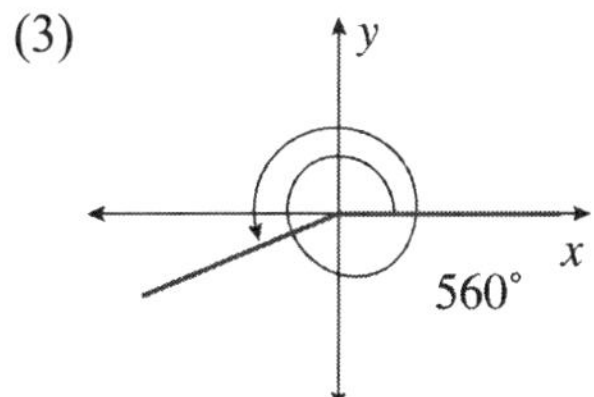

(4)

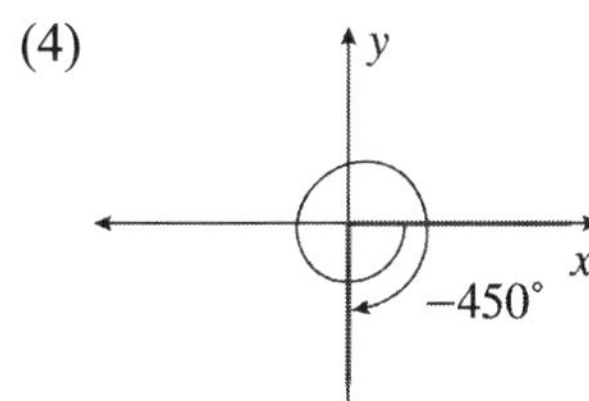

EX 1 Sketch each angle in standard position.

(1) $50°$ (2) $670°$

(3) $-310°$ (4) $-770°$

The radian measure θ of a central angle is defined as the ratio of the arc length a to the radius r.

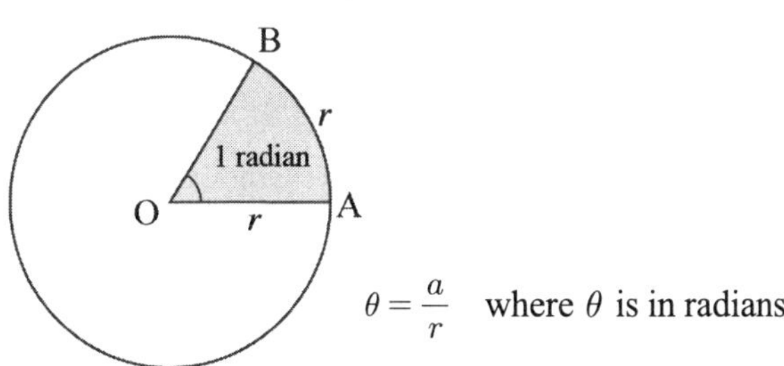

$\theta = \dfrac{a}{r}$ where θ is in radians.

- $360° = \dfrac{2\pi r}{r} = 2\pi \Rightarrow \therefore 180° = \pi$
- Radians is a real number with no units.
- One radian is the measure of a central angle subtended by an arc, whose arc length is equal to the radius of the circle ($1\,radian \approx 57.3°$).

(1) Degree $\times \dfrac{\pi}{180^\circ} \rightarrow$ Radian

Example 2 Convert the degree measure to radians.

(1) 240° (2) -420°

(3) 540° (4) $12x^\circ$

Solution

(1) $240^\circ \times \dfrac{\pi}{180^\circ} = \dfrac{4\pi}{3}$ (2) $-420^\circ \times \dfrac{\pi}{180^\circ} = -\dfrac{7\pi}{3}$

(3) $540^\circ \times \dfrac{\pi}{180^\circ} = 3\pi$ (4) $12x^\circ \times \dfrac{\pi}{180^\circ} = \dfrac{\pi x}{15}$

EX 2 Convert the degree measure to radians.

(1) 180° (2) -120°

(3) 300° (4) 75°

(2) Radian $\times \dfrac{180^\circ}{\pi} \rightarrow$ Degree

Example 3 Convert the radian measure to degree.

(1) $\dfrac{7}{6}\pi$ (2) $-\dfrac{\pi}{6}$

Solution

(1) $\dfrac{7}{6}\pi = \dfrac{7}{6}\pi \times \dfrac{180^\circ}{\pi} = 210^\circ$ (2) $-\dfrac{\pi}{6} = -\dfrac{\pi}{6} \times \dfrac{180^\circ}{\pi} = -30^\circ$

EX 3 Convert the radian measure to degrees.

(1) $\dfrac{2}{3}\pi$ (2) $\dfrac{4}{3}\pi$

(3) $-\dfrac{11}{6}\pi$ (4) $\dfrac{7}{12}\pi$

Special Angles and Solving Simple Trigonometric Equations.

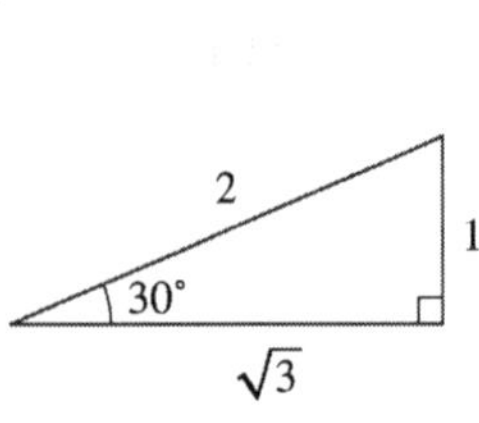

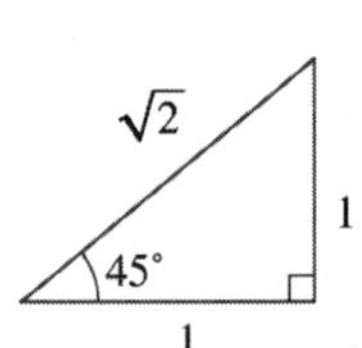

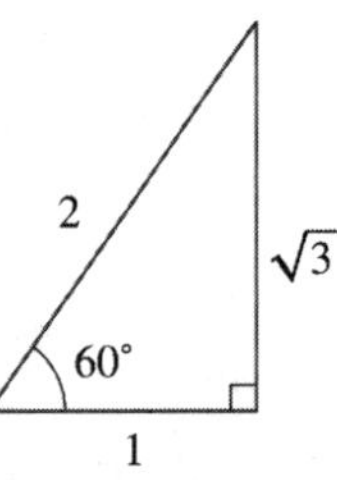

$\sin 30^\circ = \dfrac{1}{2}$ $\quad \sin 45^\circ = \dfrac{1}{\sqrt{2}}\left(\text{or } \dfrac{\sqrt{2}}{2}\right)$ $\quad \sin 60^\circ = \dfrac{\sqrt{3}}{2}$

$\cos 30^\circ = \dfrac{\sqrt{3}}{2}$ $\quad \cos 45^\circ = \dfrac{1}{\sqrt{2}}\left(\text{or } \dfrac{\sqrt{2}}{2}\right)$ $\quad \cos 60^\circ = \dfrac{1}{2}$

$\tan 30^\circ = \dfrac{1}{\sqrt{3}}\left(\text{or } \dfrac{\sqrt{3}}{3}\right)$ $\quad \tan 45^\circ = 1$ $\quad \tan 60^\circ = \sqrt{3}$

Example 4 Find the exact volue of each ratio.

(1) $\sin\frac{7}{6}\pi$ (2) $\cot\frac{2}{3}\pi$

(3) $\cos\frac{5}{6}\pi$ (4) $\tan\frac{\pi}{4}$

Solution

(1)

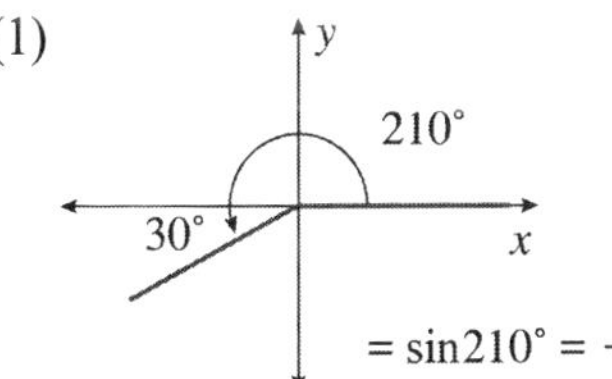

$= \sin 210° = -\frac{1}{2}$

(2)

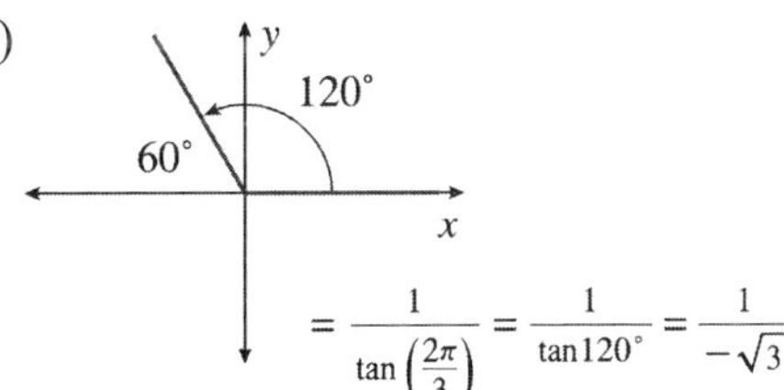

$= \frac{1}{\tan\left(\frac{2\pi}{3}\right)} = \frac{1}{\tan 120°} = \frac{1}{-\sqrt{3}}$

(3) $\cos\frac{5}{6}\pi = \cos 150° = -\cos 30° = -\frac{\sqrt{3}}{2}$

(4) $\tan\frac{\pi}{4} = \tan 45° = 1$

EX 4 Find the exact value of each ratio.

(1) $\cos\frac{\pi}{3}$ (2) $\sin\pi$

(3) $\tan\frac{3}{4}\pi$ (4) $\sin\frac{5}{4}\pi$

(5) $\csc\frac{7}{4}\pi$ (6) $\sec\frac{5}{3}\pi$

(7) $\cot\frac{3}{2}\pi$ (8) $\cos\frac{11}{3}\pi$

Example 5 Evaluate.

(1) $\sin 120°$ (2) $\sin 330°$

(3) $\sin 225°$ (4) $\sin(-450°)$

Solution

(1)

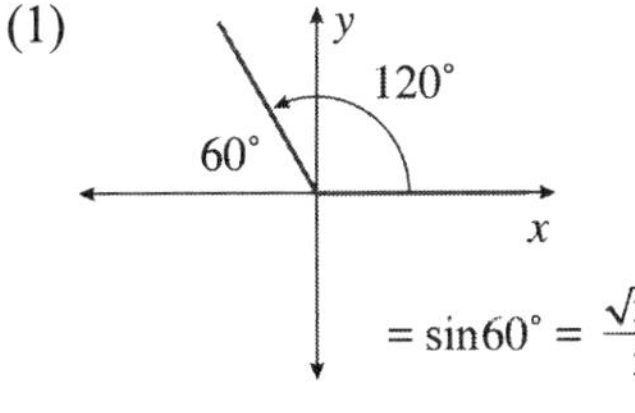

$= \sin 60° = \frac{\sqrt{3}}{2}$

(2)

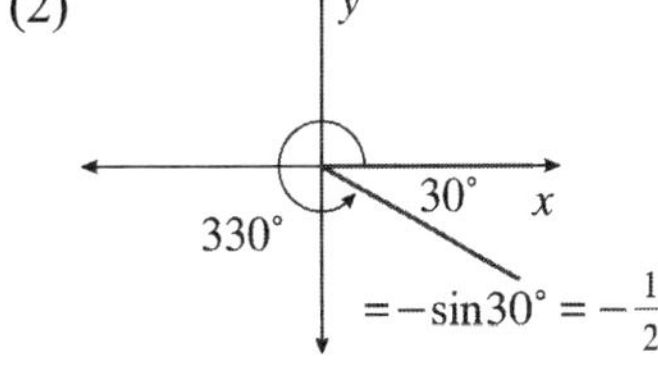

$= -\sin 30° = -\frac{1}{2}$

(3)

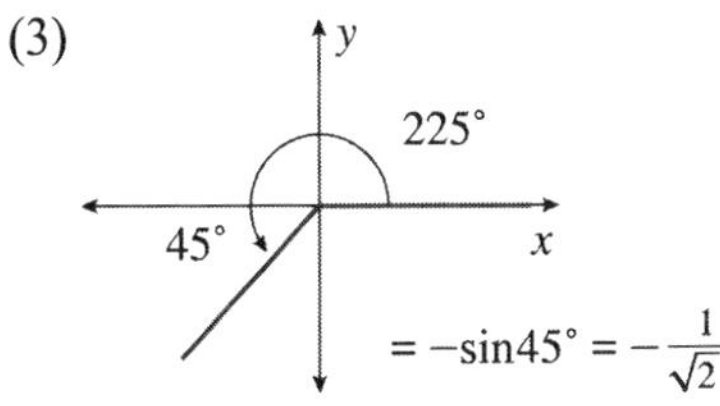

$= -\sin 45° = -\frac{1}{\sqrt{2}}$

(4)

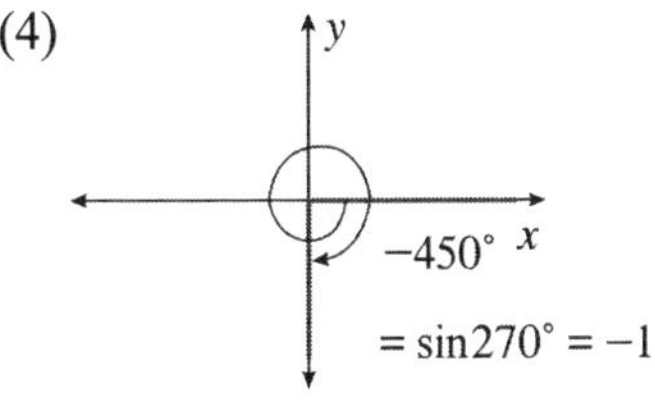

$= \sin 270° = -1$

Example 6 Evaluate.

(1) $\cos 120^\circ$　　(2) $\cos 330^\circ$

(3) $\cos 225^\circ$　　(4) $\cos(-180^\circ)$

Solution

(1)

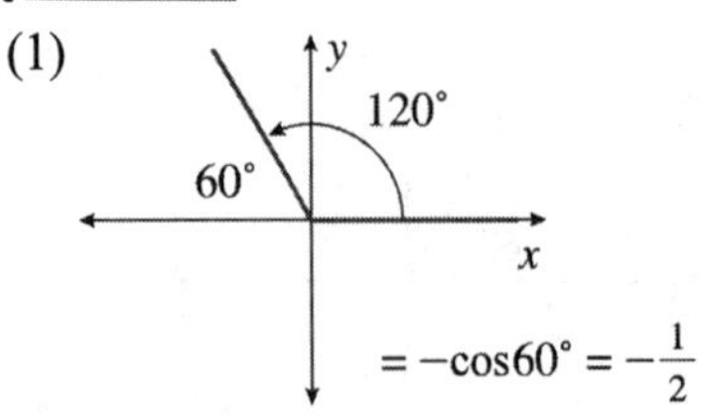

(2)

y
330°
30°
x
$= \cos 30^\circ = \frac{\sqrt{3}}{2}$

(3)

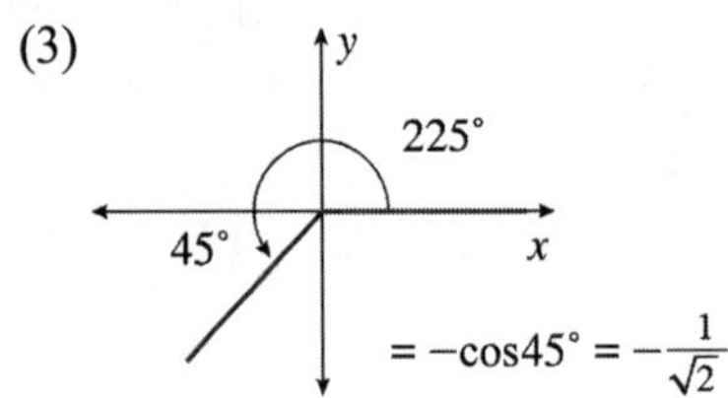

(4)

y
−180°
x
$= \cos 180^\circ = -1$

Example 7 Evaluate.

(1) $\tan 120^\circ$　　(2) $\tan 330^\circ$

(3) $\tan 225^\circ$　　(4) $\tan 270^\circ$

Solution

(1)

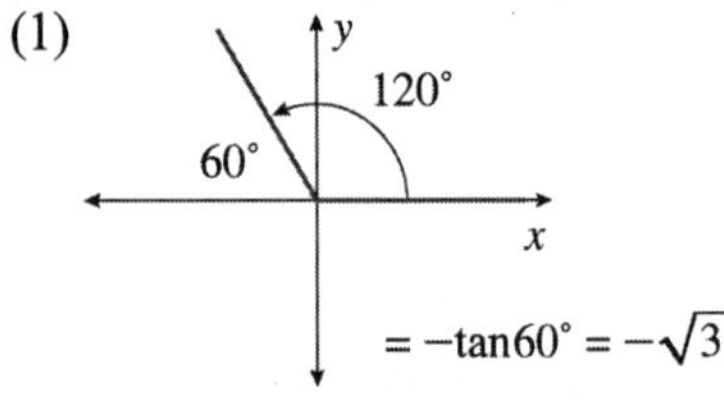

(2)

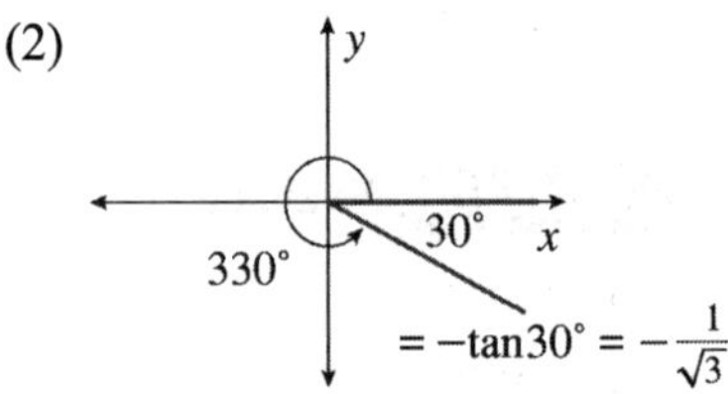

(3)

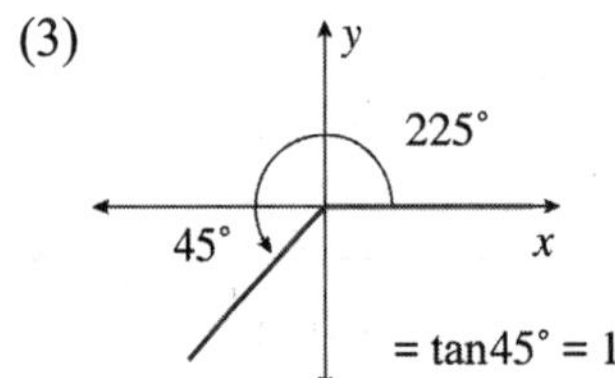

(4)

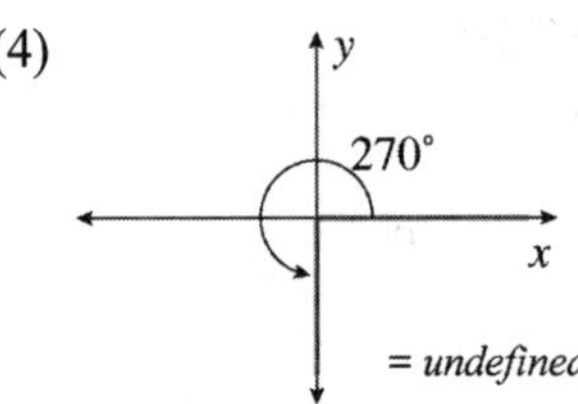

EX 5 Find the exact value of each ratio.

(1) $\cos 150^\circ$　　(2) $\cos(-210^\circ)$

(3) $\sin 150^\circ$　　(4) $\sin(-240^\circ)$

(5) $\tan 135^\circ$　　(6) $\tan 180^\circ$

Example 8 Solve for $\theta(0 \le \theta \le 2\pi)$.

(1) $\sin\theta = -\frac{1}{2}$ (2) $\cos\theta = \frac{\sqrt{3}}{2}$

Solution

(1) Reference angle of $\sin\theta = -\frac{1}{2}$; 30°.

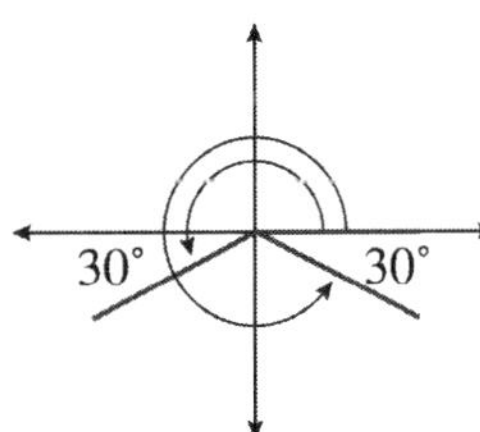

$\therefore \theta = \frac{7}{6}\pi, \frac{11}{6}\pi$

(2) $\theta = \frac{\pi}{6}, \frac{11\pi}{6}$

EX 6 Find all values of θ in terms of $(0 \le \theta < 2\pi)$.

(1) $\sin\theta = \frac{\sqrt{3}}{2}$ (2) $\cos\theta = -\frac{1}{2}$

(3) $\sin\theta = -\frac{\sqrt{2}}{2}$ (4) $\tan\theta = -1$

(5) $\sec\theta = -\sqrt{2}$ (6) $\cot\theta = -\frac{\sqrt{3}}{3}$

(7) $\cos\theta = 1$ (8) $\tan\theta = undefined$

Example 9 Solve for $\theta(0 \le \theta \le 360^\circ)$.

(1) $\sin\theta = -\frac{1}{\sqrt{2}}$ (2) $\cos\theta = \frac{1}{2}$ (3) $\tan\theta = \frac{1}{\sqrt{3}}$

Solution

(1) Reference angle of $\sin\theta = \frac{1}{\sqrt{2}}$; 45°

$\theta = 180^\circ + 45^\circ,\ 360^\circ - 45^\circ$ $\therefore 225^\circ$ or 315°

(2) Reference angle of $\cos\theta = \frac{1}{2}$; 60°.

$\theta = 60^\circ, 300^\circ$.

(3) Reference angle of $\tan\theta = \frac{1}{\sqrt{3}}$; 30°.

$\therefore 30^\circ$ or 210°

EX 7 Solve for θ. $(0° \leq \theta < 360°)$

(1) $\cos\theta = \dfrac{\sqrt{3}}{2}$

(2) $\cos\theta = -\dfrac{\sqrt{2}}{2}$

(3) $\sin\theta = \dfrac{\sqrt{3}}{2}$

(4) $\sin\theta = -\dfrac{1}{\sqrt{2}}$

(5) $\tan\theta = 1$

(6) $\tan\theta = -\dfrac{1}{\sqrt{3}}$

1.4.2 Graphs of Trigonometric Functions

A periodic function is a function whose value repeat at regual intervals, or periods. $f(x+p)=f(x)$ where the smallest number p is the period of $f(x)$.

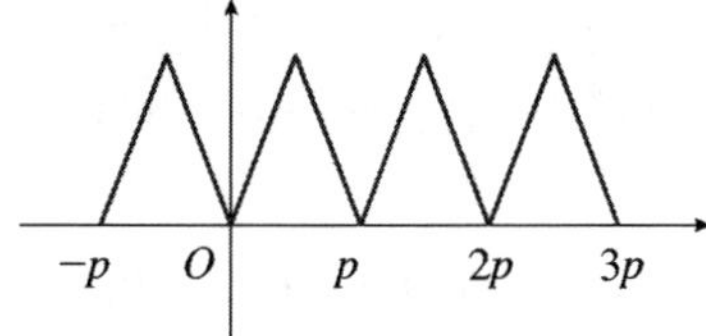

The amplitude of a periodic function is half the difference between the maximum and minimum values of the function.

$$Amplitude = \frac{Max - Min}{2}$$

(1) $y = \sin x$

x	y
$\vdots$	$\vdots$
0	0
$\frac{\pi}{6}$	0.5
$\frac{\pi}{3}$	0.87
$\frac{\pi}{2}$	1
$\vdots$	$\vdots$
π	0
$\vdots$	$\vdots$
$\frac{3\pi}{2}$	−1
$\vdots$	$\vdots$
2π	0
$\vdots$	$\vdots$

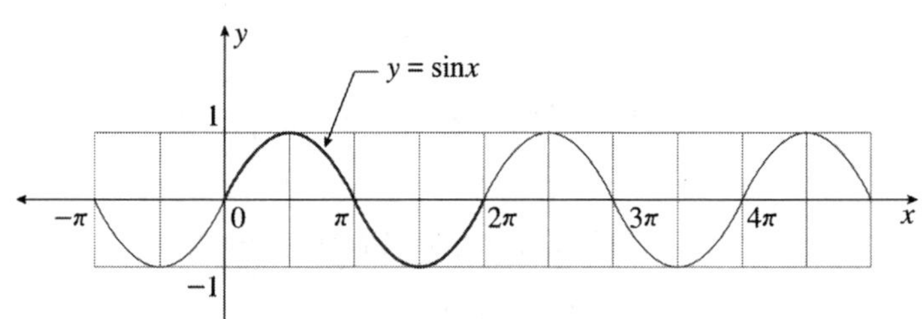

– Amplitude : 1, Period : 2π

– Domain : $\{x \mid x \in R\}$, Range : $\{\, y \mid -1 \leq 1 \leq y \leq 1\}$

(2) $y = \cos x$

x	y
$\vdots$	$\vdots$
0	0
$\frac{\pi}{6}$	0.87
$\frac{\pi}{3}$	0.5
$\frac{\pi}{2}$	0
$\vdots$	$\vdots$
π	-1
$\vdots$	$\vdots$
$\frac{3\pi}{2}$	0
$\vdots$	$\vdots$
2π	1
$\vdots$	$\vdots$

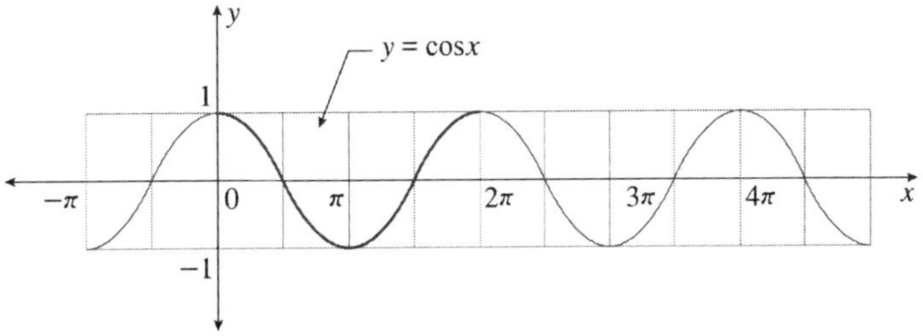

- Amplitude : 1, period : 2π
- Domain : $\{x \mid x \in R,\}$ Range : $\{\, y \mid -1 \leq 1 \leq y \leq 1\}$

(3) $y = \tan x$

x	y
$\vdots$	$\vdots$
$-\frac{\pi}{2}(0.5\pi)$	unde fined
-0.4999π	-3183
$\vdots$	$\vdots$
$-\frac{\pi}{3}$	-1.73
$-\frac{\pi}{4}$	-1
0	0
$\frac{\pi}{4}$	1
$\frac{\pi}{3}$	1.73
$\vdots$	$\vdots$
0.4999π	3183
$\frac{\pi}{2}(0.5\pi)$	unde fined
$\vdots$	$\vdots$

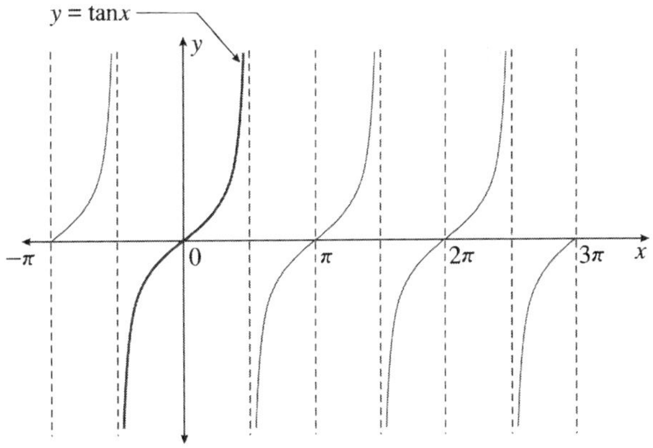

- Amplitude : none. Period : π
- Domain : $\{x \mid x \neq \frac{\pi}{2} + n\pi,\ n \in z\}$, Range : $y \mid y \in R$

(4) Amplitude and period of Trigonometric Functions

$y = a\sin bx \quad y = a\cos bx$: Amplitude $= |a|$, period $= \dfrac{2\pi}{|b|}$

$y = a\tan bx$: Amplitude = none, period $= \dfrac{\pi}{|b|}$

Example 10 Graph two periods of each function.

(1) $y = 2\sin x$

(2) $y = \sin(x)$

(3) $y = -2\sin(x)$

(4) $y = \sin\left(\dfrac{1}{2}x\right)$

Solution

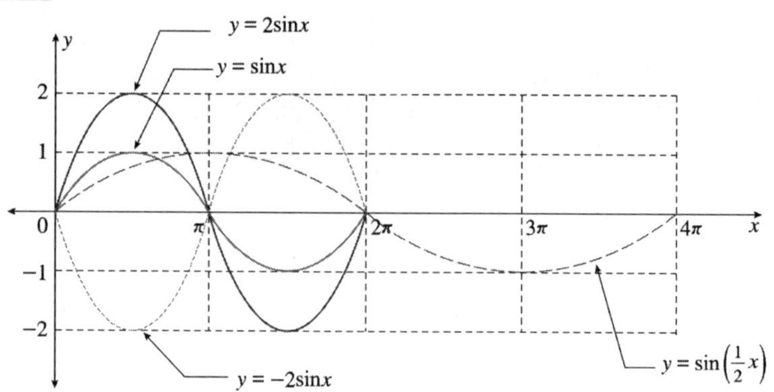

Example 11 Graph two periods of each function.

(1) $y = 4\sin 2x$

(2) $y = 2\cos\left(\dfrac{1}{2}x\right)$

(3) $y = 2\tan\left(\dfrac{1}{2}x\right)$

Solution

(1) Amp $= 4$

$P = \dfrac{2\pi}{2} = \pi$

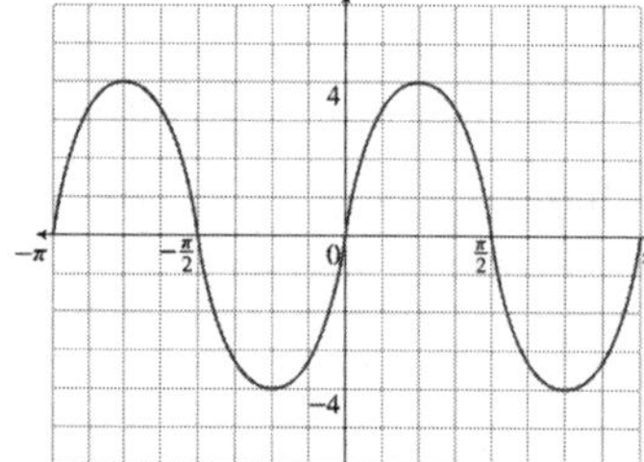

(2) Amp $= 2$

$P = \dfrac{2\pi}{1/2} = 4\pi$

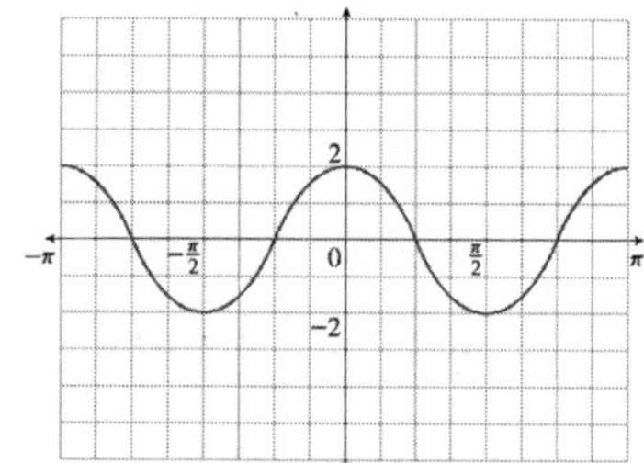

(3) $P = \frac{\pi}{1/2} = 2\pi$

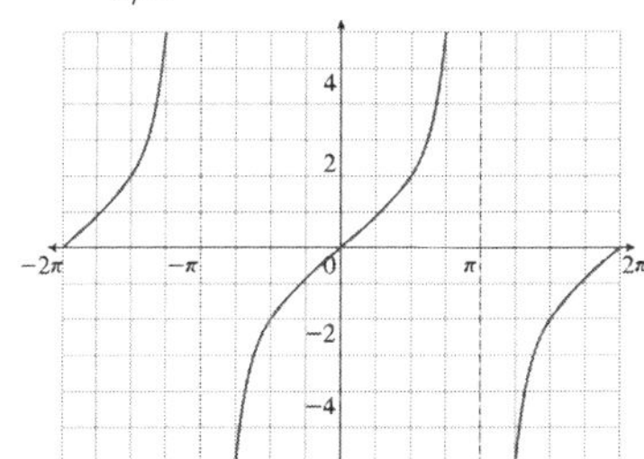

EX 8 Determin in the amplitude and period of each function.

(1) $y = 3\sin 2x$ (2) $y = -\frac{1}{2}\sin \pi x$

(3) $y = -4\sin\frac{1}{2}x$ (4) $y = 3\sin\left(\frac{\pi}{3}x\right)$

EX 9 Graph two periods of each function.

(1) $y = 3\cos 3x$ (2) $y = -2\cos(2\pi x)$

(3) $y = -6\cos\left(\frac{2}{3}x\right)$ (4) $y = -2\cos\left(\frac{\pi}{2}x\right)$

EX 10 Graph two periods of each function.

(1) $y = \tan 2x$ (2) $y = \tan\left(\frac{1}{2}x\right)$

(3) $y = 2\tan\left(\frac{x}{3}\right)$ (4) $y = -2\tan\left(\frac{\pi}{2}x\right)$

1.4.3 Inverse Trigonometric Functions

(1) $y = \sin^{-1}x = \arcsin x$

$y = \sin x$

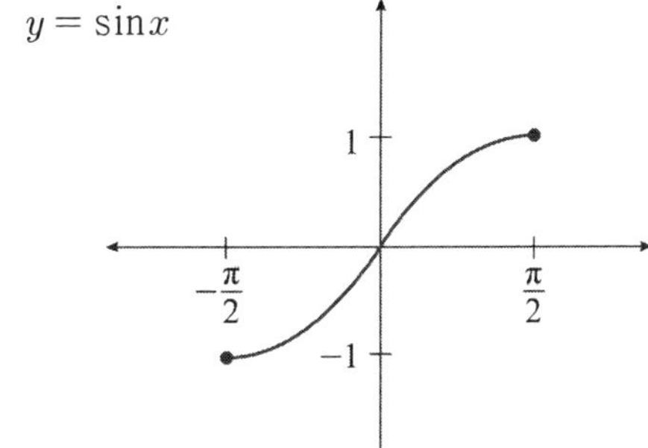

$y = \sin^{-1}x$

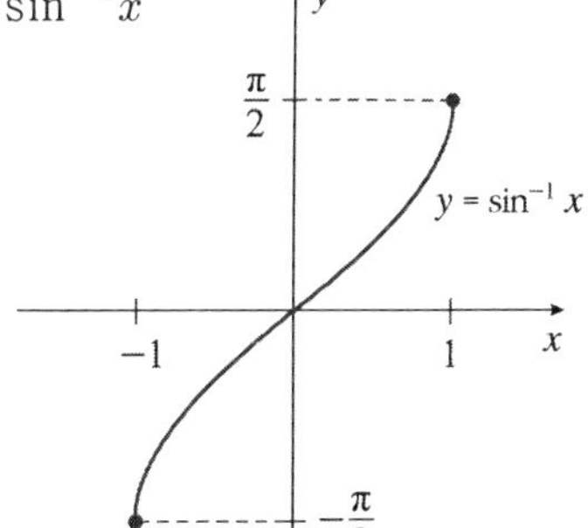

– Domain : $-\frac{\pi}{2} \le x \le \frac{\pi}{2}$

– Range : $-1 \le y \le 1$

– Domain : $-1 \le x \le 1$

– Range : $-\frac{\pi}{2} \le y \le \frac{\pi}{2}$

Example 12 Evaluate.

(1) $\sin^{-1}\left(-\frac{1}{2}\right)$ (2) $\sin^{-1}\left(\frac{\sqrt{3}}{2}\right)$

Solution

(1) $\sin^{-1}\left(-\frac{1}{2}\right)=-\frac{\pi}{6}$ (2) $\sin^{-1}\left(\frac{\sqrt{3}}{2}\right)=\frac{\pi}{3}$

Example 13 Evaluate.

(1) $\sin^{-1}\left(\sin\left(\frac{7\pi}{6}\right)\right)$ (2) $\sin^{-1}\left(\sin\left(\frac{2\pi}{3}\right)\right)$

Solution

(1) $\sin^{-1}\left(\sin\left(\frac{7\pi}{6}\right)\right)=\sin^{-1}\left(-\frac{1}{2}\right)$
$=-\frac{\pi}{6}$

(2) $\sin^{-1}\left(\sin\left(\frac{2\pi}{3}\right)\right)=\sin^{-1}\left(\frac{\sqrt{3}}{2}\right)$
$=\frac{\pi}{3}$

EX 11 Evaluate.

(1) $\sin^{-1}(-\frac{\sqrt{3}}{2})$ (2) $\sin^{-1}(1)$

(3) $\sin\left(\sin^{-1}\left(-\frac{1}{2}\right)\right)$ (4) $\cos(\sin^{-1}(-1))$

(2) $y=\cos^{-1}x=\arccos x$

$y=\cos x$

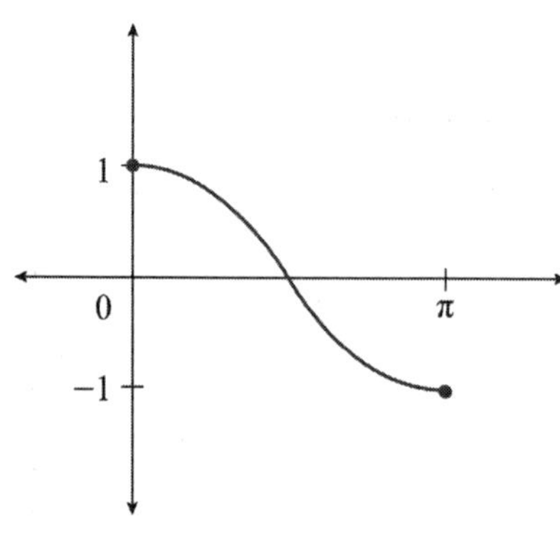

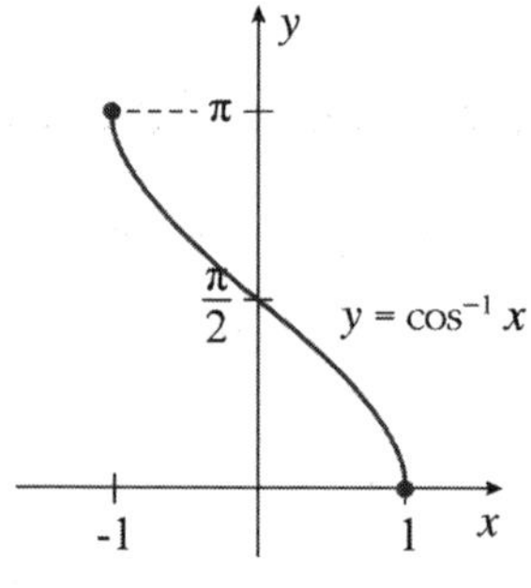

– Domain : $0 \le x \le \pi$
– Range : $-1 \le y \le 1$

– Domain : $-1 \le x \le 1$
– Range : $0 \le y \le \pi$

Example 14 Evaluate.

(1) $\cos^{-1}\left(-\frac{1}{2}\right)$ (2) $\cos^{-1}\left(\frac{\sqrt{3}}{2}\right)$

Solution

(1) $\cos^{-1}\left(-\frac{1}{2}\right)=\frac{2}{3}\pi$ (2) $\cos^{-1}\left(\frac{\sqrt{3}}{2}\right)=\frac{1}{6}\pi$

Example 15 Evaluate.

(1) $\cos^{-1}\left(\cos\left(\frac{5\pi}{6}\right)\right)$ (2) $\cos^{-1}\left(\cos\left(-\frac{\pi}{6}\right)\right)$

Solution

(1) $\cos^{-1}\left(\cos\left(\frac{5\pi}{6}\right)\right)=\cos^{-1}\left(-\frac{\sqrt{3}}{2}\right)=\frac{5\pi}{6}$ (2) $\cos^{-1}\left(\cos\left(-\frac{\pi}{6}\right)\right)=\cos^{-1}\left(\frac{\sqrt{3}}{2}\right)=\frac{\pi}{6}$

EX 12 Evaluate.

(1) $\cos^{-1}(-\frac{\sqrt{3}}{2})$ (2) $\cos^{-1}(0)$

(3) $\cos\left(\cos^{-1}\left(-\frac{1}{2}\right)\right)$ (4) $\tan\left(\cos^{-1}(1)\right)$

(3) $y=\tan^{-1}x=\arctan x$

$y=\tan x$

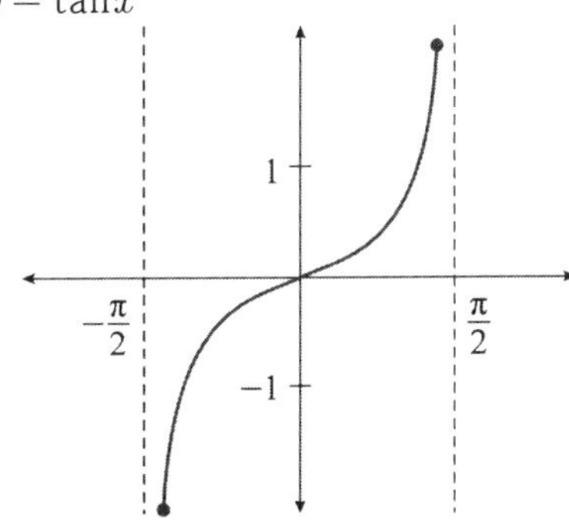

$y=\tan^{-1}x$

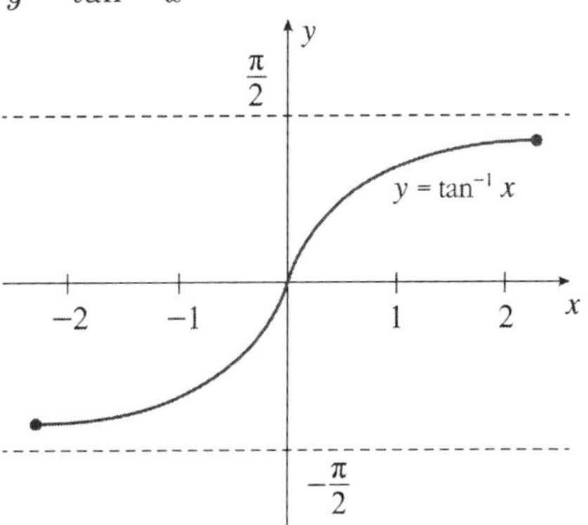

– Domain : $-\frac{\pi}{2}<x<\frac{\pi}{2}$
– Range : $y\in R$

– Domain : $\infty < x < \infty$
– Range : $-\frac{\pi}{2}<y<\frac{\pi}{2}$

Example 16 Evaluate.

(1) $\tan^{-1}(-1)$ (2) $\tan^{-1}(\sqrt{3})$

Solution

(1) $\tan^{-1}(-1)=-\dfrac{\pi}{4}$ (2) $\tan^{-1}(\sqrt{3})=\dfrac{\pi}{3}$

Example 17 Evaluate.

(1) $\tan^{-1}\left(\tan\left(\dfrac{3\pi}{4}\right)\right)$ (2) $\tan^{-1}\left(\tan\left(-\dfrac{7\pi}{6}\right)\right)$

Solution

(1) $\tan^{-1}\left(\tan\left(\dfrac{3\pi}{4}\right)\right)=\tan^{-1}(-1)$

$=-\dfrac{\pi}{4}$

(2) $\tan^{-1}\left(\tan\left(-\dfrac{7\pi}{6}\right)\right)=\tan^{-1}\left(-\dfrac{1}{\sqrt{3}}\right)$

$=-\dfrac{\pi}{6}$

EX 13 Evaluate.

(1) $\tan^{-1}(\dfrac{1}{\sqrt{3}})$ (2) $\tan^{-1}(0)$

(3) $\tan\left(\tan^{-1}\left(-\dfrac{1}{\sqrt{3}}\right)\right)$ (4) $\cos(\tan^{-1}(-\sqrt{3}))$

(4) $y=\csc^{-1}x,\ y=\sec^{-1}x,\ y=\cot^{-1}x$

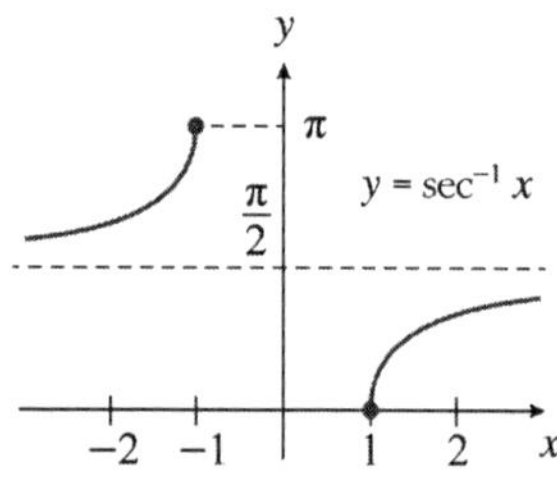

– Domain : $x\le -1$ or $x\ge 1$
– Range : $0\le y\le \pi,\ y\ne \dfrac{\pi}{2}$

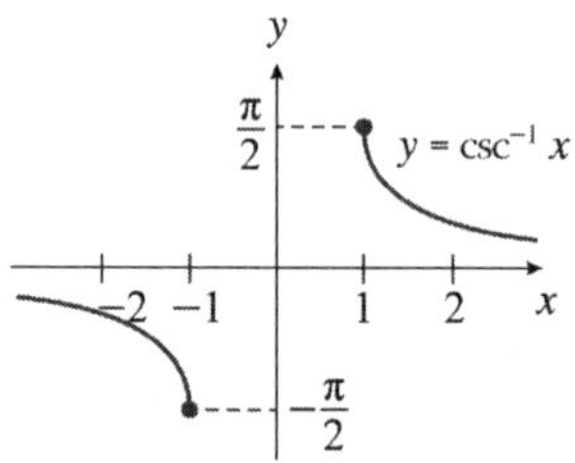

– Domain : $x\le -1$ or $x\ge 1$
– Range : $-\dfrac{\pi}{2}\le y\le \dfrac{\pi}{2},\ y\ne 0$

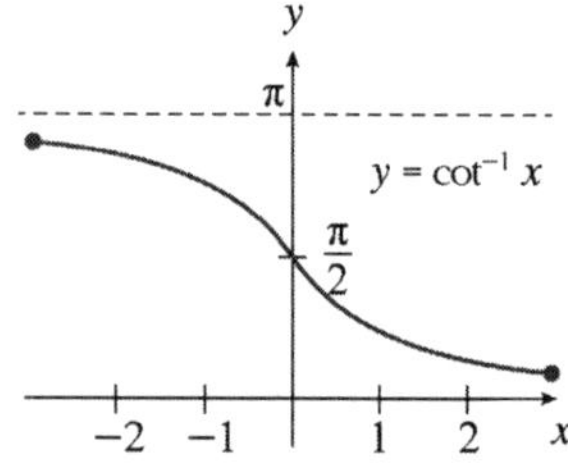

– Domain : $-\infty<x<\infty$
– Range : $0<y<\pi$

Example 18 Evaluate.

(1) $\csc^{-1}(2)$

(2) $\sec^{-1}(-\sqrt{2})$

(3) $\sec^{-1}(2)$

(4) $\cot^{-1}\left(\frac{1}{\sqrt{3}}\right)$

Solution

(1) $\csc^{-1}(2)=\sin^{-1}\left(\frac{1}{2}\right)$

$=\frac{\pi}{6}$

(2) $\sec^{-1}(-\sqrt{2})=\cos^{-1}\left(-\frac{1}{\sqrt{2}}\right)$

$=\frac{3\pi}{4}$

(3) $\sec^{-1}(2)=\cos^{-1}\left(\frac{1}{2}\right)$

$=\frac{\pi}{3}$

(4) $\cot^{-1}\left(\frac{1}{\sqrt{3}}\right)=\frac{\pi}{2}-\tan^{-1}\left(\frac{1}{\sqrt{3}}\right)$

$=\frac{\pi}{2}-\frac{\pi}{6}=\frac{\pi}{3}$

Example 19 Evaluate.

(1) $\sin\left(\cos^{-1}\left(\frac{\sqrt{2}}{2}\right)\right)$

(2) $\tan\left(\sin^{-1}\left(-\frac{1}{2}\right)\right)$

Solution

(1) $\sin\left(\cos^{-1}\frac{\sqrt{2}}{2}\right)=\sin\left(\frac{\pi}{4}\right)=\frac{1}{\sqrt{2}}$

(2) $\tan\left(\sin^{-1}\left(-\frac{1}{2}\right)\right)=\tan\left(-\frac{\pi}{6}\right)=-\frac{1}{\sqrt{3}}$

Multiple Choice

01 $(a^2)^3 + \dfrac{b^3 \cdot b^2}{b^{-4}} =$

(A) $a^5 + b$ (B) $a^6 + b^{10}$ (C) $a^5 + b^{10}$ (D) $a^6 + b^9$ (E) $a^6 \cdot b^5$

02 The domain of the function $y = \dfrac{x-3}{(x-2)(x+4)}$ is

(A) $x \neq 3$ (B) $x \neq 2$ and $x \neq -4$ (C) $x \geq -2$ (D) $x \leq 4$ (E) $-2 < x < 4$

03 In the figure below, the coordinates of points P are

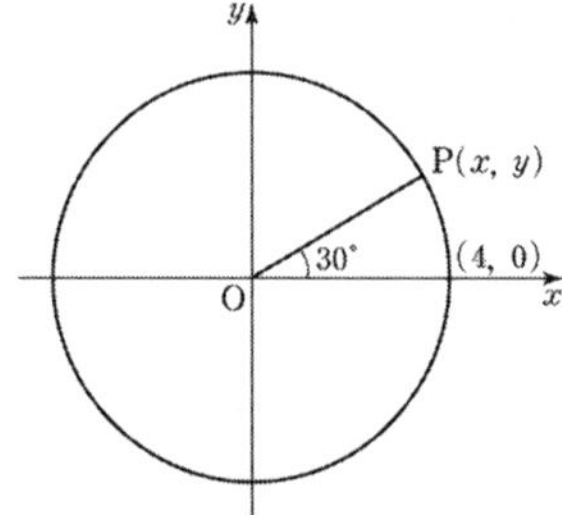

(A) (3.46, 3.46) (B) (3.46, 2) (C) (2, 2) (D) (2, 3.46) (E) (3, 1.73)

04 If $\sin^2\theta - \cos^2\theta = 0.5$, then $\sin^4\theta - \cos^4\theta =$

(A) 0.5 (B) 1.5 (C) 0.25 (D) $\sqrt{0.5}$ (E) 0

05 What is the area of the shaded region in the figure below?

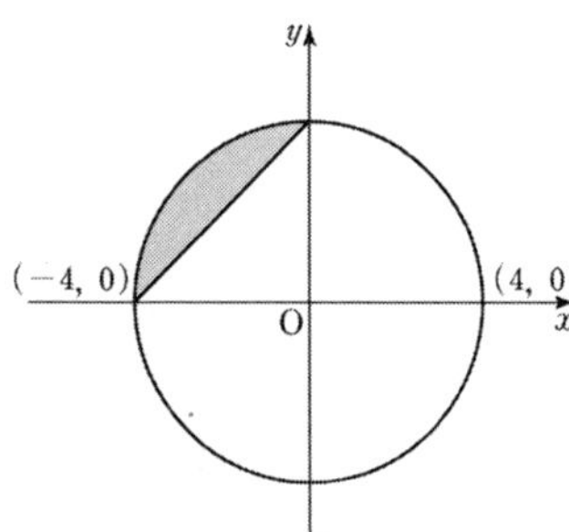

(A) 12.57 (B) 8 (C) 4.57 (D) 7.89 (E) 6.32

06 In the figure below, if $\cos\theta = 0.8$, and the length of segment $\overline{BC} = 8$, what is the perimeter of the triangle?

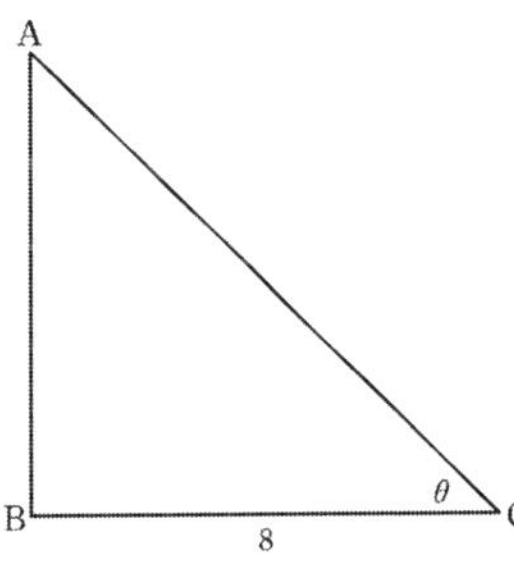

(A) 26 (B) 16 (C) 24 (D) 30 (E) 32

07 If $\arcsin(\sin x) = \frac{\pi}{6}$, and $0 \le x \le \frac{\pi}{2}$, then x could be

(A) 0 (B) $\frac{1}{2}$ (C) $\frac{\pi}{6}$ (D) $\frac{\pi}{3}$ (E) $\frac{\pi}{2}$

08 The value of $\ln e^3$ is

(A) 2.718 (B) 3.123 (C) 3 (D) 2.93 (E) 3.287

09 If the vertex of a function $f(x)$ is at (1,1), where is the vertex of the function $f(x-2)+1$?

(A) (2,3) (B) (3,2) (C) (2,−1) (D) (−1,0) (E) (−1,2)

10 If $f = (0,-3), (-1,-2), (2,-1)$ and $g = (-1,2), (-2,1), (-3,3)$, then $g \circ f(-1)$ equals

(A) −1 (B) 1 (C) 2 (D) −2 (E) Not defined

11 What are the asymptotes of the hyperbola $4x^2 - 9y^2 = 36$?

(A) $y = 9x$ and $y = -9x$ (B) $y = \frac{3x}{2}$ and $y = \frac{-3x}{2}$

(C) $y = 2x$ and $y = -2x$ (D) $y = \frac{2x}{3}$ and $y = \frac{-2x}{3}$

(E) $y = 4x$ and $y = -4x$

12 What is the exact value of $\tan(\arccos\frac{3}{4})$?

(A) $\frac{-3\sqrt{7}}{7}$ (B) $\frac{3\sqrt{7}}{7}$ (C) $\frac{-\sqrt{7}}{3}$ (D) $\frac{3}{4}$ (E) $\frac{\sqrt{7}}{3}$

13 To draw the graph of the inverse of a function $f(x)$, one must mirror the graph of $f(x)$ about the

(A) $x-axis$ (B) $y-axis$ (C) line $y=-x$

(D) line $y=x$ (E) lines $y=x$ and $y=-x$

14 The axis of symmetry for $f(x)=x^2-2x+3$ is $x=$

(A) 2 (B) 1 (C) −2 (D) −1 (E) $\frac{1}{2}$

15 Simplify $\sin^2 x+\frac{(\cos 2x)}{2}+\frac{1}{2}$

(A) $\sin 2x$ (B) 1 (C) $\sin^2 x+\sin(\frac{x}{2})$ (D) $\sin^2 x+\cos(\frac{x}{2})$ (E) −1

16 What must be the value of k if the lines $3x-y=9$ and $kx+3y=5$ are to be perpendicular?

(A) 1 (B) 3 (C) 9 (D) −1 (E) −9

17 If $\sec\theta=2$, then $\cos\theta\sec\theta=$

(A) 2 (B) 4 (C) $\frac{1}{4}$ (D) 1 (E) 0

18 $\log_2 16\sqrt{2}=$

(A) $\frac{7}{2}$ (B) 5 (C) 4 (D) 2 (E) $\frac{9}{2}$

19 For all x such that $x > 0$, $f(x) = \log_3 x$. What does $f^{-1}(x)$ equal?

(A) x^3 (B) 3^x (C) $\sqrt[3]{x}$ (D) $\log_x 3$ (E) $3^{\frac{1}{x}}$

20 The graph of $16x^2 + 8y^2 - 32x + 8y = 0$ is which of the following?

(A) a hyperbola (B) an ellipse (C) a parabola (D) a circle (E) a semicircle

21 Which of the following is the equation of a line with x-intercept (6,0) and y-intercept (0,−15)?

(A) $y = +\frac{5}{2}x - 15$ (B) $y = -\frac{5}{3}x - 15$ (C) $y = -\frac{5}{2}x - 6$
(D) $y = -\frac{2}{5}x + 15$ (E) $y = \frac{5}{2}x - 6$

22 If a function is an odd function, then $f(-x) = -f(x)$ for all values of x in the domain. Which of the following is an odd function?

(A) $f(x) = \sin x$ (B) $f(x) = \cos x$ (C) $f(x) = x^2 - 10$
(D) $f(x) = 4^x$ (E) $f(x) = \log_2 x$

23 What is the vertex of the parabola given by $f(x) = x^2 - 6x + 5$?

(A) (−3, 4) (B) (3, −4) (C) (3, 4) (D) (3, −3) (E) (4, −4)

24 If $\cos 2\theta = \frac{3}{4}$, then $\frac{1}{\cos^2\theta - \sin^2\theta} =$

(A) −1 (B) $\frac{3}{4}$ (C) $\frac{4}{3}$ (D) 4 (E) 1

25 Which of the following is a y-intercept of the hyperbola $\frac{y^2}{9} - \frac{x^2}{16} = 1$?

(A) (0, −3) (B) (0, 4) (C) (3, 0) (D) (0, 81) (E) (0, 9)

26 $\tan(\arccos 0.75) =$

(A) 0.66 (B) 0.75 (C) 0.80 (D) 0.88 (E) 41.4

27 What is the range of the piecewise function

$$f(x) = \begin{cases} x^2 & x > 3 \\ -\frac{2}{3}x + 11 & x \le 3 \end{cases}$$

(A) All real numbers (B) $y \ge 0$ (C) $-3 \le y \le 3$ (D) $y \le 3$ (E) $y \ge 9$

28 If the lines $y_1 = (n+2)x + 10$ and $y_2 = (n-4)x + 2$ are perpendicular, then n could equal which of the following?

(A) 3.45 (B) 3.83 (C) 1 (D) 0.5 (E) −6

29 If (x, y) is a point on the graph of $f(x)$, then which of the following is a point on the graph of $f^{-1}(x)$?

(A) $(-x, -y)$ (B) $(-x, y)$ (C) (y, x)
(D) $(x, -y)$ (E) $(-y, -x)$

30 If $f(x) = 6^x$, then $f(\log_7 \frac{1}{343}) =$

(A) -3 (B) -18 (C) $\frac{1}{216}$ (D) $\frac{1}{1296}$ (E) 216

02

Limits and Continuity

2.1 Limit Existence

The number L is the *limit of the function* $f(x)$ as x approaches c if, as the values of x get arbitrarily close (but not equal) to c the values of $f(x)$ approach (or equal) L. We write

$$\lim_{x \to c} f(x) = L$$

In order for $\lim_{x \to c} f(x)$ to exist, the values of f must tend to the same number L as x approaches c from either the left or the right. We write

$$\lim_{x \to c^-} f(x)$$

for the *left-hand limit of f* at c (as x approaches c through values *less* than c), and

$$\lim_{x \to c^+} f(x)$$

for the *right-hand limit of f* at c (as x approaches c through values *greater* than c)

Example 1 Given $f(x) = \dfrac{x^2 - x - 2}{x - 2}$, find the value of $\lim_{x \to 2} f(x)$.

Solution

$$\begin{aligned} y &= \frac{x^2 - x - 2}{x - 2} \\ &= \frac{(x-2)(x+1)}{x-2} \\ &= x + 1,\ x \neq 2 \end{aligned}$$

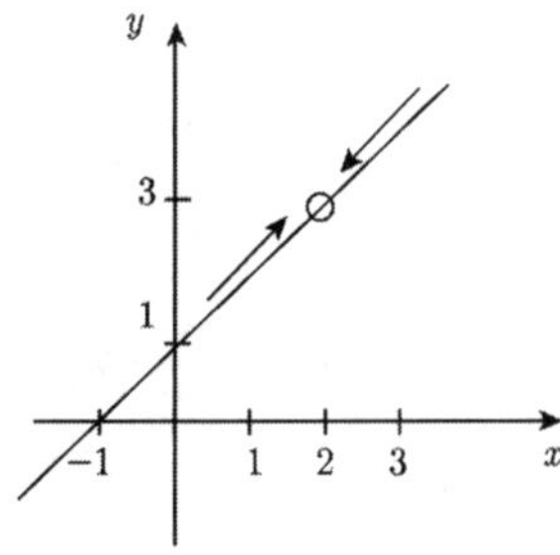

$$\lim_{x \to 2^-} f(x) = 3,\ \lim_{x \to 2^+} f(x) = 3$$

$$\lim_{x \to 2^-} f(x) = \lim_{x \to 2^+} f(x) = 3$$

$$\therefore \lim_{x \to 2} f(x) = 3$$

Example 2 Given $f(x) = \begin{cases} \dfrac{2(x-2)}{x^2-4}, & x \neq 2 \\ 2 & , x = 2 \end{cases}$, find the value of $\lim_{x \to 2} f(x)$.

Solution

$$y = \frac{2(x-2)}{x^2-4}$$

$$= \frac{2(x-2)}{(x-2)(x+2)}$$

$$= \frac{2}{x+2}, x \neq 2$$

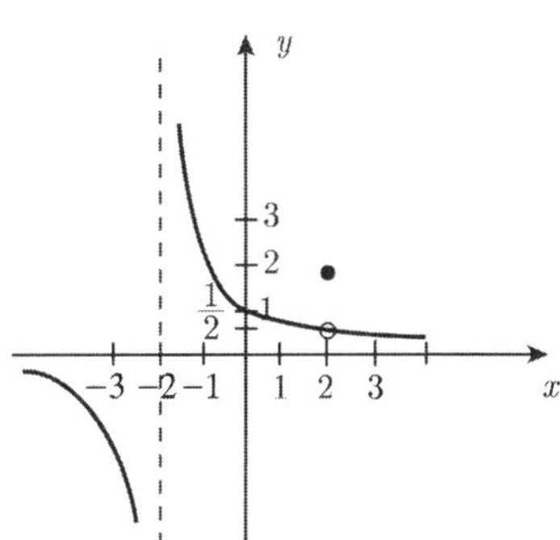

Vertical asymptote: $x = -2$

Horizontal asymptote: $y = 0$

$$\lim_{x \to 2^-} f(x) = \frac{1}{2}, \lim_{x \to 2^+} f(x) = \frac{1}{2}$$

$$\lim_{x \to 2^-} f(x) = \lim_{x \to 2^+} f(x) = \frac{1}{2}$$

$$\therefore \lim_{x \to 2} f(x) = \frac{1}{2}$$

Example 3 Given $f(x) = \begin{cases} 3-x, & x \leq 2 \\ x & , x > 2 \end{cases}$, find the value of $\lim_{x \to 2} f(x)$.

Solution

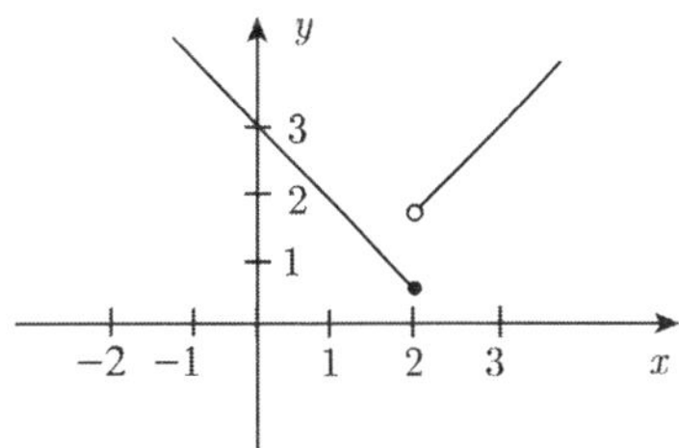

$$\lim_{x \to 2^-} f(x) = 1$$

$$\lim_{x \to 2^+} f(x) = 2$$

$$\lim_{x \to 2^-} f(x) \neq \lim_{x \to 2^+} f(x)$$

$\therefore \lim_{x \to 2} f(x) = DNE$ (Does Not Exist)

EX 1 Use the graph of f to evaluate the following limits.

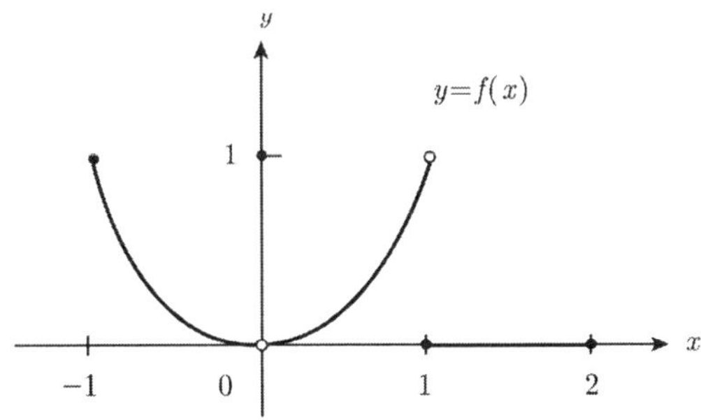

(1) $\lim_{x \to -1^+} f(x) =$

(2) $\lim_{x \to 0^+} f(x) =$

(3) $\lim_{x \to 0^-} f(x) =$

(4) $\lim_{x \to 0} f(x) =$

(5) $\lim_{x \to 1} f(x) =$

(6) $\lim_{x \to 1^+} f(x) =$

(7) $\lim_{x \to 2^-} f(x) =$

(8) $\lim_{x \to -1^-} f(x) =$

(9) $\lim_{x \to 2^+} f(x) =$

(10) $\lim_{x \to 1^-} f(x) =$

EX 2 Use the graph of f to evaluate the following limits.

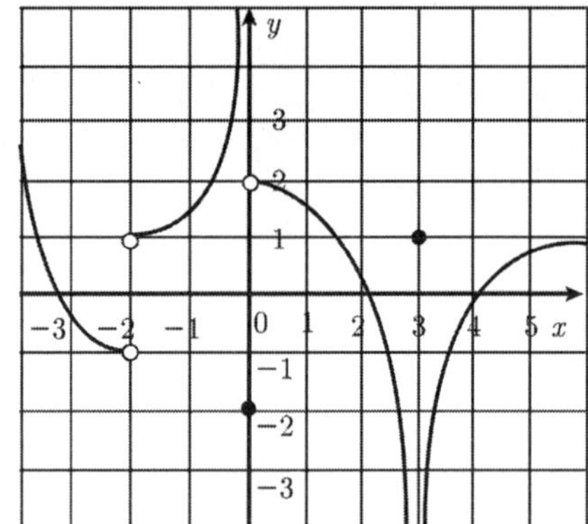

(1) (a) $\lim_{x \to -2^-} f(x) =$

(b) $\lim_{x \to -2^+} f(x) =$

(c) $\lim_{x \to -2} f(x) =$

(d) $f(-2) =$

(2) (a) $\lim_{x \to 0^-} f(x) =$

(b) $\lim_{x \to 0^+} f(x) =$

(c) $\lim_{x \to 0} f(x) =$

(d) $f(0) =$

(3) (a) $\lim_{x \to 3^-} f(x) =$

(b) $\lim_{x \to 3^+} f(x) =$

(c) $\lim_{x \to 3} f(x) =$

(d) $f(3) =$

(4) (a) $\lim_{x \to 4^-} f(x) =$

(b) $\lim_{x \to 4^+} f(x) =$

(c) $\lim_{x \to 4} f(x) =$

(d) $f(4) =$

EX 3 Sketch the graph of $y = f(x)$ and find the following limits.

(1) $f(x) = \begin{cases} 3 - x, & x < 2 \\ \dfrac{x}{2} + 1, & x > 2 \end{cases}$

(a) $\lim_{x \to 2^+} f(x) =$

(b) $\lim_{x \to 2^-} f(x) =$

(c) $\lim_{x \to 2} f(x) =$

(d) $\lim_{x \to 4^-} f(x) =$

(e) $\lim_{x \to 4^+} f(x) =$

(f) $\lim_{x \to 4} f(x) =$

(2) $f(x) = \begin{cases} 3 - x, & x < 2 \\ 2, & x = 2 \\ \dfrac{x}{2}, & x > 2 \end{cases}$

(a) $\lim_{x \to 2^+} f(x) =$

(b) $\lim_{x \to 2^-} f(x) =$

(c) $f(2) =$

(d) $\lim_{x \to 2} f(x) =$

(e) $\lim_{x \to -1^-} f(x) =$

(f) $\lim_{x \to -1^+} f(x) =$

(g) $\lim_{x \to -1} f(x) =$

(3) $f(x) = \begin{cases} 1 - x^2, & x \le 0 \\ 1/x, & x > 0 \end{cases}$

(a) $\lim_{x \to 0^-} f(x) =$

(b) $\lim_{x \to 0^+} f(x) =$

(c) $\lim_{x \to 0} f(x) =$

(d) $f(0) =$

(4) $f(x) = \begin{cases} \dfrac{x^2 - 1}{x + 1}, & x \ne -1 \\ 3, & x = -1 \end{cases}$

(a) $\lim_{x \to -1^-} f(x) =$

(b) $\lim_{x \to -1^+} f(x) =$

(c) $\lim_{x \to -1} f(x) =$

(d) $f(-1) =$

(5) $f(x) = \begin{cases} x^2 + 2x, & x \le -1 \\ x, & -1 < x < 1 \\ 1 - x, & x \ge 1 \end{cases}$

(a) $\lim_{x \to -1^-} f(x) =$

(b) $\lim_{x \to -1^+} f(x) =$

(c) $\lim_{x \to -1} f(x) =$

(d) $\lim_{x \to 1^-} f(x) =$

(e) $\lim_{x \to 1^+} f(x) =$

(f) $\lim_{x \to 1} f(x) =$

(6) $f(x) = \begin{cases} \dfrac{x-1}{x^2-3x+2}, & x \neq 1 \\ 1, & x = 1 \end{cases}$

(a) $\lim_{x\to 1^-} f(x) =$

(b) $\lim_{x\to 1^+} f(x) =$

(c) $\lim_{x\to 1} f(x) =$

(d) $f(1) =$

- **Limit Laws If L, M, c, and k are real numbers and $\lim_{x\to c} f(x)=L$ and $\lim_{x\to c} g(x)=M$, then**

1. Sum Rule: $\lim_{x\to c}(f(x)+g(x)) = L+M$
2. Difference Rule: $\lim_{x\to c}(f(x)-g(x)) = L-M$
3. Constant Multiple Rule: $\lim_{x\to c}(k\cdot f(x)) = k\cdot L$
4. Product Rule: $\lim_{x\to c}(f(x)\cdot g(x)) = L\cdot M$
5. Quotient Rule: $\lim_{x\to c}\dfrac{f(x)}{g(x)} = \dfrac{L}{M},\ M\neq 0$
6. Power Rule: $\lim_{x\to c}[f(x)]^n = L^n$, n a positive integer
7. Root Rule: $\lim_{x\to c}\sqrt[n]{f(x)} = \sqrt[n]{L} = L^{1/n}$, n a positive integer

(If n is even, we assume that $\lim_{x\to c} f(x) = L > 0$.)

EX 4 Use the properties of limits to evaluate the following limits.

(1) $\lim_{x\to c}(x^3+4x^2-3)$

(2) $\lim_{x\to c}\dfrac{x^4+x^2-1}{x^2+5}$

(3) $\lim_{x\to 3}\dfrac{\sqrt{x+1}-1}{x-1}$

(4) $\lim_{x\to -2}\sqrt{4x^2-3}$

EX 5 Suppose $\lim_{x\to c} f(x)=5$ and $\lim_{x\to c} g(x)=-2$. Find

(1) $\lim_{x\to c} f(x)g(x)$

(2) $\lim_{x\to c} 2f(x)g(x)$

(3) $\lim_{x\to c}(f(x)+3g(x))$

(4) $\lim_{x\to c}\dfrac{f(x)}{f(x)-g(x)}$

- **Indeterminate form of type $\frac{0}{0}$**

  – Rational Expressions: Step 1. Factor the numerator and denominator.
  Step 2. Cancel the common factor.

  – Radical Expressions: Step 1. Multiply the numerator and denominator by the conjugate.
  Step 2. Cancel the common factor.

Example 4 Evaluate the following limit.

$$\lim_{x\to 3}\frac{x^2-9}{x^3-27}$$

Solution

$$\lim_{x\to 3}\frac{x^2-9}{x^3-27}=\lim_{x\to 3}\frac{(x-3)(x+3)}{(x-3)(x^2+3x+9)}$$
$$=\lim_{x\to 3}\frac{x+3}{x^2+3x+9}$$
$$=\frac{3+3}{3^2+3(3)+9}=\frac{6}{27}=\frac{2}{9}$$

Example 5 Evaluate the following limit.

$$\lim_{x\to 0}\frac{\sqrt{4+x}-2}{x}$$

Solution

$$\lim_{x\to 0}\frac{\sqrt{4+x}-2}{x}=\lim_{x\to 0}\frac{(\sqrt{4+x}-2)(\sqrt{4+x}+2)}{x(\sqrt{4+x}+2)}$$
$$=\lim_{x\to 0}\frac{4+x-4}{x(\sqrt{4+x}+2)}=\lim_{x\to 0}\frac{x}{x(\sqrt{4+x}+2)}$$
$$=\lim_{x\to 0}\frac{1}{\sqrt{4+x}+2}=\frac{1}{\sqrt{4+0}+2}=\frac{1}{4}$$

EX 6 Evaluate the following limit.

(1) $\lim_{t\to 1}\frac{t^2+t-2}{t^2-1}$

(2) $\lim_{x\to 4}\frac{x^2-4x}{x^2-5x+4}$

(3) $\lim_{x\to 2}\frac{x^3-8}{x^4-16}$

(4) $\lim_{x\to 2}\frac{x^3-8}{x^2-3x+2}$

(5) $\lim_{y\to 0}\frac{5y^3+8y^2}{3y^4-16y^2}$

(6) $\lim_{x\to 2}\frac{x^3-2x^2-4x+8}{x^3-2x^2}$

(7) $\lim_{h\to 0}\dfrac{(1+h)^4-1}{h}$

(8) $\lim_{x\to 1}\dfrac{(x+1)^3-8}{x-1}$

EX7 Evaluate the following limit.

(1) $\lim_{x\to 2}\dfrac{\sqrt{x^2+12}-4}{x-2}$

(2) $\lim_{x\to 4}\dfrac{x-4}{2-\sqrt{x}}$

(3) $\lim_{x\to 1}\dfrac{x-1}{\sqrt{x+3}-2}$

(4) $\lim_{x\to 3}\dfrac{\sqrt{x+1}-2}{\sqrt{x-2}-1}$

(5) $\lim_{x\to 2}\dfrac{\sqrt{x+2}-\sqrt{3x-2}}{\sqrt{5x-1}-\sqrt{4x+1}}$

EX8 Evaluate the following limit.

(1) $\lim_{x\to 0}\dfrac{\dfrac{1}{x-1}+\dfrac{1}{x+1}}{x}$

(2) $\lim_{h\to 0}\dfrac{\dfrac{1}{(h+3)^2}-\dfrac{1}{9}}{h}$

(3) $\lim_{x\to 9}\dfrac{\dfrac{1}{\sqrt{x}}-\dfrac{1}{3}}{x-9}$

(4) $\lim_{x\to 2}\dfrac{\dfrac{1}{\sqrt{x+2}}-\dfrac{1}{2}}{x^2-4}$

- **Definition of Absolute Value**

$$|x|=\begin{cases} x & \text{if } x\ge 0 \\ -x & \text{if } x<0 \end{cases}$$

- $[x]$ **represents the greatest integer less than or equal to** x

(eg) $[3.9]=3,\ [2.1]=2,\ [-1.9]=-2,\ [-2.1]=-3$

EX9 Evaluate the following limit.

(1) (a) $\lim_{x\to -2^+}(x+3)\dfrac{|x+2|}{x+2}$ (b) $\lim_{x\to -2^-}(x+3)\dfrac{|x+2|}{x+2}$

(2) (a) $\lim_{x\to 1^+}\dfrac{\sqrt{2}(x-1)}{|x-1|}$ (b) $\lim_{x\to 1^-}\dfrac{\sqrt{2}(x-1)}{|x-1|}$

(3) (a) $\lim_{\theta\to 3^+}\dfrac{[\theta]}{\theta}$ (b) $\lim_{\theta\to 3^-}\dfrac{[\theta]}{\theta}$

(4) (a) $\lim_{t\to 4^+}(t-[t])$ (b) $\lim_{t\to 4^-}(t-[t])$

- $\lim_{x \to 0^+} \frac{1}{x} = +\infty$, $\lim_{x \to 0^-} \frac{1}{x} = -\infty$, $\lim_{x \to 0} \frac{1}{x} = DNE$ (Does Not Exist)
- $\lim_{x \to 0^+} \frac{1}{x^2} = +\infty$, $\lim_{x \to 0^-} \frac{1}{x^2} = +\infty$, $\lim_{x \to 0} \frac{1}{x^2} = +\infty$

Example 6 Evaluate the following limit.

(1) $\lim_{x \to 2^+} \frac{4}{x-2}$

(2) $\lim_{x \to 2^-} \frac{4}{x-2}$

(3) $\lim_{x \to 2} \frac{4}{x-2}$

Solution

(1) $\lim_{x \to 2^+} \frac{4}{x-2} = \frac{4}{+0.000\ldots\ldots1} = +\infty$

(2) $\lim_{x \to 2^-} \frac{4}{x-2} = \frac{4}{-0.000\ldots\ldots1} = -\infty$

(3) $\lim_{x \to 2} \frac{4}{x-2} = DNE$

EX 10 Evaluate the following limit.

(1) $\lim_{x \to 2^-} \frac{3}{x-2}$

(2) $\lim_{x \to 2^-} \frac{-1}{x-2}$

(3) $\lim_{x \to -8^+} \frac{2x}{x+8}$

(4) $\lim_{x \to 2} \frac{1}{x-2}$

(5) $\lim_{x \to 0} \frac{4}{x^{2/5}}$

(6) $\lim_{x \to 3^-} \frac{-1}{(x-3)^2}$

(7) $\lim_{x \to 7} \frac{4}{(x-7)^2}$

(8) $\lim_{x \to 3} \frac{1}{|x-3|}$

(9) $\lim_{x \to 0^+} \frac{2}{3x^{1/3}}$

(10) $\lim_{x \to 0^-} \frac{2}{3x^{(1/3)}}$

(11) $\lim_{x \to (\pi/2)^-} \tan x$

(12) $\lim_{\theta \to 0^-} (1 + \csc\theta)$

EX 11 Without sketching the graph of f. find the following limits. (Use the definition of the limit existence)

(1) Let $f(x) = \begin{cases} x+2 & , x < -1 \\ x^2+1 & , x \geq -1 \end{cases}$. Find $\lim_{x \to -1} f(x)$.

(2) Let $f(x) = \begin{cases} 4\cos(\pi \frac{x}{9}) & , x < 3 \\ \sqrt{x^2-5} & , x \geq 3 \end{cases}$. Find $\lim_{x \to 3} f(x)$.

(3) Let $f(x) = \begin{cases} e^x + 3 & , x \leq \ln 3 \\ e^{2x} - 3 & , x > \ln 3 \end{cases}$. Find $\lim_{x \to \ln 3} f(x)$.

EX 12 Find the values of a and b that satisfy the following equations.

(1) $\lim_{x\to 2}\dfrac{x^2-4x+a}{x-2}=b$ (2) $\lim_{x\to 1}\dfrac{x^2-a}{x-1}=b$

(3) $\lim_{x\to 1}\dfrac{a\sqrt{x+1}-b}{x-1}=\sqrt{2}$ (4) $\lim_{x\to 2}\dfrac{2-x}{ax^2+x-6}=b \ (b\neq 0)$

2.2 Squeezing Theorem (Sandwich Theorem)

- **The Sandwich Theorem**

Suppose that $g(x)\leq f(x)\leq h(x)$ for all x in some open interval containing c, except possibly at $x=c$ itself.

Suppose also that $\lim_{x\to c}g(x)=\lim_{x\to c}h(x)=L$,

$$\text{Then } \lim_{x\to c}f(x)=L.$$

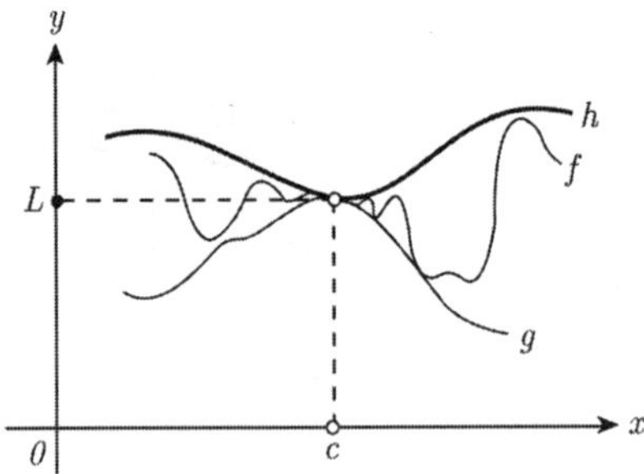

Example 1 Given that $1-\dfrac{x^2}{4}\leq u(x)\leq 1+\dfrac{x^2}{2}$ for all $x\neq 0$, find $\lim_{x\to 0}u(x)$, no matter how complicated u is.

Solution

Since

$\lim_{x\to 0}(1-(x^2/4))=1$ and

$\lim_{x\to 0}(1-(x^2/2))=1$,

the Sandwich Theorem implies that

$\lim_{x\to 0}u(x)=1$

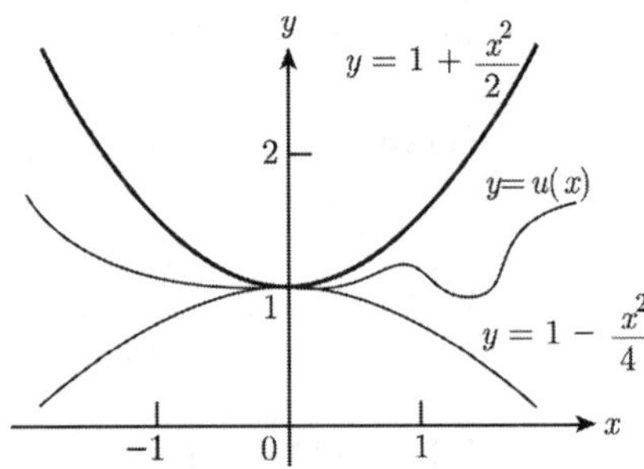

Use the squeezing Theorem to prove the following limits.

(1) $\lim_{x\to 0} x^2\cos\frac{1}{x}=0$　　(2) $\lim_{x\to 0} x\sin(\frac{1}{x^2})=0$

(3) $\lim_{x\to\infty}\frac{\sin x}{x}=0$　　(4) $\lim_{x\to\infty}\frac{\cos 2x}{x^2}=0$

EX 2

(1) Given $1-\frac{x^2}{6}<\frac{x\sin x}{2-2\cos x}<1$ for all x, find $\lim_{x\to 0}\frac{x\sin x}{2-2\cos x}$.

(2) Given $-6x-4<f(x)<x^4+x^2$ for all x, find $\lim_{x\to -1} f(x)$.

(3) Given $\sqrt{5-2x}<f(x)<\sqrt{5}$ for all x, find $\lim_{x\to 0} f(x)$.

2.3 Limit of Trigonometric Function

- $\lim_{\theta\to 0}\frac{\sin\theta}{\theta}=1$ (θ in radians)

Proof

The plan is to show that the right-hand and left-hand limits are both 1. Then we will know that the two-sided limit is 1 as well.

(i) To show that the right-hand is 1, we begin with positive values of θ less than $\pi/2$,

Area $\triangle OAP <$ area sector $OAP <$ area $\triangle OAT$

We can express these areas in terms of θ as follows.

Area $\triangle OAP = \frac{1}{2}$base $\times$ height

$= \frac{1}{2}(1)(\sin\theta) = \frac{1}{2}\sin\theta$

Area sector $OAP = \frac{1}{2}r^2\theta = \frac{1}{2}(1)^2\theta = \frac{\theta}{2}$

Area $\triangle OAT = \frac{1}{2}$base $\times$ height $= \frac{1}{2}(1)(\tan\theta)$

$= \frac{1}{2}\tan\theta$

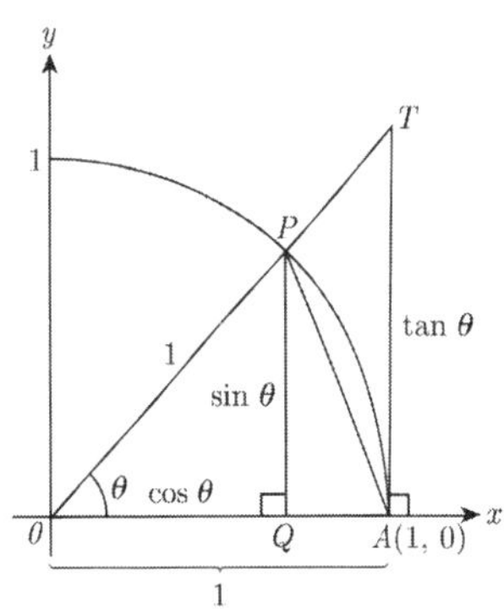

Thus,

$$\frac{1}{2}\sin\theta < \frac{1}{2}\theta < \frac{1}{2}\tan\theta.$$

This last inequality goes the same way if we divide all three terms by the number (1/2) $\sin\theta$, which is positive since $0 < \theta < \pi/2$:

$$1 < \frac{\theta}{\sin\theta} < \frac{1}{\cos\theta}$$

Taking reciprocals reverses the inequalities:

$$1 > \frac{\sin\theta}{\theta} > \frac{\cos\theta}{\theta}$$

Since $\lim_{\theta \to 0+} \cos\theta = 1$, the Sandwich Theorem gives

$$\lim_{\theta \to 0^+} \frac{\sin\theta}{\theta} = 1.$$

(ii) To show that the left-hand is 1, we begin with negative values of θ more than $-\pi/2$, let $\theta = -x$

$$\lim_{\theta \to 0^-} \frac{\sin\theta}{\theta} = \lim_{x \to 0^+} \frac{\sin(-x)}{(-x)} = \lim_{x \to 0^+} \frac{\sin x}{x} = 1.$$

by (i),(ii)

$$\lim_{\theta \to 0} \frac{\sin\theta}{\theta} = 1.$$

The formula $\lim_{\theta \to 0} \frac{\sin\theta}{\theta} = 1$ can be generalized . If $\lim_{x \to c} f(x) = 0$ and $f(x)$ is never zero in an open interval containing the point $x = c$, except possibly c itself, then $\lim_{x \to c} \frac{\sin f(x)}{f(x)} = 1.$

Example 1 Evaluate the following limits.

(a) $\lim_{x \to 0} \frac{\sin x^2}{x^2}$

(b) $\lim_{x \to 0} \frac{\sin x^2}{x}$

(c) $\lim_{x \to -1} \frac{\sin(x^2 - x - 2)}{x+1}$

(d) $\lim_{x \to 1} \frac{\sin(1 - \sqrt{x})}{x - 1}$

Solution

(a) $\lim_{x\to 0}\frac{\sin x^2}{x^2}=1$

(b) $\lim_{x\to 0}\frac{\sin x^2}{x}=\lim_{x\to 0}\frac{\sin x^2}{x^2}\lim_{x\to 0}\frac{x^2}{x}=1\cdot 0=0$

(c) $\lim_{x\to -1}\frac{\sin(x^2-x-2)}{x+1}=\lim_{x\to -1}\frac{\sin(x^2-x-2)}{x^2-x-2}\cdot\lim_{x\to -1}\frac{(x^2-x-2)}{x+1}$

$=1\cdot\lim_{x\to -1}\frac{(x+1)(x-2)}{x+1}=-3$

(d) $\lim_{x\to 1}\frac{\sin(1-\sqrt{x})}{x-1}=\lim_{x\to 1}\frac{\sin(1-\sqrt{x})}{1-\sqrt{x}}\frac{1-\sqrt{x}}{x-1}=1\cdot\lim_{x\to 1}\frac{(1-\sqrt{x})(1+\sqrt{x})}{(x-1)(1+\sqrt{x})}$

$=\lim_{x\to 1}\frac{1-x}{(x-1)(1+\sqrt{x})}=-\frac{1}{2}$

EX 1 Evaluate the following limits.

(1) $\lim_{\theta\to 0}\frac{\sin\sqrt{2}\theta}{\sqrt{2}\theta}$ (2) $\lim_{x\to 0}\frac{\sin 2x}{3x}$ (3) $\lim_{y\to 0}\frac{\sin 3y}{4y}$

(4) $\lim_{x\to 0}\frac{\sin 5x}{\sin 2x}$ (5) $\lim_{x\to 0}\frac{\sin^3 2x}{x^3}$ (6) $\lim_{x\to 0}\frac{\sin^2 3x}{\sin^2 2x}$

(7) $\lim_{x\to 0}\frac{2x}{\tan x}$ (8) $\lim_{x\to 0}\frac{\tan 3x}{2x}$ (9) $\lim_{x\to 0}\frac{\tan 3x}{\sin 8x}$

(10) $\lim_{x\to 0}\frac{\tan^3 3x}{8x^3}$ (11) $\lim_{x\to 0}\frac{\sin 2x}{x^2+3x}$ (12) $\lim_{x\to 0}\frac{\sin^3 x}{x^2}$

(13) $\lim_{x\to 0}\frac{\sin(1-\cos x)}{x}$ (14) $\lim_{x\to 0^+}\frac{\sin x}{\sin\sqrt{x}}$ (15) $\lim_{x\to 0}\frac{\sin(x^2+x)}{x}$

(16) $\lim_{x\to 0}\frac{x-x\cos x}{\sin^2 3x}$ (17) $\lim_{x\to 0}\frac{\sin(1-\cos x)}{x^2}$

2.4 Limit of Exponential and Logarithmic Functions

- **Exponential** e

$$\lim_{x\to\pm\infty}\left(1+\frac{1}{x}\right)^x=\lim_{x\to 0}(1+x)^{\frac{1}{x}}=2.718281828459045235=e$$

x	$(1+x)^{\frac{1}{x}}$	x	$(1+x)^{\frac{1}{x}}$
0.1	2.59374⋯	- 0.1	2.86797⋯
0.01	2.70481⋯	- 0.01	2.73199⋯
0.001	2.71692⋯	- 0.001	2.71964⋯
0.0001	2.71814⋯	- 0.0001	2.71841⋯
0.00001	2,71827⋯	- 0.00001	2,71829⋯
……	……	……	……

Example 1 Evaluate the following limits.

(1) $\lim_{x\to\infty}\left(1+\frac{3}{x}\right)^{\frac{x}{3}}$ (2) $\lim_{x\to\infty}(1+\frac{3}{x})^{x}$

Solution

(1) $\lim_{x\to\infty}\left(1+\frac{3}{x}\right)^{\frac{x}{3}}=\lim_{x\to 0}(1+5x)^{\frac{1}{5x}}=e$

(2) $\lim_{x\to\infty}(1+\frac{3}{x})^{x}=\lim_{x\to\infty}[(1+\frac{3}{x})^{\frac{x}{3}}]^{3}=e^{3}$

EX 1 Evaluate the following limits.

(1) $\lim_{x\to 0}(1+3x)^{\frac{1}{x}}$ (2) $\lim_{x\to\infty}\left(1+\frac{3}{x}\right)^{x}$ (3) $\lim_{x\to 0}(1+x)^{\frac{3}{x}}$

(4) $\lim_{x\to -\infty}\left(1+\frac{2}{3x}\right)^{x}$ (5) $\lim_{x\to 0}(1+3x)^{\frac{1}{2x}}$ (6) $\lim_{n\to\infty}P\left(1+\frac{r}{n}\right)^{nt}$

(7) $\lim_{x\to\infty}\left(1+\frac{3}{x}\right)^{x}$ (8) $\lim_{x\to 0}\left(1+\frac{x}{2}\right)^{\frac{1}{2x}}$ (9) $\lim_{x\to\infty}\left(1+\frac{1}{x}\right)^{3x}$

(10) $\lim_{x\to 0}\left(1-\frac{x}{2}\right)^{\frac{1}{x}}$ (11) $\lim_{x\to\infty}\left(1+\frac{3}{x}\right)^{2x}$ (12) $\lim_{x\to\infty}\left(\frac{2x+1}{x}\right)^{3x}$

EX 2 Evaluate the following limits.

(1) $\lim_{x\to 0}\frac{\ln(x+1)}{x}$

(2) $\lim_{x\to 0}\frac{e^{x}-1}{x}$ (Hint: Let $e^{x}-1=t$)

2.5 Limits Involving Infinity

• Limit of Rational Function in the form $\frac{\infty}{\infty}$.

Divide both the numerator and the denominator by the highest power of x in the denominator.

Example 1

(1) $\lim_{x\to\infty}\frac{5x^2+8x-3}{3x^2+2}$ (2) $\lim_{x\to-\infty}\frac{11x+2}{2x^3-1}$

Solution

(1) $\lim_{x\to\infty}\frac{5x^2+8x-3}{3x^2+2}=\lim_{x\to\infty}\frac{5+(8/x)-(3/x^2)}{3+(2/x^2)}=\frac{5+0-0}{3+0}=\frac{5}{3}$

(2) $\lim_{x\to-\infty}\frac{11x+2}{2x^3-1}=\lim_{x\to-\infty}\frac{(11/x^2)+(2/x^3)}{2-(1/x^3)}=\frac{0+0}{2-0}=0$

EX 1 Evaluate the following limits.

(1) $\lim_{x\to-\infty}\left(\frac{1-x^3}{x^2+7x}\right)^5$ (2) $\lim_{x\to\infty}\frac{3x^3+4}{2x^3-3x+5}$ (3) $\lim_{x\to\infty}\frac{2\sqrt{x}+x^{-1}}{3x-7}$

(4) $\lim_{x\to\infty}\frac{3x^2+1}{5x^2-2}$ (5) $\lim_{x\to-\infty}\frac{\sqrt[3]{x}-\sqrt[5]{x}}{\sqrt[3]{x}+\sqrt[5]{x}}$ (6) $\lim_{x\to-\infty}\frac{2x^4-4x}{x^5+9x-5}$

(7) $\lim_{x\to\infty}\frac{2x^{5/3}-x^{1/3}+7}{x^{8/5}+3x+\sqrt{x}}$ (8) $\lim_{x\to\infty}\frac{\sqrt{x^2+1}}{x+1}$ (9) $\lim_{x\to-\infty}\frac{\sqrt{x^2+1}}{x+1}$

(10) $\lim_{x\to\infty}\frac{x-3}{\sqrt{4x^2+25}}$ (11) $\lim_{x\to-\infty}\frac{4-3x^3}{\sqrt{x^6+9}}$

• Horizontal Asymptote

A line $y=b$ is a horizontal asymptote of the graph of a function $y=f(x)$ if either

$$\lim_{x\to\infty}f(x)=b \text{ or } \lim_{x\to-\infty}f(x)=b.$$

Example 2 Find the equation of the horizontal asymptote for thd following function

The graph of the function

$$f(x)=\frac{5x^2+8x-3}{3x^2+2}$$

Solution

sketched in Figure has the line $y=5/3$ as a horizontal asymptote on both the right and the left because

$$\lim_{x\to\infty} f(x)=\frac{5}{3} \text{ and } \lim_{x\to-\infty} f(x)=\frac{5}{3}.$$

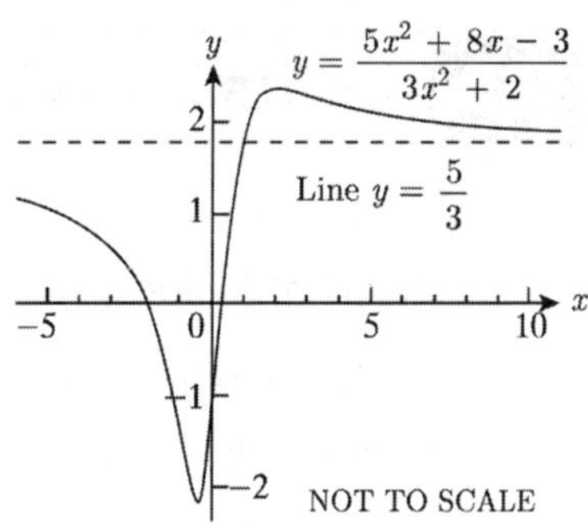

- **Horizontal Asymptote of Rational Functions**

$$y=\frac{ax^m+\cdots\cdots}{bx^n+\cdots\cdots}$$

(1) $m=n$: $y=\frac{a}{b}$ is the horizontal asymptote

(2) $m<n$: $y=0$ is the horizontal asymptote

(3) $m>n$: there is no horizontal asymptote

EX 2 Find the equation of horizontal asymptote for the following functions.

(1) $y=\frac{2x+3}{5x+7}$ (2) $y=\frac{1-x^2}{2x^2+1}$ (3) $y=\frac{2x^4+7}{x^3-x^2+x+7}$

(4) $y=\frac{x+1}{1-x^4}$ (5) $y=\frac{x+1}{x^2+3}$ (6) $y=\frac{(1-x)(2x-1)(x+2)}{(x-2)(x^2+x+2)}$

(7) $y=\frac{7}{x+4}+5$ (8) $y=\frac{2x}{x^2+4}+2$ (9) $y=-\frac{4x}{2x-3}+5$

- **Vertical Asymptote**

A line $x=a$ is a vertical asymptote of the graph of a function. $y=f(x)$ if either

$$\lim_{x\to a^+} f(x)=\pm\infty \text{ or } \lim_{x\to a^-} f(x)=\pm\infty.$$

Example 3 Find the horizontal and vertical asymptotes of graph of $f(x)=-\frac{8}{x^2-4}$

The behavior as $x\to\pm 2$. Since

$$\lim_{x\to 2^+} f(x)=-\infty \text{ and } \lim_{x\to 2^-} f(x)=\infty,$$

the line $x=2$ is a vertical asymptote both from the right and from the left. By symmetry, the line $x=-2$ is also a vertical asymptote.

There are no other asymptotes because f has a finite limit at every other point.

EX 3 (a) Find the equation(s) of the vertical asymptote(s). (b) Describe the behaviour of the function to the left and right of each horizontal asymptote.

(1) $y=\dfrac{2+x}{x-3}$ (2) $y=\dfrac{x^2}{x+2}$ (3) $y=\dfrac{1}{x^2}$

(4) $y=\dfrac{x^2}{1-x^2}$ (5) $y=\dfrac{2x}{x^2-4}$ (6) $y=\dfrac{x}{\sqrt{2-x}}+3$

EX 4 Find the equation of the (a) horizontal and (b) vertical asymptotes for the following functions.

(1) $y=\dfrac{2x}{|x|-1}$ (2) $y=\dfrac{x-1}{2|x|+1}$

(3) $y=\dfrac{5(4)^x+2}{4^x-1}$ (4) $y=\dfrac{3(2)^x+1}{2^x-3}$

EX 5 Evaluate the following limits.

(1) $\lim\limits_{x\to\left(\frac{\pi}{2}\right)^-}\tan x$ (2) $\lim\limits_{x\to\left(\frac{\pi}{2}\right)^+}\sec x$

(3) $\lim\limits_{x\to 0^-}\csc x$ (4) $\lim\limits_{x\to 0^+}\cot x$

(5) $\lim\limits_{x\to 0^+}\ln x$ (6) $\lim\limits_{x\to 2^-}\log_3(2-x)$

EX 6 Evaluate the following limits.

(1) $\lim\limits_{x\to\infty}e^{-x}$ (2) $\lim\limits_{x\to-\infty}(0.9)^x$

(3) $\lim\limits_{x\to 0^-}e^{\frac{1}{x}}$ (4) $\lim\limits_{x\to 3^+}(0.9)^{\frac{1}{x-3}}$

(5) $\lim\limits_{x\to 0^+}e^{\csc x}$ (6) $\lim\limits_{x\to 1^-}\dfrac{1}{\ln(1-x)}$

(7) $\lim\limits_{x\to-\infty}\dfrac{3^x+1}{3^x-1}$ (8) $\lim\limits_{x\to-\infty}\tan^{-1}x$

EX 7 Determine the following limits.

(1) $\lim\limits_{x\to\infty}\sin x$ (2) $\lim\limits_{x\to\infty}\dfrac{\sin x}{x}$

(3) $\lim\limits_{x\to 0}\sin\left(\dfrac{1}{x}\right)$ (4) $\lim\limits_{x\to\infty}x\sin\dfrac{1}{x}$

EX 8 Determine the following limits.

(1) $\lim_{x \to -\infty} \dfrac{-3x+1}{\sqrt{2x^2+1}}$

(2) $\lim_{x \to \infty} \dfrac{x+3}{\sqrt{x^2+1}+2}$

(3) $\lim_{x \to -\infty} \dfrac{x+3}{\sqrt{x^2+1}+2x}$

(4) $\lim_{x \to \infty} \dfrac{\sqrt{x^2-1}+3}{2x}$

EX 9 Determine the following limits.

(1) $\lim_{x \to \infty} (\sqrt{x^2+3x} - \sqrt{x^2-2x})$

(2) $\lim_{x \to -\infty} (\sqrt{x^2+x}+x)$

(3) $\lim_{x \to -\infty} (\sqrt{4x^2+3x-2}+2x)$

(4) $\lim_{x \to \infty} (2x - \sqrt{4x^2+x})$

EX 10 For the function f whose graph is given, determine the following limits.

(1) $\lim_{x \to 4} f(x)$

(2) $\lim_{x \to 2^+} f(x)$

(3) $\lim_{x \to 2^-} f(x)$

(4) $\lim_{x \to 2} f(x)$

(5) $\lim_{x \to -3^+} f(x)$

(6) $\lim_{x \to -3^-} f(x)$

(7) $\lim_{x \to -3} f(x)$

(8) $\lim_{x \to 0^+} f(x)$

(9) $\lim_{x \to 0^-} f(x)$

(10) $\lim_{x \to 0} f(x)$

(11) $\lim_{x \to \infty} f(x)$

(12) $\lim_{x \to -\infty} f(x)$

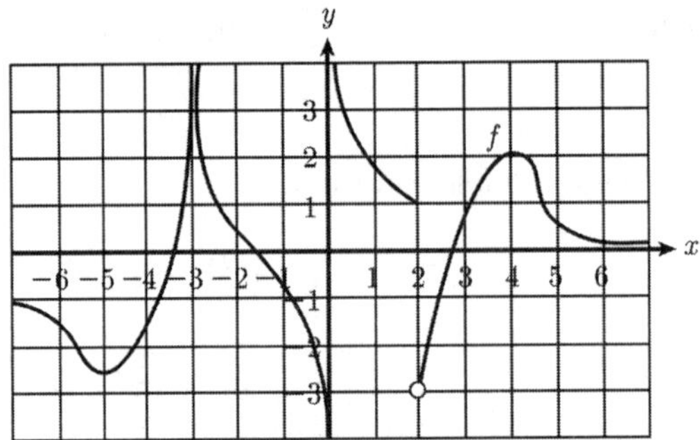

EX 11 Sketch a graph of function $f(x)$ that satisfies all of the following conditions.

(1) The domain of f is $(-\infty, 3)$.

The range of f is $(-\infty, 1)$.

$\lim_{x \to 3^-} f(x) = -\infty$.

$\lim_{x \to -\infty} f(x) = 1$.

$f(-2) = f(-1) = f(1) = 0$.

$\lim_{x \to -1} f(x) = -1$.

(2) The domain of f is all real numbers.

The range of f is $(-3, \infty)$.

$\lim_{x\to\infty} f(x) = 1$ and $\lim_{x\to-\infty} f(x) = -3$.

$\lim_{x\to 0} f(x) = +\infty$.

$\lim_{x\to-2^-} f(x) = -1$ and $\lim_{x\to-2^+} f(x) = 0$.

$f(-2) = f(0) = -2$.

2.6 Continuity

• Continuity Test

A function $f(x)$ is continuous at an interior point $x = c$ of its domain if and only if it meets the following three conditions.

1. $f(c)$ exists (c lies in the domain of f).
2. $\lim_{x\to c} f(x)$ exists (f has a limit as $x\to c$).
3. $\lim_{x\to c} f(x) = f(c)$ (the limit equals the function value).

Example 1 Explain why the function f below is discontinuous at $x = -2, 0, 3$ and 4.

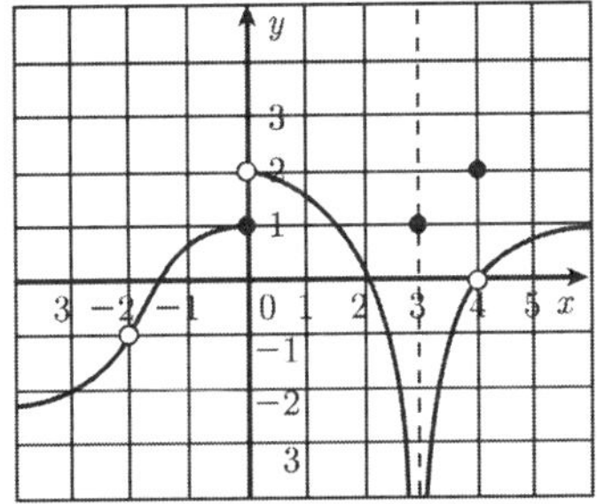

Solution

At $x = -2$, $f(-2)$ does not exist. (Removable Discontinuity)

At $x = 0$, $\lim_{x\to 0} f(x)$ does not exist. (Jump Discontinuity)

At $x = 3$, $\lim_{x\to 3} f(x) = -\infty$. (Infinite Discontinuity)

At $x = 4$, $0 = \lim_{x\to 4} f(x) \neq f(4) = 2$. (Removable Discontinuity)

EX 1 (a) Explain why the function f below is discontinuous at x = − 3, −1 , 1, 3 and 5

(b) State the type of discontinuity at each point.

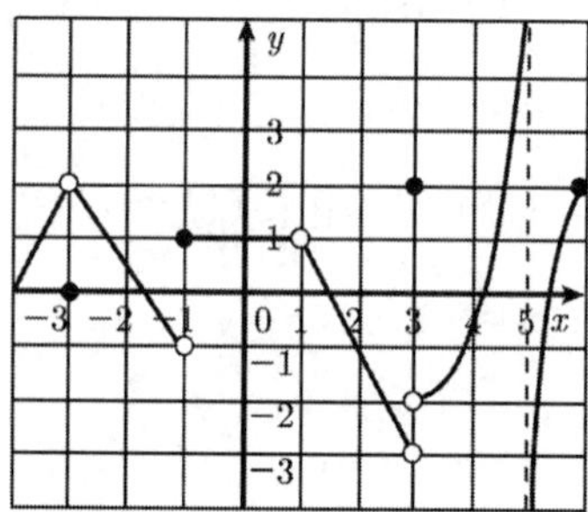

(1) At $x = -3$ (2) At $x = -1$ (3) At $x = 1$

(4) At $x = 3$ (5) At $x = 5$

EX 2 Sketch the graph of the following function and determine any values of x at which f is discontinuous.

(1) $f(x)=\begin{cases} x+3 & , x \le -2 \\ x^2 & , -2 < x < 1 \\ 1/x & , x \ge 1 \end{cases}$ (2) $f(x)=\begin{cases} |x+3|-1 & , x < -2 \\ \sqrt{4-x^2} & , |x| \le 2 \\ \sqrt{x-2} & , x > 2 \end{cases}$

(3) $f(x)=\begin{cases} x+4, & x \le -1 \\ \dfrac{|x|}{x}, & |x|<1 \\ 1, & x=1 \\ (x-2)^2, & x>1 \end{cases}$ (4) $f(x)=\begin{cases} \sqrt{1-x^2}, & |x|<1 \\ \dfrac{x+1}{x-1}, & |x|>1 \\ 0, & x=-1 \\ 1, & x=1 \end{cases}$

EX 3 Determine any values of x of which f is continuous

(1) Define $f(3)$ in a way that extends $f(x)=(x^2-9)/(x-3)$ to be continuous at $x=3$.

(2) Define $f(2)$ in a way that extends $f(t)=(t^2+3t-10)/(t-2)$ to be continuous at $t=2$.

(3) Define $f(1)$ in a way that extends $f(s)=(s^3-1)/(s^2-1)$ to be continuous at $s=1$.

(4) Define $f(4)$ in a way that extends $f(x)=(x^2-16)/(x^2-3x-4)$ to be continuous at $x=4$.

EX 4 Use the three−part definition of continuity to prove f in continuous

(1) $f(x)=\begin{cases} x+3 & , x \le -1 \\ \sqrt{x+5} & , x > -1 \end{cases}$ (2) $f(x)=\begin{cases} 3-\ln x & , 0 < x < 1 \\ 2e^{\sin(\pi x)}+1 & , x \ge 1 \end{cases}$

EX 5 Find the constant k that makes $f(x)$ continuous everywhere.

(1) $f(x)=\begin{cases}\dfrac{x^2-4}{x-2}, & x\neq 2\\ k, & x=2\end{cases}$

(2) $f(x)=\begin{cases}\dfrac{x^2-2x}{x}, & x\neq 0\\ k, & x=0\end{cases}$

(3) $f(x)=\begin{cases}\dfrac{x^3+1}{x+1}, & x\neq -1\\ k+3, & x=-1\end{cases}$

(4) $f(x)=\begin{cases}\dfrac{\sin x}{x}, & x\neq 0\\ k, & x=0\end{cases}$

EX 6 Find the constants a, b that makes $f(x)$ continuous everywhere.

(1) For what values of a is $f(x)=\begin{cases}a^2x-2a, & x\geq 2\\ 12, & x<2\end{cases}$ continuous at every x?

(2) For what values of b is $g(x)=\begin{cases}\dfrac{x-b}{b+1}, & x<0\\ x^2+b, & x>0\end{cases}$ continuous at every x?

(3) For what values of a, b is $f(x)=\begin{cases}-2, & x\leq -1\\ ax-b, & -1<x<1\\ 3, & x\geq 1\end{cases}$ continuous at every x?

(4) For what values of a, b is $g(x)=\begin{cases}ax+2b, & x\leq 0\\ x^2+3a-b, & 0<x\leq 2\\ 3x-5, & x\geq 2\end{cases}$ continuous at every x?

2.7 Average Rate of Change and Instantaneous Rate of Change

The average rate of change of $y=f(x)$ with respect to x over the interval $[x_1, x_2]$ is

$$\frac{\Delta y}{\Delta x}=\frac{f(x_2)-f(x_1)}{x_2-x_1}=\frac{f(x_1+h)-f(x_1)}{h},\ h\neq 0.$$

A secant to the graph $y=f(x)$. Its slope is $\Delta y/\Delta x$, the average rate of change of f over the interval $[x_1, x_2]$.

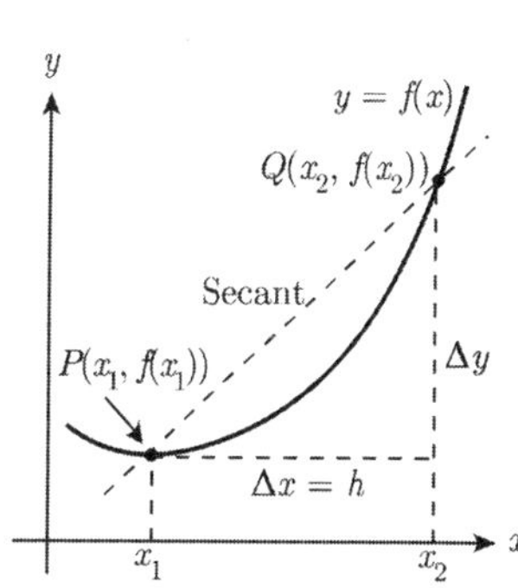

Example 1 Find the average rate of change for $f(x) = x^2 + 1$, with respect to x over the interval $[1, 3]$.

Solution

$$\frac{f(3) - f(1)}{3 - 1} = \frac{10 - 2}{3 - 1} = 4$$

Ex 1 Find the average rate of change for the function over the indicated interval.

(1) $f(x) = x^3$ $\quad [a, b]$

(2) $f(x) = x^2 + 2x$ $\quad [2, 4]$

(3) $f(x) = \cot x$ $\quad \left[\frac{\pi}{4}, \frac{3\pi}{4}\right]$

(4) $f(x) = \sqrt{2x - 1}$ $\quad [1, 5]$

(5) $f(x) = \frac{1}{x}$ $\quad [a, a + h]$

(6) $f(x) = \sin x + \cos x$ $\quad \left[-\frac{\pi}{6}, \frac{\pi}{3}\right]$

- **The Instantaneous Rate of Change of f with respect to x at $x = a$: Tangent Slope**

$$f'(a) = \frac{dy}{dx}\Big|_{x=a} = \lim_{x \to a} \frac{f(x) - f(a)}{x - a} \text{ or } \lim_{h \to 0} \frac{f(a+h) - f(a)}{h}$$

The tangent slope is the slope of the line tangent to the curve at a point $x = a$.
As point B approaches point A, the secant slope of $\overline{AB}$ becomes the slope of the tangent at point A.

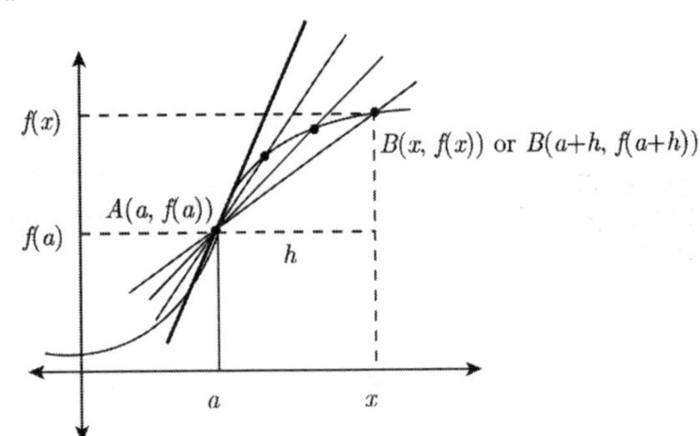

Example 2 Find the slope of the tangent for $f(x) = x^2 + 1$ at $x = 2$.

Solution

(Method 1)

$$f'(2) = \lim_{x \to 2} \frac{f(x) - f(2)}{x - 2} = \lim_{x \to 2} \frac{x^2 + 1 - 5}{x - 2} = \lim_{x \to 2} \frac{x^2 - 4}{x - 2} = \lim_{x \to 2}(x + 2) = 4$$

(Method 2)

$$f'(2) = \lim_{h \to 0} \frac{f(2+h) - f(2)}{h} = \lim_{h \to 0} \frac{(2+h)^2 + 1 - 5}{h} = \lim_{h \to 0} \frac{h^2 + 4h}{h} = \lim_{h \to 0}(h + 4) = 4$$

EX 2 (a) Find the slope of the tangent to each curve at the point indicated by using two methods.

(Method 1) $f'(a) = \lim_{x \to a} \dfrac{f(x) - f(a)}{x - a}$

(Method 2) $f'(a) = \lim_{h \to 0} \dfrac{f(a+h) - f(a)}{h}$

(b) Write the equation of the tangent line in point-slope form $y - y_1 = m(x - x_1)$ $(m = f'(x_1))$

(1) $f(x) = x^2 + 2x$ $at\ (1, 3)$

(Method 1) (Method 2)

(2) $f(x) = x^2 - 6x$ $at\ (2, -8)$

(Method 1) (Method 2)

(3) $f(x) = x^3$ $at\ (-1, -1)$

(Method 1) (Method 2)

(4) $f(x) = \dfrac{1}{x+1}$ $at\ (1, \dfrac{1}{2})$

(Method 1) (Method 2)

(5) $f(x) = \sqrt{x-2}$ $at\ x = 6$

(Method 1) (Method 2)

(6) $f(x) = \dfrac{1}{\sqrt{x+1}}$ $at\ x = 3$

(Method 1) (Method 2)

2.8 Intermediate Value Theorem (I.V.T.)

If f is a continuous function on a closed interval $[a, b]$, and if y_0 is any value between $f(a)$ and $f(b)$, then $f(c) = y_0$ for some c in $[a, b]$.

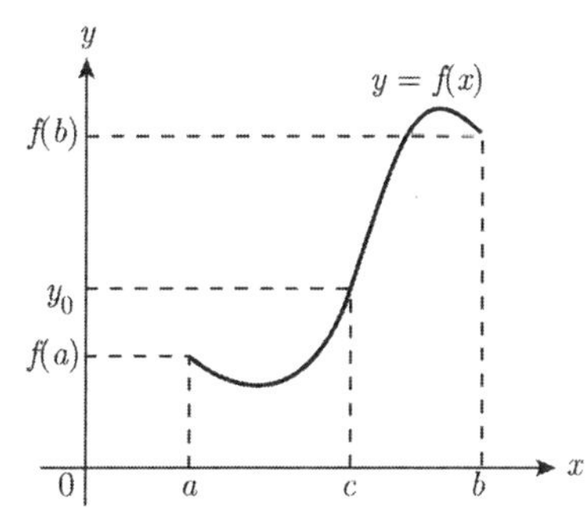

Example 1 Show that the equation $x^3-2x-2=0$ has a solution in the interval $[1, 2]$.

Solution

Let $f(x)=x^3-2x-2$.

Since f is a polynomial function, it is continuous on the interval $[1, 2]$.

$$f(1)=1-2-2=-3 \text{ and } f(2)=8-4-2=2, \text{ so } f(1)<0<f(2).$$

By the Intermediate Value Theorem, there must be at least one solution c, $1<c<2$, such that $f(c)=0$.

(1) Use the Intermediate Value Theorem to show that the equation $x^3-3x=8$ has at least one solution in the interval $[2, 3]$.

(2) Use the Intermediate Value Theorem to show that the equation $x^3+x^2-2x-6=0$ has at least one solution in the interval $[1, 2]$.

EX 2 A particle is moving along a horizontal line. The particle's velocity v and acceleration a are continuous functions of time t. The values are shown below.

t (seconds)	0	5	10	15	20
$v(t)$ (meters/second)	−3	−1	5	10	13
$a(t)$ (meters/second^2)	1	3	2	6	5

(1) Explain why there must be a value t for $5<t<10$, such that $v(t)=0$.

(2) Explain why there must be a value t for $10<t<20$, such that $a(t)=4$.

(1) A function f is continuous on the closed interval $[1, 3]$ and has values that are shown in the table. The equation $f(x)=0$ has at least two roots on the interval $[1, 3]$. Find the values of M.

x	1	2	3
$f(x)$	−2	M	−3

(2) A function f is continuous on the closed interval $[-2, 2]$ and has values that are shown in the table. The equation $f(x)=0$ has at least one positive roots in the interval $[-2, 2]$. Find the values of M.

x	−2	0	2
$f(x)$	−2	M	3

Multiple Choice

01 If f is a function such that $\lim_{x\to2}\frac{f(x)-f(2)}{x-2}=0$, which of the following must be true?

(A) The limit of $f(x)$ as x approaches 2 does not exist.

(B) f is not defined at $x=2$.

(C) The derivative of f at $x=2$ is 0.

(D) f is continuous at $x=0$.

(E) $f(2)=0$.

02 $\lim_{x\to\infty}(1+5e^x)^{\frac{1}{x}}$ is

(A) 0 (B) 1 (C) e (D) e^5 (E) nonexistent

03 If $f(x)=2x^2+1$, then $\lim_{x\to0}\frac{f(x)-f(0)}{x^2}$ is

(A) 0 (B) 1 (C) 2 (D) 4 (E) nonexistent

04 $\lim_{x\to0}(1+2x)^{\csc x}=$

(A) 0 (B) 1 (C) 2 (D) e (E) e^2

05 $\lim_{h\to0}\frac{\ln(e+h)-1}{h}$ is

(A) $f'(e)$, where $f(x)=\ln x$

(B) $f'(e)$, where $f(x)=\frac{\ln x}{x}$

(C) $f'(1)$, where $f(x)=\ln x$

(D) $f'(1)$, where $f(x)=\ln(x+e)$

(E) $f'(0)$, where $f(x)=\ln x$

06 If k is a positive integer, the $\lim_{x\to+\infty}\frac{x^k}{e^x}$ is

(A) 0 (B) 1 (C) e (D) $k!$ (E) nonexistent

07 Let f be the function given by $f(x)=\dfrac{(x-1)(x^2-4)}{x^2-a}$. For what positive values of a is f continuous for all real numbers x?

(A) None (B) 1 only (C) 2 only (D) 4 only (E) 1 and 4 only

08 $\lim\limits_{x\to0}\dfrac{1-\cos^2(2x)}{x^2}=$

(A) −2 (B) 0 (C) 1 (D) 2 (E) 4

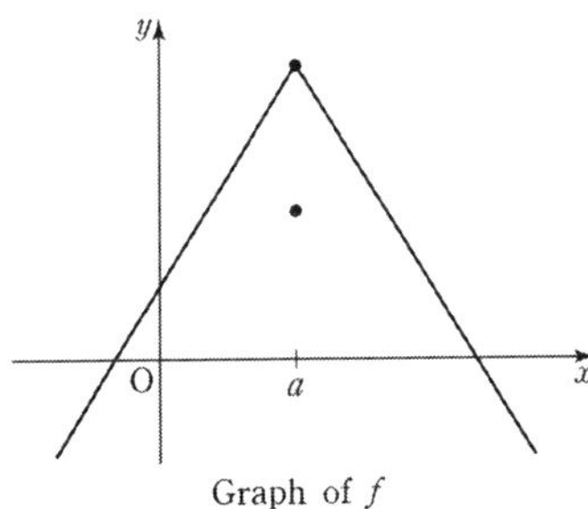

Graph of f

09 The graph of the function f is shown above. Which of the following statements must be false?

(A) $f(a)$ exists.

(B) $f(x)$ is defined for $0<x<a$.

(C) f is not continuous at $x=a$.

(D) $\lim\limits_{x\to a}f(x)$ exists.

(E) $\lim\limits_{x\to a}f'(x)$ exists

10 What is $\lim\limits_{h\to0}\dfrac{8\left(\frac{1}{2}+h\right)^8-8\left(\frac{1}{2}\right)^8}{h}$?

(A) 0

(B) $\dfrac{1}{2}$

(C) 1

(D) The limit does not exist.

(E) It cannot be determined from the information given

11 The line $y = 5$ is a horizontal asymptote to the graph of which of the following functions?

(A) $y = \dfrac{\sin(5x)}{x}$ (B) $y = 5x$ (C) $y = \dfrac{1}{x-5}$

(D) $y = \dfrac{5x}{1-x}$ (E) $y = \dfrac{20x^2 - x}{1+4x^2}$

12 $\displaystyle\lim_{x\to 3} \frac{x-3}{x^2-2x-3}$ is

(A) 0 (B) 1 (C) 1/4 (D) ∞ (E) none of these

13 $\displaystyle\lim_{x\to\infty} \frac{4-x^2}{4x^2-x-2}$ is

(A) -2 (B) $-1/4$ (C) 1 (D) 2 (E) nonexistent

14. $\displaystyle\lim_{x\to\infty} \frac{3x^2+27}{x^3-27}$ is

(A) 3 (B) ∞ (C) 1 (D) -1 (E) 0

15 $\displaystyle\lim_{x\to -\infty} \frac{2^{-x}}{2^x}$ is

(A) -1 (B) 1 (C) 0 (D) ∞ (E) none of these

16 $\displaystyle\lim_{x\to 0} \frac{\sin x}{x^2+3x}$ is

(A) 1 (B) $\dfrac{1}{3}$ (C) 3 (D) ∞ (E) $\dfrac{1}{4}$

17 $\displaystyle\lim_{x\to\infty} \frac{2x^2+1}{(2-x)(2+x)}$ is

(A) -4 (B) -2 (C) 1 (D) 2 (E) nonexistent

18 $\displaystyle\lim_{x\to\infty} x\sin\frac{1}{x}$ is

(A) 0 (B) ∞ (C) nonexistent (D) -1 (E) 1

19 If $\begin{cases} f(x) = \dfrac{x^2 - x}{2x} & \text{for } x \neq 0, \\ f(0) = k & \text{for } x = 0 \end{cases}$, when f is continuous at $x = 0$, $k = ?$

(A) -1 (B) $-\dfrac{1}{2}$ (C) 0 (D) $\dfrac{1}{2}$ (E) 1

20 The graph of y=$\dfrac{2x^2 + 2x + 3}{4x^2 - 4x}$ has

(A) a horizontal asymptote at y=$\pm\dfrac{1}{2}$ but no vertical asymptote

(B) no horizontal asymptote but two vertical asymptotes,at x=0 and x=1

(C) a horizontal asymptote at y=$\dfrac{1}{2}$ and two vertical asymptotes,at x=0 and x=1

(D) a horizontal asymptote at x=2 but no vertical astmptote

(E) a horizontal asymptote at y=$\dfrac{1}{2}$ and two vertical asymptotes,at x=± 1

21 Let $f(x) = \begin{cases} \dfrac{x^2 + x}{x} & \text{if } x \neq 0 \\ 1 & \text{if } x = 0 \end{cases}$

Which of the following statements is(are) true ?

Ⅰ. $f(0)$ exists

Ⅱ. $\lim\limits_{x \to 0} f(x)$ exists

Ⅲ. f is continuous at $x = 0$

(A) Ⅰ only (B) Ⅱ only (C) Ⅰ and Ⅱ only

(D) Ⅰ,Ⅱ,and Ⅲ (E) none of Ⅰ,Ⅱ,or Ⅲ

22 $\lim\limits_{x \to 0} \sqrt{3 + \arctan\dfrac{1}{x}}$ is

(A) $-\infty$ (B) $\sqrt{3 - \dfrac{\pi}{2}}$ (C) $\sqrt{3 + \dfrac{\pi}{2}}$ (D) ∞ (E) nonexistent

23 If $y = \dfrac{1}{2 + 10^{1/x}}$, then $\lim\limits_{x \to 0} y$ is

(A) 0 (B) $\dfrac{1}{12}$ (C) $\dfrac{1}{2}$ (D) $\dfrac{1}{3}$ (E) nonexistent

03

Derivatives

3.1 Definition of Derivative

The derivative of the function $f(x)$ with respect to the variable x is the function f' whose value at x is

$$f'(x) = \lim_{h \to 0} \frac{f(x+h) - f(x)}{h}$$

provided the limit exists

f' is the tangent slope function f. As point B approaches point A, the secant slope of $\overline{AB}$ becomes slope of the tangent at point A

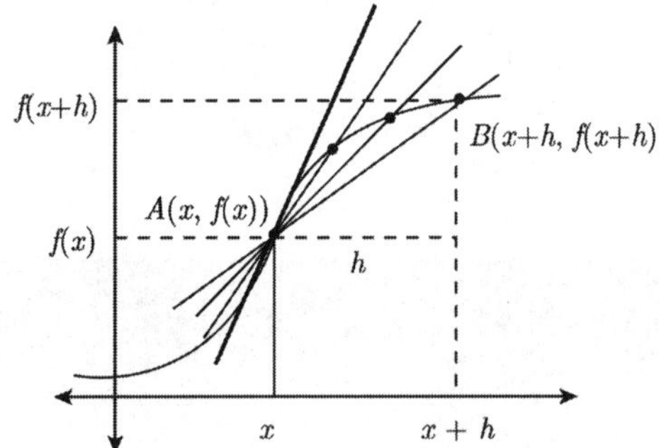

- **Derivative Notation**

$$f'(x) = y' = \frac{dy}{dx} = \frac{df}{dx} = \frac{d}{dx} f(x) = D(f)(x) = D_x f(x)$$

The symbols d/dx and D indicate the operation of differentiation

Example 1 Use the formula $f'(a) = \lim_{h \to 0} \frac{f(a+h) - f(a)}{h}$ to find the slope of the tangent to the curve $f(x) = x^2 + x$ at $x = 1$, 2, and 3.

Solution

$$f'(1) = \lim_{h \to 0} \frac{f(1+h) - f(1)}{h} = \lim_{h \to 0} \frac{(1+h)^2 + f(1+h) - 2}{h}$$

$$= \lim_{h \to 0} \frac{h^2 + 3h}{h} = \lim_{h \to 0} (h+3) = 3$$

$$f'(2) = \lim_{h \to 0} \frac{f(2+h) - f(2)}{h} = \lim_{h \to 0} \frac{(2+h)^2 + (2+h) - 6}{h}$$

$$= \lim_{h \to 0} \frac{h^2 + 5h}{h} = \lim_{h \to 0} (h+5) = 5$$

$$f'(3) = \lim_{h \to 0} \frac{f(3+h) - f(3)}{h} = \lim_{h \to 0} \frac{(3+h)^2 + (3+h) - 12}{h}$$

$$= \lim_{h \to 0} \frac{h^2 + 7h}{h} = \lim_{h \to 0} (h+7) = 7$$

Example 2 Use the of the definition to find the slope of the tangent to the curve $f(x)=(x-1)^2+1$ at $x=-1$, 0, and 2.

Solution

$f(x)=(x-1)^2+1$ and $f(x+h)=(x+h-1)^2+1$

$$\Rightarrow f'(x)=\lim_{h\to 0}\frac{[(x+h-1)^2+1]-[(x-1)^2+1]}{h}$$

$$=\lim_{h\to 0}\frac{(x^2+2xh+h^2-2x-2h+1+1)-(x^2-2x+1+1)}{h}$$

$$=\lim_{h\to 0}\frac{2xh+h^2-2h}{h}=\lim_{h\to 0}(2x+h-2)$$

$$=2(x-1);\ f'(-1)=-4,\ f'(0)=-2,\ f'(2)=2$$

EX 1 Use the definition of the derivative to find the derivative of each function.

(1) $f(x)=\dfrac{1}{\sqrt{3x-2}}$

(2) $f(x)=x^3-2x$

(3) $f(x)=x-\dfrac{1}{x}$

(4) $f(x)=\dfrac{1}{x^2}$

(5) $y=\dfrac{3x}{x+1}$

(6) $f(x)=\sqrt{x^2+1}$

3.2 Differentiability

If f is continuous at $x=a$, the left-hand derivative at $x=a$ is equal to the right-hand derivative at $x=a$

$$\left(\lim_{h\to 0^-}\frac{f(a+h)-f(a)}{h}=\lim_{h\to 0^+}\frac{f(a+h)-f(a)}{h}\right)$$

then $f(x)$ is differentiable at $x=a$.

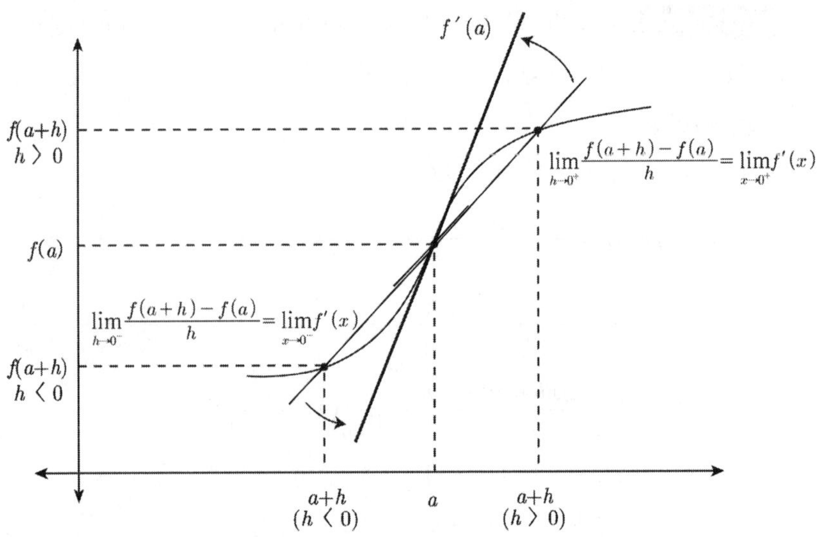

For the tangent slope to exist at x=a, the function must be continuous, and the left-hand slope must equal to the right-hand slope

- **Functions that are not differentiable**

① Discontinuity	② Corner	③ Cusp	④ Vertical tangent
$y=\frac{\lvert x\rvert}{x}$	$y=\lvert x\rvert$	$y=\sqrt[3]{x^2}$	$y=\sqrt[3]{x}$

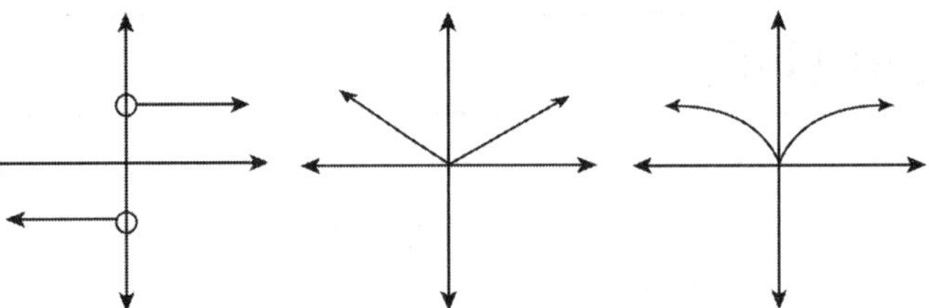

discontinuous at $x=0$	$\lim_{x\to0^-}f'(x)=-1$	$\lim_{x\to0^-}f'(x)=-\infty$	$\lim_{x\to0^-}f'(x)=+\infty$
	$\lim_{x\to0^+}f'(x)=1$	$\lim_{x\to0^+}f'(x)=\infty$	$\lim_{x\to0^+}f'(x)=+\infty$

EX 1 Each figure in (1)~(6) shows the graph of a function over a closed interval D. At what domain points does the function appear to be

a. differentiable?

b. continuous but not differentiable?

c. neither continuous nor differentiable?

Give reasons for your answers.

(1)

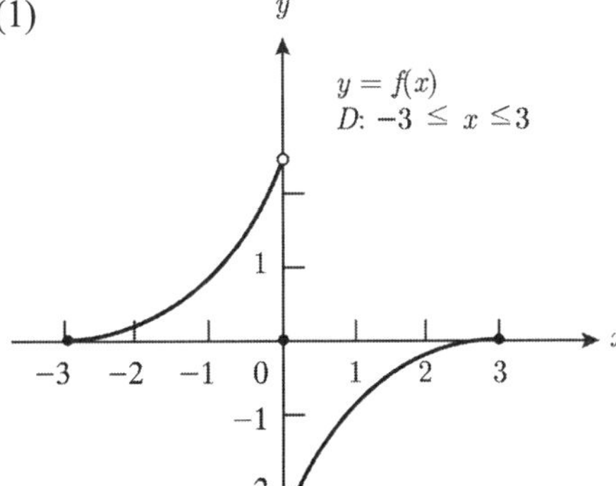

(2)

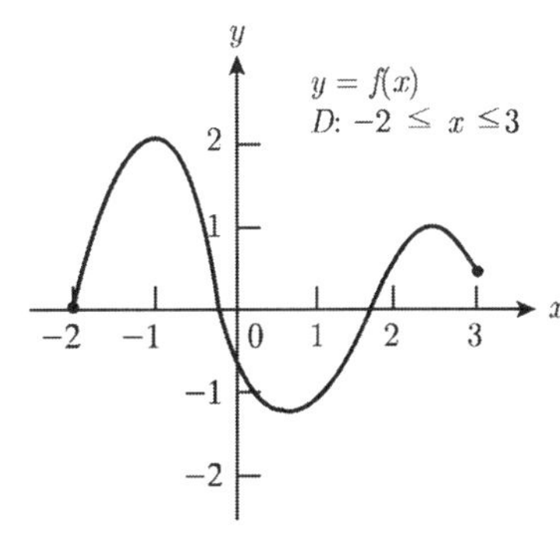

(3)

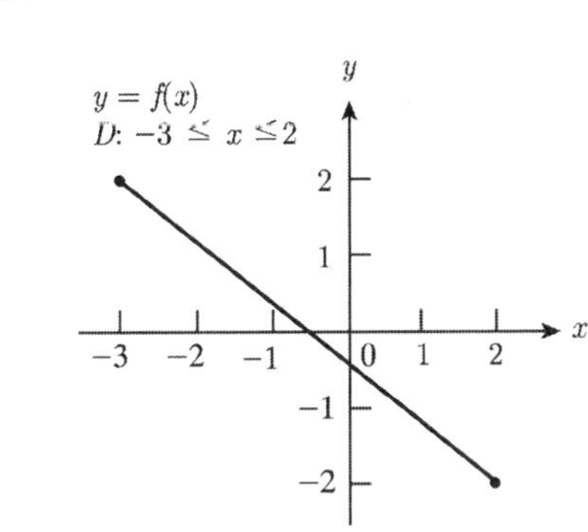

(4)

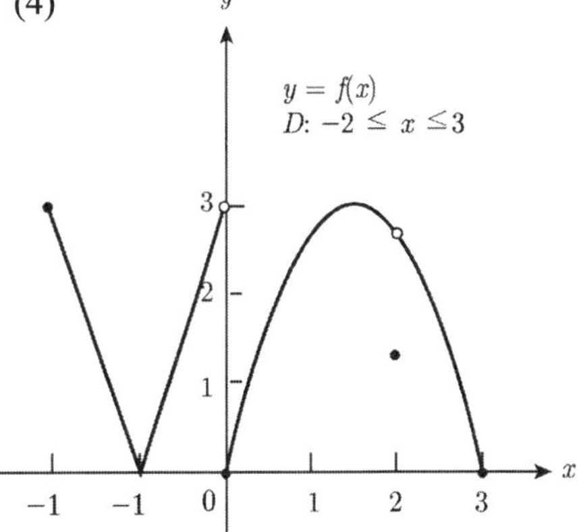

(5)

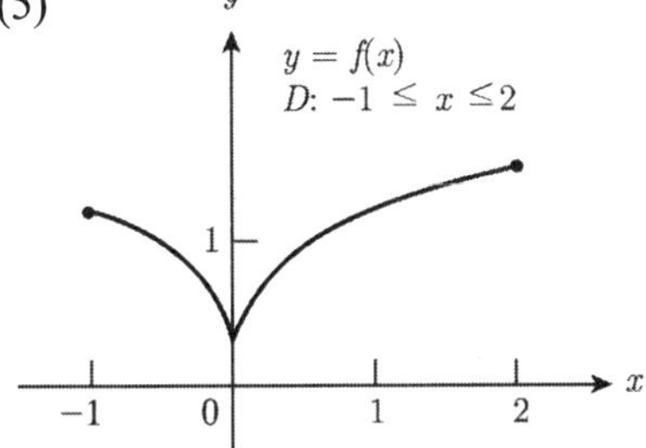

(6)

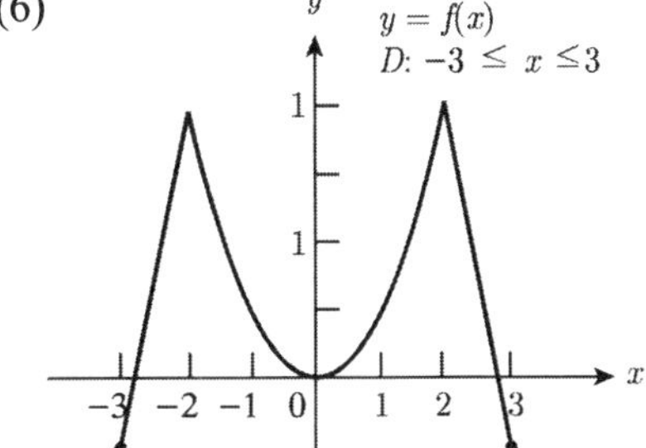

EX 2 The graph of f is shown in the figure below. Determine whether each of the following statements is true or false.

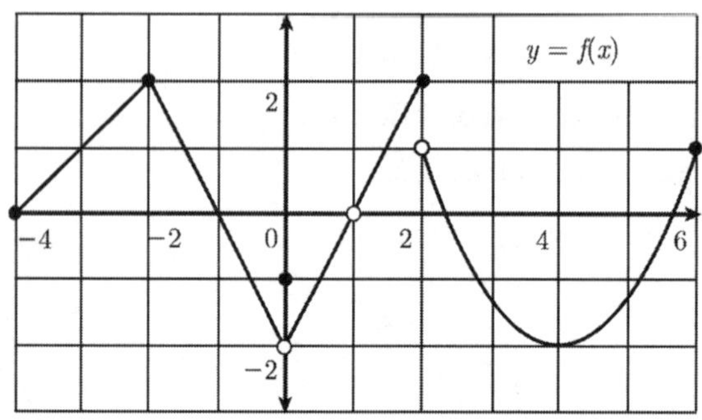

(1) $\lim_{x \to -2^-} f(x) = \lim_{x \to -2^+} f(x)$

(2) $\lim_{x \to -2} f(x) = f(-2)$

(3) $\lim_{x \to -2^-} f'(x) = \lim_{x \to -2^+} f'(x)$

(4) $\lim_{x \to 0} f(x)$

(5) $\lim_{x \to 0} f(x) = f(0)$

(6) $\lim_{x \to 0^-} f'(x) = \lim_{x \to 0^+} f'(x)$

(7) $f(1)$ is not defined

(8) $\lim_{h \to 0} \frac{f(1+h) - f(1)}{h}$ exists

(9) $\lim_{h \to 0^+} \frac{f(h) - f(0)}{h} = 2$

(10) $\lim_{x \to 1} \frac{f(x) - f(1)}{x - 1}$ does not exist

(11) $\lim_{x \to 1^-} f'(x) = \lim_{x \to 1^+} f'(x)$

(12) $\lim_{x \to 2} f(x)$ does not exist

(13) $\lim_{x \to 2^-} f(x) \neq f(2)$

(14) $\lim_{h \to 0} \frac{f(3+h) - f(3)}{h}$ exists

(15) $\lim_{x \to 4} f(x) = f(4)$

(16) $\lim_{x \to 4} \frac{f(x) - f(4)}{x - 4} = -2$

(17) $\lim_{x \to c} f(x)$ exists at every c in the interval (2,6)

(18) $\lim_{h \to 0^+} \frac{f(c+h) - f(c)}{h} = \lim_{h \to 0^-} \frac{f(c+h) - f(c)}{h}$ at every c in the interval (2, 6)

(19) $f'(-3) = 1$

(20) $f'(-1) = 0$

(21) $f'(3) < f'(4) < f'(5) < f'(5.5)$

Example 1 Sketch the graph of $f'(x)$ using the graph of $f(x)$ given below.

(1) $y = f(x)$

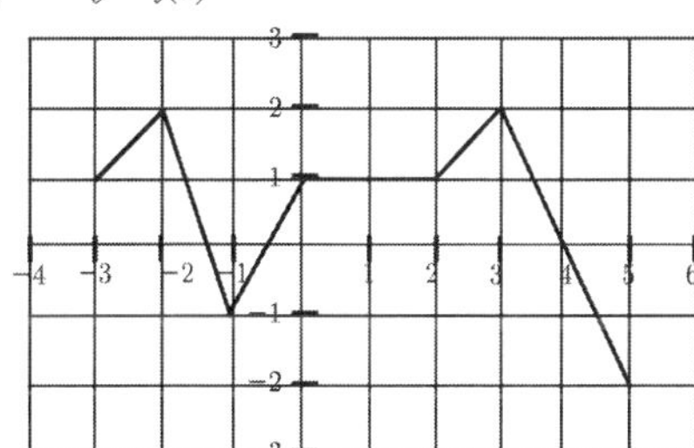

(2) $y = f(x)$

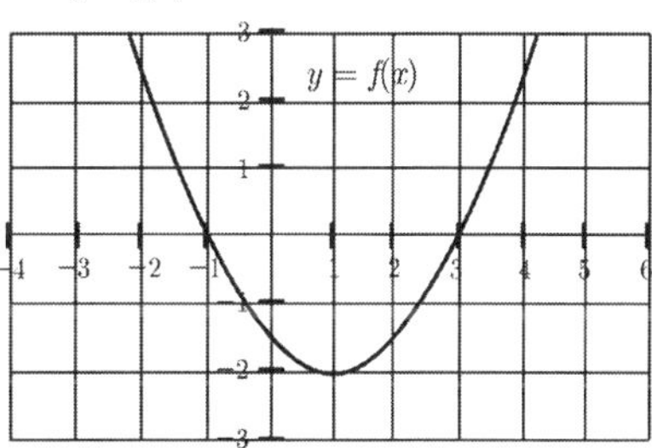

(3) $y = f'(x) = slope\ of\ f$

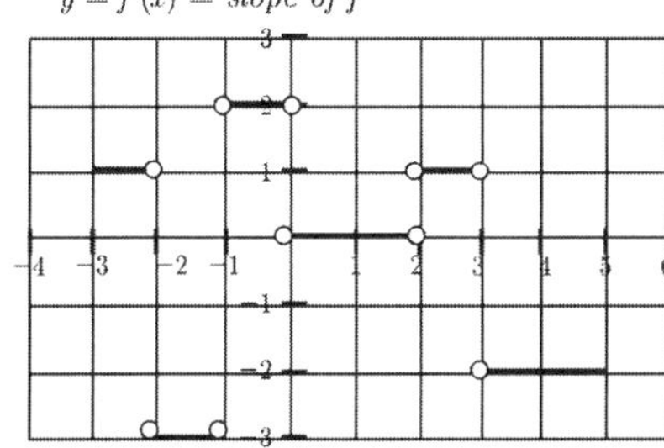

(4) $y = f'(x) = slope\ of\ f$

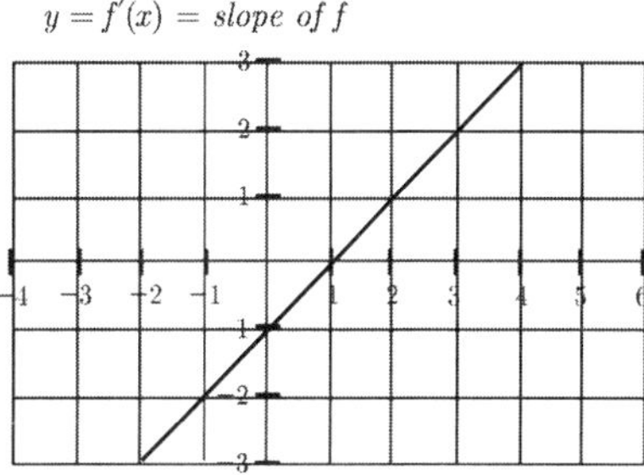

EX 3 Sketch the graph of $f'(x)$ using the graph of $f(x)$ given below.

(1) $y = f(x)$

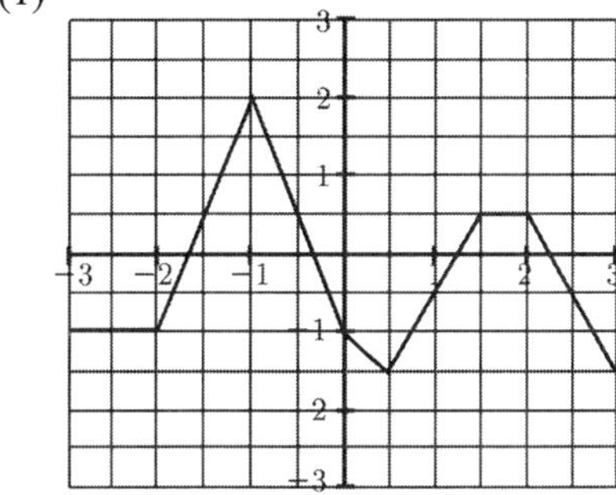

$y = f'(x)$

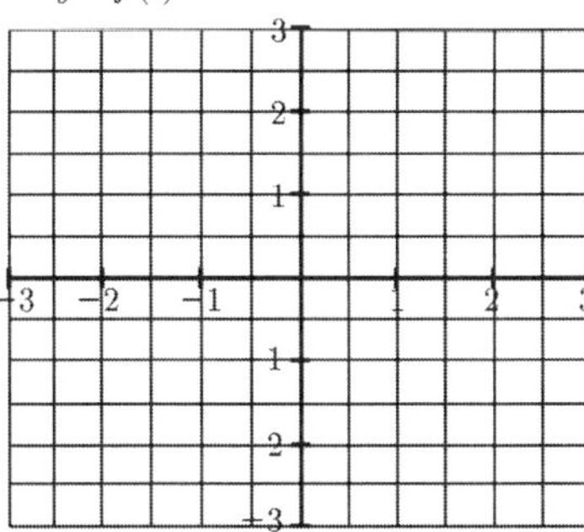

(2) $y = f(x)$

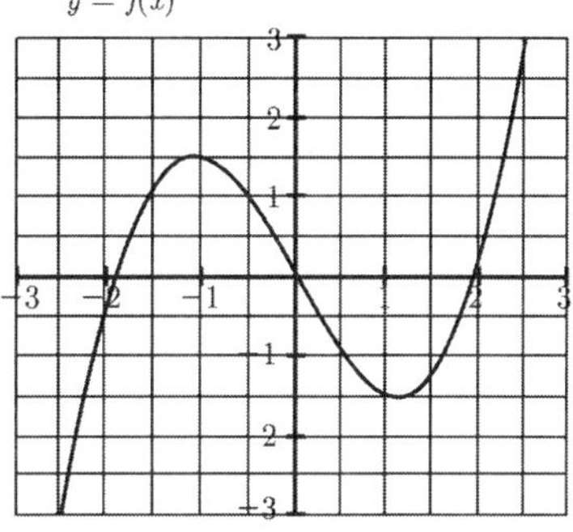

$y = f'(x)$

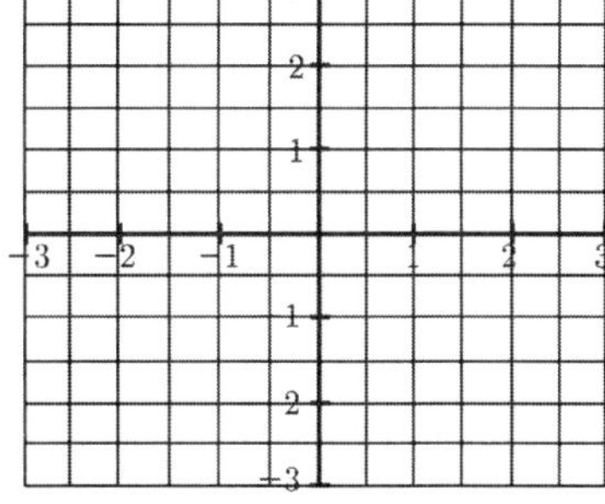

(3)

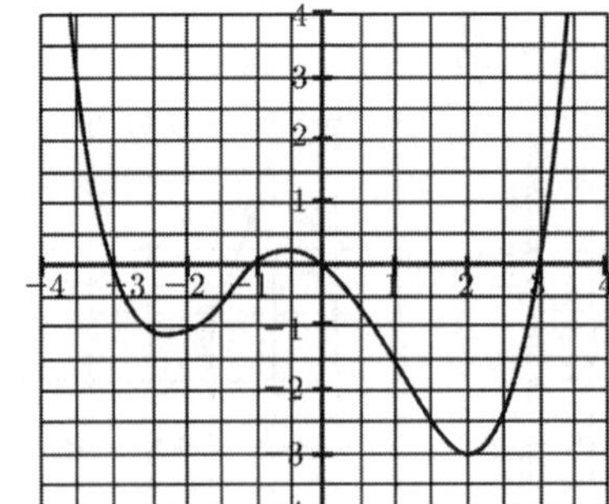

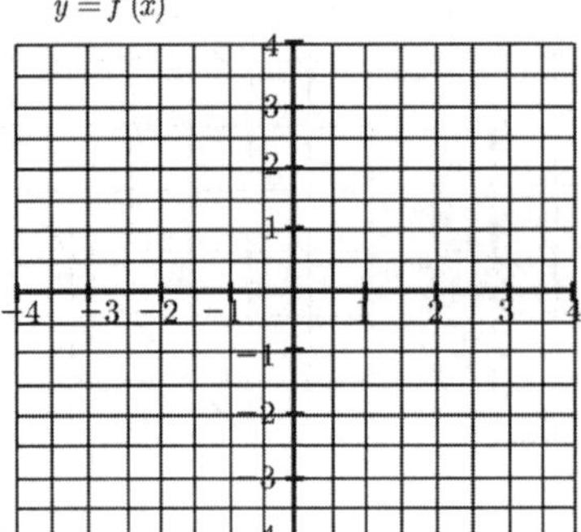

(4)

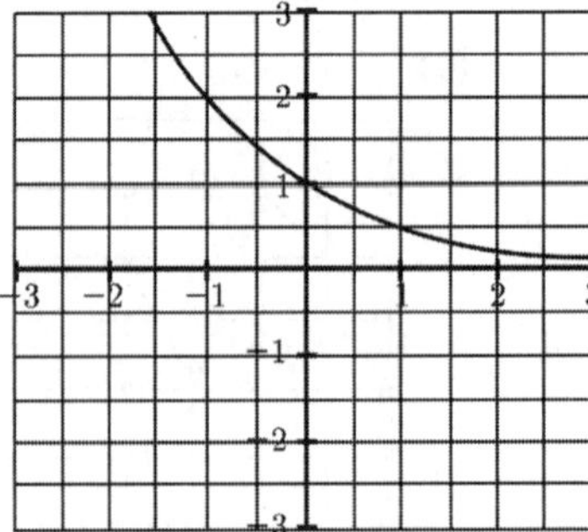

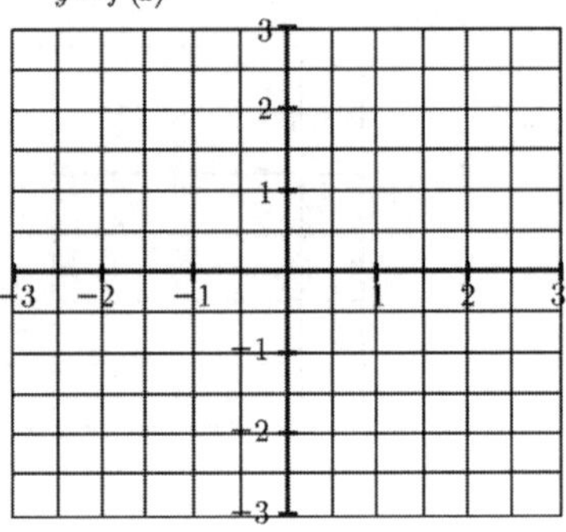

(5)

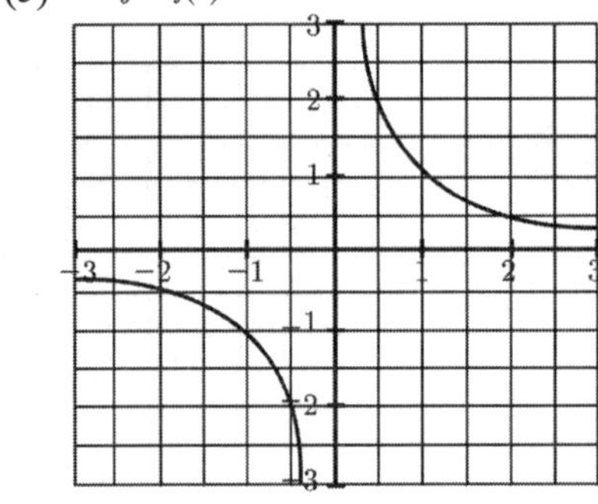

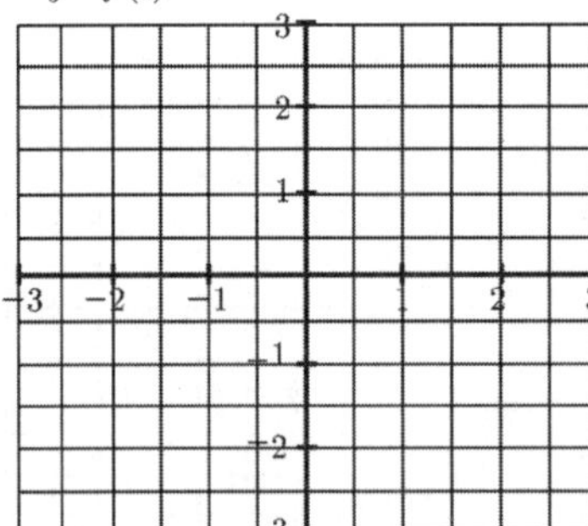

(6)

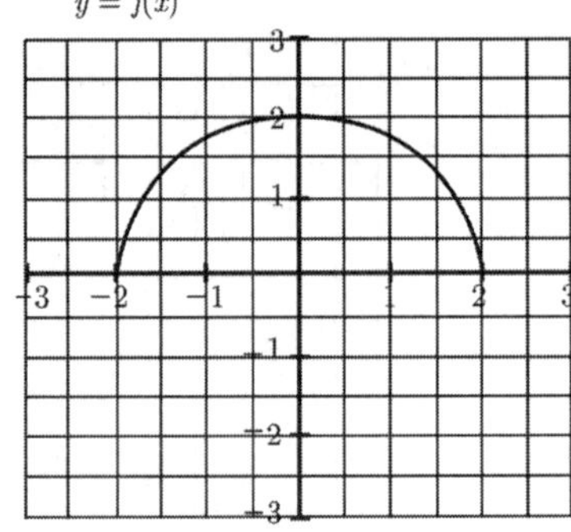

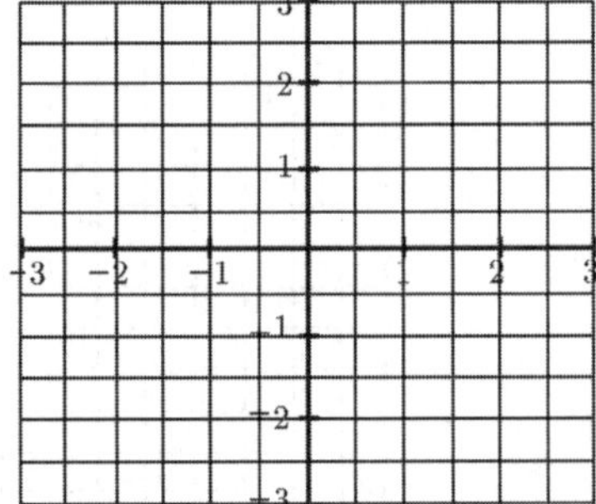

3.3 Power Rule

If n is any real number,

$$\frac{d}{dx}(x^n) = nx^{n-1}$$

Proof

Let $f(x) = x^n$

$$f'(x) = \frac{d}{dx}(x^n) = \lim_{h\to 0}\frac{f(x+h)-f(x)}{h}$$

$$= \lim_{h\to 0}\frac{(x+h)^n - x^n}{h}$$

$$= \lim_{h\to 0}[[{}_nC_0x^n + {}_nC_1x^{n-1}h + {}_nC_2x^{n-2}h^2 + {}_nC_3x^{n-3}h^3$$

$$+ \ldots\ldots + {}_nC_{n-1}xh^{n-1} + {}_nC_nh^n] - x^n]/h$$

$$= \lim_{h\to 0}[\left[x^n + nx^{n-1}h + \frac{n(n-1)}{2!}x^{n-2}h^2 + \frac{n(n-1)(n-2)}{3!}x^{n-3}h^3\right.$$

$$+ \ldots\ldots + nxh^{n-1} + h^n] - x^n]/h$$

$$= \lim_{h\to 0}\frac{nx^{n-1}h + \frac{n(n-1)}{2!}x^{n-2}h^2 + \frac{n(n-1)(n-2)}{3!}x^{n-3}h^3 + \ldots\ldots + nxh^{n-1} + h^n}{h}$$

$$= \lim_{h\to 0}\left[nx^{n-1} + \frac{n(n-1)}{2!}x^{n-2}h + \frac{n(n-1)(n-2)}{3!}x^{n-3}h^2 + \ldots\ldots + nxh^{n-2} + h^{n-1}\right]$$

$$= nx^{n-1} + 0 + 0 + 0 + \ldots\ldots + 0$$

$$= nx^{n-1}$$

Example 1 Find the derivative of the following functions.

(1) $y = x^3$ (2) $y = \frac{1}{x^4}$ (3) $y = x^{\frac{2}{3}}$

Solution

(1) $y = x^3$

$y' = 3x^2$

(2) $y = \frac{1}{x^4}$

$y' = -4x^{-5}$

(3) $y = x^{\frac{2}{3}}$

$y' = \frac{2}{3}x^{-\frac{1}{3}}$

• Meaning of the notation $\frac{d}{dx}(x^3)$: Given $y=x^3$, find $\frac{dy}{dx}$ or y'.

Example 2

(1) $\frac{d}{dx}(x^{\sqrt{2}})$ (2) $\frac{d}{dx}(x^{10})$ (3) $\frac{d}{dx}(x^{-\frac{4}{3}})$

Solution

(1) $\frac{d}{dx}(x^{\sqrt{2}})=\sqrt{2}x^{\sqrt{2}-1}$ (2) $\frac{d}{dx}(x^{10})=10x^9$ (3) $\frac{d}{dx}(x^{-\frac{4}{3}})=-\frac{4}{3}x^{-\frac{7}{3}}$

• $\frac{d}{dx}(c)=0$

Proof $y'=\lim_{h\to 0}\frac{c-c}{h}=0$

Example 3 $\frac{d}{dx}(7)$

Solution

$\frac{d}{dx}(7)=0$

• $\frac{d}{dx}(cf(x))=cf'(x)$

Proof

$$y'=\lim_{h\to 0}\frac{cf(x+h)-cf(x)}{h}$$
$$=c\lim_{h\to 0}\frac{f(x+h)-f(x)}{h}=cf'(x)$$

Example 4

(1) $\frac{d}{dx}(3x^2)$ (2) $\frac{d}{dx}(-4x^{-\frac{1}{2}})$

Solution

(1) $\frac{d}{dx}(3x^2)=6x$ (2) $\frac{d}{dx}(-4x^{-\frac{1}{2}})=2x^{-\frac{3}{2}}$

• $\frac{d}{dx}(f(x)\pm g(x))=\frac{d}{dx}f(x)\pm\frac{d}{dx}g(x)=f'(x)\pm g'(x)$

Proof

If $y=f(x)+g(x)$,

$$y'=\lim_{h\to 0}\frac{[f(x+h)+g(x+h)]-[f(x)+g(x)]}{h}$$
$$=\lim_{h\to 0}\frac{[f(x+h)-f(x)]+[g(x+h)-g(x)]}{h}$$
$$=\lim_{h\to 0}\frac{f(x+h)-f(x)}{h}+\lim_{h\to 0}\frac{g(x+h)-g(x)}{h}=f'(x)+g'(x)$$

Example 5

(1) $\frac{d}{dx}(x^3 - \frac{4}{3}x^2 - 5x)$ (2) $\frac{d}{dx}(3x^3 - 6x^{\frac{1}{2}} + 4)$

Solution

(1) $\frac{d}{dx}(x^3 - \frac{4}{3}x^2 - 5x) = 3x^2 - \frac{8}{3}x - 5$

(2) $\frac{d}{dx}(3x^3 - 6x^{\frac{1}{2}} + 4) = 9x^2 - 3x^{-\frac{1}{2}} = 9x^2 - \frac{3}{\sqrt{x}}$

EX 1 Find the derivative of the following functions.

(1) $y = 8$ (2) $y = k$ (k : constant) (3) $y = -5x^4$

(4) $y = \sqrt[5]{x^4}$ (5) $y = -\frac{2}{x^3}$ (6) $y = \frac{2}{\sqrt{x}}$

EX 2 Find the derivative of the following functions.

(1) $y = 3x^2 - 5x + 2$ (2) $y = 5x^3 - 2\sqrt{x}$ (3) $y = 1 + \frac{1}{x} + \frac{2}{x^2}$

(4) $y = 2(\sqrt{x} + \frac{1}{x})$ (5) $y = -2x^{-1} + \frac{4}{x^2}$ (6) $y = \frac{12}{x} - \frac{4}{x^3} + \frac{1}{x^4}$

EX 3 Find the derivative of the following functions.

(1) $y = \frac{x^2 + 5x - 1}{x^2}$ (2) $y = \frac{2x^3 - 3x\sqrt{x} - 1}{\sqrt{x}}$ (3) $y = \frac{3x^{-3} + 2x^{-1} - 1}{x^{-2}}$

(4) $f(x) = 6x^{\frac{2}{3}}(2x - 1)$ (5) $f(x) = \sqrt{x^3}(x - 3x^2)$

• The point-slope form of the equation of a line : $y - y_1 = m(x - x_1)$ ($m = f'(x_1)$).

EX 4

(1) Find the equation of the tangent line to the curve $y = f(x)$ at $x = -2$, if $f(-2) = 3$ and $f'(-2) = 2$.

(2) Find the equation of the tangent line to the curve $y = g(x)$ at $x = 1$, if $g(1) = -2$ and $g'(1) = 5$.

EX 5 Find the equation of the tangent line to the curves at given point.

(1) $y = 4 - x^2$ *at* $(-1, 3)$ (2) $f(x) = 4\sqrt{x} - \frac{1}{x}$ *at* $x = 1$

(3) $g(x) = \frac{8}{x^2} + \frac{2}{\sqrt{x}}$ *at* $x = 4$

(1) Find equations for the horizontal tangents to the curve $y = x^3 - 3x - 2$. Also find equations for the lines that are perpendicular to these tangents at the points of tangency.

(2) Find all points $(x,\ y)$ on the graph of $g(x) = \frac{1}{3}x^3 - \frac{3}{2}x^2 + 1$ with tangent lines parallel to the line $8x - 2y = 1$.

EX7

(1) The curve $y = ax^2 + bx + c$ passes through the point (1,2) and is tangent to the line $y = x$ at the origin. Find a, b, and c.

(2) The curves $y = x^2 + ax + b$ and $y = cx - x^2$ have a common tangent line at the point (1,0). Find a, b, and c.

EX8 Find the values of a and b that make $f(x)$ differentiable everywhere.

(1) $f(x) = \begin{cases} ax, & x < 0 \\ x^2 - 3x, & x \geq 0 \end{cases}$

(2) $f(x) = \begin{cases} ax + b, & x > -1 \\ bx^2 - 3, & x \leq -1 \end{cases}$

(3) $f(x) = \begin{cases} 4\sqrt{x}, & x \leq 4 \\ ax^2 - bx, & x > 4 \end{cases}$

(4) $f(x) = \begin{cases} ax + b, & x \leq -1 \\ ax^3 + x + 2b, & x > -1 \end{cases}$

3.4 Product Rule

If f and g are differentiable at x, then $f \cdot g$ is differentiable and

$$\frac{d}{dx}(f(x) \cdot g(x)) = f'(x)g(x) + f(x)g'(x)$$

Proof

$$\begin{aligned}
&\frac{d}{dx}(f(x) \cdot g(x)) \\
&= \lim_{h \to 0} \frac{f(x+h)g(x+h) - f(x)g(x)}{h} \\
&= \lim_{h \to 0} \frac{[f(x+h)g(x+h) - f(x)g(x+h)] + [f(x)g(x+h) - f(x)g(x)]}{h} \\
&= \lim_{h \to 0} \frac{[f(x+h) - f(x)]g(x+h) + f(x)[g(x+h) - g(x)]}{h}
\end{aligned}$$

$$= \lim_{h \to 0} \frac{[f(x+h) - f(x)]g(x+h)}{h} + \lim_{h \to 0} \frac{f(x)[g(x+h) - g(x)]}{h}$$

$$= \lim_{h \to 0} \frac{[f(x+h) - f(x)]}{h} \cdot \lim_{h \to 0} g(x+h) + \lim_{h \to 0} f(x) \cdot \lim_{h \to 0} \frac{[g(x+h) - g(x)]}{h}$$

$$= f'(x)g(x) + f(x)g'(x)$$

Example 1 Use the product rule to find the derivative.

$$y = (x^2 - 2)(3x^2 - 2x)$$

Solution

$$y' = 2x(3x^2 - 2x) + (x^2 - 2)(6x - 2)$$

$$= 6x^3 - 4x^2 + 6x^3 - 2x^2 - 12x + 4$$

$$= 12x^3 - 6x^2 - 12x + 4$$

EX 1 Use the product rule to find the derivative.

(1) $y = (3 - x^2)(x^3 - x + 1)$ (2) $y = (1 + x^2)(x^{\frac{3}{4}} - x^{-3})$

(3) $y = (3 - 2x)\sqrt{x}$

EX 2

(1) Find the equation of the tangent line the curve $y = (x^2 - 2)(x^2 - 3x - 1)$ at $x = -1$.

(2) Find the equation of the tangent line the curve $y = (x - 2\sqrt{x})(x^2 - 6x + 2)$ at $x = 4$.

EX 3 Suppose u and v are function of x that are differentiable at $x = 0$ and that $u(0) = 5$, $u'(0) = -3$, $v(0) = -1$, $v'(0) = 2$.

Find the values of the following derivatives at $x = 0$.

(1) $\frac{d}{dx}(uv)$ (2) $\frac{d}{dx}(\frac{u}{v})$ (3) $\frac{d}{dx}(\frac{v}{u})$ (4) $\frac{d}{dx}(7v - 2u)$

EX 4 Show that $\frac{d}{dx}(f \cdot g \cdot h) = f'gh + fg'h + fgh'$

3.5 Quotient Rule

If f and g are differentiable at x, then $\frac{f}{g}$ is differentiable and

$$\frac{d}{dx}\left(\frac{f(x)}{g(x)}\right)=\frac{f'(x)g(x)-f(x)g'(x)}{g^2(x)}.$$

Proof

$$\frac{d}{dx}\left(\frac{f(x)}{g(x)}\right)$$

$$=\lim_{h\to0}\frac{\frac{f(x+h)}{g(x+h)}-\frac{f(x)}{g(x)}}{h}=\lim_{h\to0}\frac{\left[\frac{f(x+h)}{g(x+h)}-\frac{f(x)}{g(x)}\right]\cdot g(x+h)g(x)}{h\cdot g(x+h)g(x)}$$

$$=\lim_{h\to0}\frac{f(x+h)g(x)-f(x)g(x+h)}{hg(x+h)g(x)}$$

$$=\lim_{h\to0}\frac{[f(x+h)g(x)-f(x)g(x)]+[f(x)g(x)-f(x)g(x+h)]}{hg(x+h)g(x)}$$

$$=\lim_{h\to0}\frac{[f(x+h)g(x)-f(x)g(x)]-[f(x)g(x+h)-f(x)g(x)]}{hg(x+h)g(x)}$$

$$=\lim_{h\to0}\frac{[f(x+h)-f(x)]g(x)-f(x)[g(x+h)-g(x)]}{hg(x+h)g(x)}$$

$$=\lim_{h\to0}\frac{[f(x+h)-f(x)]g(x)}{hg(x+h)g(x)}-\lim_{h\to0}\frac{f(x)[g(x+h)-g(x)]}{hg(x+h)g(x)}$$

$$=\lim_{h\to0}\frac{[f(x+h)-f(x)]}{h}\cdot\lim_{h\to0}\frac{g(x)}{g(x+h)g(x)}-\lim_{h\to0}\frac{f(x)}{g(x+h)g(x)}\cdot\lim_{h\to0}\frac{[g(x+h)-g(x)}{h}$$

$$=\frac{f'(x)g(x)}{g(x)g(x)}-\frac{f(x)g'(x)}{g(x)g(x)}=\frac{f'(x)g(x)-f(x)g'(x)}{g^2(x)}$$

EX1 Find derivative of the function $y=\frac{2x}{x+1}$.

EX2 Differentiable the following functions.

(1) $y=\frac{2x+5}{3x-2}$

(2) $y=\frac{4-3x}{3x^2+x}$

(3) $y=\frac{(x+1)(x+2)}{(x-1)(x-2)}$

(4) $y=\frac{\sqrt{x}-1}{\sqrt{x}+1}$

(1) Find the equation of the tangent line to the curve $y=\dfrac{1+x}{2x-1}$ at $(1,2)$.

(2) Find the equation of the normal line to the curve $y=\dfrac{x}{\sqrt{x}-3}$ at $x=4$.

EX 4 At what point does the function $y=\dfrac{4x}{x^2+1}$ have a horizontal tangent line?

EX 5 Suppose that functions f and g and their derivatives with respect to x have the following values at $x=2$ and$x=3$.

x	$f(x)$	$g(x)$	$f'(x)$	$g'(x)$
2	8	2	1/3	−3
3	3	−4	2π	5

Find the derivatives with respect to of the following combinations at the given value of x.

(1) $2f(x),\ x=2$

(2) $f(x)+g(x),\ x=3$

(3) $f(x)\cdot g(x),\ x=3$

(4) $f(x)/g(x),\ x=2$

EX 6 If $f(3)=2, g(3)=-2, f'(3)=4$ and $g'(3)=-5$, find $\left(\dfrac{f}{g}\right)'(3)$.

(1) Find the tangents to newton's serpentine (graphed here) at the origin and he point (1,2).

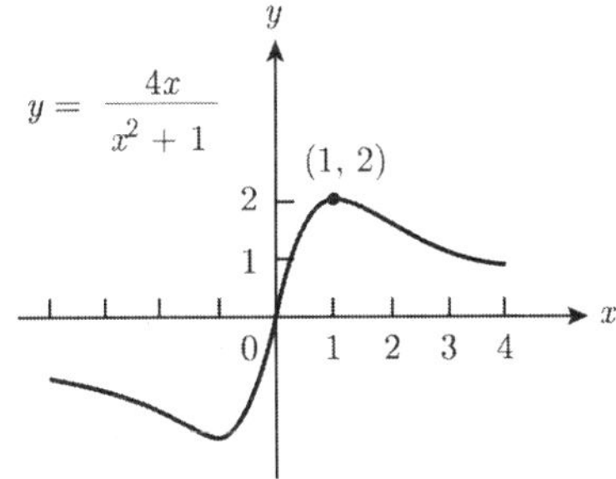

(2) Find the tangent to the *Witch of Agnesi* (graphed here)at the point (2,1).

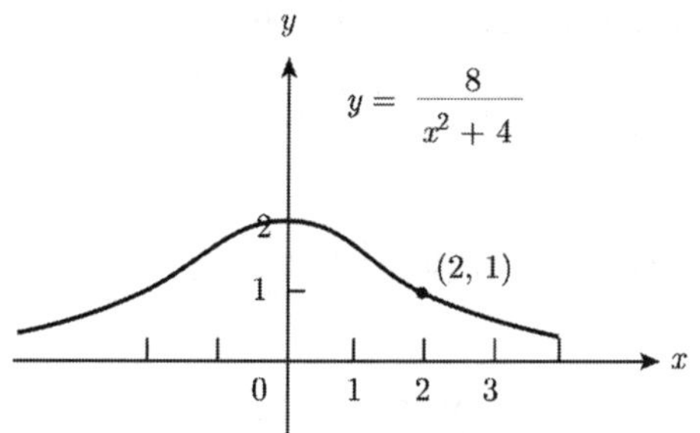

3.6 Chain Rule

If $f(u)$ is differentiable at the point $u = g(x)$ and $g(x)$ is differentiable at x, then the composite function $(f \circ g)(x) = f(g(x))$ is differentiable at x, and

$$(f \circ g)'(x) = f'(g(x)) \cdot g'(x).$$

In Leibniz's notation, if $y = f(u)$ and $u = g(x)$, then

$$\frac{dy}{dx} = \frac{dy}{du} \cdot \frac{du}{dx}$$

where dy/du is evaluated $u = g(x)$.

Example 1 Given $y = (x^3 - 4x + 1)^{10}$, find $\dfrac{dy}{dx}$.

Solution

$y' = 10(x^3 - 4x + 1)^9(3x^2 - 4)$

EX 1 Find the derivative of the following functions.

(1) $y = (4 - 3x)^9$

(2) $y = (3x^2 - \sqrt{x})^4$

(3) $y = (\dfrac{x^2}{8} + x - \dfrac{1}{x})^4$

(4) $y = \sqrt[3]{2x - x^2}$

EX 2 Find the derivative functions.

(1) $y=\dfrac{1}{(2-3x)^4}$ (2) $y=(2x+1)^3(3x-4)^5$

(3) $y=(2-5x)^4(3x^2+1)^3$ (4) $y=\dfrac{(x+1)^2}{(2x+1)^3}$

• Square Root Rule $\dfrac{d}{dx}(\sqrt{f(x)})=\dfrac{1}{2\sqrt{f(x)}}\cdot f'(x)$

Proof $\dfrac{d}{dx}(\sqrt{f(x)})=\dfrac{d}{dx}[f(x)]^{\frac{1}{2}}=\dfrac{1}{2}[f(x)]^{-\frac{1}{2}}\cdot f'(x)$

Example 2 $\dfrac{d}{dx}(\sqrt{3x^2-1})$.

Solution

$$\frac{d}{dx}(\sqrt{3x^2-1})=\frac{1}{2\sqrt{3x^2-1}}\cdot 6x=\frac{3x}{\sqrt{3x^2-1}}$$

EX 3 Find the derivative of the following functions.

(1) $y=\sqrt{x^2-4x}$ (2) $y=3x^2\sqrt{x^2+1}$

(3) $y=\dfrac{1-2x}{\sqrt{4x+1}}$ (4) $y=\dfrac{\sqrt{x^2+1}}{x-1}$

(5) $y=\left(\dfrac{x+2}{2x-1}\right)^5$ (6) $y=\left(\dfrac{\sqrt{x}-1}{\sqrt{x}+1}\right)^4$

EX 4

(1) Find the tangent to $y=\sqrt{x^2-x+7}$ at $x=2$.

(2) Find the tangent to $y=((x-1)/(x+1))^2$ at $x=0$.

EX 5 Differentiate the following functions.

(1) $y=[f(x)]^3$ (2) $y=f(x^3)$ (3) $y=[f(x^3)]^3$

(4) $y=\sqrt{f(x)-3x}$ (5) $y=\dfrac{1}{\sqrt{f(x)}}$ (6) $y=3x^2f(g(\sqrt{x}))$

(7) $y=3[f(x^2)+2x]^4$

EX 6 Suppose that functions f and g and their derivatives with respect to x have the following values at $x=0$ and $x=1$.

x	$f(x)$	$g(x)$	$f'(x)$	$g'(x)$
0	1	1	5	1/3
1	3	−4	−1/3	−8/3

Find the derivatives with respect to of the following combinations at the given value of x.

(1) $5f(x)-g(x),\ x=1$

(2) $f(x)g^3(x),\ x=0$

(3) $\dfrac{f(x)}{g(x)+1},\ x=1$

(4) $f(g(x)),\ x=0$

(5) $g(f(x)),\ x=0$

(6) $(x^{11}+f(x))^{-2},\ x=1$

(7) $f(x+g(x)),\ x=0$

3.7 Implicit Differentiation

Most of the function we have dealt with so far have beem described by an equation of the form $y=f(x)$ that expresses y explicitly in terms of the variable x. We have learned rules for differentiating functions defines in this way. Another situation occurs when we encounter equations like

$$x^3+y^3-9xy=0, \quad y^2-x=0, \quad \text{or} \quad x^2+y^2-25=0$$

These equations define an *implicit* relation between the variables x and y, In some cases we may be able to solve such an equation for y as an explicit function (or even several functions) of x. When we cannot put an equation $F(x,y)=0$ in the form $y=f(x)$ to differentiable it in the usual way, we may still be able to find dy/dx by *implicit differentiation.*

1. Differentiation both sides of the equation with repect to x treating y as a differentiation function of x.
2. Collect the terms with dy/dx on one side of the equation and solve for dy/dx.

Example 1 Find $\dfrac{dy}{dx}$ for the equation $xy = 2y^3 + x$.

Solution

Step 1. $\dfrac{d}{dx}(xy) = \dfrac{d}{dx}(2y^3 + x)$

$1 \cdot y + x \cdot y' = 6y^2y' + 1$

$y + xy' = 6y^2y' + 1$

Step 2. $xy' - 6y^2y' = 1 - y$

$(x - 6y^2)y' = 1 - y$

$\therefore \dfrac{dy}{dx} = \dfrac{1-y}{x-6y^2}$

EX 1 If y is a function of x, find

(1) $\dfrac{d}{dx}(3y^3)$

(2) $\dfrac{d}{dx}(-\sqrt{y})$

(3) $\dfrac{d}{dx}(\dfrac{4}{y})$

(4) $\dfrac{d}{dx}(xy^2)$

(5) $\dfrac{d}{dx}(\dfrac{7x}{y})$

EX 2 Use implicit differentiation to find $\dfrac{dy}{dx}$.

(1) $x^3 + y^3 = 16$

(2) $x^{\frac{2}{3}} + y^{\frac{2}{3}} = 1$ (astroid)

(3) $x^2 - xy + y^2 = 6$ (ellipse)

(4) $x^3 + y^3 = 18xy$

(5) $2\sqrt{y} = x - y$

(6) $x^2y^3 - x^3y^2 = -5$

EX 3 Find the slope of the tangent to the curves as the given point.

(1) $x^2 + xy - y^2 = 1,\ (2,3)$

(2) $(x^2 + y^2)^2 = (x - y)^2,\ (1,-1)$

(3) $\sqrt{4x - y^2} - 4\sqrt{xy} = -6,\ (2,2)$

(4) $x = y^2\sqrt{y-1},\ (4,2)$

EX 4

(1) Find the equation of the tangent line to the curve $x^3 + y^3 - 9xy = 0$ at the point $(4,2)$.

(2) The line that is normal to the curve $x^2 + 2xy - 3y^2 = 0$ at $(1,1)$ intersect the curve at what other point?

(1) Consider the curve defined by $xy^2 - x^2y = 16$. Find the coordinates of the point on the curve where the tangent line is horizontal.

(2) Consider the curve defined by $x^2y - xy^3 = 6$. Find the coordinates of the point on the curve where the tangent line is vertical.

3.8 Higher Order Derivatives

- First derivative $: y' = \dfrac{dy}{dx}$
- Second derivative $: y'' = \dfrac{dy'}{dx} = \dfrac{d}{dx}\left(\dfrac{dy}{dx}\right) = \dfrac{d^2y}{dx^2}$
- Third derivative $: y''' = \dfrac{dy''}{dx} = \dfrac{d}{dx}\left(\dfrac{d^2y}{dx^2}\right) = \dfrac{d^3y}{dx^3}$
- Fourth derivative $: y^{(4)} = \dfrac{dy'''}{dx} = \dfrac{d}{dx}\left(\dfrac{d^3y}{dx^3}\right) = \dfrac{d^4y}{dx^4}$
- Fifth derivative $: y^{(5)} = \dfrac{dy^{(4)}}{dx} = \dfrac{d}{dx}\left(\dfrac{d^4y}{dx^4}\right) = \dfrac{d^5y}{dx^5}$

  ⋮ ⋮

- nth derivative $: y^{(n)} = \dfrac{dy^{(n-1)}}{dx} = \dfrac{d}{dx}\left(\dfrac{d^{n-1}y}{dx^{n-1}}\right) = \dfrac{d^ny}{dx^n}$

Example 1 Find the first six derivatives of $y = \dfrac{1}{4}x^4 - \dfrac{2}{3}x^3 + x^2 - 6x + 2$.

Solution

$y' = x^3 - 2x^2 - 2x - 6$

$y'' = 3x^2 - 4x - 2$

$y''' = 6x - 4$

$y^{(4)} = 6$

$y^{(5)} = 0$

$y^{(6)} = 0$

EX 1 Find the second derivative of following functions.

(1) $y=\dfrac{1}{2x-1}$ (2) $y=(x^2-2x)^6$

(3) $y=\sqrt{x^2+2}$ (4) $x^2-y^2=4$

EX 2

(1) If $f(x)=\dfrac{3x-2}{2x+3}$, find $f''(-2)$ (2) If $y=\dfrac{1}{\sqrt{1-2x}}$, find $\left.\dfrac{d^2y}{dx^2}\right|_{x-0}$

(3) If $g(x)=(3-2x^2)^5$, find $g''(1)$ (4) Find $\left.\dfrac{d^3}{dx^3}(x^n-1)\right|_{x=1}$

EX 3

(1) Given $G(x)=xf(x^2)$, $\dfrac{d^2G(x)}{dx^2}$.

(2) Let $f(2)=-1$, $f'(2)=-2$ and $f''(2)=3$. For the function $F(x)=[f(x)]^2$, find $F''(2)$.

(3) Let $f(1)=2$, $f'(1)=3$ and $f''(1)=-2$. Find the value of $\dfrac{d^2}{dx^2}[f(x^3)]_{x=1}$.

EX 4

(1) Consider the curve defined by $y^2+xy=2$. Find the value of $\left.\dfrac{d^2y}{dx^2}\right|_{(1,1)}$.

(2) Consider the curve defined by $x^2-xy+y^2=3$. Find the value of $\left.\dfrac{d^2y}{dx^2}\right|_{(1,-1)}$.

EX 5 Let f be twice-differentiable function defined on interval $-1.2<x<3.2$ with $f(1)=2$. The graph of f' crosses the derivative of f, is shown above. The graph of f' crosses the x-axis at $x=-1$ and $x=3$ and has a horizontal tangent at $x=2$. Let g be the function given by $g(x)=e^{f(x)}$.

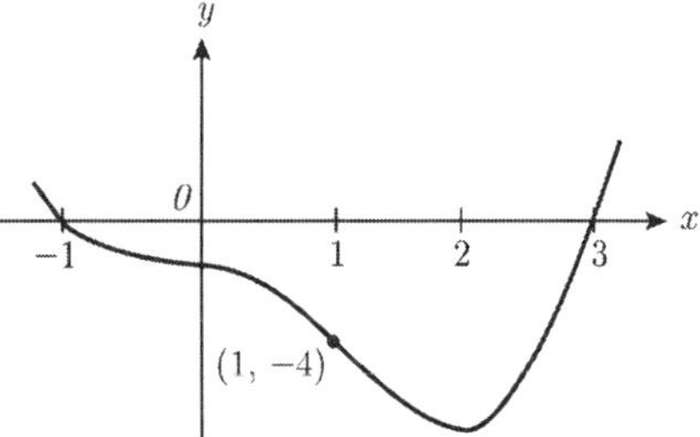

Graph of f'

(1) Write an equation for the line tangent to the graph of g at $x=1$.

(2) For $-1.2<x<3.2$, find all values of x at which g has a local maximum. justify your answer.

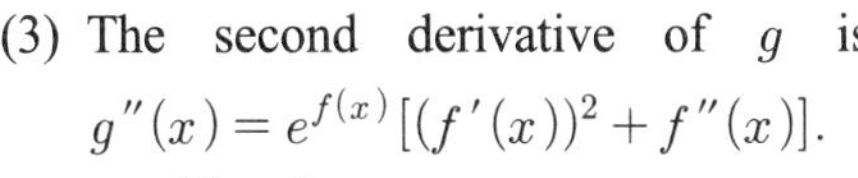

(3) The second derivative of g is $g''(x)=e^{f(x)}[(f'(x))^2+f''(x)]$.
Is $g''(-1)$ positive, negative, or zero? justify your answer.

(4) Find the average rate of change of change of g', the derivative of g, over the interval [1,3].

3.9 Derivatives of Trigonometric Functions

$$\frac{d}{dx}(\sin x) = \cos x \qquad \frac{d}{dx}(\cos x) = -\sin x$$

$$\frac{d}{dx}(\tan x) = \sec^2 x \qquad \frac{d}{dx}(\cot x) = -\csc^2 x$$

$$\frac{d}{dx}(\sec x) = \sec x \tan x \qquad \frac{d}{dx}(\csc x) = -\csc x \cot x$$

Proof

① Let $f(x) = \sin x$.

$$\begin{aligned} f'(x) &= \lim_{h\to 0}\frac{f(x+h)-f(x)}{h} = \lim_{h\to 0}\frac{\sin(x+h)-\sin x}{h} \\ &= \lim_{h\to 0}\frac{[\sin x\cos h + \cos x \sin h]-\sin x}{h} \\ &= \lim_{h\to 0}\frac{\sin x(\cos h-1)}{h} + \lim_{h\to 0}\frac{\cos x \sin h}{h} \\ &= \lim_{h\to 0}\sin x \cdot \lim_{h\to 0}\frac{\cos h-1}{h} + \lim_{h\to 0}\cos x \cdot \lim_{h\to 0}\frac{\sin h}{h} \\ &= \sin x \cdot 0 + \cos x \cdot 1 \\ &= \cos x \end{aligned}$$

② Let $f(x) = \cos x$.

$$\begin{aligned} f'(x) &= \lim_{h\to 0}\frac{f(x+h)-f(x)}{h} = \lim_{h\to 0}\frac{\cos(x+h)-\cos x}{h} \\ &= \lim_{h\to 0}\frac{[\cos x\cos h + \sin x \sin h]-\cos x}{h} \\ &= \lim_{h\to 0}\frac{\cos x(\cos h-1)}{h} + \lim_{h\to 0}\frac{\sin x \sin h}{h} \\ &= \lim_{h\to 0}\cos x \cdot \lim_{h\to 0}\frac{\cos h-1}{h} - \lim_{h\to 0}\sin x \cdot \lim_{h\to 0}\frac{\sin h}{h} \\ &= \cos x \cdot 0 - \sin x \cdot 1 \\ &= -\sin x \end{aligned}$$

③ Let $f(x) = \tan x$.

$$\begin{aligned} f'(x) &= \frac{d}{dx}(\tan x) = \frac{d}{dx}\left(\frac{\sin x}{\cos x}\right) = \frac{(\sin x)'\cos x - \sin x(\cos x)'}{(\cos x)^2} \\ &= \frac{\cos x\cos x - \sin x(-\sin x)}{(\cos x)^2} \end{aligned}$$

$$= \frac{\cos^2 x + \sin^2 x}{\cos^2 x} = \frac{1}{\cos^2 x} = \sec^2 x$$

④ Let $f(x) = \cot x$.

$$f'(x) = \frac{d}{dx}(\cot x) = \frac{d}{dx}\left(\frac{\cos x}{\sin x}\right) = \frac{(\cos x)'\sin x - \cos x(\sin x)'}{(\sin x)^2}$$

$$= \frac{(-\sin x)\sin x - \cos x(\cos x)}{(\sin x)^2}$$

$$= \frac{-\sin^2 x - \cos^2 x}{\sin^2 x} = \frac{-1}{\sin^2 x} = -\csc^2 x$$

⑤ Let $f(x) = \sec x$.

$$f'(x) = \frac{d}{dx}(\sec x) = \frac{d}{dx}\left(\frac{1}{\cos x}\right) = \frac{(1)'\cos x - 1(\cos x)'}{(\cos x)^2} = \frac{0-(-\sin x)}{(\cos x)^2}$$

$$= \frac{\sin x}{\cos^2 x} = \frac{1}{\cos x}\frac{\sin x}{\cos x} = \sec x \tan x$$

⑥ Let $f(x) = \csc x$.

$$f'(x) = \frac{d}{dx}(\csc x) = \frac{d}{dx}\left(\frac{1}{\sin x}\right) = \frac{(1)'\sin x - 1(\sin x)'}{(\sin x)^2} = \frac{0-(\cos x)}{(\sin x)^2}$$

$$= \frac{-\cos x}{\sin^2 x} = \frac{-1}{\sin x}\frac{\cos x}{\sin x} = -\csc x \cot x$$

Example 1 (1) $\frac{d}{dx}(\cos(x^2+1))$ (2) $\frac{d}{dx}(\sin\sqrt{x^2+1})$

Solution

(1) $\frac{d}{dx}(\cos(x^2+1)) = -\sin(x^2+1) \cdot 2x = -2x\sin(x^2+1)$

(2) $\frac{d}{dx}(\sin\sqrt{x^2+1}) = \cos(\sqrt{x^2+1}) \cdot \frac{1}{2\sqrt{x^2+1}} \cdot 2x = \frac{x\cos(\sqrt{x^2+1})}{\sqrt{x^2+1}}$

EX 1 Find the derivative of each of the following trigonometric functions.

(1) $y = \sin(3x+1)$ (2) $y = \cos(-3x+2)$

(3) $y = 2\sin(x^3 - \frac{1}{x})$ (4) $y = \cos(\sin x)$

(5) $y = \tan(10x-5)$ (6) $y = \cot(3x^2+2x)$

(7) $y = \sec(\frac{1}{2}x)$ (8) $y = -\frac{1}{5}\csc(5x)$

(9) $y = \sec(\tan x)$ (10) $y = \sqrt{1+2\cot x}$

(11) $y = \cot(\pi - \frac{1}{x})$ (12) $y = \sin(x^3-1)^4$

EX 2 Differentiable each of the following trigonometric functions.

(1) $y=(1+\cos 2t)^{-4}$

(2) $y=\sin^3 x$

(3) $y=3\cos^2(x^2)$

(4) $y=\tan x^3+\tan^3 x$

(5) $y=2\sec^3(\sqrt[3]{x})$

(6) $y=\sin^3(2x)+\cos^3(2x)$

(7) $y=\tan^2(\sin^3 t)$

EX 3 ifferentiable each of the following trigonometric functions.

(1) $y=x\sin x$

(2) $y=x^2\cos\frac{1}{x}$

(3) $y=3x\sqrt{\csc x}$

(4) $y=\frac{\sin x}{1+\cos x}$

(5) $y=\sin 4x\cos^2 x$

(6) $y=\sec\sqrt{x}\tan(\frac{1}{x})$

(7) $y=\frac{\cos^2 x}{\sin x}$

EX 4 Use implicit differentiation to find $\frac{dy}{dx}$.

(1) $x\cos(2x+3y)=y\sin x$

(2) $\cos(x+y)+\sin(x-y)=1$

(3) $x+\tan(xy)=0$

(4) $x^2+y^2=\sec(xy)$

EX 5 Find the equation of the tangent line to the curves at the given point.

(1) $2xy+\pi\sin y=2\pi$ at $(1,\frac{\pi}{2})$

(2) $y=x^2\cos x$ at $x=\pi$

(3) $y=\sin x\cos 2x$ at $x=\frac{\pi}{6}$

(4) $y=2\cot^2 x-\cos^2 x$ at $x=\frac{\pi}{4}$

(5) $x\sin 2y=y\cos 2x$ at $(\frac{\pi}{4},\frac{\pi}{2})$

EX 6 Find the second derivative of the following functions.

(1) $y=\sin^3 2x$

(2) $y=\cos^3 3x$

3.10 Derivatives of Exponential Functions

$$\frac{d}{dx}(e^x) = e^x \qquad \frac{d}{dx}(e^{f(x)}) = e^{f(x)} \cdot f'(x)$$

$$\frac{d}{dx}(a^x) = a^x \cdot \ln a \ (a \neq 1, a > 0)$$

$$\frac{d}{dx}(a^{f(x)}) = a^{f(x)} \cdot f'(x) \cdot \ln a \ (a \neq 1, a > 0)$$

Proof

① $f(x) = e^x$

$$f'(x) = \lim_{h\to 0}\frac{f(x+h)-f(x)}{h} = \lim_{h\to 0}\frac{e^{x+h}-e^x}{h}$$

$$= \lim_{h\to 0} e^x \cdot \frac{e^{h-1}}{h} = e^x \cdot \lim_{h\to 0}\frac{e^h-1}{h}$$

Let $e^h - 1 = t \Rightarrow e^h = 1+t$

$h = \ln(1+t)$

if $h \to 0$, $then\ t \to 0$.

$$f'(x) = e^x \cdot \lim_{h\to 0}\frac{e^h-1}{h} = e^x \cdot \lim_{h\to 0}\frac{t}{\ln(1+t)} = e^x \cdot \lim_{h\to 0}\frac{1}{\frac{1}{t}\ln(1+t)}$$

$$= e^x \cdot \lim_{t\to 0}\frac{1}{\ln(1+t)^{1/t}} = e^x \cdot \frac{1}{\ln e} = e^x \cdot 1 = e^x$$

② $y = e^{f(x)}$

Let $y = e^u$ and $u = f(x)$

$\frac{dy}{du} = e^u$ and $\frac{du}{dx} = f'(x)$

$$\frac{dy}{dx} = \frac{dy}{du} \cdot \frac{du}{dx} = e^u \cdot f'(x) = e^{f(x)} \cdot f'(x)$$

③ $y = a^x = (e^{\ln a})^x = e^{x\ln a}$ (where $\ln a$ is a constant)

$$y' = e^{x\ln a} \cdot \ln a$$

$$= a^x \cdot \ln a$$

④ $y=a^{f(x)}=(e^{\ln a})^{f(x)}=e^{f(x)\ln a}$ (where $\ln a$ is a constant)

$$y'=e^{f(x)\ln a}\cdot f'(x)\ln a$$
$$=a^{f(x)}\cdot f'(x)\ln a$$

Example 1

(1) $\dfrac{d}{dx}(e^{4x})$ (2) $\dfrac{d}{dx}(e^{-3x})$ (3) $\dfrac{d}{dx}(2^{x^2})$

Solution

(1) $\dfrac{d}{dx}(e^{4x})=e^{4x}\cdot 4$
$=4e^{4x}$

(2) $\dfrac{d}{dx}(e^{-3x})=e^{-3x}\cdot(-3)$
$=-3e^{-3x}$

(3) $\dfrac{d}{dx}(2^{x^2})=2^{x^2}\cdot 2x\cdot\ln 2$
$=2x(2)^{x^2}\ln 2$

EX 1 Find the derivative of each of the following functions.

(1) $y=e^{5-7x}$ (2) $y=e^{\sec x}$ (3) $y=\cos(e^{-x^2})$

(4) $y=3^{-x^2}$ (5) $y=xe^{-x}+e^{3x}$ (6) $y=a^{\cos x}$

(7) $y=x^3e^{2x}$ (8) $y=(9x^2-6x+2)e^{x^3}$ (9) $y=\dfrac{e^{-2x}}{x}$

(10) $y=(e^{\sin\frac{x}{2}})^3$ (11) $y=\pi^{x\cos x}$ (12) $y=\dfrac{e^{-x}+1}{e^x-1}$

(13) $e^{x^2y}=2x+2y$ (14) $y=\sin(x^2e^x)$ (15) $y=\dfrac{1}{\sqrt{4e^{-x}+2}}$

(16) $e^{x+y}=3x^2y$ (17) $e^{2x}=\sin(x+3y)$

EX 2 Find the equation of the tangent line to the curves at the given point.

(1) $y=\dfrac{e^{-x}}{\sqrt{x}}$ at $x=1$ (2) $y=x^2e^{\frac{1}{x}}$ at $x=1$

EX 3

(1) Find the equation of the tangent line to the curves $y=e^{3x}$ from the external point $(0,0)$.

(2) Find the equation of the tangent line to the curves $y=xe^{2x}$ from the external point $(-2,0)$.

EX 4 Given $f(x)=e^x\cos x$, the equation of $f''(x)-pf'(x)+qf(x)=0$ is satisfied for all real number x. Find the values of p and q.

3.11 Derivatives of Logarithmic Functions

$$\frac{d}{dx}(\ln x) = \frac{1}{x} \qquad \frac{d}{dx}(\log_a x) = \frac{1}{x} \cdot \frac{1}{\ln a}$$

$$\frac{d}{dx}(\ln f(x)) = \frac{1}{f(x)} \cdot f'(x) \qquad \frac{d}{dx}(\log_a f(x)) = \frac{1}{f(x)} \cdot f'(x) \cdot \frac{1}{\ln a}$$

Proof

① $f(x) = \ln x$

$$f'(x) = \lim_{h\to 0}\frac{f(x+h)-f(x)}{h} = \lim_{h\to 0}\frac{\ln(x+h)-\ln x}{h} = \lim_{h\to 0}\frac{\ln\left(\frac{x+h}{x}\right)}{h}$$

$$= \lim_{h\to 0}\frac{1}{h}\ln\left(\frac{x+h}{x}\right) = \lim_{h\to 0}\ln\left(1+\frac{h}{x}\right)^{\frac{1}{h}}$$

$$= \lim_{h\to 0}\frac{1}{h}\ln\left[\left(\frac{x+h}{x}\right)^{\frac{x}{h}}\right]^{\frac{1}{x}} = \ln e^{\frac{1}{x}} = \frac{1}{x}\ln e = \frac{1}{x}$$

② $f(x) = \log_a x$.

$f(x) = \log_a x = \dfrac{\ln x}{\ln a}$ (where $\ln a$ is a constant)

$$f'(x) = \frac{(\ln x)'}{\ln a} = \frac{\frac{1}{x}}{\ln a} = \frac{1}{x} \cdot \frac{1}{\ln a}$$

③ $f(x) = \ln f(x)$

Let $y = \ln u$ and $u = f(x)$

$$\frac{dy}{du} = \frac{1}{u},\ \frac{du}{dx} = f'(x)$$

$$\frac{dy}{dx} = \frac{dy}{du} \cdot \frac{du}{dx} = \frac{1}{u} \cdot f'(x) = \frac{1}{f(x)} \cdot f'(x)$$

④ $f(x) = \log_a f(x) = \dfrac{\ln f(x)}{\ln a}$ (where $\ln a$ is a constant)

$$\frac{dy}{dx} = \frac{1}{f(x)} \cdot f'(x) \cdot \frac{1}{\ln a}$$

- $y = \ln|f(x)|$

If $f(x) > 0$, $y = \ln|f(x)| = \ln f(x)$

$$y = \frac{1}{f(x)} \cdot f'(x)$$

If $f(x) < 0, y = \ln|f(x)| = \ln(-f(x))$

$$y' = \frac{1}{-f(x)} \cdot -f'(x) = \frac{1}{f(x)} \cdot f'(x)$$

$$\therefore \frac{d}{dx}(\ln|f(x)|) = \frac{1}{f(x)} \cdot f'(x)$$

Example 1 (1) $\frac{d}{dx}(\ln(x^2+1))$ (2) $\frac{d}{dx}(\ln(\ln(x^2+\sin x)))$

Solution

(1) $\frac{d}{dx}(\ln(x^2+1)) = \frac{1}{x^2+1} \cdot 2x = \frac{2x}{x^2+1}$

(2) $\frac{d}{dx}(\ln(\ln(x^2+\sin x))) = \frac{1}{\ln(x^2+\sin x)} \cdot \frac{1}{x^2+\sin x} \cdot (2x+\cos x)$

$$= \frac{2x+\cos x}{(x^2+\sin x)\ln(x^2+\sin x)}$$

EX 1 Find the derivative of each of the following functions.

(1) $y = \ln 3x$ (2) $y = \ln(\ln x)$ (3) $y = \ln(\sin x)$

(4) $y = \ln(\tan x)$ (5) $y = (\ln x)^5$ (6) $y = \log_3(\sin x)$

(7) $y = \log_2 5x$ (8) $y = \ln|x\cos x|$ (9) $y = 3^{\log_2 x}$

(10) $y = x^2 \ln x$ (11) $y = x\sqrt{\ln x}$ (12) $y = \ln(\sec x + \tan x)$

(13) $y = \tan[\log(x^2+4)]$ (14) $y = (x^2\ln x)^4$ (15) $y = \frac{\ln x}{x^2}$

(16) $\ln(xy) = x - y$ (17) $y+1 = x\ln^3 y$ (18) $y = \ln(\sec(\ln x))$

EX 2 Find the derivative of each of the following functions.

(1) $y = \ln((2x+1)^3(x^2+1)^4)$ (2) $y = \ln(3xe^{-x})$ (3) $y = \ln\left(\frac{(x^2+1)^5}{\sqrt{1-x}}\right)$

(4) $y = \ln\sqrt{\frac{(x+1)^5}{(x+2)^{20}}}$ (5) $y = \ln\left(\frac{\sqrt{\sin x \cos x}}{1+2\ln x}\right)$ (6) $y = \log_7\left(\frac{\sin x \cos x}{e^x 2^x}\right)$

- Logarithmic Differentiation

Step 1. Take the natural logarithm of both sides.

Step 2. Use the logarithmic properties to simplify the right side of the equation

Step 3. Differentiate both sides implicitly

Step 4. Solve for y'

Example 2 Use logarithmic differentiation to find the derivative of $y = x^x$.

Solution

Step 1. $\ln y = \ln x^x$

Step 2. $\ln y = x \ln x$

Step 3. $\frac{1}{y} \cdot y' = 1 \cdot \ln x + x \cdot \frac{1}{x}$

step 4. $y' = y(\ln x + 1)$ or $x^x(\ln x + 1)$

EX 3 Use logarithmic differentiation to find the derivative of each of the following functions.

(1) $y = x^{\ln x}$ (2) $y = (\sin x)^x$ (3) $y = (x+1)^x$

(4) $y = (\sin x)^{\sin x}$ (5) $y = \sqrt{(x^2+1)(x-1)^2}$ (6) $y = (\ln x)^{\ln x}$

(7) $y = \dfrac{x\sqrt{x^2+1}}{(x+1)^{\frac{2}{3}}}$ (8) $y = \dfrac{x \sin x}{\sqrt{\sec x}}$ (9) $y = \dfrac{e^{-x}\cos x}{\sqrt[3]{x^2 - x}}$

EX 4

(1) Find the slope of the tangent line to the curve $y = \dfrac{(x+1)^2}{x^3(x-2)}$ at $x = 1$.

(2) Find the slope of the tangent line to the curve $y = \dfrac{\sqrt[4]{2x-1}\sqrt{x^2+1}}{\sqrt{x+1}\sqrt{5-x}}$ at $x = 1$.

EX 5 Given $f(2) = 1$ and $f'(2) = 4$, find value of $\frac{d}{dx}[\sqrt{2x + \ln(f(x))}]$ at $x = 2$.

3.12 Inverse Trigonometric Functions

[1] $y = \sin^{-1} x = \arcsin x$

• $y = \sin x$

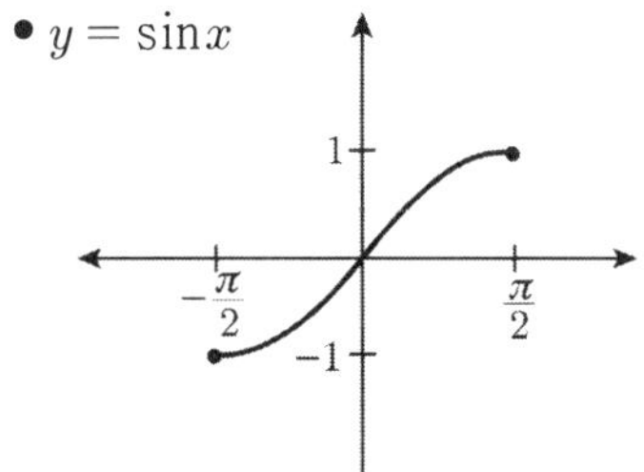

Domain : $-\frac{\pi}{2} \le x \le \frac{\pi}{2}$

Range : $-1 \le y \le 1$

• $y = \sin^{-1} x$

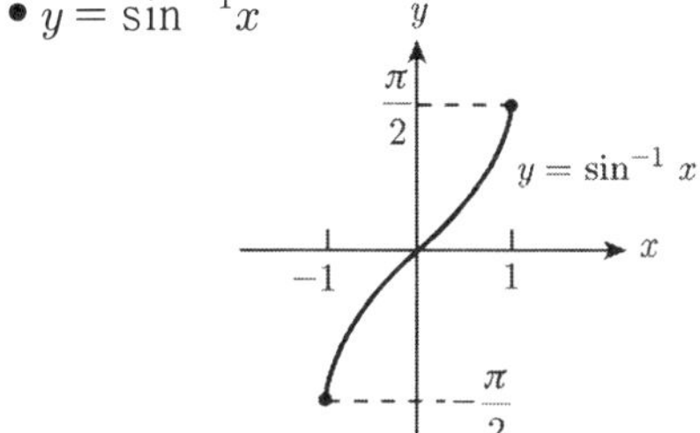

Domain : $-1 \le x \le 1$

Range : $-\frac{\pi}{2} \le y \le \frac{\pi}{2}$

EX1 Evaluate

(1) $\sin^{-1}(-\frac{1}{2})$

(2) $\sin^{-1}(-\frac{\sqrt{3}}{2})$

(3) $\sin^{-1}(\frac{\sqrt{3}}{2})$

(4) $\sin^{-1}(1)$

EX2 Evaluate

(1) $\sin^{-1}\left(\sin\left(\frac{7\pi}{6}\right)\right)$

(2) $\sin\left(\sin^{-1}\left(-\frac{1}{2}\right)\right)$

(3) $\sin^{-1}\left(\sin\left(\frac{2\pi}{3}\right)\right)$

(4) $\cos(\sin^{-1}(-1))$

[2] $y=\cos^{-1}x=\arccos x$

- $y=\cos x$

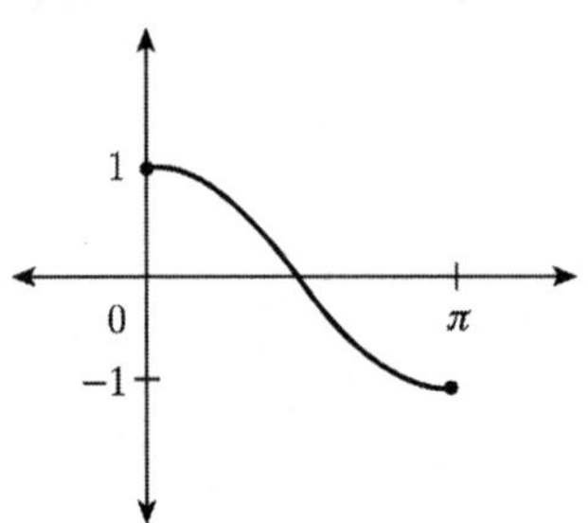

Domain : $0 \le x \le \pi$

Range : $-1 \le y \le 1$

- $y=\cos^{-1}x$

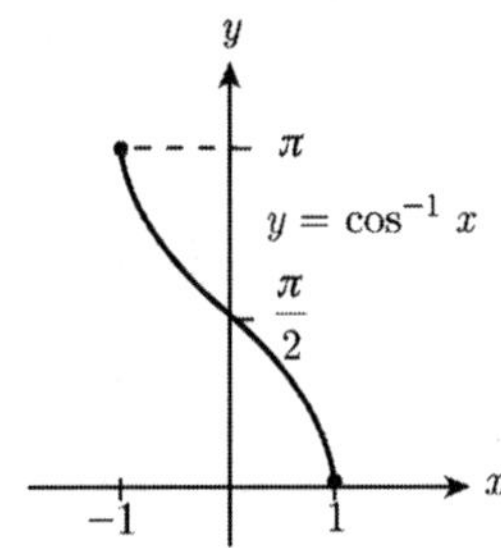

Domain : $-1 \le x \le 1$

Range : $0 \le y \le \pi$

EX3 Evaluate

(1) $\cos^{-1}(-\frac{1}{2})$

(2) $\cos^{-1}(-\frac{\sqrt{3}}{2})$

(3) $\cos^{-1}(\frac{\sqrt{3}}{2})$

(4) $\cos^{-1}(0)$

EX4 Evaluate

(1) $\cos^{-1}\left(\cos\left(\frac{5\pi}{6}\right)\right)$

(2) $\cos\left(\cos^{-1}\left(-\frac{1}{2}\right)\right)$

(3) $\cos^{-1}\left(\cos\left(-\frac{\pi}{6}\right)\right)$

(4) $\tan(\cos^{-1}(1))$

[3] $y = \tan^{-1}x = \arctan x$

- $y = \tan x$

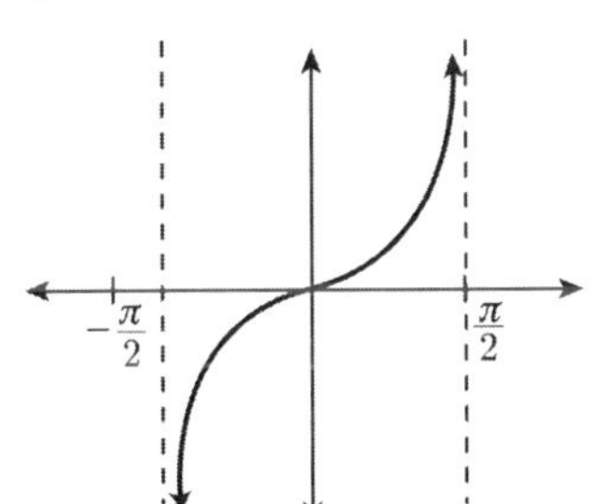

Domain : $-\dfrac{\pi}{2} < x < \dfrac{\pi}{2}$

Range : $y \in R$

- $y = \tan^{-1}x$

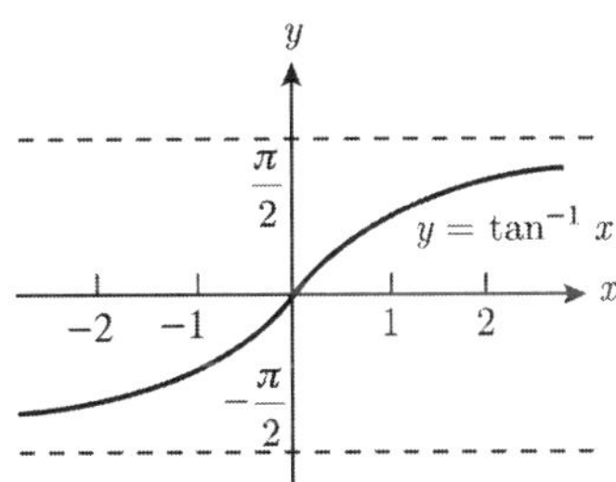

Domain : $\infty < x < \infty$

Range : $-\dfrac{\pi}{2} < y < \dfrac{\pi}{2}$

EX 5 Evaluate

(1) $\tan^{-1}(-1)$

(2) $\tan^{-1}(\frac{1}{\sqrt{3}})$

(3) $\tan^{-1}(\sqrt{3})$

(4) $\tan^{-1}(0)$

EX 6 Evaluate

(1) $\tan^{-1}\left(\tan\left(\frac{3\pi}{4}\right)\right)$

(2) $\tan\left(\tan^{-1}\left(-\frac{1}{\sqrt{3}}\right)\right)$

(3) $\tan^{-1}\left(\tan\left(-\frac{7\pi}{6}\right)\right)$

(4) $\cos\left(\tan^{-1}(-\sqrt{3})\right)$

[4] $y = \csc^{-1}x$, $y = \sec^{-1}x$, and $y = \cot^{-1}x$

Domain: $x \le -1$ or $x \ge 1$

Range: $0 \le y \le \pi,\ y \ne \frac{\pi}{2}$

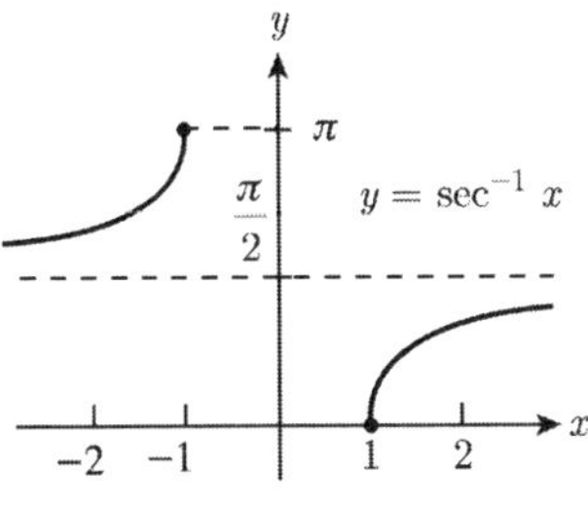

Domain: $x \le -1$ or $x \ge 1$

Range: $-\frac{\pi}{2} \le y \le \frac{\pi}{2},\ y \ne 0$

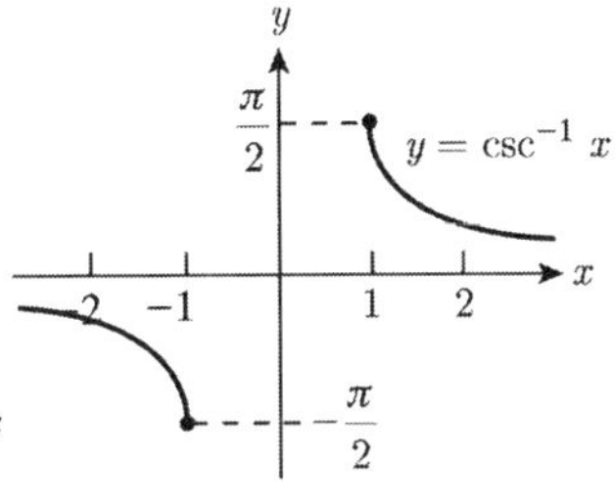

Domain: $-\infty < x < \infty$

Range: $0 < y < \pi$

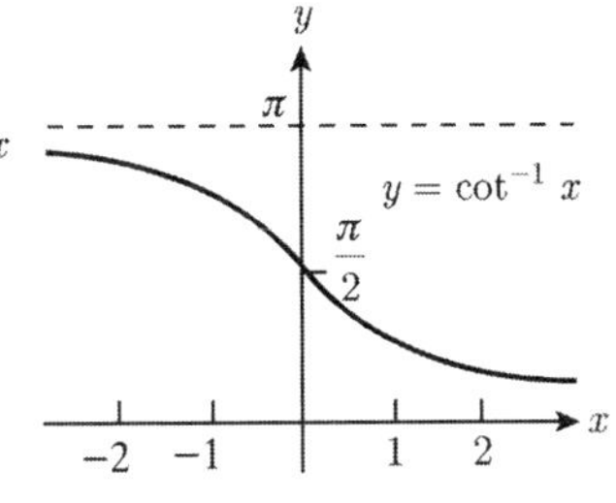

EX 7 Evaluate

(1) $\csc^{-1}(2)$

(2) $\csc^{-1}\left(-\dfrac{2}{\sqrt{3}}\right)$

(3) $\sec^{-1}(-\sqrt{2})$

(4) $\csc^{-1}(undefined)$

(5) $\sec^{-1}(2)$

(6) $\sec^{-1}(undefined)$

(7) $\cot^{-1}\left(\dfrac{1}{\sqrt{3}}\right)$

(8) $\cot^{-1}(-\sqrt{3})$

EX 8 Evaluate

(1) $\sin\left(\cos^{-1}\left(\dfrac{\sqrt{2}}{2}\right)\right)$

(2) $\sec\left(\cos^{-1}\left(\dfrac{1}{2}\right)\right)$

(3) $\tan\left(\sin^{-1}\left(-\dfrac{1}{2}\right)\right)$

(4) $\cot\left(\sin^{-1}\left(-\dfrac{\sqrt{3}}{2}\right)\right)$

3.13 Derivatives of Inverse Trigonometric Functions

$$\frac{d}{dx}(\sin^{-1}x)=\frac{1}{\sqrt{1-x^2}} \qquad \frac{d}{dx}(\cos^{-1}x)=-\frac{1}{\sqrt{1-x^2}}$$

$$\frac{d}{dx}(\tan^{-1}x)=\frac{1}{1+x^2} \qquad \frac{d}{dx}(\cot^{-1}x)=-\frac{1}{1+x^2}$$

$$\frac{d}{dx}(\sec^{-1}x)=\frac{1}{|x|\sqrt{x^2-1}} \qquad \frac{d}{dx}(\csc^{-1}x)=-\frac{1}{|x|\sqrt{x^2-1}}$$

Proof

① $y=\sin^{-1}x$ where $-\dfrac{\pi}{2}\le y\le\dfrac{\pi}{2}$

$$\sin y=x$$

$$\frac{d}{dx}(\sin y)=\frac{d}{dx}(x)$$

$$\cos y\cdot y'=1$$

$$y'=\frac{1}{\cos y}$$

$$=\frac{1}{\sqrt{1-x^2}}$$

$$\sin^2 y+\cos^2 y=1$$

$$\cos y=\pm\sqrt{1-\sin^2 y}=\pm\sqrt{1-x^2}$$

For $-\dfrac{\pi}{2}\le y\le\dfrac{\pi}{2}$, $\cos y>0$

$$\therefore \cos y=\sqrt{1-x^2}$$

② $y=\cos^{-1}x$ where $0 \le y \le \pi$

$$\cos y = x$$

$$\frac{d}{dx}(\cos y)=\frac{d}{dx}(x)$$

$$-\sin y \cdot y' = 1$$

$$y' = -\frac{1}{\sin y}$$

$$= -\frac{1}{\sqrt{1-x^2}}$$

$$\sin^2 y + \cos^2 y = 1$$

$$\sin y = \pm\sqrt{1-\cos^2 y} = \pm\sqrt{1-x^2}$$

For $0 \le y \le \pi$, $\sin y > 0$

$$\therefore \sin y = \sqrt{1-x^2}$$

③ $y=\tan^{-1}x$

$$\tan y = x$$

$$\frac{d}{dx}(\tan y)=\frac{d}{dx}(x)$$

$$\sec^2 y \cdot y' = 1$$

$$y' = \frac{1}{\sec^2 y}$$

$$= \frac{1}{1+x^2}$$

$$\tan^2 y + 1 = \sec^2 y$$

④ $y=\cot^{-1}x$

$$\cot y = x$$

$$\frac{d}{dx}(\cot y)=\frac{d}{dx}(x)$$

$$-\csc^2 y \cdot y' = 1$$

$$y' = -\frac{1}{\csc^2 y}$$

$$= -\frac{1}{1+x^2}$$

$$\cot^2 y + 1 = \csc^2 y$$

⑤ $y=\sec^{-1}x$ where $0 \le y \le \pi$, $y =\ne \frac{\pi}{2}$

$$\sec y = x$$

$$\frac{d}{dx}(\sec y)=\frac{d}{dx}(x)$$

$$\sec y \tan y \cdot y' = 1$$

$$y' = \frac{1}{\sec y \tan y}$$

$$= \pm \frac{1}{x\sqrt{x^2-1}}$$

$$= \frac{1}{|x|\sqrt{x^2-1}}$$

$\tan^2 y + 1 = \sec^2 y$

$\tan y = \pm\sqrt{\sec^2 y - 1}$

$\tan y = \pm\sqrt{x^2-1}$ and $\sec y = x$

(6) $y = \csc^{-1}x$ where, $0 < y < \pi, y = \frac{\pi}{2}$

$$\csc y = x$$

$$\frac{d}{dx}(\csc y) = \frac{d}{dx}(x)$$

$$-\csc y \cot y \cdot y' = 1$$

$$y' = -\frac{1}{\csc y \cot y}$$

$$= \pm\frac{-1}{x\sqrt{x^2-1}}$$

$$= \frac{-1}{|x|\sqrt{x^2-1}}$$

$\cot^2 y + 1 = \csc^2 y$

$\cot y = \pm\sqrt{\csc^2 y - 1}$

← $\cot y = \pm\sqrt{x^2-1}$, $\csc y = x$

Example 1

(1) $\frac{d}{dx}(\sin^{-1}x^2)$ (2) $\frac{d}{dx}(\tan^{-1}(\sin x))$

Solution

(1) $\frac{d}{dx}(\sin^{-1}x^2) = \frac{1}{\sqrt{1-(x^2)^2}} \cdot 2x = \frac{2x}{\sqrt{1-x^4}}$

(2) $\frac{d}{dx}(\tan^{-1}(\sin x)) = \frac{1}{1+(\sin x)^2} \cdot \cos x = \frac{\cos x}{1+\sin^2 x}$

EX 1 Find the derivative of each of the following inverse trigonometric functions.

(1) $y = \sin^{-1}(\frac{3}{x^2})$ (2) $y = \cos^{-1}(-3x+2)$

(3) $y = \sin^{-1}(1-x)$ (4) $y = 2\cos^{-1}(\sqrt{x-1})$

(5) $y = \tan^{-1}(\ln x)$ (6) $y = \cot^{-1}\left(\frac{1}{x}\right)$

(7) $y=\ln(x^2+4)-x\tan^{-1}\left(\frac{x}{2}\right)$ (8) $y=\cot^{-1}(\csc x^2)$

(9) $y=\sec^{-1}(2x+1)$ (10) $y=\sin^{-1}((1-x^2)^2)$

(11) $y=\tan^{-1}\sqrt{x^2-1}+\csc^{-1}x,\ x>1$ (12) $y=\sqrt{x^2-1}+\csc^{-1}x,\ x>1$

EX 2 Find the slope of the tangent line to the curves at the given point.

(1) $y=(\cos^{-1}x)^4\ \ at\ x=0$ (2) $y=x\tan^{-1}(x^2-1)\ \ at\ x=\sqrt{2}$

EX 3 Find the equation of the tangent line to the curves at the given point.

(1) $y=\sec^{-1}(\sqrt{x})\ at\ x=4$ (2) $f(x)=(6\tan^{-1}x)^3\ \ at\ \ x=\sqrt{3}$

3.14 Derivatives of Inverse Functions

Let f be differentiable and a one-to-one function on a domain.

$$\frac{d}{dx}[f^{-1}(x)]=(f^{-1})'(x)=\frac{1}{f'(f^{-1}(x))}$$

Proof

if f has an inverse function, then $f(f^{-1}(x))=x$ for every x in the domain.

$$f(f^{-1}(x))=x$$
$$\frac{d}{dx}f(f^{-1}(x))=1$$
$$f'(f^{-1}(x))\cdot\frac{d}{dx}f^{-1}(x)=1$$
$$\frac{d}{dx}f^{-1}(x)=\frac{1}{f'(f^{-1}(x))}$$

If f has an interval I as domain and $f'(x)$ exist and is never zero on I. then f^{-1} is differentiable at every point in its domain (the range of f). The value of the $(f^{-1})'$ at a point b in the domain of f^{-1} is the reciprocal of the value of f' at the point $a=f^{-1}(b)$:

$$(f^{-1})'(b)=\frac{1}{f'(f^{-1}(b))}\quad\text{or}\quad\left.\frac{df^{-1}}{dx}\right|_{x=b}=\frac{1}{\left.\frac{df}{dx}\right|_{x=f^{-1}(b)}}$$

Example 1 Let $y = f(x) = x^3 + x - 2$, and let g be the inverse function. Evaluate $g'(0)$.

Solution

Since $f'(x) = 3x^2 + 1$, $g'(y) = \dfrac{1}{3x^2+1}$. To find x when $y = 0$, we must solve the equation $x^3 + x - 2 = 0$. Note by inspection that $x = 1$, so

$$g'(0) = \frac{1}{3(1)^2+1} = \frac{1}{4}.$$

EX 1

(1) A function f and its derivative take on the values shown in the table. If g is the inverse of f, find $g'(6)$.

x	$f(x)$	$f'(x)$
2	6	$\frac{1}{3}$
6	8	$\frac{3}{2}$

(2) If $f(3) = 8$ and $f'(3) = 5$, what do we know about f^{-1}?

EX 2

(1) Suppose $y = f(x) = 2x^3 - 3x$. If $h(x)$ is the inverse function of f, then $h'(-1) =$

(2) Let $f(x) = x^3 - 3x^2 - 1, x \geq 2$. Find the value of df^{-1}/dx at the point $x = -1 = f(3)$.

(3) Let $f(x) = x^2 - 4x - 5, x > 2$. Find the value of df^{-1}/dx at the point $x = 0 = f(5)$.

(4) Suppose that the differentiable function $y = f(x)$ has an inverse and that the graph of f passes through the point $(2,4)$ and has a slope of 1/3 there. Find the value of $df^{-1}/dx \ at \ x = 4$.

(5) Suppose that the differentiable function $y = g(x)$ has an inverse and that the graph of g passes through the origin with slope 2. Find the slope of the graph of g^{-1} at the origin.

(6) Let $f(x) = e^x + x + 1$ and $f(g(x)) = x$. Find the value of $g'(2)$.

- **Basic Derivative Formulas**

$$\frac{d}{dx}(x^n) = nx^{n-1}, \qquad \frac{d}{dx}(\sqrt{x}) = \frac{1}{2\sqrt{x}}, \qquad \frac{d}{dx}\left(\frac{1}{x}\right) = -\frac{1}{x^2}$$

$$\frac{d}{dx}(f(x) \cdot g(x)) = f'(x)g(x) + f(x)g'(x), \qquad \frac{d}{dx}\left(\frac{f(x)}{g(x)}\right) = \frac{f'(x)g(x) - f(x)g'(x)}{g^2(x)}$$

$$\frac{d}{dx}f(g(x)) = f'(g(x)) \cdot g'(x), \qquad \frac{d}{dx}f(g(h(x))) = f'(g(h(x))) \cdot g'(h(x)) \cdot h'(x)$$

$$\frac{d}{dx}(\sqrt{f(x)}) = \frac{1}{2\sqrt{f(x)}} \cdot f'(x)$$

- **Derivatives of Trigonometric Formulas**

$$\frac{d}{dx}(\sin x) = \cos x \qquad \frac{d}{dx}(\cos x) = -\sin x$$

$$\frac{d}{dx}(\tan x) = \sec^2 x \qquad \frac{d}{dx}(\cot x) = -\csc^2 x$$

$$\frac{d}{dx}(\sec x) = \sec x \tan x \qquad \frac{d}{dx}(\csc x) = -\csc x \cot x$$

- **Derivatives of Inverse Trigonometric Formulas**

$$\frac{d}{dx}(\sin^{-1}x) = \frac{1}{\sqrt{1-x^2}} \qquad \frac{d}{dx}(\cos^{-1}x) = -\frac{1}{\sqrt{1-x^2}}$$

$$\frac{d}{dx}(\tan^{-1}x) = \frac{1}{1+x^2} \qquad \frac{d}{dx}(\cot^{-1}x) = -\frac{1}{1+x^2}$$

$$\frac{d}{dx}(\sec^{-1}x) = \frac{1}{|x|\sqrt{x^2-1}} \qquad \frac{d}{dx}(\csc^{-1}x) = -\frac{1}{|x|\sqrt{x^2-1}}$$

- **Derivatives of Exponential and Logarithmic Functions Formulas**

$$\frac{d}{dx}(e^x) = e^x \qquad \frac{d}{dx}(a^x) = a^x \ln a \quad (a \neq 1,\ a > 0)$$

$$\frac{d}{dx}(e^{f(x)}) = e^{f(x)} \cdot f'(x) \qquad \frac{d}{dx}(a^{f(x)}) = a^{f(x)} \cdot f'(x) \cdot \ln a$$

$$\frac{d}{dx}(\ln x) = \frac{1}{x} \qquad \frac{d}{dx}(\log_a x) = \frac{1}{x} \cdot \frac{1}{\ln a}$$

$$\frac{d}{dx}(\ln f(x)) = \frac{1}{f(x)} \cdot f'(x) \qquad \frac{d}{dx}(\log_a f(x)) = \frac{1}{f(x)} \cdot f'(x) \cdot \frac{1}{\ln a}$$

- **Derivatives of Inverse Formulas**

$$(f^{-1})'(x) = \frac{1}{f'(f^{-1}(x))}$$

Multiple Choice

01 If $xy^2+2xy=8$, then, at the point (1,2), y' is

(A) $-\frac{5}{2}$ (B) $-\frac{4}{3}$ (C) -1 (D) $-\frac{1}{2}$ (E) 0

02 If f and g are twice differentiable functions such that $g(x)=e^{f(x)}$ and $g''(x)=h(x)e^{f(x)}$, then $h(x)=$

(A) $f'(x)+f''(x)$
(B) $f'(x)+(f''(x))^2$
(C) $(f'(x)+f''(x))^2$
(D) $(f'(x))^2+f''(x)$
(E) $2f'(x)+f''(x)$

03 If $f(x)=\ln(\sqrt{x})$, then $f''(x)=$

(A) $-\frac{2}{x^2}$ (B) $-\frac{1}{2x^2}$ (C) $-\frac{1}{2x}$ (D) $-\frac{1}{2x^{\frac{3}{2}}}$ (E) $\frac{2}{x^2}$

04 If $y^2-2xy=16$, then $\frac{dy}{dx}=$

(A) $\frac{x}{y-x}$ (B) $\frac{y}{x-y}$ (C) $\frac{y}{y-x}$ (D) $\frac{y}{2y-x}$ (E) $\frac{2y}{x-y}$

05 $\lim_{h\to 0}\frac{\sin(x+h)-\sin x}{h}$ is

(A) 0 (B) 1 (C) $\sin x$ (D) $\cos x$ (E) nonexistent

06 $\frac{d}{dx}\ln\left|\cos\left(\frac{\pi}{x}\right)\right|$ is

(A) $\frac{-\pi}{x^2\cos\left(\frac{\pi}{x}\right)}$
(B) $-\tan\left(\frac{\pi}{x}\right)$
(C) $\frac{1}{\cos\left(\frac{\pi}{x}\right)}$
(D) $\frac{\pi}{x}\tan\left(\frac{\pi}{x}\right)$
(E) $\frac{\pi}{x^2}\tan\left(\frac{\pi}{x}\right)$

07 If $x = t^2 + 1$ and $y = t^3$, then $\dfrac{d^2y}{dx^2} =$

(A) $\dfrac{3}{4t}$ (B) $\dfrac{3}{2t}$ (C) $3t$ (D) $6t$ (E) $\dfrac{3}{2}$

08 If $f(x) = 1 + x^{\frac{2}{3}}$, which of the following is NOT true?

(A) f is continuous for all real numbers.
(B) f has a minimum at $x = 0$.
(C) f is increasing for $x > 0$.
(D) $f'(x)$ exists for all x.
(E) $f''(x)$ is negative for $x > 0$.

09 If $e^{f(x)} = 1 + x^2$, then $f'(x) =$

(A) $\dfrac{1}{1+x^2}$ (B) $\dfrac{2x}{1+x^2}$ (C) $2x(1+x^2)$
(D) $2x(e^{1+x^2})$ (E) $2x\ln(1+x^2)$

10 Let f and g be functions that are differentiable for all real numbers, with $g(x) \neq 0$ for $x \neq 0$.

If $\lim_{x\to 0} f(x) = \lim_{x\to 0} g(x) = 0$ and $\lim_{x\to 0} \dfrac{f'(x)}{g'(x)}$ exists, then $\lim_{x\to 0} \dfrac{f(x)}{g(x)}$ is

(A) 0 (B) $\dfrac{f'(x)}{g'(x)}$ (C) $\lim_{x\to 0} \dfrac{f'(x)}{g'(x)}$
(D) $\dfrac{f'(x)g(x) - f(x)g'(x)}{(f(x))^2}$ (E) nonexistent

11 If $x = e^{2t}$ and $y = \sin(2t)$, then $\dfrac{dy}{dx} =$

(A) $4e^{2t}\cos(2t)$ (B) $\dfrac{e^{2t}}{\cos(2t)}$ (C) $\dfrac{\sin(2t)}{2e^{2t}}$
(D) $\dfrac{\cos(2t)}{2e^{2t}}$ (E) $\dfrac{\cos(2t)}{e^{2t}}$

12 If $f(x) = (x-1)^{\frac{3}{2}} + \dfrac{e^{x-2}}{2}$, then $f'(2) =$

(A) 1 (B) $\dfrac{3}{2}$ (C) 2 (D) $\dfrac{7}{2}$ (E) $\dfrac{3+e}{2}$

13 The function f given by $f(x) = 3x^5 - 4x^3 - 3x$ has a relative maximum at $x =$

(A) -1 (B) $-\frac{\sqrt{5}}{5}$ (C) 0 (D) $\frac{\sqrt{5}}{5}$ (E) 1

14 If f and g are twice differentiable and if $h(x) = f(g(x))$, then $h''(x) =$

(A) $f''(g(x))[g'(x)]^2 + f'(g(x))g''(x)$

(B) $f''(g(x))g'(x) + f'(g(x))g''(x)$

(C) $f''(g(x))[g'(x)]^2$

(D) $f''(g(x))g''(x)$

(E) $f''(g(x))$

15 If $\frac{dy}{dx} = (1 + \ln x)y$ and if $y = 1$ when $x = 1$, then $y =$

(A) $e^{\frac{x^2-1}{x^2}}$ (B) $1 + \ln x$ (C) $\ln x$ (D) $e^{2x + x\ln x - 2}$ (E) $e^{x\ln x}$

16 If $f(x) = \frac{x-1}{x+1}$ for all $x \neq -1$, then $f'(1) =$

(A) -1 (B) $-\frac{1}{2}$ (C) 0 (D) $\frac{1}{2}$ (E) 1

17 If $y = \sin x$ and $y^{(n)}$ means "the nth derivative of y with respect to x," then the smallest positive integer n for which $y^{(n)} = y$ is

(A) 2 (B) 4 (C) 5 (D) 6 (E) 8

18 If $f(x) = \ln(\ln x)$, then $f'(x)$ is

(A) $\frac{1}{x}$ (B) $\frac{1}{\ln x}$ (C) $\frac{\ln x}{x}$ (D) x (E) $\frac{1}{x\ln x}$

19 Suppose that f is an odd function; i.e., $f(-x) = -f(x)$ for all x. Suppose that $f'(x_0)$ exists. Which of the following must necessarily be equal to $f'(-x_0)$?

(A) $f'(x_0)$ (B) $-f'(x_0)$ (C) $\frac{1}{f'(x_0)}$

(D) $-\frac{1}{f'(x_0)}$ (E) None of the above

20 If $y = \sin(3x)$, then $\frac{dy}{dx} =$

(A) $-3\cos(3x)$ (B) $-\cos(3x)$ (C) $-\frac{1}{3}\cos(3x)$

(D) $\cos(3x)$ (E) $3\cos(3x)$

21 Let f be the function defined by $f(x) = x + \ln x$. What is the value of c for which the instantaneous rate of change of f at $x = c$ is the same as the average rate of change of f over [1, 4]?

(A) 0.456 (B) 1.244 (C) 2.164 (D) 2.342 (E) 2.452

22 If $h(x) = f^2(x) - g^2(x)$, $f'(x) = -g(x)$, and $g'(x) = f(x)$, then $h'(x) =$

(A) 0 (B) 1 (C) $-4f(x)g(x)$

(D) $(-g(x))^2 - (f(x))^2$ (E) $-2(-g(x) + f(x))$

23 If $\sin x = e^y$, $0 < x < \pi$, what is $\frac{dy}{dx}$ in terms of x?

(A) $-\tan x$ (B) $-\cot x$ (C) $\cot x$

(D) $\tan x$ (E) $\csc x$

24 If $y = \tan u$, $u = v - \frac{1}{v}$, and $v = \ln x$, what is the value of $\frac{dy}{dx}$ at $x = e$?

(A) 0 (B) $\frac{1}{e}$ (C) 1 (D) $\frac{2}{e}$ (E) $\sec^2 e$

25 If $\arcsin x = \ln y$, then $\frac{dy}{dx} =$

(A) $\frac{y}{\sqrt{1-x^2}}$ (B) $\frac{xy}{\sqrt{1-x^2}}$ (C) $\frac{y}{1+x^2}$

(D) $e^{\arcsin x}$ (E) $\frac{e^{\arcsin x}}{1+x^2}$

04

Application of Derivatives

4.1 Critical Points

An interior point of the domain of a function f where f' is zero or undefined is a critical point of f.

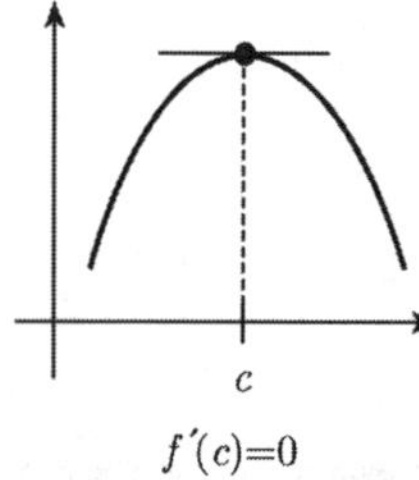

$f'(c)=0$

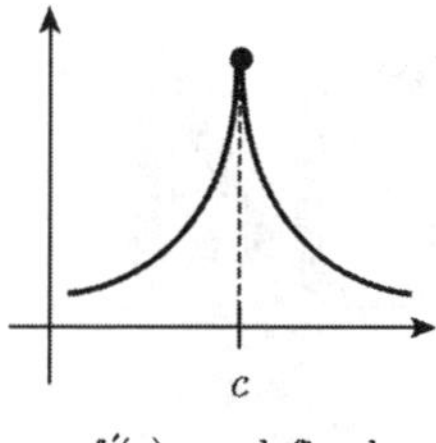

$f'(c)=$undefined

Example 1 Determine the critical points of the function $y = x^2 - 32\sqrt{x}$.

Solution

$y = x^2 - 32\sqrt{x} \Rightarrow y' = 2x - \dfrac{16}{\sqrt{x}} = \dfrac{2x^{3/2} - 16}{\sqrt{x}} = 0 \Rightarrow \ 2x^{3/2} - 16 = 0$

$\Rightarrow x = 4;\ \dfrac{2x^{3/2} - 16}{\sqrt{x}} =$ undefined $\Rightarrow \sqrt{x} = 0 \Rightarrow x = 0$. The critical points are $x = 4$ and $x = 0$

EX1 Determine the critical points of the following functions.

(1) $y = 6x^2 - x^3$

(2) $y = x^2 + \dfrac{2}{x}$

(3) $y = \sqrt{2x - x^2}$

(4) $y = \sqrt[3]{x}\,(4 + x)$

(5) $y = x^2 \ln x$

(6) $y = x + 2\sin x,\ [0, 2\pi]$

4.2 Extrema: Maximum & Minimum value

Let f be a function with domain D. Then f has an absolute maximum value on D at a point c if

$$f(x) \le f(c) \qquad \text{for all } x \text{ in } D$$

and an absolute minimum value on D at c if

$$f(x) \geq f(c) \qquad \text{for all } x \text{ in } D.$$

A function f has a local maximum value at a point c within its domain D if $f(x) \leq f(c)$ for all $x \in D$ lying in some open interval containing c. A function f has a local minimum value at a point c within its domain D if $f(x) \geq f(c)$ for all $x \in D$ lying in some open interval containing c.

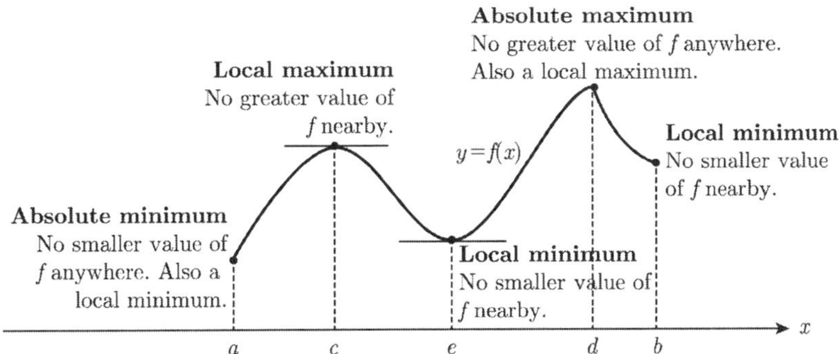

EX 1 Determine whether the function has an absolute maximum or minimum, a relative maximum or minimum, or neither a maximum nor a minimum for each point.

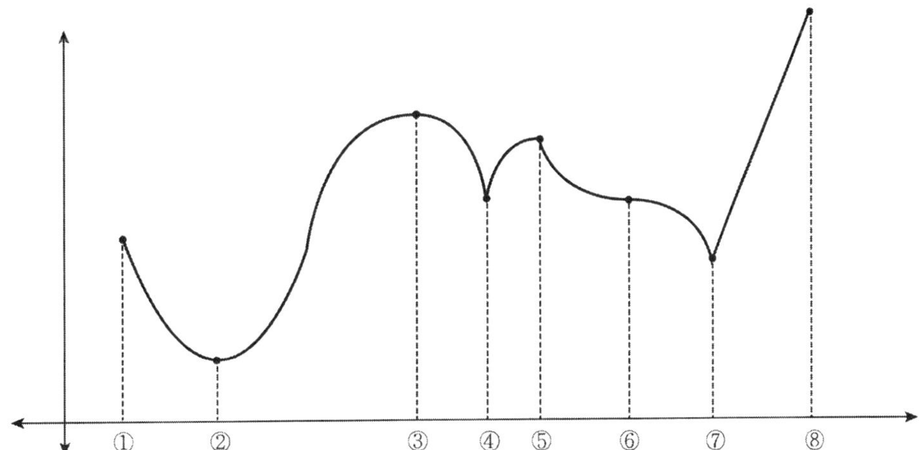

- **Increasing & Decreasing Functions**

Suppose that f is continuous on [a, b] and differentiable on (a, b)

If $f'(x) > 0$ at each point $x \in (a, b)$, then f is increasing on [a, b].

If $f'(x) < 0$ at each point $x \in (a, b)$, then f is decreasing on [a, b].

$f'(x) > 0 \quad \Leftrightarrow$ The graph of $f(x)$ is

$f'(x) < 0 \quad \Leftrightarrow$ The graph of $f(x)$ is

Example 1 Find the intervals on which the function $h(x) = -x^3 + 2x^2$ is increasing/decreasing.

Solution

$h(x) = -x^3 + 2x^2 \Rightarrow h'(x) = -3x^2 + 4x = x(4-3x)$

$\Rightarrow$ critical points at $x = 0, \dfrac{4}{3}$

$\Rightarrow h' = \underset{0}{---|} \underset{4/3}{+++|} ---$, increasing on $(0, \frac{4}{3})$, decreasing or $(-\infty, 0)$ or $(\frac{4}{3}, \infty)$

EX 2 Find the intervals on which the function is increasing/decreasing.

(1) $y = x^4 - 4x^3 + 4x^2$

(2) $h(x) = \ln x + \dfrac{1}{x}$

(3) $y = x + 2\cos x\ ,\ [0, \pi]$

(4) $y = x^3 e^{-x}$

(5) $y = \dfrac{2x}{x^2 + 4}$

(6) $h(x) = x^2 \ln x$

• First Derivative Test for Local Extrema

Suppose that c is a critical point of a continuous function f, and that f is differentiable at every point in some interval containing c except possibly at c itself. Moving across this interval from left to right,

1. if f' changes from negative to positive at c, then f has a local minimum at c;
2. if f' changes from positive to negative at c, then f has a local maximum at c;
3. if f' does not change sign at c (that is, f' is positive on both sides of c or negative on both sides), then f has no local extremum at c.

Relative Minimum

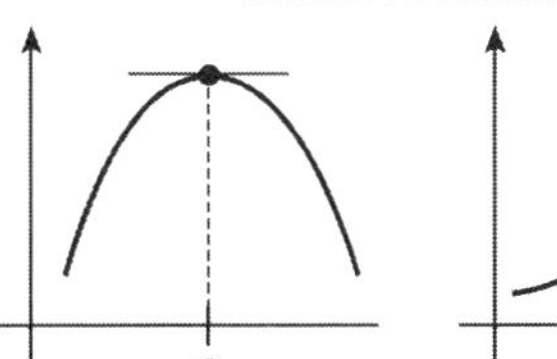

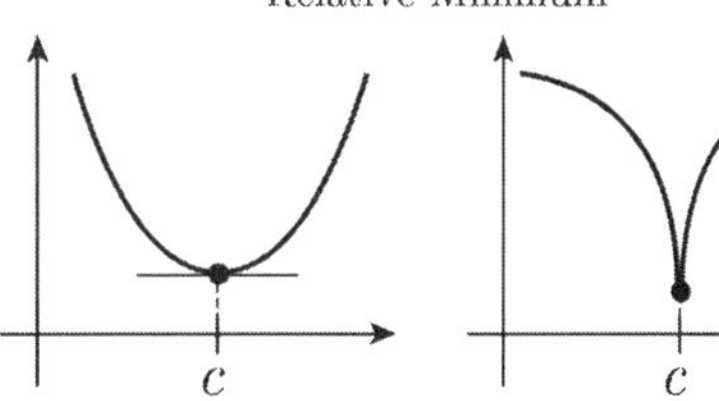

Example 2 Find the relative extreme values of the function $f(\theta)=3\theta^2-4\theta^3$.

Solution

$f(\theta)=3\theta^2-4\theta^3 \Rightarrow f'(\theta)=6\theta-12\theta^2=6\theta(1-2\theta) \Rightarrow$ critical points at $\theta=$ $0,\ \frac{1}{2} \Rightarrow f' = ---|+++|---,$

$\qquad\qquad 0 \quad 1/2$

increasing on $(0, \frac{1}{2})$, decreasing on $(-\infty, 0)$ and $(\frac{1}{2}, \infty)$

a local maximum is $f(\frac{1}{2})=\frac{1}{4}$ at $\theta=\frac{1}{2}$, a local minimum is $f(0)=0$ at $\theta=0$, no absolute extrema

EX 3 Use the first derivative test to find the relative extreme values of the function and where they occur.

(1) $k(t)=15t^3-t^5$

(2) $y=\frac{e^x}{x}$

(3) $y=x^2e^{-x}$

(4) $f(x)=x^{\frac{1}{3}}(x+8)$

(5) $g(x)=x^{\frac{2}{3}}(x+5)$

- **The Extreme Value Theorem**

If f is continuous on a closed interval [a, b], then f attains both an absolute maximum value M and an absolute minimum value m in [a, b]. That is, there are numbers x_1 and x_2 in [a, b] with $f(x_1)=m$, $f(x_2)=M$, and $m \le f(x) \le M$ for every other x in [a, b].

Example 3 Find the absolute extreme values of the function given by $g(x)=e^{-x^2}$, $-2 \le x \le 1$.

Solution

$g(x) = e^{-x^2} \Rightarrow g'(x) = -2xe^{-x^2} \Rightarrow$ a critical point at $x = 0$; $g(-2) = e^{-4}$, $g(0) = 1$, and $g(1) = e^{-1}$ $\Rightarrow$ the absolute maximum is 1 at $x = 0$ and the absolute minimum is e^{-4} at $x = -2$

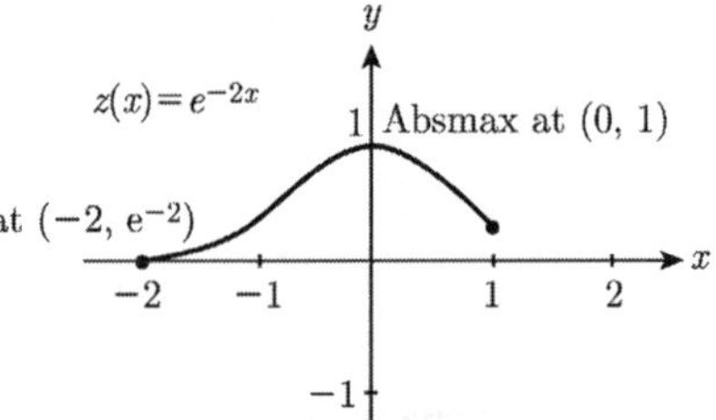

EX 4 Find the absolute extreme values of the function and where they occur.

(1) $h(x) = -3x^{\frac{2}{3}}$, $[-1, 1]$

(2) $f(x) = 12x - x^3$, $[-3, 4]$

(3) $y = \dfrac{\ln x}{x}$, $[1, e]$

(4) $y = xe^{-x}$, $[-1, 1]$

(5) $y = x\sqrt{4 - x^2}$, $[0, 2]$

(6) $g(x) = 4x - 6\sqrt[3]{x^2}$, $[-1, 8]$

(7) $f(x) = 2\cos x - \cos 2x$, $[0, 2\pi]$

4.3 Concavity

The graph of a differentiable function $y = f(x)$ is

(a) concave up on an open interval I if f' is increasing on I;

(b) concave down on an open interval I if f' is decreasing on I.

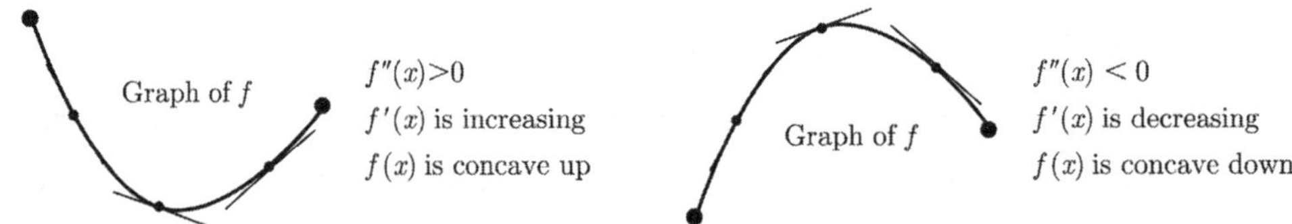

- **Point of Inflection**

A point where the graph of a function has a tangent line and where the concavity changes is a point of inflection.

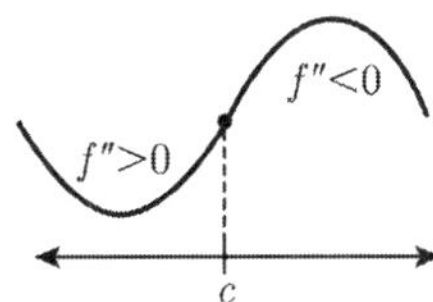

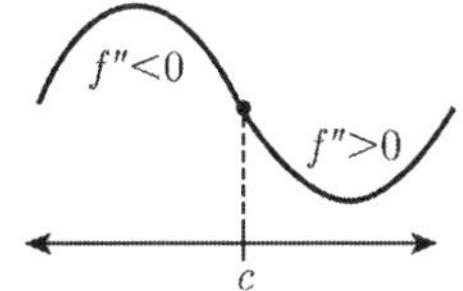

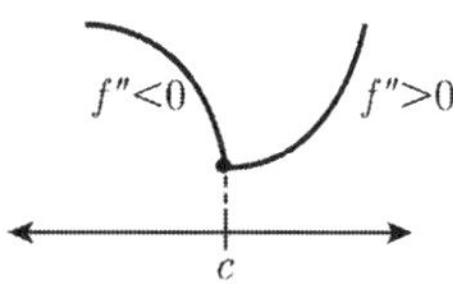

Example 1 Find the coordinated of the inflection point of the function $f(x)=x^3-6x^2+12$.

Solution

$f'(x)=3x^2-12x$

$f''(x)=6x-12$

$=6(x-2)$

$f''(x)=0$ at $x=2$

$\therefore$ The point of inflection is (2, -4)

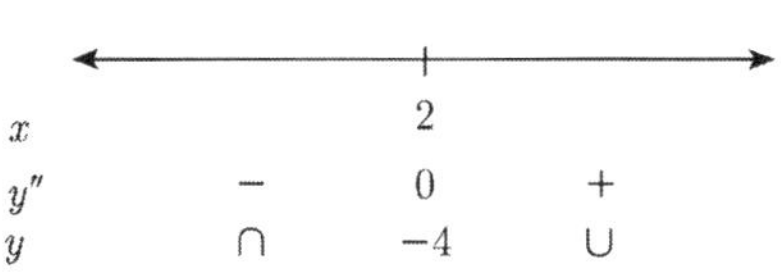

EX 1 Find (a) all intervals on which the function is concave up/down, (b) all values of x at which the function has a point of inflection.

(1) $y=4x^3-x^4$

(2) $y=x^5-5x^4$

(3) $y=x^2\ln x$

(4) $y=\dfrac{e^x}{x}$

(1) The cubic function $y=x^3+ax^2+b$ has a point of inflection at (2, 8). Find the values of a and b.

(2) The cubic function $y=x^3+ax^2+bx+c$ has a relative maximum at point (-1, 3) and a point of inflection at $x=-3$. Find the values of a, b, and c.

(3) The quartic function $f(x)=x^4+ax^2-b\ln x$ has a horizontal tangent and a point of inflection at $x=1$. Find the values of a and b.

(4) The cubic function $y=ax^3+bx^2+c$ has a tangent slope of -9 at $x=1$ and has a point of inflection at $(2, -12)$. Find the values of a, b, and c.

EX 3 The graph of f is shown below. Determine whether f' and f'' are positive, negative or zero and describe the behaviour of the graph of f at each point given in the table below.

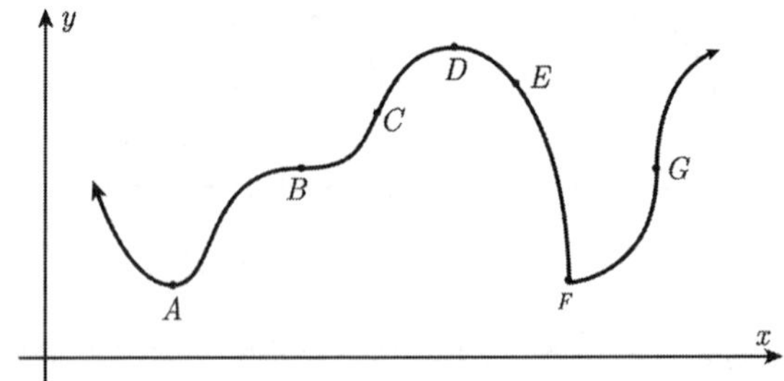

	$\frac{dy}{dx}$	$\frac{d^2y}{dx^2}$	Behaviour of the graph of f
A			
B			
C			
D			
E			
F			
G			

• Second Derivative Test for Local Extrema

Suppose f'' is continuous on an open interval that contains $x = c$.

1. If $f'(c) = 0$ and $f''(c) < 0$, then f has a local maximum at $x = c$.
2. If $f'(c) = 0$ and $f''(c) > 0$, then f has a local minimum at $x = c$.
3. If $f'(c) = 0$ and $f''(c) = 0$, then the test fails. The function f may have a local maximum, a local minimum, or neither.

$f'=0,\ f''>0$
⇒ local max

$f'=0,\ f''>0$
⇒ local min

Example 2 Find the relative extreme values of the function $y = x^3 - 3x + 2$.

Solution

$$f'(x) = 3x^2 - 3 = 3(x^2 - 1) = 3(x+1)(x-1)$$

$f'(x) = 0$ at $x = -1, 1$

$f''(x) = 6x$

$f''(-1) = -6 < 0$

f has a relative maximum at $x = -1$.

$\therefore$ Relative Maximum $= f(-1) = -1 + 3 + 2 = 4$

$f''(1) = 6 > 0$

f has a relative minimum at $x = 1$.

$\therefore$ Relative Minimum $= f(1) = 1 - 3 + 2 = 0$

EX 4 Use the second derivative test to find the relative extreme values of the function.

(1) $f(x) = x^3 - 3x^2 - 9x$ (2) $f(x) = 2xe^x$

EX 5 Consider the curve defined by $3x^2 - 2y^3 + 12x + 28 = 0$.

(1) Show that $\dfrac{dy}{dx} = \dfrac{x+2}{y^2}$.

(2) Find the coordinates of point P on the curve where $\dfrac{dy}{dx} = 0$.

(3) Find the value of $\dfrac{d^2y}{dx^2}$ at point P.

(4) Determine whether f has a relative maximum, relative minimum, or neither at point P. Explain.

EX 6 Use the second derivative test to find the relative extrema of the curve given by $x^2 + y^2 - 2y = 0$.

4.4 Curve Sketching

Step 1. Find y'.

Step 2. Find y''.

Step 3. Make a sign diagram.

(Identify the critical points, extreme values, concavity, points of inflection and y-intercept).

Step 4. Sketch the graph.

Example 1 Sketch the graph of the function $y = -2x^3 + 6x^2 - 3$. Label the relative extrema, point of inflection, and y-intercept.

Solution

When $y = -2x^3 + 6x^2 - 3$, then $y' = -6x^2 + 12x = -6x(x-2)$ and $y'' = -12x + 12 = -12(x-1)$. The curve rises on $(0, 2)$ and falls on $(-\infty, 0)$ and $(2, \infty)$.

At $x = 0$ there is a local minimum and at $x = 2$ a local maximum. The curve is concave up on $(-\infty, 1)$ and concave down on $(1, \infty)$. At $x = 1$ there is a point of inflection.

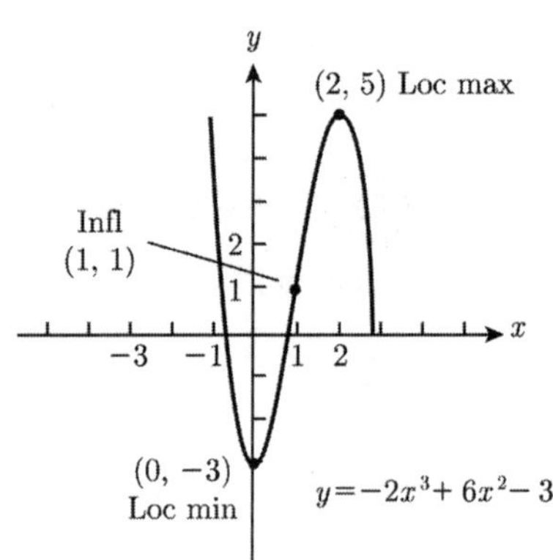

Example 2 Sketch the graph of the function $y = \cos x + \sqrt{3}\sin x$. Label the relative extrema, points of inflection, and y-intercept.

Solution

When $y = \cos x + \sqrt{3}\sin x$, then $y' = -\sin x + \sqrt{3}\cos x$ and $y'' = -\cos x - \sqrt{3}\sin x$. The curve is increasing on $(0, \frac{\pi}{3})$ and $(\frac{4\pi}{3}, 2\pi)$, and decreasing on $(\frac{\pi}{3}, \frac{4\pi}{3})$. At $x = 0$ there is a local minimum, at $x = \frac{\pi}{3}$ there is a local and absolute maximum, and at $x = \frac{4\pi}{3}$, there is a local and absolute minimum, and at $x = 2\pi$ there is a local maximum.

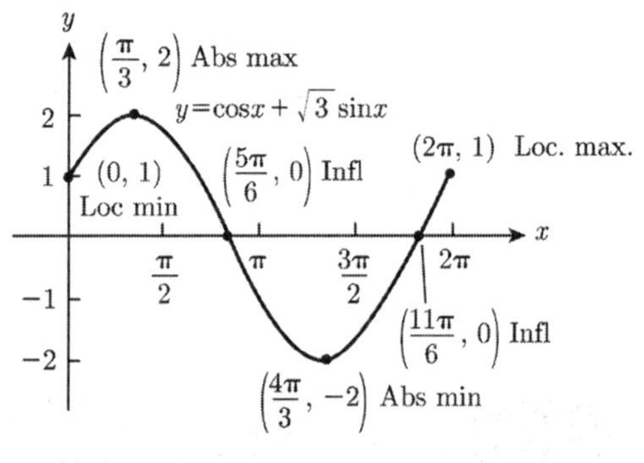

The curve is concave down on $(0, \frac{5\pi}{6})$ and $(\frac{11\pi}{6}, 2\pi)$, and is concave up on $(\frac{5\pi}{6}, \frac{11\pi}{6})$. At $x = \frac{5\pi}{6}$ and $x = \frac{11\pi}{6}$ there are points of inflection.

EX 1 Sketch the graph on a separate sheet of paper and label the relative extrema, points of inflection, and y-intercept of the following functions.

(1) $f(x)=x^3+3x^2-9x+5$ (2) $y=-2x^3-3x^2+12x+4$

(3) $y=x^4-4x^3+5$ (4) $f(x)=-x^4+6x^2+1$

(5) $y=x(x+2)^3$

EX 2 Sketch the graph on a separate sheet of paper and label the relative extrema, points of inflection, y-intercept and asymptotes for the following functions.

(1) $y=2x-3x^{2/3}$ (2) $y=xe^{\frac{1}{x}}$

(3) $y=xe^{-x}$ (4) $y=\dfrac{\ln x}{\sqrt{x}}$

(5) $y=\ln(3-x^2)$ (6) $y=\dfrac{4}{3}x-\tan x \quad (-\dfrac{\pi}{2}, \dfrac{\pi}{2})$

EX 3 Sketch the graph on a separate sheet of paper and label the relative extrema, point of inflection, y-intercept and asymptotes for the following functions.

(1) $y=\dfrac{8x}{x^2+4}$ (2) $f(x)=\dfrac{x^2-3}{x-2}$

(3) $y=\dfrac{e^x}{1+e^x}$ (4) $y=\dfrac{x}{\sqrt{x^2+1}}$

4.5 Relation of $f'(x)$, $f''(x)$ to the Graph of $f(x)$

- $f'(x)>0 \Leftrightarrow f(x)$ is increasing.
- $f'(x)$ is increasing $\Leftrightarrow f''(x)>0 \Leftrightarrow f(x)$ is concave up.

f'	+	+	−	−	+	−	0
f''	+	−	+	−	0	0	0
graph of f	↗ (concave up)	↗ (concave down)	↘ (concave up)	↘ (concave down)	↗	↘	→

EX 1 Each of Exercises (1) ~ (4) shows the graphs of the first and second derivatives of a function $y = f(x)$. Copy the picture and add to it a sketch of the approximate graph of f, given that the graph passes through the point P.

(1)

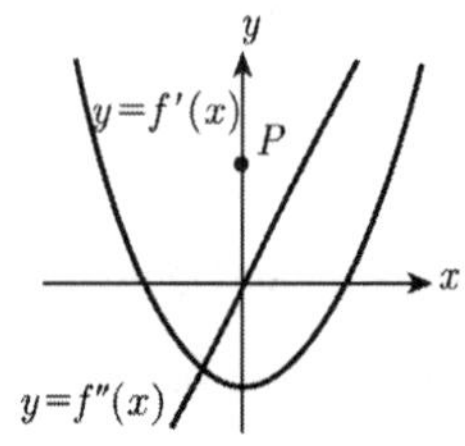

(2)

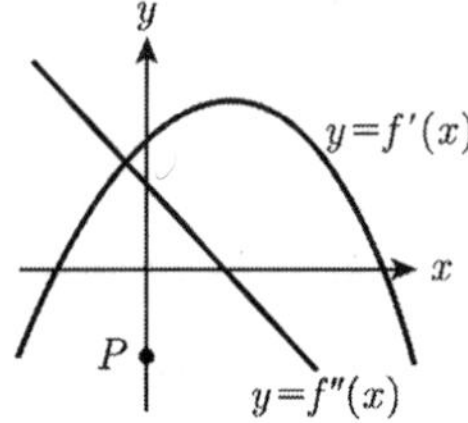

(3)

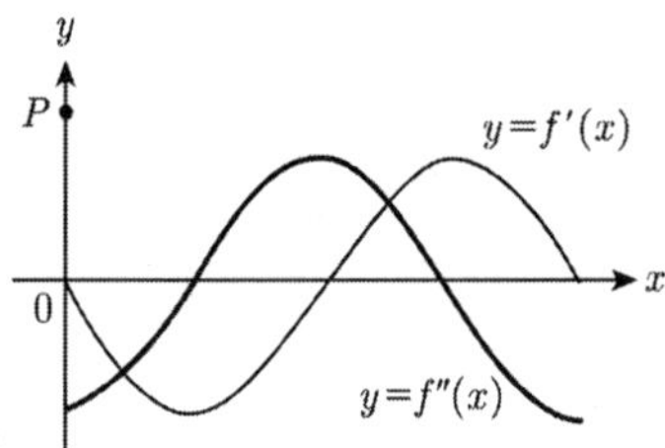

(4)

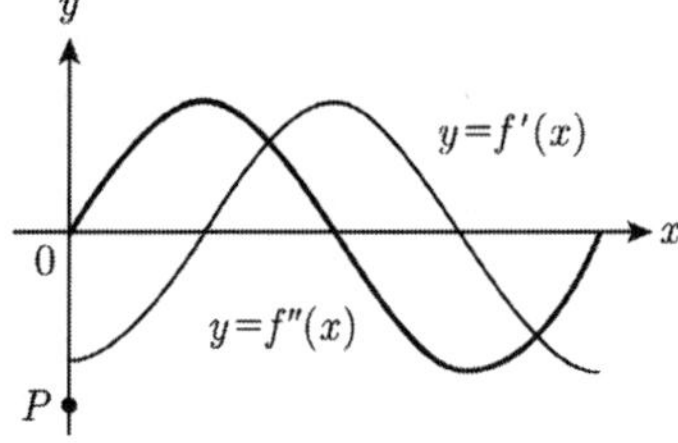

EX 2 The accompanying figure shows a portion of the graph of a twice-differentiable function $y = f(x)$. At each of the five labeled points, classify y' and y'' as positive, negative, or zero.

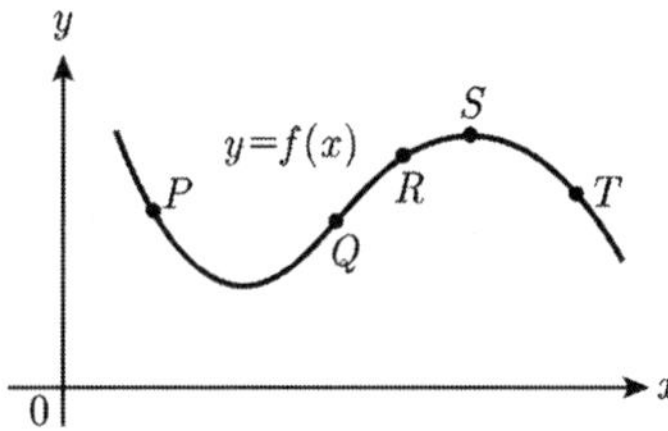

EX 3

(1) Sketch a smooth connected curve $y = f(x)$ with

$f(-2) = 8,$	$f'(2) = f'(-2) = 0,$
$f(0) = 4,$	$f'(x) < 0$ for $\lvert x\rvert < 2,$
$f(2) = 0,$	$f''(x) < 0$ for $x < 0,$
$f'(x) > 0$ for $\lvert x\rvert > 2,$	$f''(x) > 0$ for $x > 0.$

(2) Sketch the graph of a twice-differentiable function $y = f(x)$ with the following properties. Label coordinates where possible.

x	y	Derivatives
$x < 2$		$y' < 0,\ y'' > 0$
2	1	$y' = 0,\ y'' > 0$
$2 < x < 4$		$y' > 0,\ y'' > 0$
4	4	$y' > 0,\ y'' = 0$
$4 < x < 6$		$y' > 0,\ y'' < 0$
6	7	$y' = 0,\ y'' < 0$
$x > 6$		$y' < 0,\ y'' < 0$

(3) Sketch the graph of a twice-differentiable function $y = f(x)$ that passes through the points $(-2, 2)$, $(-1, 1)$, $(0, 0)$, $(1, 1)$, $(2, 2)$ and whose first two derivatives have the following sign patterns.

y' : + on $x < -2$, − on $-2 < x < 0$, + on $0 < x < 2$, − on $x > 2$

y'' : − on $x < -1$, + on $-1 < x < 1$, − on $x > 1$

EX 4 The graph of f' is shown below.

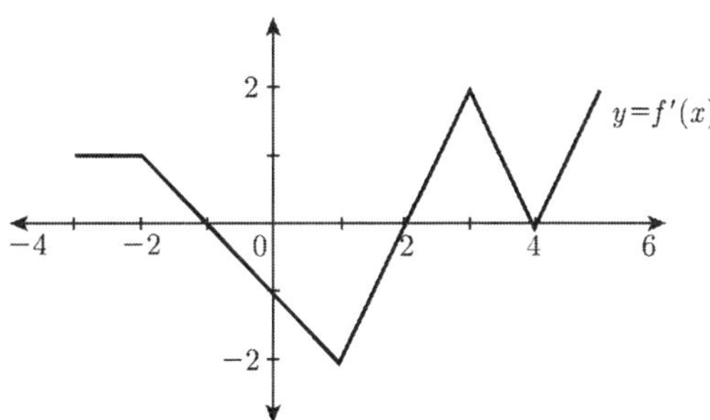

(1) Find the values of $f'(-1)$, $f'(1)$, $f'(5)$, $f''(-1)$, $f''(1)$, $f''(2)$ and $f''(3)$.
(2) Find the average rate of change of f' over the interval $-1 \le x \le 3$.
(3) Find all intervals on which the graph of f is increasing. Justify your answer.
(4) Find all intervals on which the graph of f is concave down. Justify your answer.
(5) Find all values of x at which the graph of f has a relative minimum/maximum. Explain.
(6) Find all values of x at which the graph of f has a point of inflection. Explain.

The graph of f is shown below.

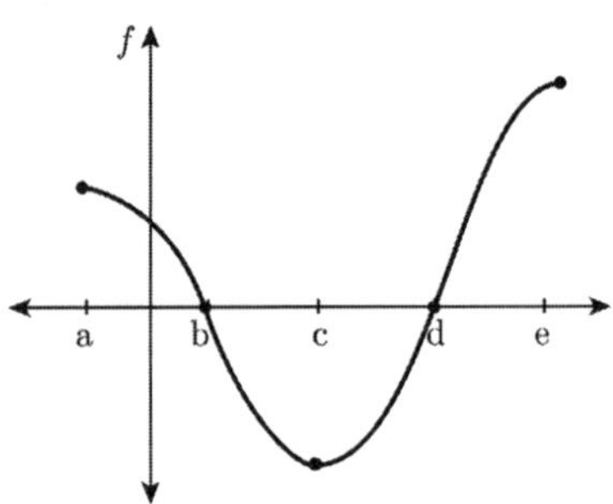

(1) Find all intervals on which the graph of f is increasing.

(2) Find all intervals on which the graph of f' is increasing.

(3) Find all values of x at which the graph of f has a relative minimum.

(4) Find all values of x at which the graph of f has a point of inflection.

EX6 The figure below shows the graph of f', the derivative of a twice-differentiable function f, on the closed interval $0 \le x \le 8$. The graph of f' has horizontal tangent lines at $x=1$, $x=3$, and $x=5$. The areas of the regions between the graph of f' and the x-axis are labeled in the figure.

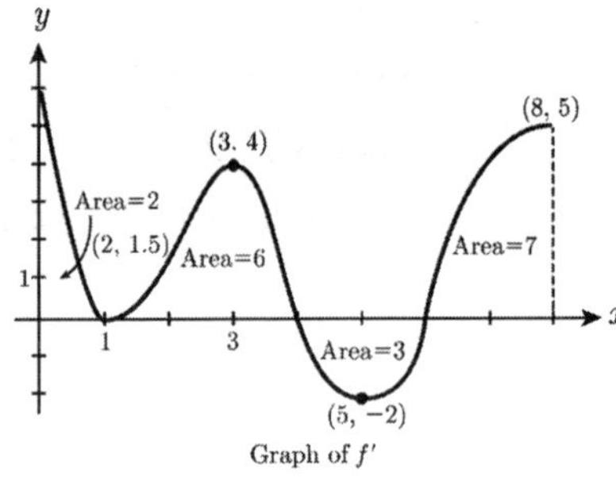

Graph of f'

The function f is defined for all real numbers and satisfies $f(8)=4$.

(1) Find all values of x on the open interval $0 < x < 8$ for which the function f has a local minimum. Justify your answer.

(2) Determine the absolute minimum value of f on the closed interval $0 \le x \le 8$. Justify your answer.

(3) On what open intervals contained in $0 < x < 8$ is graph of f both concave down and increasing? Explain your reasoning.

(4) The function g is defined by $g(x)=(f(x))^3$. If $f(x)=-\dfrac{5}{2}$, find the slope of the line tangent to the graph of g at $x=3$.

EX 7 The graph of f'' is shown below.

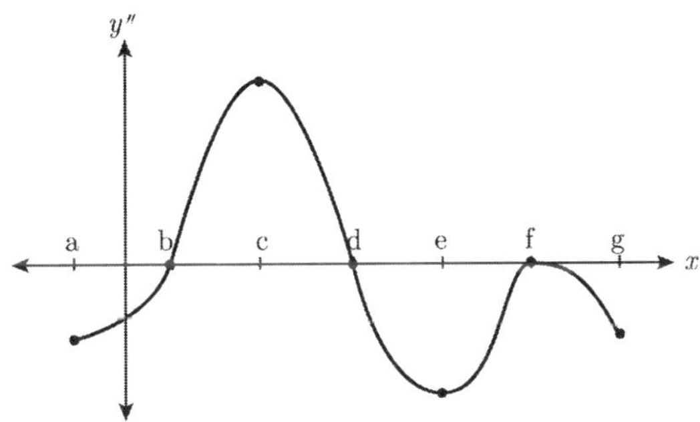

(1) Find all values of x at which the graph of f has a point of inflection. Explain briefly.

(2) Find all intervals on which the graph of f is concave up. Explain briefly.

(3) Find all intervals on which the graph of f' is increasing. Explain briefly.

(4) Find all values of x at which the graph of f' has a relative maximum. Justify your answer.

(5) Find all intervals on which the graph of f' is concave down. Justify your answer.

4.6 Motion Along a Line (1-Dimension)

If a particle moves along a line according to the law $s = f(t)$, where s represents the position of the particle P on the line at time t, then the velocity v of P at time t is given by $\frac{ds}{dt}$ and its acceleration a by $\frac{dv}{dt}$ or by $\frac{d^2s}{dt^2}$. The speed of the particle is $|v|$, the magnitude of v. If the line of motion is directed positively to the right, then the motion of the particle P is subject to the following: at any instant,

(1) If $v > 0$, then P is moving to the right and its distance s is increasing; If $v < 0$, then P is moving to the left and its distance s is decreasing;

(2) If $a > 0$, then v is increasing; if $a < 0$, then v is decreasing;

(3) If a and v are both positive or both negative, then (1) and (2) imply that the speed of P is increasing or that P is accelerating; if a and v have opposite signs, then the speed of P is decreasing or P is decelerating;

(4) If s is a continuous function of t, then P reverses direction whenever v is zero and a is different from zero; note that zero velocity does not necessarily imply a reversal in direction.

Example 1 A particle moves along a line according to the law $s = 2t^3 - 9t^2 + 12t - 4$, where $t \geq 0$.

(1) Find all t for which the distance s is increasing.

(2) Find all t for which the velocity is increasing.

(3) Find all t for which the speed of the particle is increasing.

(4) Find the speed when $t = \frac{3}{2}$.

(5) Find the total distance travelled between $t = 0$ and $t = 4$.

Solution

We have

$$\begin{aligned} v = \frac{ds}{dt} &= 6t^2 - 18t + 12 \\ &= 6(t^2 - 3t + 2) \\ &= 6(t-2)(t-1) \end{aligned}$$

and

$$a = \frac{dv}{dt} = \frac{d^2s}{dt^2} = 12t - 18 = 12\left(t - \frac{3}{2}\right).$$

The sign of v behaves as follows:

$$\begin{array}{llll} \text{if} & t < 1, & \text{then} & v > 0, \\ & 1 < t < 2, & & v < 0, \\ & t > 2, & & v > 0. \end{array}$$

For a, We have:

$$\begin{array}{llll} \text{if} & t < \frac{3}{2}, & \text{then} & a < 0, \\ & t > \frac{3}{2}, & & a > 0. \end{array}$$

These signs of v and a immediately yield the answers, as follows:

(1) s increases when $0 \leq t < 1$ or $t > 2$.

(2) v increases when $t > \frac{3}{2}$.

(3) The speed $|v|$ is increasing when v and a are both positive, that is, for $t > 2$, and when v and a are both negative, that is for $1 < t < \frac{3}{2}$.

(4) The speed when $t = \frac{3}{2}$ equals

$$|v| = \left|-\frac{3}{2}\right| = \frac{3}{2}.$$

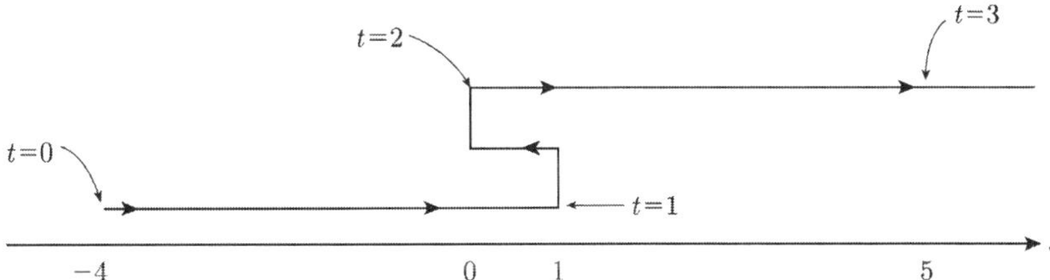

(5) P's motion can be indicated as shown in Figure. P moves to the right if $t < 1$, reverses its direction at $t = 1$, moves to the left when $1 < t < 2$, reversed again at $t = 2$, and continues to the right for all $t > 2$.

The position of P at certain times t are shown in the following table:

t:	0	1	2	4
s:	-4	1	0	28

Thus P travels a total of 34 units between times $t = 0$ and $t = 4$.

EX 1 A particle is moving along a horizontal line. Its position as a function of time is given by $s(x) = t^3 - 12t + 1$, $t \geq 0$ where s is in meters and t is in seconds.

(1) Find the velocity and acceleration functions of thc particle.

(2) At what time does the particle change direction?

(3) For what time intervals is the velocity of the particle increasing?

(4) For what time intervals is the speed of the particle decreasing? Justify your answer.

EX 2 A particle is moving along a horizontal line. Its position as a function of time is given by $s(t) = t^3 - 6t^2 + 9t - 3$, $t \geq 0$ where s is in meters and t is in seconds.

(1) Find the displacement of the particle from $t = 1$ to $t = 2$.

(2) Find the average velocity of the particle from $t = 1$ to $t = 2$.

(3) Find the velocity and acceleration functions of the particle.

(4) At what time(s) is the particle at rest?

(5) Find the position, velocity, speed, and acceleration of the particle at time $t = 4$.

(6) Is the particle moving away from the origin or toward the origin at time $t = 4$? Justify your answer.

(7) Find the average acceleration of the particle from $t=0$ to $t=2$.

(8) For what time intervals is the particle moving in the positive direction?

(9) For what time intervals is the velocity of the particle decreasing?

(10) For what time intervals is the speed of the particle increasing? Justify your answer.

(11) Find the total distance travelled by the particle during the first 4 seconds. Justify your answer.

EX 3 The velocity, in m/sec, of a particle moving along the x-axis is shown in the graph below.

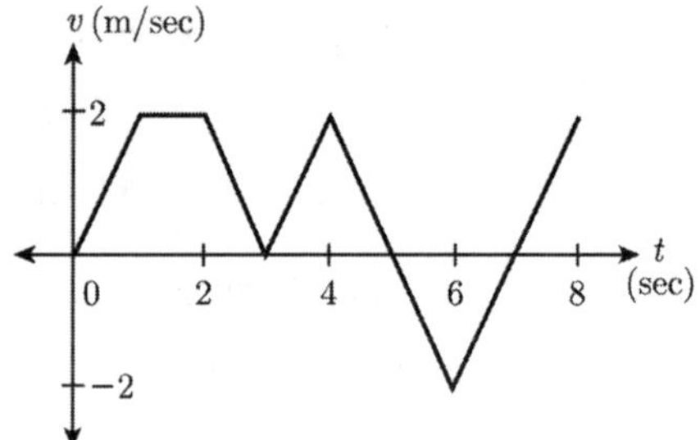

(1) At what time(s) is the particle at rest?

(2) At what time(s) does the particle change direction?

(3) Find the speed of the particle at time $t=6$.

(4) Find the acceleration of the particle at time $t=5$.

(5) Find the average acceleration of the particle from $t=1$ to $t=7$.

(6) For what time intervals is the particle moving in the negative direction?

(7) For what time intervals is the speed of particle decreasing? Justify your answer.

(8) Find the total distance travelled by the particle from $t=0$ to $t=8$.

(9) How many seconds will it take the particle to reach the farthest point from the origin? Explain.

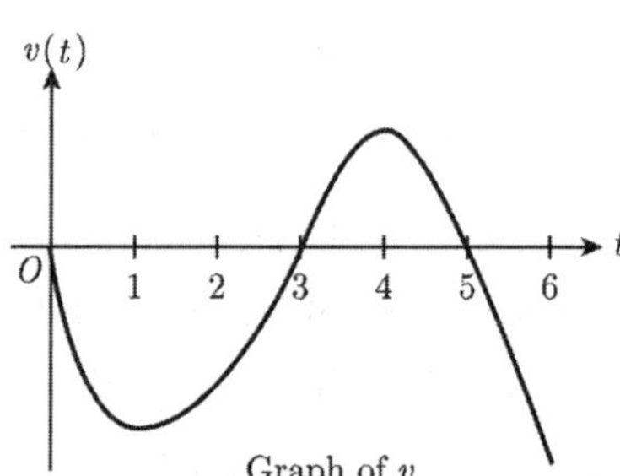

Graph of v

A particle moves along the x-axis so that its velocity at time t, for $0 \le t \le 6$, is given by a differentiable function v whose graph is shown above. The velocity is 0 at $t=0$, $t=3$, and $t=5$, and the graph has horizontal tangents at $t=1$ and $t=4$.

The areas of the regions bounded by the t-axis and the graph of v on the intervals $[0,3]$, $[3,5]$, and $[5,6]$ are 8, 3, and 2, respectively. At time $t=0$, the particle is at $x=-2$.

(1) For $0 \le t \le 6$, find both the time and the position of the particle when the particle is farthest to the left.
Justify your answer.

(2) For how many values of t, where $0 \le t \le 6$, is the particle at $x=-8$? Explain your reasoning.

(3) On the interval $2 < t < 3$, is the speed of the particle increasing or decreasing? Give a reason for your answer.

(4) During what time intervals, if any, is the acceleration of the particle negative? Justify your answer.

4.7 Motion Along a Curve: Velocity and Acceleration Vectors

If a point P moves along a curve defined parametrically by $P(t)=(x(t),y(t))$, where t represents time, then the vector from the origin to P is called the *position vector* with x as its *horizontal component* and y as its *vertical component*. The set of position vectors for all values of t in the domain common to $x(t)$ and $y(t)$ is called *vector function*.

The position vector, then, may be written as $\vec{R}(t)=(x,y)$ or as $R=xi+yj$, where i is the horizontal unit vector from $(0,0)$ to $(1,0)$ and j is the vertical unit vector from $(0,0)$ to $(0,1)$.

The *velocity vector* is the derivative of the vector function (the position vector):

$$\mathrm{v}=\frac{\mathrm{dR}}{\mathrm{dt}}=\frac{\mathrm{dx}}{\mathrm{dt}}\mathrm{i}+\frac{\mathrm{dy}}{\mathrm{dt}}\mathrm{j} \text{ or } \vec{v}(t)=\left(\frac{dx}{dt},\frac{dy}{dt}\right).$$

Alternative notations for $\frac{dx}{dt}$ and $\frac{dy}{dt}$ are v_x and v_y respectively; these are the components of v in the horizontal and vertical directions, respectively.

$$\text{The slope of v is } \frac{\frac{dy}{dt}}{\frac{dx}{dt}} = \frac{dy}{dx},$$

(speed) the *magnitude* of v is

$$|\text{v}| = \sqrt{(\frac{dx}{dt})^2 + (\frac{dy}{dt})^2} = \sqrt{v_x^2 + v_y^2}$$

The *acceleration vector* a is $\frac{dv}{dt}$ or $\frac{d^2R}{dt^2}$, and can be obtained by a second differentiation of the components of R. Thus

$$\text{a} = \frac{d^2x}{dt^2}i + \frac{d^2y}{dt^2}j;$$

the direction of **a** is

$$\tan^{-1}\frac{\frac{d^2y}{dt^2}}{\frac{d^2x}{dt^2}};$$

and its magnitude is

$$|\mathbf{a}| = \sqrt{(\frac{d^2x}{dt^2})^2 + (\frac{d^2y}{dt^2})^2} = \sqrt{a_x^{\ 2} + a_y^{\ 2}},$$

where we have used a_x and a_y for $\frac{d^2x}{dt^2}$ and $\frac{d^2y}{dt^2}$, respectively.

Example 1 A particle moves according to the equations $x = 3\cos t$, $y = 2\sin t$.

(1) Find a single equation in x and y for the path of the particle and sketch the curve.

(2) Find the velocity and acceleration vectors at any time t, and show that a $= -R$ at all time.

(3) Find R, v and a when $t_1 = \frac{\pi}{6}$, $t_2 = \pi$,

(4) Find the speed of the particle and the magnitude of its acceleration at each instant in (3).

(5) When is the speed a maximum? A minimum?

Solution

(1) Since $\frac{x^2}{9}=\cos^2 t$ and $\frac{y^2}{4}=\sin^2 t$, therefore

$$\frac{x^2}{9}+\frac{y^2}{4}=1$$

and the particle moves in a counterclockwise direction along an ellipse, starting, when $t=0$, at $(3,0)$ and $t_2=\pi$, returning to this point when $t=2\pi$.

(2) We have

$$R=3\cos t i+2\sin t j,$$
$$v=-3\sin t i+2\cos t j$$
$$a=-3\cos t i-2\sin t j=-R.$$

The acceleration, then, is always directed toward the center of the ellipse.

(3) At $t_1=\frac{\pi}{6}$,

$$R_1=\frac{3\sqrt{3}}{2}i+j$$
$$v_1=-\frac{3}{2}i+\sqrt{3}j$$
$$a_1=-\frac{3\sqrt{3}}{2}i-j.$$

At $t_2=\pi$

$$R_2=-3i$$
$$v_2=-2j$$
$$a_2=3i.$$

(4) At $t_1=\frac{\pi}{6}$,

$$|v_1|=\sqrt{\frac{9}{4}+3}=\frac{\sqrt{21}}{2},$$
$$|a_1|=\sqrt{\frac{27}{4}+1}=\frac{\sqrt{31}}{2}.$$

At $t_2=\pi$

$$|v_2|=\sqrt{0+4}=2,$$
$$|a_2|=\sqrt{9+0}=3$$

(5) For the speed $|v|$ at any time t we have

$$\begin{aligned} v &= \sqrt{9\sin^2 t+4\cos^2 t} \\ &= \sqrt{4\sin^2 t+4\cos^2 t+5\sin^2 t} \\ &= \sqrt{4+5\sin^2 t}. \end{aligned}$$

We see immediately that the speed is a maximum when $t=\frac{\pi}{2}$ or $\frac{3\pi}{2}$, and a minimum when $t=0$ or π.

EX 1 In Question (1)~(4) $R=3\cos\frac{\pi}{3}ti + 2\sin\frac{\pi}{3}tj$ is the (position) vector $xi+yj$ from the origin to a moving point $P(x,y)$ at time t,

(1) A single equation in x and y for the path of the point is

(A) $x^2+y^2=13$
(B) $9x^2+4y^2=36$
(C) $2x^2+3y^2=13$
(D) $4x^2+9y^2=1$
(E) $4x^2+9y^2=36$

(2) When $t=3$, the speed of the particle is

(A) $\frac{2\pi}{3}$
(B) 2
(C) 3
(D) π
(E) $\frac{\sqrt{13}}{3}\pi$

(3) The magnitude of the accelerating when $t=3$ is

(A) 2
(B) $\frac{\pi^2}{3}$
(C) 3
(D) $\frac{2\pi^2}{9}$
(E) π

(4) At the point where $t=\frac{1}{2}$, the slope of the curve along which the particle moves is

(A) $-\frac{2\sqrt{3}}{9}$
(B) $-\frac{\sqrt{3}}{2}$
(C) $\frac{2}{\sqrt{3}}$
(D) $-\frac{2\sqrt{3}}{3}$
(E) none of these

EX 2 In Question (1)~(4) the motion of a particle in a plane is given by the pair of equations $x=2t$ and $y=4t-t^2$.

(1) The particle moves along

(A) an ellipse
(B) a circle
(C) a hyperbola
(D) a line
(E) a parabola

(2) The speed of the particle at any time t is

(A) $\sqrt{6-2t}$
(B) $2\sqrt{t^2-4t+5}$
(C) $2\sqrt{t^2-2t+5}$
(D) $\sqrt{8}(|t-2|)$
(E) $2(|3-t|)$

(3) The minimum speed of the particle is

(A) 2 (B) $2\sqrt{2}$

(C) 0 (D) 1

(E) 4

(4) The acceleration of the particle

(A) depends on t

(B) is always directed upward

(C) is constant both in magnitude and in direction

(D) never exceeds 1 in magnitude

(E) is non of these

EX 3 For $t \geq 0$, a particle is moving along a curve so that its position at time t is $(x(t), y(t))$. At time $t=2$, the particle is at position (1,5). It is known that $\frac{dx}{dt} = \frac{\sqrt{t+2}}{e^t}$ and $\frac{dy}{dt} = \sin^2 t$.

(1) Is the horizontal movement of the particle to the left or to the right at time $t=2$? Explain your answer.

(2) Find the slope of the path of the particle at time $t=2$.

(3) Find the x-coordinate of the particle's position at time $t=4$.

(4) Find the speed of the particle at time $t=4$. Find the acceleration vector of the particle at time $t=4$.

EX 4 An object moving along a curve in the xy-plane is at position $(x(t), y(t))$ at time t with

$$\frac{dx}{dt} = \arctan\left(\frac{t}{1+t}\right) \text{ and } \frac{dy}{dt} = \ln(t^2+1)$$

for $t \geq 0$. At time $t=0$, the object is at position $(-3, -4)$. (Note: $\tan^{-1}x = \arctan x$)

(1) Find the speed of the object at time $t=4$.

(2) For $t>0$, there is a point on the curve where the line tangent to the curve has slope 2. At what time t is the object at this point? Find the acceleration vector at this points.

4.8 Modeling and Optimization

Step 1. Draw a large, clear diagram with the variable and the constants. Identify the quantity to be maximized or minimized.

Step 2. Write an equation connecting the variables and the constants.
Express the quantity to be maximized or minimized as a function of a single variable.

Step 3. Differentiate and simplify the function. Consider the domain to the function. Find the critical pont.

Step 4. Make a sign diagram and find the answers.

Example 1 What is the smallest perimeter possible for a rectangle whose area is 16 in^2, and what are its dimensions?

Solution

Let ℓ and w represent the length and width of the rectangle, respectively. With an area of 16 in^2, we have that $(\ell)(w) = 16 \Rightarrow w = 16\ell^{-1} \Rightarrow$ the perimeter is $P = 2\ell + 2w = 2\ell + 32\ell^{-1}$ and

$$p'(\ell) = 2 - \frac{32}{\ell^2} = \frac{2(\ell^2 - 16)}{\ell^2}.$$

Solving $p'(\ell) = 0 \Rightarrow \dfrac{2(\ell+4)(\ell-4)}{\ell^2} = 0 \Rightarrow \ell = -4, 4$. Since $\ell > 0$ for the length of a rectangle ℓ must be 4 and $w = 4 \Rightarrow$ The perimeter is 16 in., a minimum since $p''(\ell) = \dfrac{64}{\ell^3} > 0$.

EX 1 A rectangular plot of farmland will be bounded on one side by a river and on the other three sides by a single-strand electric fence. With 800 m of wire at your disposal, what is the largest area you can enclose, and what are its dimensions?

(1) Two number have a difference of 10 and their product is a minimum. Determine the numbers.

(2) The sum of two positive numbers is 18. The product of one and the square of the other is a maximum. Determine the numbers.

(1) You are planning to make an open rectangular box from an 8 in by 15 in piece of cardboard by cutting congruent squares from the corners and folding up the sides. What are the dimensions of the box of largest volume you can make this way, and what is its volume?

(2) An open-top, rectangular box is to be made from a 10 in by 16 in sheet of tin. It is made by cutting out equal squares from the four corners measuring x cm, and folding up the sides to create the box. What dimensions will yield a box of maximum volume?

(1) A container company is designing an open-top, square based, rectangular box that will have a volume of 500 ml. What is the minimum surface area the of the box?

(2) What are the dimensions of the lightest open-top right circular cylindrical can that will hold a volume of 1000 cm^3?

(3) A container company is designing a square based, rectangular box with top that will have a volume of 256 ft^3. The cost of the top is 15 ₡/ft^2, the bottom is 5 ₡/ft^2 and the sides is 2.5 ₡/ft^2. Find the dimensions of the box that will minimize the total cost.

(4) An oil can shaped like a right circular cylinder has a volume of 1000 ml. Find the dimensions of the can that will minimize the material required to build it.

(1) A piece of string measuring 24 cm is cut into two pieces. One is bent into a circle and the other is bent into a square. The sum of the areas of the circle and square is a minimum. What is the length of the piece bent into the circle?

(2) A piece of string measuring 20 cm is cut into two pieces. One is bent into an equilateral triangle and the other is bent into a square. The sum of the areas of the triangle and square is a minimum. What is the length of the piece bent into the equilateral triangle?

(1) Find the dimensions of the largest rectangle that can be inscribed in a circle with a radius of 10 cm.

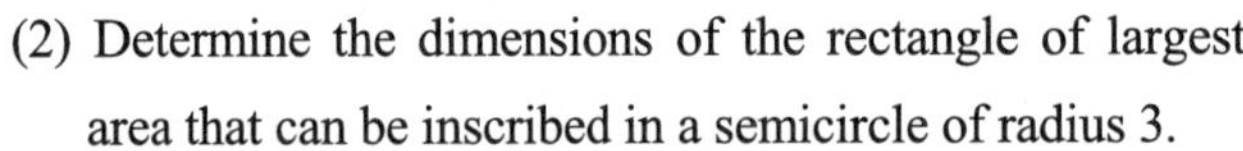

(2) Determine the dimensions of the rectangle of largest area that can be inscribed in a semicircle of radius 3.

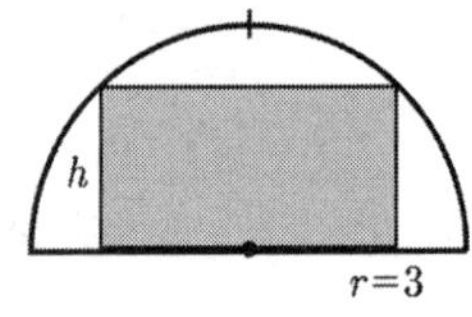

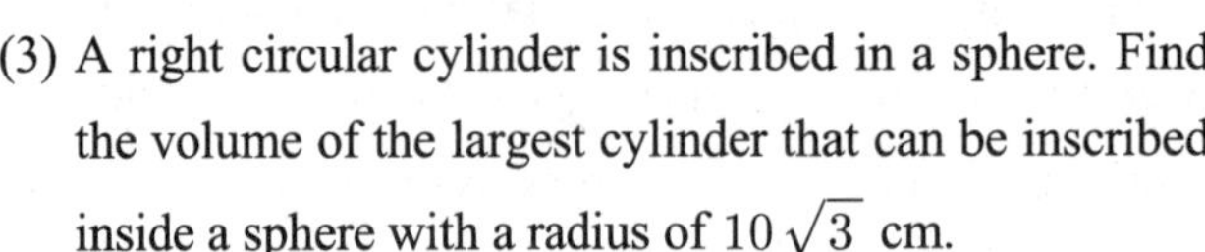

(3) A right circular cylinder is inscribed in a sphere. Find the volume of the largest cylinder that can be inscribed inside a sphere with a radius of $10\sqrt{3}$ cm.

(4) A right circular cone is inscribed in a sphere. Find the dimensions of the largest cone that can be inscribed inside a sphere with a radius of 3 cm.

(1) A utilities company is designing a power line from point A to point C, which is located directly across a straight river 1 km from point B. Point B is 4 km down shore from point A. It costs \$50 per meter to lay the line under the water but only \$30 per meter to lay the line underground. How far from B should P be located to minimize the cost of the line?

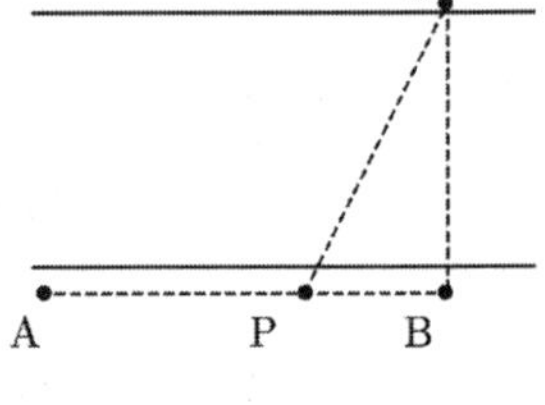

(2) An oil company is construction a pipeline from their refinery to storage tanks. The storage tanks are located 10 km away from the point A, which is situated directly across the 3 km wide river from the refinery. The cost of lying pipe is \$20 per miter underground and \$25 per meter under the river. how far from A should P be located to minimize the cost of the pipeline?

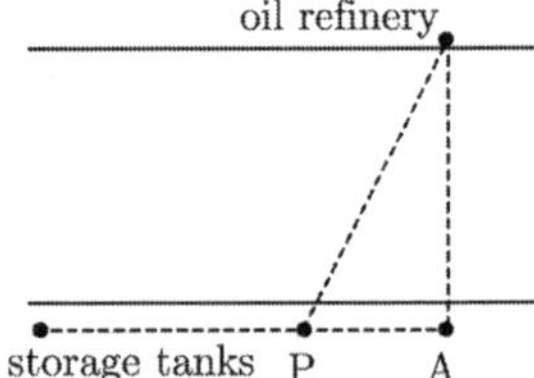

(3) Jane is 2 mi offshore in a boat and wishes to reach coastal village 6 mi down a straight shoreline from the point nearest the boat. She can row 2 mph and can walk 5 mph. Where should she land her boat to reach the village in the least amount of time.

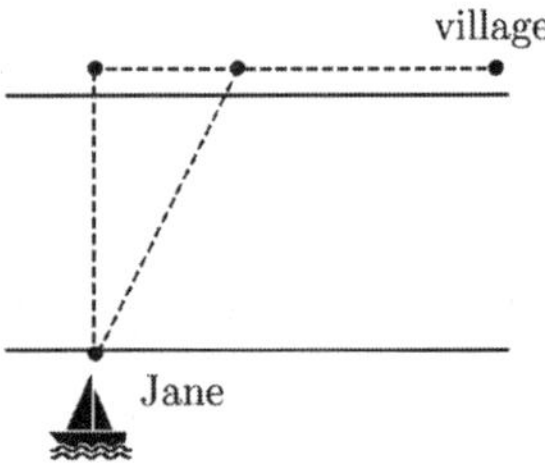

(1) Find the coordinated of the point (x, y), on the curve $y = 2\sqrt{x}$, which is closest to the point $A(6, 0)$.

(2) Find the coordinates of the point (x, y), on the curve $y = \frac{1}{4}x^2 - 1$, which is closest to $(0, 3)$.

(3) Find the shortest distance from the origin to a point on the curve $xy^2 = 16$.

(4) Find the shortest distance from the pont $(4,0)$ to a point on the curve $x^2 - y^2 = -1$.

(5) Find the coordinates of the point (x,y), on the curve $y = x^2 - 3$, which is closest to $(6,0)$.

EX 9

(1) An isosceles triangle is inscribed in the parabola of $y = 9 - x^2$ above the x-axis. One vertex is at the origin and the other two vertices are on the curve. Find the maximum area of the isosceles triangle.

(2) A rectangle is inscribed in the parabola of $y = x^2 - 12$ below the x-axis. Two vertices are on the x-axis and the other two vertices are on the curve. Find the maximum area of the rectangle.

(3) A rectangle with two vertices on the x-axis is inscribed under the curve $y = e^{-2x^2}$ as shown in the figure. Find the maximum area of the rectangle.

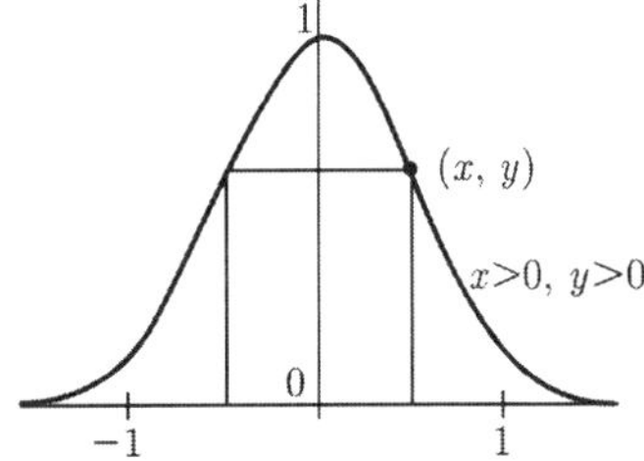

4.9 Related Rates

1. Draw a picture and name the variables and constants. Use t for time.
2. Write down the numerical information
3. Write down what you are asked to find
4. Write and equation that relates the variables.
5. Differentiate with respect to t.
6. Evaluate. Use known values to find the unknown rate.

Example 1 If edge length x of a cube decreases at the rate of 5 m/min, when $x = 3$ m at what rate does the cube's.

(1) Surface area change?

(2) Volume change?

Solution

(1) $S = 6x^2,\ \dfrac{dx}{dt} = -5\dfrac{m}{\text{min}} \Rightarrow \dfrac{ds}{dt} = 12x\dfrac{dx}{dt};$

when $x = 3 \Rightarrow \dfrac{dS}{dt} = 12(3)(-5) = -180\dfrac{m^2}{\text{min}}$

(2) $V = x^3,\ \dfrac{dx}{dt} = -5\dfrac{m}{\text{min}} \Rightarrow \dfrac{dV}{dt} = 3x^2\dfrac{dx}{dt};$

when $x = 3 \Rightarrow \dfrac{dV}{dt} = 3(3)^2(-5) = -135\dfrac{m^3}{\text{min}}$

EX 1 The voltage V(volts), current I(amperes), and resistance R(ohms) of an electric circuit like the one shown here are related by the equation $V = IR$. Suppose that V is increasing at the rate of 1 volt/sec while I is decreasing at the rate of 1/3 amp/sec. Let t denote time in seconds.

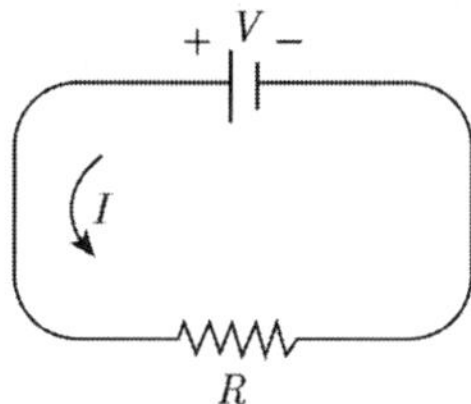

(1) What is the value of dV/dt?

(2) What is the value of dI/dt?

(3) What equation relates dR/dt to dV/dt and dI/dt?

(4) Find the rate at which R is changing when $V = 12$ volts and $I = 2$ amp. Is R increasing, or decreasing?

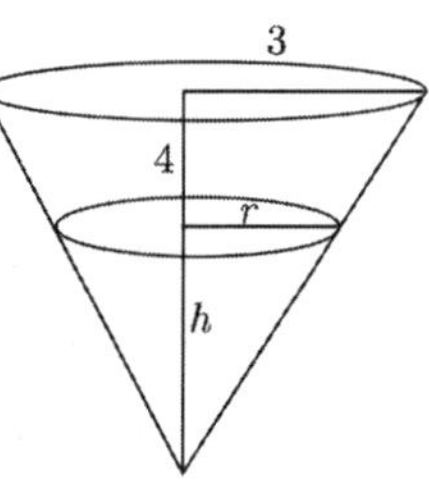

(1) A conical tank with a radius of 3 m and a height of 4 m is being filled with water at a rate of 18π m^3/min. How fast is the water level rising when the water is 2 m deep?

(2) A conical tank with a radius of 45 m and a height of 6 m is filled with water. The water is leaking out at a constant rate of 16π m^3/min. How fast is the radius of the water's surface decreasing when it has a 5 m deep.

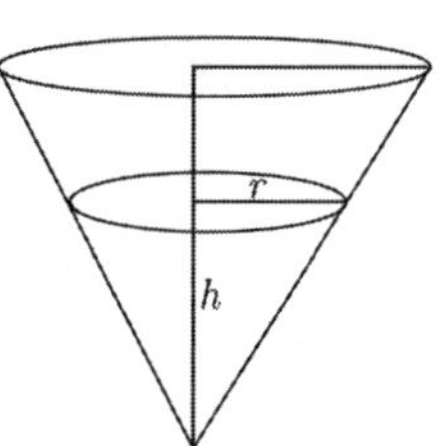

(3) A spherical balloon is inflated with helium at the rate of 100π ft^3/min. How fast is the balloon's radius increasing at the instant the radius is 5 ft? How fast is the surface area increasing?

EX 3 A 13 ft ladder is leaning against a house when its base starts to slide away (see accompanying figure). By the time the base is 12 ft from the house, the base is moving at the rate of 5 ft/sec.

(1) How fast is the top of the ladder sliding down the wall then?

(2) At what rate is the area of the triangle formed by the ladder, wall, and ground changing then?

(3) At what rate is the angle θ between the ladder and the ground changing then?

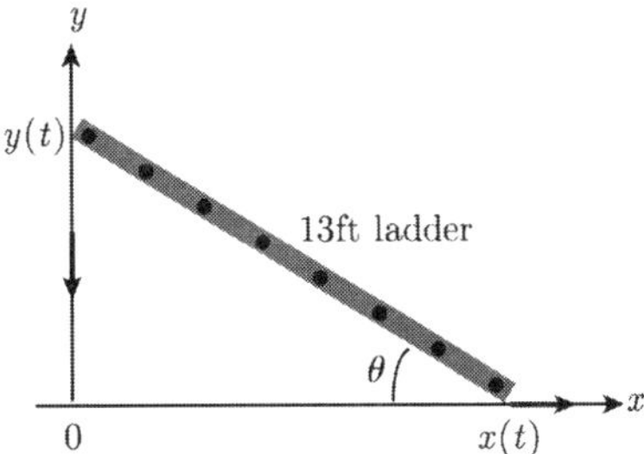

EX 4

(1) A dinghy is pulled toward a dock by a rope from the bow through a ring on the dock 6 ft above the bow. The rope is hauled in at the rate of 2 ft/sec.

(a) How fast is the boat approaching the dock when 10 ft of rope are out?

(b) At what rate is the angle θ changing at this instant (see the figure)?

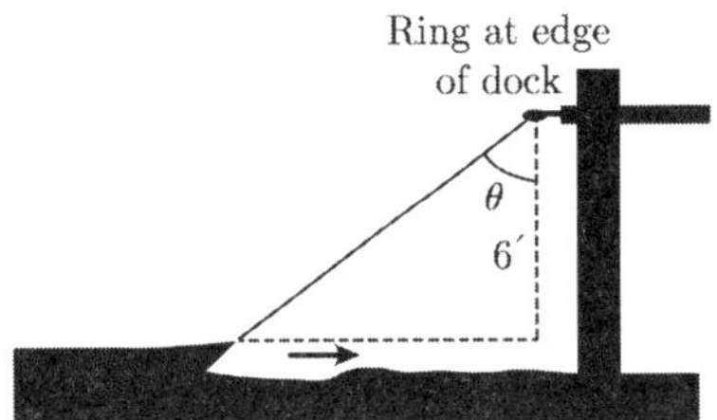

(2) You are videotaping a race from a stand 132 ft from the track, following a car that is moving at 180 mi/h(264 ft/sec), as shown in the accompanying figure. How fast will your camera angle θ be changing when the car is right in front of you? A half second later?

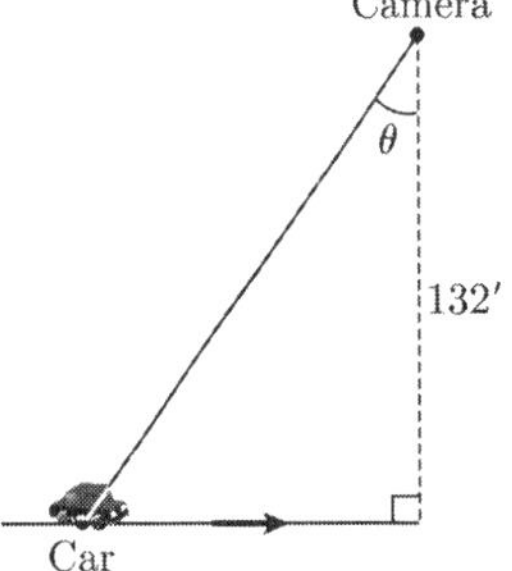

(1) A 6 ft tall man is walking man toward a 20 ft lamp post at a rate of 3 ft/sec. (a) How fast is the length of his shadow changing? (b) How fast is the tip of the shadow moving?

(2) A balloon is rising vertically above a level, straight road at a constant rate of 1 ft/sec. Just when the balloon is 65 ft above the ground, a bicycle moving at a costant rate of 17 ft/sec passes under it. How fast is the distance $s(t)$ between the bicycle and balloon increasing 3 sec later?

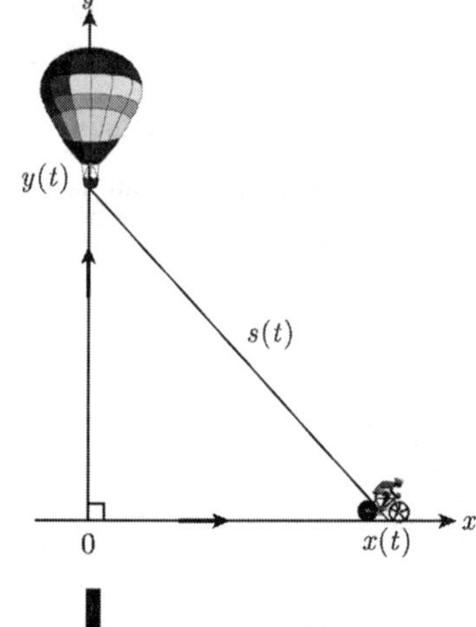

(3) Car A travels north at a rate 30 mph and car B travels east at 50 mph, as shown in the figure. When car A is 80 miles south from the intersection, car B is 70 miles east from the intersection. How fast is the distance between the cars changing 1 hour later?

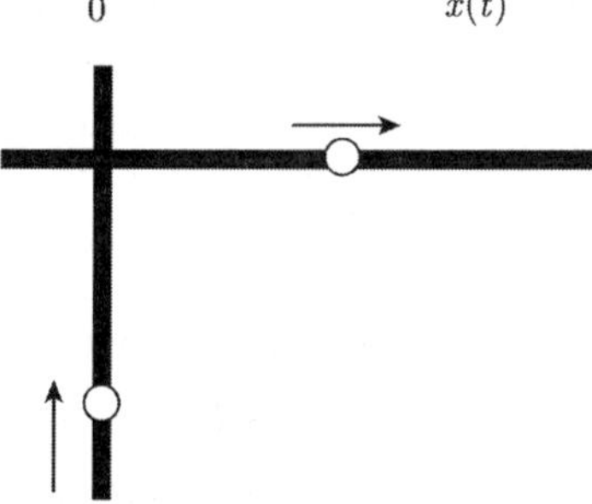

(1) One side of a triangle is 7 cm long and another side is 8 cm long. The angle between the two sides is decreasing at a rate of $\frac{\pi}{14}$ rad/sec. How fast is the area of the triangle changing when the angle is $\frac{\pi}{3}$?

(2) The height of a triangle is increasing at a rate of 2 cm/sec and the base of the triangle is decreaing in length at a rate of 3 cm/sec. How fast is the area of the triangle changing when the base is 8 cm and the area is 40 cm^2?

(3) The height of a cylindrically shaped rubber tube, with a volume of 240π cm^3, is stretched at a constant rate of 6 cm/min. The volume of the rubber tube does not change as it stretches. Find the rate of change in radius when the height is 15 cm.

EX 7 A point is moving along the curve $y = \sqrt{x}$ and x is increasing at a rate of 4 units/sec.

(1) How fast is y changing when x is 4?

(2) How fast is the distance from the origin changing when x is 4?

(3) Find the rate of change in the angle of elevation of the point from the origin when x is 4.

EX 8 Two ships are steaming straight away from a point 0 along routes that make a 120° angle. Ship A moves at 14 knots (nautical miles per hour; a nautical mile is 2000 yd). Ship B moves at 21 knots. How fast are the ships moving apart when $OA = 5$ and $OB = 3$ nautical miles?

4.10 Mean Value Theorem (M.V.T)

• The Mean Value Theorem

Suppose $y = f(x)$ is continuous on a closed interval $[a, b]$ and differentiable on the interval's interior (a, b). Then there is at least one point c in (a, b) at which

$$\frac{f(b) - f(a)}{b - a} = f'(c).$$

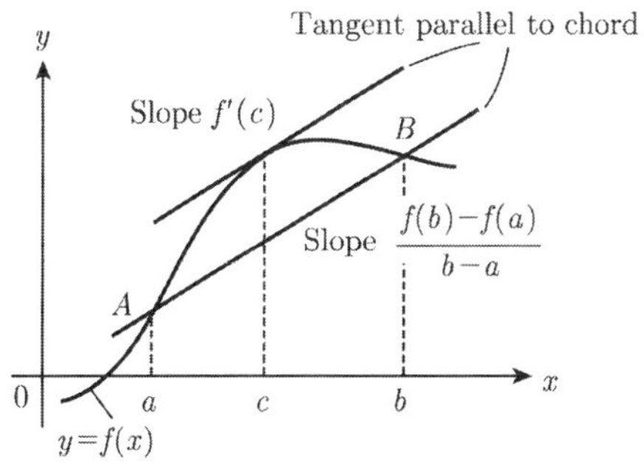

• Rolle's Theorem

Suppose that $y = f(x)$ is continuous at every point of the closed interval $[a, b]$ and differentiable at every point of its interior (a, b). It $f(a) = f(b)$, then there is at least one number c in (a,b) at which $f'(c)=0$.

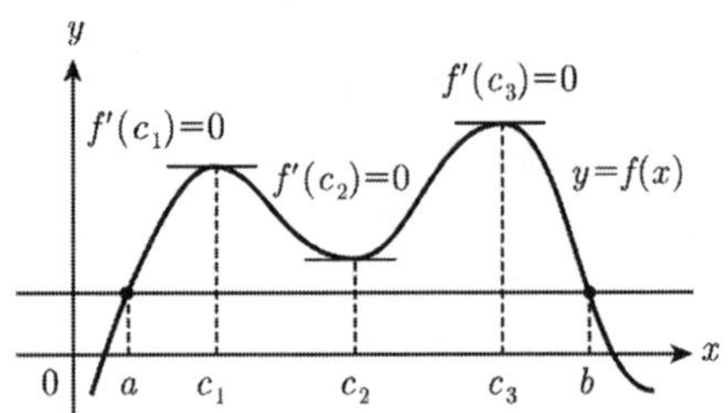

Example 1 Let $f(x)=x^2-2x$. Find all numbers c, $1<c<3$, such that $f'(c)=\dfrac{f(3)-f(1)}{3-1}$.

Solution

$$f'(x)=2x-2,\quad \frac{f(3)-f(1)}{3-1}=\frac{3-(-1)}{3-1}=2 \Rightarrow 2c-2=2$$

$$\therefore\ c=2$$

EX 1 Find the value or values of c that satisfy the equation

$$\frac{f(b)-f(a)}{b-a}=f'(c)$$

in the conclusion of the Mean Value Theorem for the functions and intervals in Exercises 1~8.

(1) $f(x)=x^2+3x-1,\ [0,1]$

(2) $f(x)=x^{2/3},\ [0,1]$

(3) $f(x)=x+\dfrac{1}{x},\ [\dfrac{1}{2},2]$

(4) $f(x)=\sqrt{x-1},\ [1,3]$

(5) $f(x)=\sin^{-1}x,\ [-1,1]$

(6) $f(x)=\ln(x-1),\ [2,4]$

(7) $f(x)=x^3-x^2,\ [-1,2]$

(8) $g(x)=\begin{cases} x^3, & -2\le x\le 0 \\ x^2, & 0\le x\le 2 \end{cases}$

EX 2 Explain why the Mean Value Theorem can not be applied to the following functions on the given interval.

(1) $y=\dfrac{1}{x},\ [-2,4]$

(2) $y=\dfrac{x}{|x|},\ [-1,1]$

(3) $y=|x-1|+2,\ [0,2]$

(4) $y=\sqrt[3]{x^2},\ [-1,8]$

EX 3 Use a graphing calculator to find all numbers c in the interval $[a,b]$ that satisfy the Mean Value Theorem for the following functions.

(1) $f(x) = 2^x - \sin x,\ [0, \pi]$

(2) $f(x) = 3x\cos^{-1}x,\ [0, \frac{1}{2}]$

EX 4 If $f(x) = 2x^3 - 6x$, at what point on the interval $0 \le x \le \sqrt{3}$, is any, is the tangent to the curve parallel to the secant line?

EX 5 Let $f(x) = 3^x - x^3$. Find all numbers c, $0 < c < 3$, such that the tangent to the curve is parallel to the secant through $(0,1)$ and $(3,0)$ for x equal.

4.11 Local Linear Approximation

If f is differentiable at $x = a$, then the approximating function

$$L(x) = f(a) + f'(a)(x-a)$$

is the linearization of f at a. The approximation

$$f(x) \approx L(x)$$

of f by L is the standard linear approximation of f at a. The point $x = a$ is the center of the approximation.

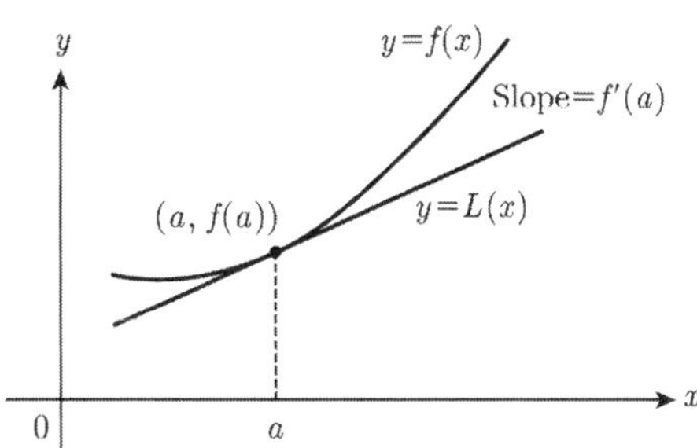

Example 1 Find the linearization of $f(x) = \sqrt{1+x}$ at $x=0$.

Solution

Since

$$f'(x) = \frac{1}{2}(1+x)^{-1/2},$$

we have $f(0)=1$ and $f'(0) = 1/2$, giving the linearization

$$L(x) = f(a) + f'(a)(x-a) = 1 + \frac{1}{2}(x-0) = 1 + \frac{x}{2}.$$

Approximation	True value
$\sqrt{12} \approx 1 + \frac{0.2}{2} = 1.10$	1.095445
$\sqrt{1.05} \approx 1 + \frac{0.05}{2} = 1.025$	1.024695
$\sqrt{1.005} \approx 1 + \frac{0.005}{2} = 1.00250$	1.002497

EX 1 Use local linear approximation to estimate the values of the following.

(1) $\sqrt{16.08}$ (2) $\sqrt[3]{7.4}$

(3) $\sqrt{63}$ (4) $(1.01)^{10}$

(5) $2^{3.99}$ (6) $\tan^{-1}(1.01)$

EX 2 Find the linearization $L(x)$ of $f(x)$ at $x=a$.

(1) $f(x) = x^3 - 2x + 3,\ a=2$ (2) $f(x) = \sqrt{x^2+9},\ a=-4$

(3) $f(x) = x + \frac{1}{x},\ a=1$ (4) $f(x) = \sqrt[3]{x},\ a=-8$

(5) $f(x) = \tan x,\ a=\pi$

EX 3 The volume of a spherical hot air balloon expands as the air inside the balloon is heated. The radius of the balloon, in feet, is modeled by a twice-differentiable function r of time t, where t is measured in minutes.

t (minutes)	0	2	5	7	11	12
$r'(t)$ (feet per minute)	5.7	4.0	2.0	1.2	0.6	0.5

For $0 < t < 12$, the graph of r is concave down. The table above gives selected values of the rate of change, $r'(t)$, of the radius of the balloon over the time interval $0 \leq t \leq 12$. The radius of the balloon is 30 feet when $t = 5$.

(Note: The volume of a sphere of radius r is given by $V = \frac{4}{3}\pi r^3$.)

(1) Estimate the radius of the balloon when $t = 5.4$ using the tangent line approximation at $t = 5$. Is your estimate grater than or less than the true value? Give a reason for your answer.

(2) Find the rate of change of the volume of the balloon with respect to time when $t = 5$. Indicate units of measure.

EX 4 A table of values of a differentiable function f and its derivative f' is shown below.

x	0	1	2	3
f	1	2	4	7
f'	0.5	1	3	4

The second derivative of f has the property that $f'' > 0$ on the interval $0 < x < 3$.

(1) Use local linear approximation to estimate $f(2.1)$.

(2) Is this approximate value greater than or less than the actual value of $f(2.1)$? Give a reason to support your answer.

4.12 Newton's Method

1. Guess a first approximation to a solution of the equation $f(x) = 0$. A graph of $y = f(x)$ may help.
2. Use the first approximation to get a second, the second to get a third, and so on, using thc formula

$$x_{n+1} = x_n - \frac{f(x_n)}{f'(x_n)}, \qquad \text{if } f'(x_n) \neq 0,$$

Given the approximation x_n, the point

slope equation for the tangent to the curve at $(x_n, f(x_n))$ is

$$y = f(x_n) + f'(x_n)(x - x_n).$$

We can find where it crosses the x-axis by setting $y = 0$(Figure):

$$0 = f(x_n) + f'(x_n)(x - x_n)$$

$$-\frac{f(x_n)}{f'(x_n)} = x - x_n$$

$$x = x_n - \frac{f(x_n)}{f'(x_n)} \text{ it } f(x_n) \neq 0.$$

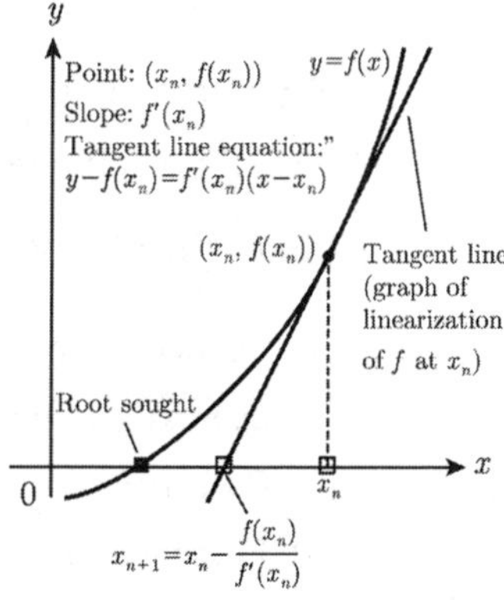

The value of x is the next approximation x_{n+1}.

Example 1 Use Newton's method to approximate the solution of the equation $x^3 - 5 = 0$. Give the first four iterations starting with $x_1 = 2$.

Solution

Let $f(x) = x^3 - 5$.

$f'(x) = 3x^2$

$x_1 = 2$

$$x_2 = x_1 - \frac{f(x_1)}{f'(x_1)} = 2 - \frac{f(2)}{f'(2)} = 2 - \frac{(2)^3 - 5}{3(2)^2} = \frac{7}{4} = 1.75$$

$$x_3 = x_2 - \frac{f(x_2)}{f'(x_2)} = 1.75 - \frac{f(1.75)}{f'(1.75)} = 1.75 - \frac{(1.75)^3 - 5}{3(1.75)^2} = 1.7108844$$

$$x_4 = x_3 - \frac{f(x_3)}{f'(x_3)} = 1.7108844 - \frac{f(1.7108844)}{f'(1.7108844)}$$

$$= 1.7108844 - \frac{(1.7108844)^3 - 5}{3(1.7108844)^2} = 1.7099764$$

$$x_5 = x_4 - \frac{f(x_4)}{f'(x_4)} = 1.7099764 - \frac{f(1.7099764)}{f'(1.7099764)}$$

$$= 1.7099764 - \frac{(1.7099764)^3 - 5}{3(1.7099764)^2} = 1.7099759$$

EX 1 Use Newton's method to approximate the solution of the following equation correct to six decimal places.

(1) $x^3 - x - 2 = 0 \ (x_1 = 2)$

$x_1 = 2$

$x_2 =$

$x_3 =$

$x_4 =$

$x_5 =$

(2) $x^5 + 2x - 5 = 0 \ (x_1 = 2)$

$x_1 = 2$

$x_2 =$

$x_3 =$

$x_4 =$

$x_5 =$

(3) $x - 2\cos x = 0 \ (x_1 = 2)$

$x_1 = 2$

$x_2 =$

$x_3 =$

$x_4 =$

$x_5 =$

(4) $\sin x = 1 - x \ (x_1 = 0)$

$x_1 = 0$

$x_2 =$

$x_3 =$

$x_4 =$

$x_5 =$

EX 2 Use Newton's method to estimate the solution of the equation $x^2 + x - 1 = 0$. Start with $x_0 = -1$ for the left-hand solution and with $x_0 = 1$ for the solution on the right. Then, in each case, find x_2.

EX 3 Use Newton's method to estimate the one real solution of $x^3 + 3x + 1 = 0$. Start with $x_0 = 0$ and then find x_2.

EX 4 Use Newton's method to estimate the two zeros of the function $f(x) = x^4 + x - 3$. Start with $x_0 = -1$ for the left-hand zero and with $x_0 = 1$ for the zero on the right. Then, in each case, find x_2.

4.13 Approximating Derivatives Numerically

Let f be differentiable on an open interval containing a.

- Forward Difference Quotient: $f'(a) \approx \dfrac{f(a+h)-f(a)}{h}$
- Backward Difference Quotient: $f'(a) \approx \dfrac{f(a)-f(a-h)}{h}$
- Symmetric Difference Quotient :
  - The average of the forward and backward difference quotient.
  - A better approximation of the derivative than the forward/backward difference quotient.

$$f'(a) \approx \frac{f(a+h)-f(a-h)}{2h}$$

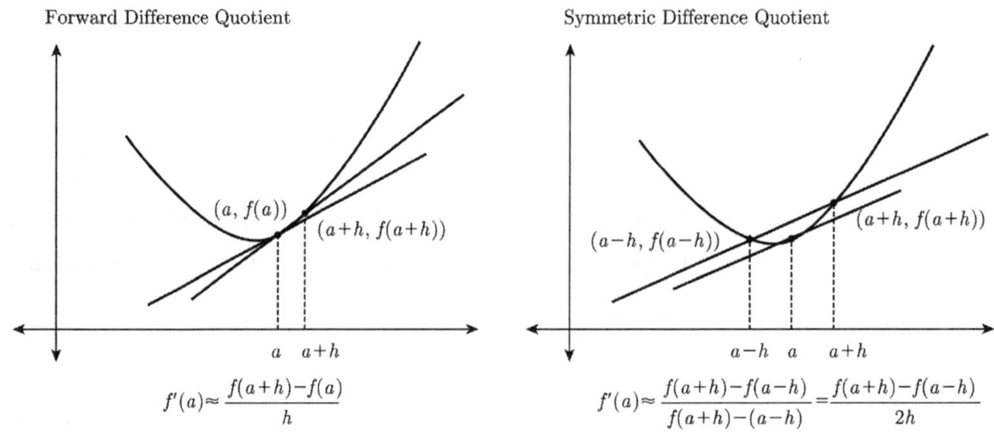

Example 1 Figure is a radioactive substance. The amount of mass remaining is given by a differentiable function f of time t. The table below shows the amount of the substance remaining as measured for a 40 minute period.

t(minutes)	0	10	20	30	40
$f(t)$(mg)	100	51	24	13	6

Approximate $f'(0)$, $f'(20)$, and $f'(40)$.

Solution

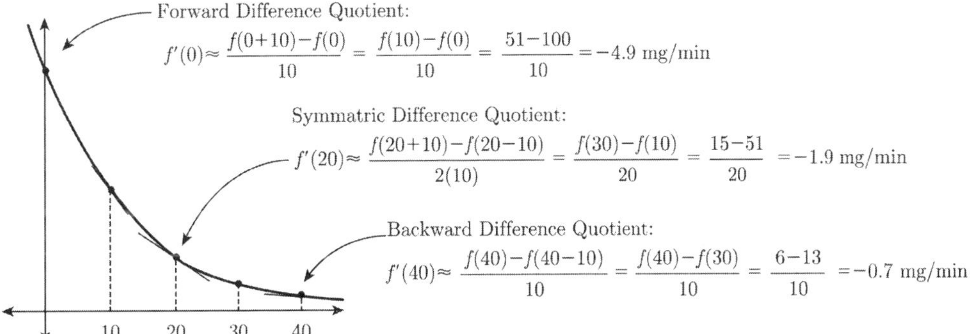

EX 1 The table gives the values of a function f that is differentiable on the interval $[0,1]$:

x	0.10	0.20	0.30	0.40	0.50	0.60
$f(x)$	0.171	0.288	0.357	0.384	0.375	0.336

According to this table, the best approximation of $f'(0.10)$ is

EX 2 A differentiable function f has the values shown. Estimate $f'(1.5)$.

x	1.0	1.2	1.4	1.6
$f(x)$	8	10	14	22

EX 3 For $f(x)=5^x$, what is the estimate of $f'(2)$ obtained by using the symmetric difference quotient with $h=0.03$?

EX 4 From the values of f shown, estimate $f'(2)$.

x	1.92	1.94	1.96	1.98	2.00
$f(x)$	6.00	5.00	4.40	4.10	4.00

EX 5 Using the values shown in the table for Question [EX4] estimate $(f^{-1})'(4)$.

 The values of differentiable function f are shown below.

x	0	2	4	6	8	10	12
$f(x)$	9	13	19	28	37	43	45

(1) Find the average rate of change of f over the interval [0,12].

(2) Approximate $f'(0)$, $f'(4)$, $f'(9)$ and $f'(12)$.

(3) Approximate $(f^{-1})'(19)$.

t (minutes)	0	1	2	3	4	5	6
$C(t)$ (ounces)	0	5.3	8.8	11.2	12.8	13.8	14.5

How water is dripping through a coffee maker, filling a large cup with coffee. The amount of coffee in the cup at time t, $0 \le t \le 6$, is by a differentiable function C, where t is measured in minutes. Selected values of $C(t)$, measured in ounces, are given in the table above.

(1) Use the date in the table to approximate $C'(3.5)$, Show the computations that lead to your answer, and indicate units of measure.

(2) Is there a time t, $2 \le t \le 4$ at which $C'(t) = 2$? Justify your answer.

(3) The amount of coffee in the cup in ounces is modeled b $B(t) = 16 - 16e^{-0.4t}$. Using this model, find the rate at which the amount of coffee in the cup is changing when $t = 5$.

4.14 Indeterminate Forms and L'Hospital's Rule

To find the limit of an indeterminate form of the type $\frac{0}{0}$ or $\frac{\infty}{\infty}$, we apply L'Hospital's rule, which involves taking derivatives. In the following, a is a finite number. The rule has several parts:

(a) If $\lim_{x\to a} f(x) = \lim_{x\to a} g(x) = 0$ and if $\lim_{x\to a}\frac{f'(x)}{g'(x)}$ exists, then

$$\lim_{x\to a}\frac{f(x)}{g(x)} = \operatorname{Lim}_{x\to a}\frac{f'(x)}{g'(x)};$$

if $\lim_{x\to a}\frac{f'(x)}{g'(x)}$ does not exist, then L'Hopital's rule can not be applied.

(b) If $\lim_{x\to a} f(x) = \lim_{x\to a} g(x) = \infty$, the same consequences follow as in case (a). The rules in (a) and (b) both hold for one-sides limits.

(c) $\lim_{x\to\infty} f(x) = \lim_{x\to\infty} g(x) = 0$ and if $\lim_{x\to\infty}\frac{f'(x)}{g'(x)}$ exists, then

$$\lim_{x\to\infty}\frac{f(x)}{g(x)} = \lim_{x\to\infty}\frac{f'(x)}{g'(x)};$$

if $\lim_{x\to\infty}\frac{f'(x)}{g'(x)}$ does not exist, then L'Hopital's rule can not be applied. (Here the notation "$x\to\infty$" represents either "$x\to+\infty$" or "$x\to-\infty$.")

(d) If $\lim_{x\to\infty} f(x) = \lim_{x\to\infty} g(x) = \infty$, the same consequences follow as in case (c).

In applying any of the above rules, if we obtain $\frac{0}{0}$ or $\frac{\infty}{\infty}$ again, we can apply the rule once more, repeating the process until the form we obtain is no longer indeterminate.

Example 1 Find in $\lim_{x\to-2}\frac{x^3+8}{x^2-4}$.

Solution

$\lim_{x\to-2}\frac{x^3+8}{x^2-4}$ is of type $\frac{0}{0}$ and thus equals $\lim_{x\to-2}\frac{3x^2}{2x} = -3$.

Example 2 Find in $\lim_{x\to\infty} \dfrac{x^3-4x^2+7}{3-6x-2x^3}$.

Solution

$\lim_{x\to\infty} \dfrac{x^3-4x^2+7}{3-6x-2x^3}$ is of type $\dfrac{\infty}{\infty}$, so that it equals $\lim_{x\to\infty} \dfrac{3x^2-8x}{-6-6x^2}$, which is again of type $\dfrac{\infty}{\infty}$, Apply L'Hopital's rule twice more:

$$\lim_{x\to\infty} \frac{6x-8}{-12x} = \lim_{x\to\infty} \frac{6}{-12} = -\frac{1}{2}$$

(e) L'Hopital's rule can be applied also to indeterminaue forms of the types $0 \cdot \infty$ and $\infty - \infty$, if the forms can be transformed to either $\dfrac{0}{0}$ or $\dfrac{\infty}{\infty}$.

Example 3 Find in $\lim_{x\to\infty} x \sin \dfrac{1}{\mathrm{x}}$.

Solution

$\lim_{x\to\infty} x\sin\dfrac{1}{\mathrm{x}}$ is of the type $\infty \cdot 0$. Since $x \sin\dfrac{1}{x} = \dfrac{\sin 1/x}{1/x}$ and, as $x\to\infty$, the latter is the indeterminate form $\dfrac{0}{0}$, we see that

$$\lim_{x\to\infty} x\sin\frac{1}{\mathrm{x}} = \lim_{\mathrm{x}\to\infty} \frac{-\frac{1}{\mathrm{x}^2}\cos\frac{1}{\mathrm{x}}}{-\frac{1}{\mathrm{x}^2}} = \lim_{\mathrm{x}\to\infty} \cos\frac{1}{x} = 1.$$

Example 4 Find in $\lim_{x\to 0^+} (\ln x - \ln \sin \mathrm{x})$.

Solution

$$\lim_{x\to 0^+} (\ln x - \ln\sin\mathrm{x}) = \lim_{\mathrm{x}\to 0^+} \ln\left(\frac{x}{\sin x}\right) = \ln\left(\lim_{x\to 0^+} \frac{x}{\sin x}\right)$$
$$= \ln\left(\lim_{x\to 0^+} \frac{1}{\cos x}\right) = \ln 1 = 0.$$

(f) Other indeterminate forms, such as 0^0, 1^∞ and ∞^0, may be resolved by taking the natural logarithm and then applying L'Hopital's rule.

Example 5 Find in $\lim_{x\to 0} (1+x)^{1/x}$.

Solution

$\lim_{x\to 0} (1+x)^{1/x}$ is of type 1^∞. Let $y = (1+x)^{1/x}$, so that $\ln y = \dfrac{1}{x}\ln(1+x)$. Then

$\lim_{x\to 0}\ln y=\lim_{x\to 0}\frac{\ln(1+x)}{x}$, which is of type $\frac{0}{0}$. Thus,

$$\lim_{x\to 0}\ln y=\lim_{x\to 0}\frac{\frac{1}{1+x}}{1}=\frac{1}{1}=1,$$

and since $\lim_{x\to 0}\ln y=1$, $\lim_{x\to 0}y=e^1=e$.

Example 6 Find in $\lim_{x\to\infty}(\ln x)^{1/x}$.

Solution

The limit leads to the indeterminate form ∞^0.

Let $f(x)=(\ln x)^{1/x}\Rightarrow\ln f(x)=\ln(\ln x)^{1/x}=\frac{\ln(\ln x)}{x}$.

Now $\lim_{x\to\infty}\ln f(x)=\lim_{x\to\infty}\frac{\ln(\ln x)}{x}=\lim_{x\to\infty}\frac{(\frac{1}{x\ln x})}{1}=0$.

Therefore $\lim_{x\to\infty}(\ln x)^{1/x}=\lim_{x\to\infty}f(x)=\lim_{x\to\infty}e^{\ln f(x)}=e^0=1$.

EX 1 Use L'Hopital's rule to find the limits in Exercises.

(1) $\lim_{x\to 2}\frac{x-2}{x^2-4}$

(2) $\lim_{x\to\infty}\frac{5x^3-2x}{7x^3+3}$

(3) $\lim_{x\to 0}\frac{8x^2}{\cos x-1}$

(4) $\lim_{\theta\to\pi/2}\frac{1-\sin\theta}{1+\cos 2\theta}$

(5) $\lim_{t\to 0}\frac{t(1-\cos t)}{t-\sin t}$

(6) $\lim_{\theta\to 0}\frac{3^{\sin\theta}-1}{\theta}$

(7) $\lim_{x\to\infty}\frac{\ln(x+1)}{\log_2 x}$

(8) $\lim_{y\to 0}\frac{\sqrt{5y+25}-5}{y}$

(9) $\lim_{x\to\infty}(\ln 2x-\ln(x+1))$

(10) $\lim_{x\to 1^+}(\frac{1}{x-1}-\frac{1}{\ln x})$

(11) $\lim_{x\to 0}\frac{x-\sin x}{x\tan x}$

(12) $\lim_{\theta\to 0}\frac{\theta-\sin\theta\cos\theta}{\tan\theta-\theta}$

(13) $\lim_{x\to 1^+}x^{1/(x-1)}$

(14) $\lim_{x\to e^+}(\ln x)^{1/(x-e)}$

(15) $\lim_{x\to\infty}x^{1/\ln x}$

(16) $\lim_{x\to 0}(e^x+x)^{1/x}$

(17) $\lim_{x\to 0^+}(1+\frac{1}{x})^x$

(18) $\lim_{x\to 0^+}x(\ln x)^2$

(19) $\lim_{x\to 0^+}\sin x\cdot\ln x$

Multiple Choice

01 At what values of x does $f(x) = 3x^5 - 5x^3 + 15$ have a relative maximum?

(A) -1 only (B) 0 Only (C) 1 Only

(D) -1 and 1 only (E) -1, 0 and 1

02 The graph of $y = f(x)$ on the closed interval $[2, 7]$ is shown below. How many points of inflection does this graph have on this interval?

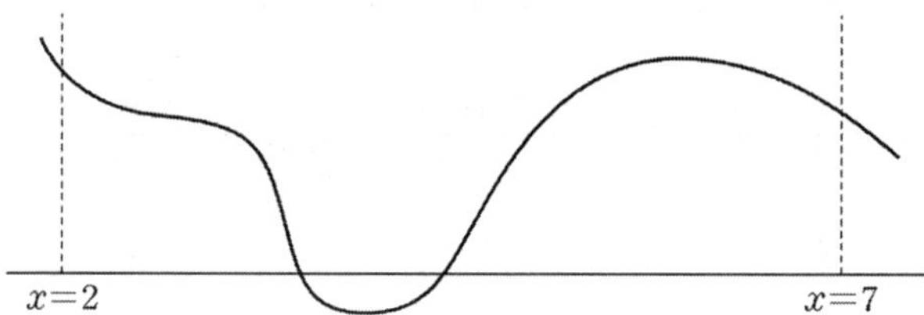

(A) One (B) Two (C) Three

(D) Four (E) Five

03 A particle moves along the x-axis so that at any time t its position is given by $x(t) = te^{-2t}$. For what values of t is the particle at rest?

(A) No values (B) 0 only (C) $\frac{1}{2}$ only

(D) 1 only (E) 0 and $\frac{1}{2}$

04 If $x + 7y = 29$ is an equation of the line normal to the graph of f at the point $(1, 4)$, then $f'(1) =$

(A) 7 (B) $\frac{1}{7}$ (C) $-\frac{1}{7}$

(D) $-\frac{7}{29}$ (E) -7

05 If c is number that satisfies the conclusion of the Mean Value Theorem for $f(x)=x^3-2x^2$ on the interval $0 \le x \le 2$, then $c=$

(A) 0 (B) $\frac{1}{2}$ (C) 1

(D) $\frac{4}{3}$ (E) 2

06 A curve in the plane is defined parametrically by the equations $x=t^3+t$ and $y=t^4+2t^2$. An equation of the line tagent to the curve at $t=1$ is

(A) $y=2x$ (B) $y=8x$ (C) $y=2x-1$

(D) $y=4x-5$ (E) $y=8x+13$

07 What is the area of the largest rectangle that can be inscribed in the ellipse $4x^2+9y^2=36$?

(A) $6\sqrt{2}$ (B) 12 (C) 24

(D) $24\sqrt{2}$ (E) 36

08 The position of a particle moving along the $x-$axis is $x(t)=\sin(2t)-\cos(3t)$ for time $t \ge 0$. When $t=\pi$, the acceleration of the particle is

(A) 9 (B) $\frac{1}{9}$ (C) 0

(D) $-\frac{1}{9}$ (E) -9

09 The slope of the line tangent to the graph of $\ln(xy)=x$ at the point where $x=1$ is

(A) 0 (B) 1 (C) e

(D) e^2 (E) $1-e$

10 If a particle moves in the $xy-$plane so that at time $t>0$ its position vector is $(\ln(t^2+2t), 2t^2)$ then at time $t=2$, its velocity vector is

(A) $\left(\frac{3}{4}, 8\right)$ (B) $\left(\frac{3}{4}, 4\right)$ (C) $\left(\frac{1}{8}, 8\right)$

(D) $\left(\frac{1}{8}, 4\right)$ (E) $\left(-\frac{5}{16}, 4\right)$

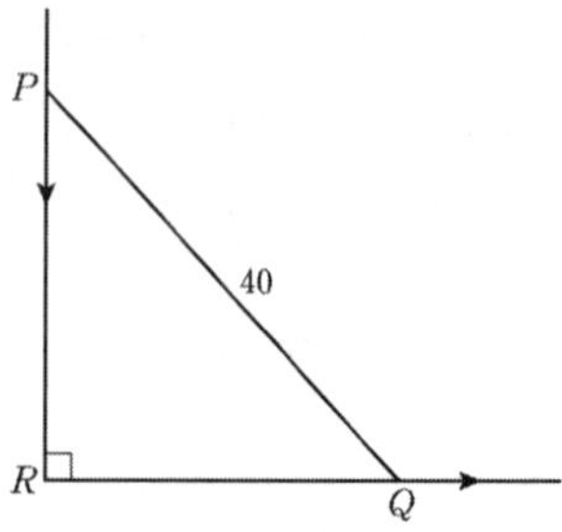

11 In the figure above, PQ represents a 40-foot ladder with end P against a vertical wall and Q on level ground. If the ladder is slipping down wall, what is the distance RQ at the instant when Q is moving along the ground $\frac{3}{4}$ as fast as P is moving down the wall?

(A) $\frac{6}{5}\sqrt{10}$ (B) $\frac{8}{5}\sqrt{10}$ (C) $\frac{80}{\sqrt{7}}$

(D) 24 (E) 32

12 The line normal to the curve $y=\sqrt{16-x}$ at the point $(0,\ 4)$ has slope

(A) 8 (B) 4 (C) $\frac{1}{8}$

(D) $-\frac{1}{8}$ (E) -8

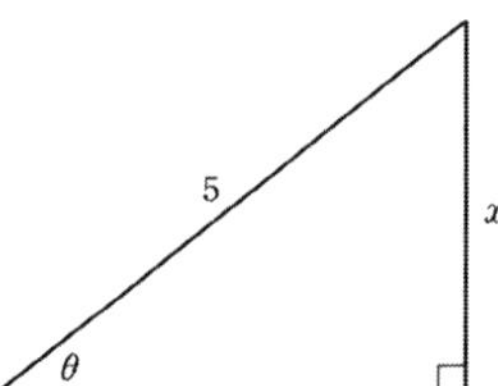

13 In the triangle shown above, if θ increases at a constant rate of 3 radians per minute, at what rate is x increasing in units per minute when x equals 3 units?

(A) 3 (B) $\frac{15}{4}$ (C) 4

(D) 9 (E) 12

14 Which of the following is an equation of the line tangent to the graph of $f(x) = x^4 + 2x^2$ at the point where $f'(x) = 1$?

(A) $y = 8x - 5$ (B) $y = x + 7$ (C) $y = x + 0.763$

(D) $y = x - 0.122$ (E) $y = x - 2.146$

15 The radius of a circle is decreasing at a constant rate of 0.1 centimeter per second. In terms of the circumference C, what is the rate of change of the area of the circle, in square centimeters per second?

(A) $-(0.2)\pi C$ (B) $-(0.1)C$ (C) $-\dfrac{(0.1)C}{2\pi}$

(D) $(0.1)^2 C$ (E) $(0.1)^2 \pi C$

16 In the xy−plane, the graph of the parametric equations $x = 5t + 2$ and $y = 3t$, for $-3 \le t \le 3$, is a line segment with slope

(A) $\dfrac{3}{5}$ (B) $\dfrac{5}{3}$ (C) 3

(D) 5 (E) 13

17 A particle move on a plane curve so that at any time $t > 0$ its x−coordinate is $t^3 - t$ and its y−coordinate is $(2t-1)^3$. The acceleration vector of the particle at $t = 1$ is

(A) (0,1) (B) (2,3) (C) (2,6)

(D) (6,12) (E) (6,24)

18 If f is the function defined by $f(x) = 3x^5 - 5x^4$, what are all the x−coordinates of points of inflection for the graph of f?

(A) −1 (B) 0 (C) 1

(D) 0 and 1 (E) −1,0, and 1

19 Which of the following is true about the graph of $y = \ln|x^2 - 1|$ in the interval $(-1,1)$?

(A) It is increasing.

(B) It attains a relative minimum at (0,0).

(C) It has a range of all real numbers.

(D) It is concave down.

(E) If has an asymptote of $x = 0$.

20 For $0 \le t \le 13$, an object travels along an elliptical path by the parametric equations $x = 3\cos t$ and $y = 4\sin t$. At point where $t = 13$, the object leaves the path and travels along the line tangent to the path at that point. What is the slope of the line on which the object travels?

(A) $-\dfrac{4}{3}$ (B) $-\dfrac{3}{4}$ (C) $-\dfrac{4\tan 13}{3}$

(D) $-\dfrac{4}{3\tan 13}$ (E) $-\dfrac{3}{4\tan 13}$

21 The rate of change of the volume, V, of water in a tank with respect to time, t, is directly proportional to the square root of the volume. Which of the following is a differential equation that describes this relationship?

(A) $V(t) = k\sqrt{t}$ (B) $V(t) = k\sqrt{V}$ (C) $\dfrac{dV}{dt} = k\sqrt{t}$

(D) $\dfrac{dV}{dt} = \dfrac{k}{\sqrt{V}}$ (E) $\dfrac{dV}{dt} = k\sqrt{V}$

22 A curve C is defined by the parametric equations $x = t^2 - 4t + 1$ and $y = t^3$. Which of the following is an equation of the line tangent to the graph of C at the point $(-3, 8)$?

(A) $x = -3$ (B) $x = 2$ (C) $y = 8$

(D) $y = -\dfrac{27}{10}(x+3)+8$ (E) $y = 12(x+3)+8$

23 The radius of a circle is increasing at a constant rate of 0.2 meters per second. What is the rate of increase in the area of the circle at the instant when circumference of the circle is 20π meters?

(A) $0.04\pi\ m^2/\mathrm{sec}$ (B) $0.4\pi\ m^2/\mathrm{sec}$ (C) $4\pi\ m^2/\mathrm{sec}$
(D) $20\pi\ m^2/\mathrm{sec}$ (E) $100\pi\ m^2/\mathrm{sec}$

24 The graph of f', the derivative of the function f, is shown above. If $f(0)=0$, which of the following must be true?

Ⅰ. $f(0) > f(1)$ Ⅱ. $f(2) > f(1)$ Ⅲ. $f(1) > f(3)$

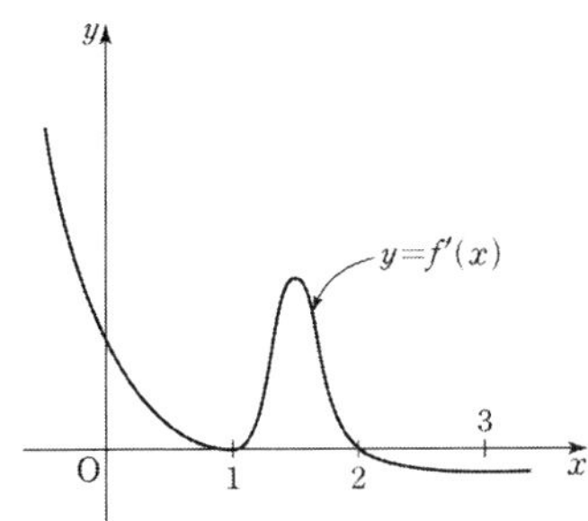

(A) Ⅰ only (B) Ⅱ only (C) Ⅲ only
(D) Ⅰ and Ⅱ only (E) Ⅱ and Ⅲ only

25 What are the coordinates of the inflection point on the graph of $y=(x+1)\ \arctan x$?

(A) (−1,0) (B) (0,0) (C) (0,1)
(D) $\left(1, \frac{\pi}{4}\right)$ (E) $\left(1, \frac{\pi}{2}\right)$

26 The Mean Value Theorem guarantees the existence of a special point on the graph of $y=\sqrt{x}$ between (0,0) and (4,2). What are the coordinates of this point?

(A) $(2,1)$ (B) $(1,1)$ (C) $(2, \sqrt{2})$
(D) $\left(\frac{1}{2}, \frac{1}{\sqrt{2}}\right)$ (E) None of the above

05

Integration

5.1 Indefinite Integrals

The antiderivative or indefinite integral of a function $f(x)$ is a function $F(x)$ whose derivative is $f(x)$. Since the derivative of a constant equals zero, the antiderivative of $f(x)$ is not unique; that is, if $F(x)$ is an integral of $f(x)$, then so is $F(x)+$C, where C is any constant. The arbitrary constant C is called the constant of integration. the indefinite integral of $f(x)$ is written as $\int f(x)\,dx$; thus

$$\int f(x)\,dx = F(x) + C \quad \text{if} \quad \frac{dF(x)}{dx} = f(x)$$

The function $f(x)$ is called the integrand.

- **Basic Integration Formulas**

$$\int kf(x)\,dx = k\int f(x)\,dx \quad (k \neq 0)$$

$$\int [f(x) \pm g(x)]\,dx = \int f(x)\,dx \pm \int g(x)dx$$

$$\int x^n\,dx = \frac{x^{n+1}}{n+1} + C \qquad (n \neq -1)$$

$$\int \frac{dx}{x} = \ln|x| + C$$

Example 1 Evaluate the following indefinite integrals.

(1) $\int 7\,dx = 7x + C$

(2) $\int 4x^3\,dx = x^4 + C$

(3) $\int \left(x^6 + \frac{1}{x}\right)dx = \frac{1}{7}x^7 + \ln|x| + C$

EX 1 Evaluate the following indefinite integrals.

(1) $\int (3-4x+2x^3)\,dx$

(2) $\int 6x^2\sqrt[3]{x}\,dx$

(3) $\int x^4 + \sqrt[3]{x^2} - \frac{2}{x^2} - \frac{1}{3\sqrt[3]{x}}\,dx$

(4) $\int \left(\frac{2}{x} - \frac{1}{x^2} + \frac{1}{2\sqrt{x}}\right)dx$

(5) $\int \frac{1}{\ln 2}\frac{dx}{x}$

(6) $\int (x-3)(x^2+3x+9)dx$

(7) $\int \frac{x^4}{x^2+1}dx - \int \frac{1}{x^2+1}dx$

(8) $\int (2\sin x + \cos x)^2 dx + \int (\sin x - 2\cos x)^2 dx$

(9) $\int (t+\sqrt{t})(t^2+3)dt$

EX 2 Evaluate the following indefinite integrals.

(1) $\int \frac{(x-2)^3}{x^2}dx$

(2) $\int \sqrt{x}(x-2)dx$

(3) $\int \frac{x^3-x-1}{x^2}dx$

(4) $\int \left(x-\frac{1}{2x}\right)^2 dx$

(5) $\int (\sqrt{x}+x)\sqrt[3]{x^2}\,dx$

(6) $\int \frac{(2-y)^2}{4\sqrt{y}}dy$

• Integration of Exponential Functions

$$\int e^x\,dx = e^x + C$$

$$\int a^x\,dx = \frac{a^x}{\ln a} + C \qquad (a>0,\ a\neq 1)$$

EX 3 Evaluate the following indefinite integrals.

(1) $\int (e^x+4)dx$

(2) $\int (5^x-7^x)dx$

(3) $\int \left(10^x + \frac{10}{x} - \frac{7}{x^3}\right)dx$

• Integration of Trigonometric Functions

$$\int \cos x\,dx = \sin x + C \qquad \int \sin x\,dx = -\cos x + C$$

$$\int \tan x\,dx = \ln|\sec x| + C \text{ or } -\ln|\cos x| + C \qquad \int \cot x\,dx = \ln|\sin x| + C \text{ or } -\ln|\csc x| + C$$

$$\int \sec^2 x\,dx = \tan x + C \qquad \int \csc^2 x\,dx = -\cot x + C$$

$$\int \sec x \tan x\,dx = \sec x + C \qquad \int \csc x \cot x\,dx = -\csc x + C$$

$$\int \sec x\,dx = \ln|\sec x + \tan x| + C \qquad \int \csc x\,dx = \ln|\csc x - \cot x| + C$$

EX 4 Evaluate the following indefinite integrals.

(1) $\int (5\cos x + \sin x)dx$

(2) $\int (2\tan x + 3\cot x)dx$

(3) $\int \frac{1}{\sin x}dx$

(4) $\int (\tan x + 3\sin x)dx$

(5) $\int (\sec x \tan x + 5\cos x)dx$

(6) $\int (\sec^2 x - 4\csc x)dx$

(7) $\int \frac{1}{1-\cos^2 x}dx$

(8) $\int \frac{\sin^2 x}{1-\cos x}dx$

(9) $\int \frac{\sin x}{\cos^2 x}dx$

(10) $\int \frac{\cos x}{\sin^2 x}dx$

(11) $\int \tan^2 x\, dx$

(12) $\int \cot^2 x dx$

(13) $\int \frac{e^x \sin^2 x + 1}{\sin^2 x}dx$

(14) $\int \frac{10^x \cot x - 1}{\cot x}dx$

- **Integration of Inverse Trigonometric Functions**

$$\int \frac{dx}{\sqrt{1-x^2}} = \sin^{-1}x + C$$

or $\arcsin x + C$

$$\int \frac{dx}{1+x^2} = \tan^{-1}x + C$$

or $\arctan x + C$

$$\int \frac{dx}{x\sqrt{x^2-1}} = \sec^{-1}|x| + C$$

or $\mathrm{arcsec}|x| + C$

EX 5 Evaluate the following indefinite integrals.

(1) $\int \frac{5}{\sqrt{1-x^2}}dx$

(2) $\int \frac{3}{x^2+1}dx$

(3) $\int \frac{-7}{x\sqrt{x^2-1}}dx$

(4) $\int \frac{4}{1+x^2}dx$

- **Integration Using Reverse Chain Rule (1)**

$$\int (ax+b)^n dx = \frac{1}{a(n+1)}(ax+b)^{n+1} + C \quad (a \text{ and } b \text{ are constants})$$

EX 6 Evaluate the following indefinite integrals.

(1) $\int (2x+1)^3 dx$

(2) $\int (3x-2)^8 dx$

(3) $\int \left(\frac{1}{2}x+3\right)^5 dx$ (4) $\int \sqrt{1-3x}\,dx$

(5) $\int \frac{7}{\sqrt{2x+1}}dx$ (6) $\int \frac{1}{(1-x)^2}dx$

• Integration Using Reverse Chain Rule (2)

$$\int e^{kx}dx = \frac{1}{k}e^{kx}+C \qquad \int a^{kx}dx = \frac{1}{k}\frac{a^{kx}}{\ln a}+C \quad (k \text{ and } a \text{ are constants})$$

$$\int \cos kx\,dx = \frac{1}{k}\sin kx + C \qquad \int \sin kx\,dx = -\frac{1}{k}\cos kx + C$$

EX 7 Evaluate the following indefinite integrals.

(1) $\int (e^{7x}+e^{3x})dx$ (2) $\int (e^{-2x}+\cos 2x)dx$

(3) $\int \left(e^{-2x}+e^{-\frac{1}{3}x}\right)dx$ (4) $\int (4e^{-x}+3^{x-1})dx$

(5) $\int (10^{5x}+\sin 2x)dx$ (6) $\int 4^x(4^{2x}+1)dx$

(7) $\int \sqrt{3^{x+1}}\,dx$ (8) $\int \left[\cos 3x+\sin\left(\frac{x}{2}\right)\right]dx$

(9) $\int \cos(3x+5)dx$ (10) $\int \sec 5x \tan 5x\,dx$

(11) $\int (e^x+e^{-x})^2 dx$ (12) $\int (\sec^2 2x+\tan 2x)dx$

• Integration Using double Angle Identities

$$\sin 2x = 2\sin x\cos x \Rightarrow \sin x\cos x = \frac{1}{2}\sin 2x$$

$$\cos 2x = 2\cos^2 x - 1 \Rightarrow \cos^2 x = \frac{1+\cos 2x}{2} = \frac{1}{2}+\frac{1}{2}\cos 2x$$

$$\cos 2x = 1-2\sin^2 x \Rightarrow \sin^2 x = \frac{1-\cos 2x}{2} = \frac{1}{2}-\frac{1}{2}\cos 2x$$

EX 8 Evaluate the following indefinite integrals.

(1) $\int 3\sin x\cos x\,dx$ (2) $\int \sin\frac{\theta}{2}\cos\frac{\theta}{2}d\theta$

(3) $\int 8\sin 2x\cos 2x\,dx$ (4) $\int 12\sin 3x\cos 3x\,dx$

(5) $\int \sin^2\frac{1}{2}x\,dx$ (6) $\int \sin^2 3x\,dx$

(7) $\int \cos^2 x\,dx$ (8) $\int \cos^2 4x\,dx$

• Integration of Special Fractions

$$\int \frac{f'(x)}{f(x)}dx = \ln|f(x)| + C$$

EX 9 Evaluate the following indefinite integrals.

(1) $\int \frac{5\cos x}{\sin x}dx$

(2) $\int \frac{3x^2+1}{x^3+x}dx$

(3) $\int \frac{7}{x+1}dx$

(4) $\int \frac{x}{x^2+1}dx$

(5) $\int \frac{6x}{x^2-4}dx$

(6) $\int \frac{2}{1-3x}dx$

(7) $\int \frac{5\sin x}{1+\cos x}dx$

(8) $\int \frac{x}{2x^2+1}dx$

(9) $\int \frac{e^x - e^{-x}}{e^x + e^{-x}}dx$

EX 10 Determine $f(x)$.

(1) $f'(x) = 1 + \frac{1}{x}$, $f(1) = 3$

(2) $\frac{dy}{dx} = 2^x - x^2\sqrt{x} + \sin x$, $y(0) = \frac{1}{\ln 2}$

(3) $f'(x) = \frac{8}{1+x^2} + \sec^2 x$, $f(0) = 1$

(4) $y' = \frac{4}{2x-1} + 8(2x+1)^3$, $y(0) = 3$

(5) $f'(x) = \frac{2x}{x^2+4} + \cos^2 x$, $f(0) = 0$

(6) $f''(x) = e^{\frac{1}{2}x} - 12x + 2$, $f(0) = 1$, $f'(0) = -1$

(7) $\frac{d^2y}{dx^2} = \frac{2}{x^3}$, $y(1) = 1$, $y'(1) = 1$

(8) $\frac{d^2y}{dx^2} = e^{-x} + 6x$, $y(0) = 2$, $y(1) = \frac{1}{e} - 1$

Integration Formulas

◆ Basic Integration Formulas

$$\int k\,dx = kx + C \quad (k : a\,constant)$$

$$\int \frac{1}{x}dx = \ln|x| + C$$

$$\int x^n\,dx = \frac{1}{n+1}x^{n+1} + C,\ n \neq -1$$

◆ Integration of Exponential Functions

$$\int e^x dx = e^x + C$$

$$\int a^x dx = \frac{a^x}{\ln a} + C \quad (a > 0,\ a \neq 1)$$

◆ Integration of Trigonometric Functions

$$\int \cos x\,dx = \sin x + C$$

$$\int \sin x\,dx = -\cos x + C$$

$$\int \sec^2 x\,dx = \tan x + C$$

$$\int \csc^2 x\,dx = -\cot x + C$$

$$\int \sec x \tan x\,dx = \sec x + C$$

$$\int \csc x \cot x\,dx = -\csc x + C$$

$$\int \tan x\,dx = \ln|\sec x| + C = -\ln|\cos x| + C$$

$$\int \cot x\,dx = \ln|\sin x| + C = -\ln|\csc x| + c$$

$$\int \sec x\,dx = \ln|\sec x + \tan x| + C$$

$$\int \csc x\,dx = \ln|\csc x - \cot x| + C$$

$$\int \sin^2 x\,dx = \int\left(\frac{1}{2} - \frac{1}{2}\cos 2x\right)dx = \frac{1}{2}x - \frac{1}{4}\sin 2x + C$$

$$\int \cos^2 x\,dx = \int\left(\frac{1}{2} + \frac{1}{2}\cos 2x\right)dx = \frac{1}{2}x + \frac{1}{4}\sin 2x + C$$

◆ Integration Using Reverse Chain Rule

$$\int e^{kx} = \frac{1}{k}e^{kx} + C$$

$$\int \sin kx\,dx = -\frac{1}{k}\cos kx + C$$

$$\int \cos kx\,dx = \frac{1}{k}\sin kx + C$$

$$\int a^{kx}\,dx = \frac{1}{k}\frac{a^{kx}}{\ln a} + C$$

$$\int (kx+b)^n dx = \frac{1}{n+1} \cdot \frac{1}{k}(kx+b)^{n+1} + C$$

(k, a and b are constants)

◆ Integration of Special Fractions

$$\int \frac{f'(x)}{f(x)}dx = \ln|f(x)| + C$$

◆ Integration of Inverse Trigonometric Functions

$$\int \frac{dx}{\sqrt{1-x^2}} = \sin^{-1}x + C$$

$$\int \frac{dx}{\sqrt{a^2-x^2}} = \sin^{-1}\left(\frac{x}{a}\right) + C$$

$$\int \frac{dx}{x^2+1} = \tan^{-1}x + C$$

$$\int \frac{dx}{x^2+a^2} = \frac{1}{a}\tan^{-1}\left(\frac{x}{a}\right) + C$$

$$\int \frac{dx}{x\sqrt{x^2-1}} = \sec^{-1}|x| + C$$

5.2 Integration by Substitution

Substitute $u = g(x)$ and $du = g'(x)dx$, integrate with respect to u, and replace u with $g(x)$.

$$\int f(g(x))g'(x)dx = \int f(u)du = F(u) + C = F(g(x)) + C$$

Example 1

(1) Find $\int \sec^2(5t+1) \cdot 5\,dt$.

Solution

We substitute $u = 5t+1$ and $du = 5dt$. Then,

$$\begin{aligned}\int \sec^2(5t+1) \cdot 5\,dt &= \int \sec^2 u\,du \\ &= \tan u + C \\ &= \tan(5t+1) + C\end{aligned}$$

(2) Find $\int \cos(7\theta+3)d\theta$.

Solution

We substitute $u = 7\theta+3$ and $du = 7d\theta$. Then,

$$\begin{aligned}\int \cos(7\theta+3)d\theta &= \frac{1}{7}\int \cos(7\theta+3) \cdot 7\,d\theta \\ &= \frac{1}{7}\int \cos u\,du \\ &= \frac{1}{7}\sin u + C \\ &= \frac{1}{7}\sin(7\theta+3) + C\end{aligned}$$

EX 1 Evaluate the following indefinite integrals.

(1) $\int x\sqrt[4]{1-x^2}\,dx$

(2) $\int (x^2-2x+3)^{10}(x-1)dx$

(3) $\int \frac{4x^3}{(x^4+1)^2}dx$

(4) $\int \frac{x-3}{\sqrt[4]{x^2-6x}}dx$

(5) $\int \frac{9x^2}{\sqrt{1-x^3}}dx$

(6) $\int \frac{x-2}{\sqrt{4x-x^2}}dx$

(7) $\int 3x\sqrt{7-3x^2}\,dx$

(8) $\int \frac{(1+\sqrt{x})^5}{\sqrt{x}}dx$

(9) $\int \frac{2^{\sqrt{x}}}{\sqrt{x}}dx$

(10) $\int \frac{1}{\sqrt{x}(\sqrt{x}-1)}dx$

EX 2 Evaluate the following indefinite integrals.

(1) $\int \frac{(\ln x)^3}{2x}dx$

(2) $\int \frac{1}{x\ln x}dx$

(3) $\int \frac{\ln x}{x\sqrt{\ln^2 x+1}}dx$

(4) $\int \frac{1}{x\ln x \cdot \ln(\ln x)}dx$

(5) $\int \sin^5\frac{x}{3}\cos\frac{x}{3}dx$

(6) $\int \cos^3 3x\sin 3x\,dx$

(7) $\int e^{\sec\pi x}\sec\pi x\tan\pi x\,dx$

(8) $\int \sec^2\theta\tan^2\theta\,d\theta$

(9) $\int \frac{e^{\sin^{-1}x}}{\sqrt{1-x^2}}dx$

(10) $\int \sec^4\theta\,d\theta$

(11) $\int \frac{1}{(\tan^{-1}x)(1+x^2)}dx$

(12) $\int \frac{\sec^2 2\theta}{1+\tan 2\theta}d\theta$

(13) $\int \frac{\sin(2x+1)}{\cos^2(2x+1)}dx$

(14) $\int \frac{\cos 2x}{\sqrt{1+\sin 2x}}dx$

(15) $\int \sin 2x\,e^{\sin^2 x}\,dx$

(16) $\int \frac{\sin x}{1+\cos^2 x}dx$

EX 3 Evaluate the following indefinite integrals.

(1) $\int \frac{e^x}{1+e^x}dx$

(2) $\int \frac{e^{2x}}{e^{2x}+1}dx$

(3) $\int 2x e^{-x^2}dx$

(4) $\int \frac{x}{e^{x^2+2}}dx$

(5) $\int \frac{\sqrt{\tan^{-1}x}}{1+x^2}dx$

(6) $\int \frac{\sin^{-1}(2x)}{\sqrt{1-4x^2}}dx$

(7) $\int \frac{e^x}{1+e^{2x}}dx$

(8) $\int \frac{e^{2x}}{1+e^{4x}}dx$

(9) $\int \frac{x^2+3}{x+1}dx$

(10) $\int \frac{e^{2x}}{e^x+1}dx$

• Integration of Inverse Trigonometric Functions

$$\int \frac{dx}{\sqrt{a^2-x^2}} = \sin^{-1}\left(\frac{x}{a}\right) + C \qquad \int \frac{dx}{a^2+x^2} = \frac{1}{a}\tan^{-1}\left(\frac{x}{a}\right) + C$$

Proof

(1) $$\int \frac{dx}{\sqrt{a^2-x^2}} = \int \frac{\frac{1}{a}dx}{\frac{1}{a}\sqrt{a^2-x^2}} = \int \frac{\frac{1}{a}dx}{\sqrt{1-\frac{x^2}{a^2}}} = \sin^{-1}\left(\frac{x}{a}\right) + C$$

(2) $$\int \frac{dx}{a^2+x^2} = \int \frac{\frac{1}{a^2}dx}{\frac{1}{a^2}(a^2+x^2)} = \int \frac{\frac{1}{a}\cdot\frac{1}{a}dx}{1+\frac{x^2}{a^2}} = \frac{1}{a}\tan^{-1}\left(\frac{x}{a}\right) + C$$

EX4 Evaluate the following indefinite integrals.

(1) $\int \frac{1}{1+4x^2}dx$

(2) $\int \frac{1}{\sqrt{16-x^2}}dx$

(3) $\int \frac{1}{x^2+2x+2}dx$

(4) $\int \frac{3}{9+x^2}dx$

(5) $\int \frac{1}{\sqrt{9-x^2}}dx$

(6) $\int \frac{1}{9+4x^2}dx$

(7) $\int \frac{e^x}{9+e^{2x}}dx$

(8) $\int \frac{2x+1}{4+x^2}dx$

(9) $\int \frac{x+1}{\sqrt{4-x^2}}dx$

(10) $\int \frac{1}{x^2-2x+5}dx$

EX5 Given that $f'(x) = \frac{1}{\sqrt{9-4x^2}}$ and $f\left(-\frac{3}{2}\right) = \frac{\pi}{12}$, find $f(x)$.

5.3 Integration by Parts

$$\int uv'\,dx = uv - \int u'v\,dx$$

Proof

$$\frac{d}{dx}(uv) = u'v + uv' \quad \int \frac{d}{dx}(uv)dx = \int (u'v + uv')dx$$

$$\int \frac{d}{dx}(uv)dx = \int (u'v)dx + \int (uv')dx$$

$$uv = \int u'vdx + \int uv'dx$$

$$\therefore \int uv'dx = uv - \int u'vdx$$

Example 1 Evaluate the integral.

(1) $\int x\ e^{x}\ dx$ (2) $\int x\ e^{-2x}\ dx$

Solution

(1) $\int x \cdot e^{x}dx = x \cdot e^{x} - \int 1 \cdot e^{x}dx = xe^{x} - e^{x} + C$

(2) $\int x \cdot c^{-2x}dx = x \cdot (-\frac{1}{2}e^{-2x}) - \int 1 \cdot (-\frac{1}{2}e^{-2x})dx$

$= -\frac{1}{2}xe^{-2x} - \frac{1}{4}e^{-2x} + C$

EX 1 Evaluate the following indefinite integrals.

(1) $\int x\cos x\,dx$ (2) $\int xe^{-\frac{1}{2}x}\,dx$

(3) $\int x\cos 3x\,dx$ (4) $\int x\sin 2x\,dx$

(5) $\int x^{2}e^{x}\,dx$ (6) $\int \ln x\,dx$

(7) $\int x^{3}\ln x\,dx$ (8) $\int \frac{\ln x}{x^{2}}dx$

(9) $\int \frac{\ln x}{\sqrt{x}}dx$ (10) $\int \tan^{-1}x\,dx$

(11) $\int x\sec^{-1}x\,dx,\ x > 0$

EX2 Evaluate the following indefinite integrals.

(1) $\int x(\ln x)^2 dx$　　(2) $\int \frac{(\ln x)^2}{x^2} dx$

- **Circular Integration by Parts:**
  $\int$ **(Exponential Function) × (Trigonometric Function)** dx

Example 2 Evaluate $\int e^x \cos x\, dx$.

Solution

$$\begin{aligned}\int e^x \cos x\, dx &= e^x \sin x - \int e^x \sin x\, dx \\ &= e^x \sin x - \left[-e^x \cos x - \int -e^x \cos x\, dx\right] \\ &= e^x \sin x + e^x \cos x - \int e^x \cos x\, dx\end{aligned}$$

Let $\int e^x \cos x dx = I$

$I = e^x \sin x + e^x \cos x - I$

$2I = e^x (\sin x + \cos x)$

$\therefore I = \frac{e^x(\sin x + \cos x)}{2} + C$

EX3 Evaluate the following indefinite integrals.

(1) $\int e^{2x} \sin x\, dx$　　(2) $\int e^x \cos 2x\, dx$

(3) $\int e^{2x} \cos 3x\, dx$　　(4) $\int e^{-2x} \sin 2x\, dx$

EX4 Evaluate the following indefinite integrals.

(1) $\int x^2 e^x dx$　　(2) $\int x^3 e^{-x} dx$

(3) $\int x^3 e^{-\frac{x}{2}} dx$　　(4) $\int x^2 \cos 3x$

(5) $\int x^3 \sin \frac{1}{2} x\, dx$

EX5 Evaluate the following indefinite integrals.

(1) $\int e^{\sqrt{x}} dx$　　(2) $\int \cos \sqrt{x}\, dx$

(3) $\int x^3 e^{-x^2} dx$　　(4) $\int x^3 \sin(x^2) dx$

5.4 Rectangular Approximation Method (RAM)

We may approximate $\int_a^b f(x)\,dx$ by any of the following sums:

(1) Left sum: $f(x_0)\triangle x + f(x_1)\triangle x + \cdots + f(x_{n-1})\triangle x$, using the value of f at the left endpoint of each subinterval.

(2) Right sum: $f(x_1)\triangle x + f(x_2)\triangle x + \cdots + f(x_n)\triangle x$, using the value of f at the right end of each subinterval.

(3) Midpoint sum:

$f(\frac{x_0+x_1}{2})\triangle x + f(\frac{x_1+x_2}{2})\triangle x + \cdots + f(\frac{x_{n-1}+x_n}{2})\triangle x$, using the value of f at the midpoint of each subinterval.

Example 1 Approximate $\int_0^2 x^3 dx$ by using four subintervals and calculating (a) the left sum, (b) the right sum, and (c) the midpoint sum. Evaluate the Integral exactly. Here $\triangle x = \frac{2-0}{4} = \frac{1}{2}$.

(a) LRAM

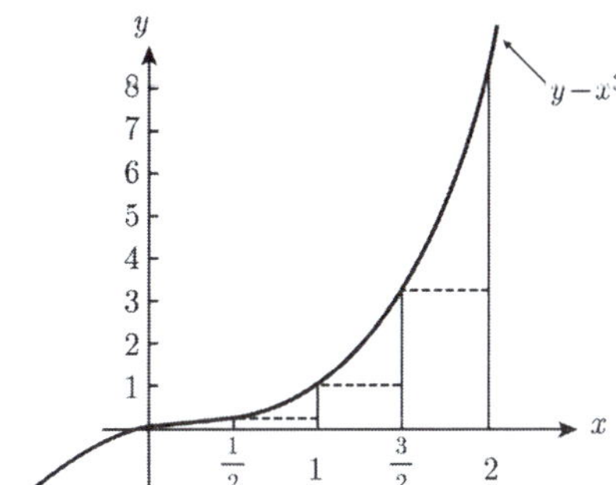

(b) RRAM

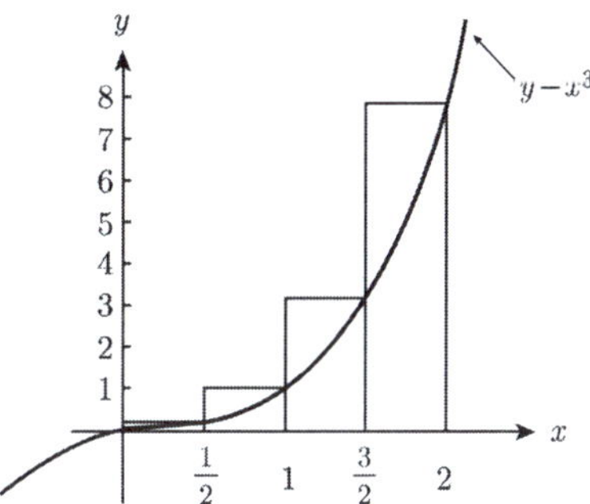

(c) MRAM

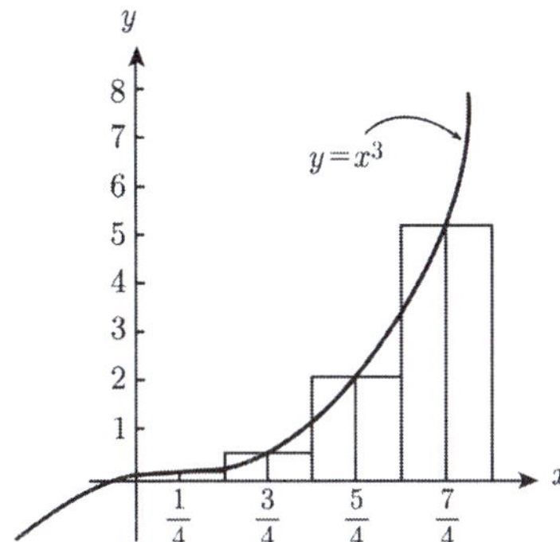

Solution

(1) For a left sum we use the left-hand altitudes at $x=0,\ \frac{1}{2},\ 1$ and $\frac{3}{2}$. The approximating sum is

$$(0)^3 \cdot \frac{1}{2}+\left(\frac{1}{2}\right)^3 \cdot \frac{1}{2}+(1)^3 \cdot \frac{1}{2}+\left(\frac{3}{2}\right)^3 \cdot \frac{1}{2}$$
$$=\frac{9}{4}.$$

The dashed lines in Figure show the inscribed rectangles used.

(2) For the right sum we use right-hand altitudes at $x=\frac{1}{2},\ 1,\ \frac{3}{2}$ and 2. The approximating sum is

$$\left(\frac{1}{2}\right)^3 \cdot \frac{1}{2}+(1)^3 \cdot \frac{1}{2}+\left(\frac{3}{2}\right)^3 \cdot \frac{1}{2}+(2)^3 \cdot \frac{1}{2}$$
$$=\frac{25}{4}.$$

(3) The midpoint sum uses the heights at the midpoints of the subintervals, as shown in Figure. The approximating sum is

$$\left(\frac{1}{4}\right)^3 \cdot \frac{1}{2}+\left(\frac{3}{4}\right)^3 \cdot \frac{1}{2}+\left(\frac{5}{4}\right)^3 \cdot \frac{1}{2}+\left(\frac{7}{4}\right)^3 \cdot \frac{1}{2}$$
$$=\frac{31}{8}.$$

EX 1 Approximate the area under the curve $y=4-x^2$ from $x=0$ to $x=2$ using (1) LRAM (2) RRAM and (3) MRAM respectively with 4 equal subintervals.

(1) LRAM

$f(x)=4-x^2$

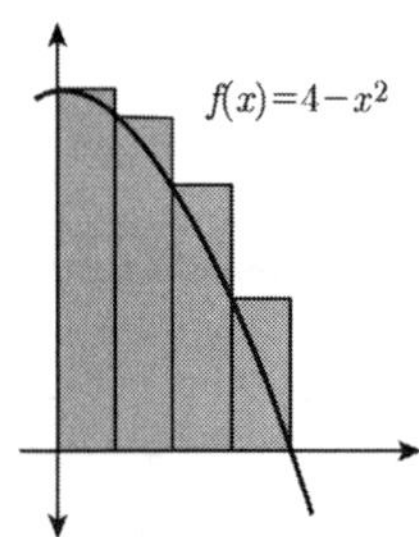

(2) RRAM

$f(x)=4-x^2$

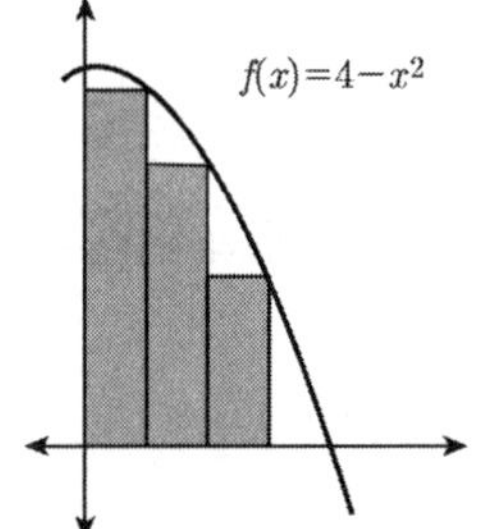

(3) MRAM

$f(x)=4-x^2$

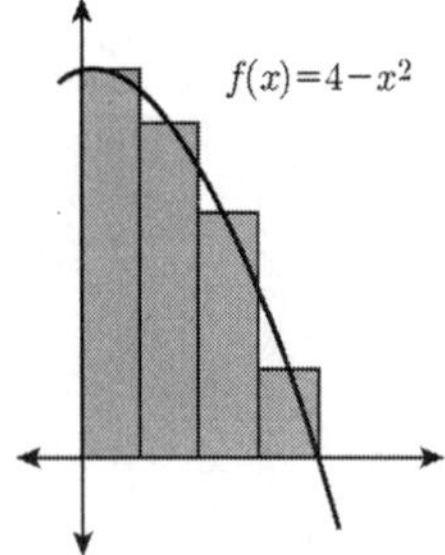

5.5 Trapezoidal Approximation

To approximate the area under a curve on the interval $[a, b]$,

$$T(n) = \frac{h}{2}(y_0 + 2y_2 + \ldots\ldots + 2y_{n-1} + y_n)$$
$$= \frac{LRAM + RRAM}{2}, \qquad where\ \ h = \frac{b-a}{n}$$

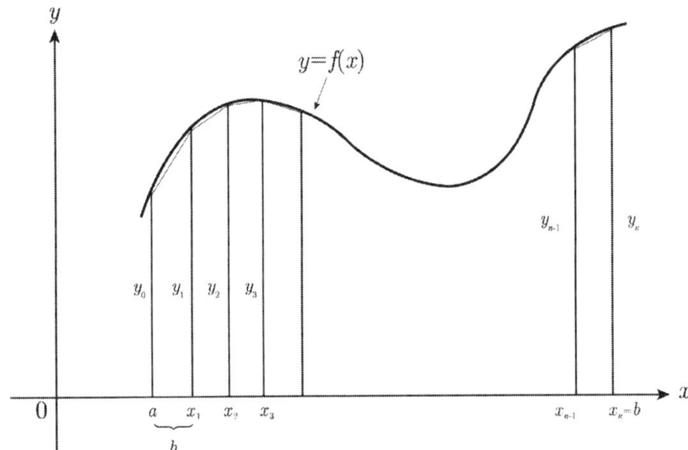

$$T(n)$$
$$= \frac{(y_0 + y_1)h}{2} + \frac{(y_1 + y_2)h}{2} + \frac{(y_2 + y_3)h}{2} \cdots + \frac{(y_{n-1} + y_n)h}{2}$$
$$= \frac{h[(y_0 + y_1) + (y_1 + y_2) + (y_2 + y_3) + (y_{n-1} + y_n)]}{2}$$
$$= \frac{h}{2}(y_0 + 2y_1 + 2y_2 + \cdots\cdots + 2y_{n-1} + y_n)$$

Example 1 Approximate the area under the curve $y = x^3$ from $x = 0$ to $x = 2$ using the trapezoidal rule with 4 equal subintervals.

Solution

$$T(4) = \frac{h}{2}[y_0 + 2y_1 + 2y_2 + 2y_3 + y_4]$$
$$= \frac{1/2}{2}\left(0^3 + 2\left(\frac{1}{2}\right)^3 + 2(1)^3 + 2\left(\frac{3}{2}\right)^3 + 2^3\right) = \frac{17}{4}$$

EX 1 Find $L(5)$, $R(5)$, and $T(5)$ for $\int_1^6 \frac{120}{x}dx$.

$f(x)=\frac{120}{x}$

EX 2 Approximate the area under the curve $y = \sin x$ form $x = 0$ to $x = \pi$ using the trapezoidal rule with 6 equal subintervals.

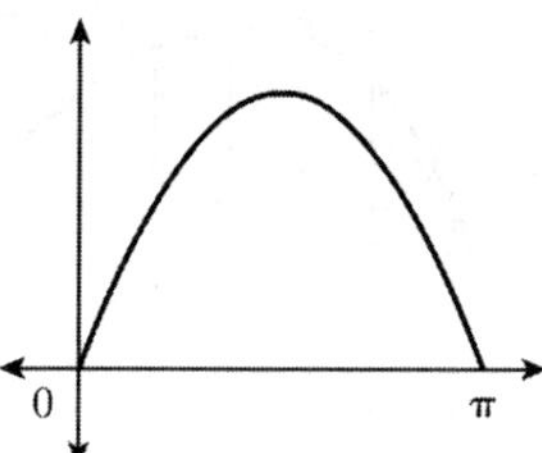

EX 3 Approximate the area under the curve $y = x^3 + 1$ from $x = -1$ to $x = 3$ using (1) LRAM (2) RRAM (3) the trapezoidal rule, with 4 equal subintervals.

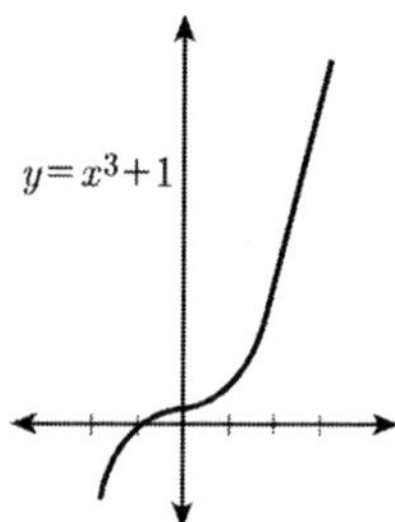

EX 4 Readings from a car's speedometer at 10 minute intervals during a 1-hour period are given in the table; t = time in minutes, v = speed in miles per hour:

t	0	10	20	30	40	50	60
v	26	40	55	10	60	32	45

(1) Draw a graph that could represent the car's speed during the hour.

(2) Approximate the distance traveled, using L(6), R(6), and T(6).

EX 5 The rate at which oil flows out of a tank, in liters per minute, is shown in the table below. Approximate the total amount of oil flowing out of the tank during 10 minutes using (1) LRAM (2) RRAM and (3) the trapezoidal rule, with 5 subintervals indicated by the data.

t (min)	R(t) (L/min)
0	4
1	6
3	8
6	9
9	7
10	5

5.6 Definition of Area as a Limit

Let f be continuous on the interval $[a, b]$ and the interval be divided into n subintervals of equal length. The area under the curve $y = f(x)$ over the interval $[a, b]$ is defined by $Area = \lim_{n \to \infty} \sum_{k=1}^{n} f(x_k)\Delta x$, where $\Delta x = \dfrac{b-a}{n}$ and $x_k = a + k\Delta x$.

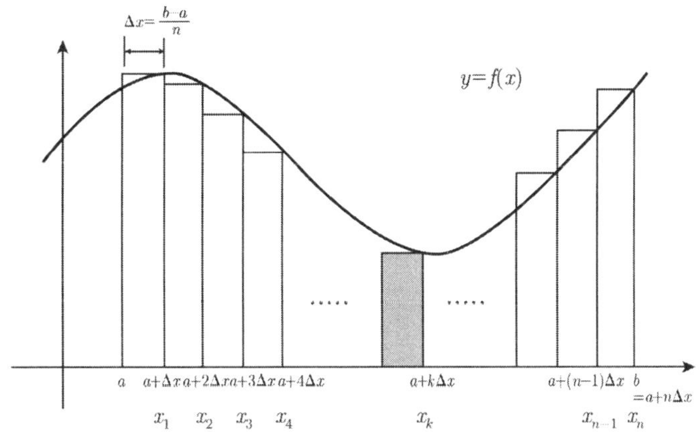

$Area \approx f(x_1)\Delta x + f(x_2)\Delta x + f(x_3)\Delta x + \cdots\cdots + f(x_k)\Delta x + \cdots\cdots f(x_n)\Delta x$

: RRAM

$$= \sum_{k=1}^{n} f(x_k)\Delta x$$

$n = 6$

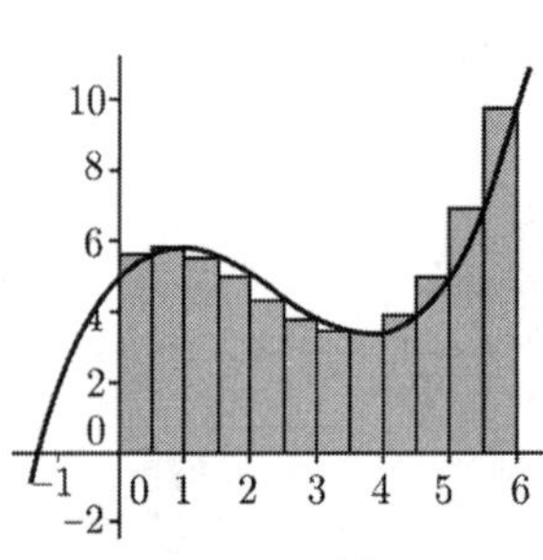

$n = 12$

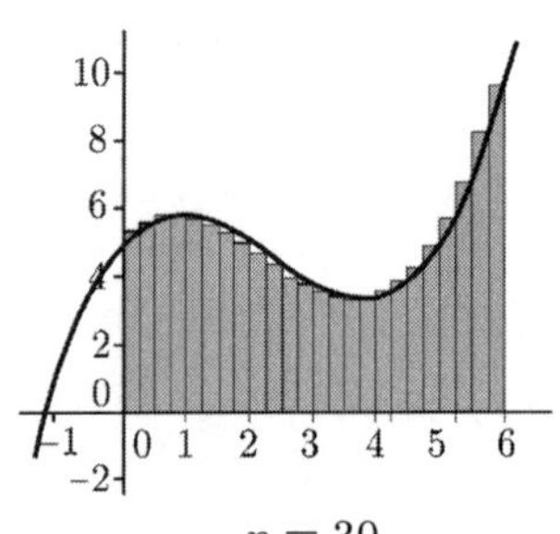

$n = 30$

The approximation becomes more accurate as the number of strips increases.

Therefore, $Area = \lim_{n\to\infty} \sum_{k=1}^{n} f(x_k)\Delta x$.

This method can also be applied to LRAM and MRAM.

$Area = \lim_{n\to\infty} \sum_{k=1}^{n} f(x_k^*)\Delta x$ for RRAM: $x_k^* = a + k\Delta x$,

LRAM: $x_k^* = a + (k-1)\Delta x$,

MRAM: $x_k^* = a + (k - \frac{1}{2})\Delta x$,

Example 1 Find the area under the curve $y = 16 - x^2$ from $x = 1$ to $x = 3$ by summing the areas of the rectangles.

Solution

$y = 16 - x^2$, $[1, 3]\left(\Delta x = \frac{2}{n}\right)$

$$s(n) = \sum_{i=1}^{n} f\left(1 + \frac{2i}{n}\right)\left(\frac{2}{n}\right) = \sum_{i=1}^{n}\left[16 - \left(1 + \frac{2i}{n}\right)^2\right]\left(\frac{2}{n}\right)$$

$$= \sum_{i=1}^{n}\left[15 - \frac{4i^2}{n^2} - \frac{4i}{n} - \right]\left(\frac{2}{n}\right)$$

$$= \frac{2}{n}\left[15n - \frac{4}{n^2}\frac{n(n+1)(2n+1)}{6} - \frac{4}{n}\frac{n(n+1)}{2}\right]$$

$$= 30 - \frac{8}{6n^2}(n+1)(2n+1) - \frac{4}{n}(n+1)$$

$$Area = \lim_{n\to\infty} s(n) = 30 - \frac{8}{3} - 4 = \frac{70}{3} = 23\frac{1}{3}$$

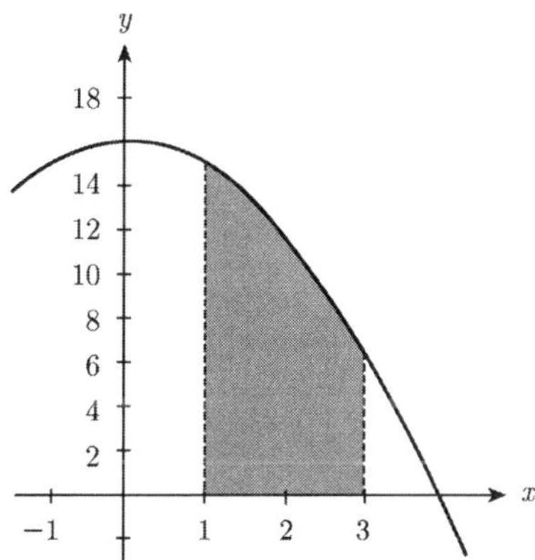

EX 1 Find the area under the curve $y = -2x + 3$ from $x = 0$ to $x = 1$ by summing the areas of the rectangles.

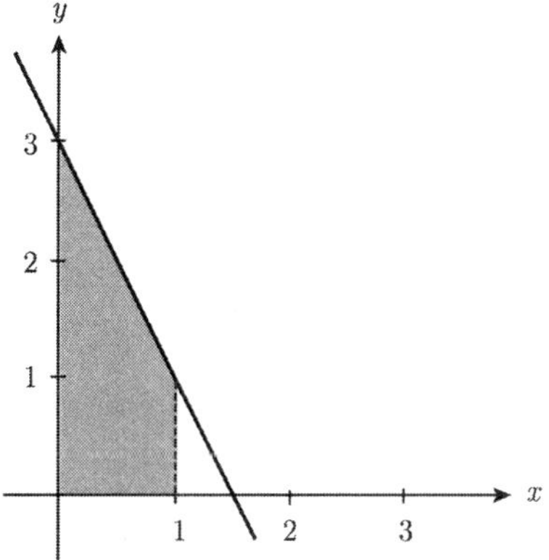

EX 2 Find the area under the curve $y = x^2 + 1$ from $x = 0$ to $x = 3$ by summing the areas of the rectangles.

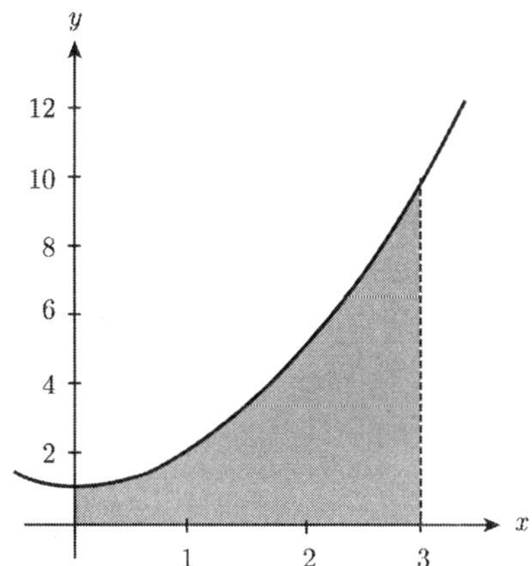

EX 3 Find the area under the curve $y = x^3$ from $x = 0$ to $x = 2$ by summing the areas of the rectangles.

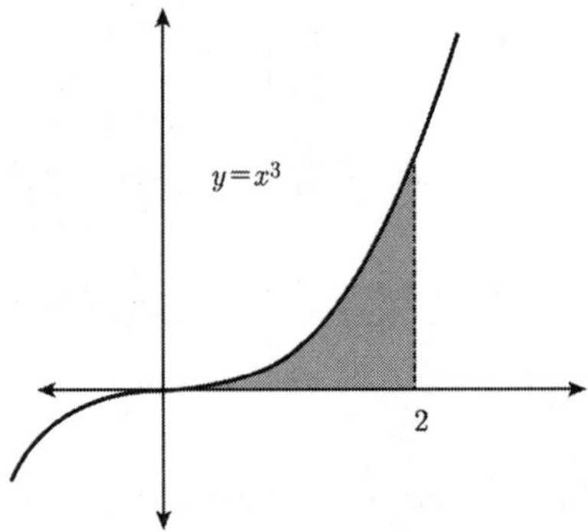

EX 4 Find the area under the curve $y = x^2$ from $x = 1$ to $x = 3$ by summing the areas of the rectangles.

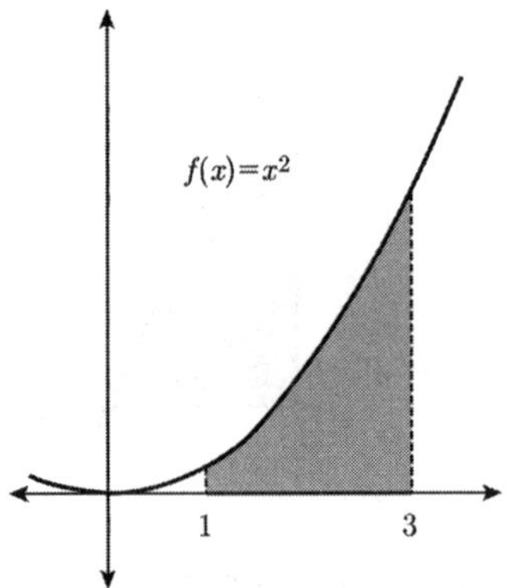

EX 5 Find the area under the curve $y = 9 - x^2$ from $x = 2$ to $x = 3$ by summing the areas of the rectangles.

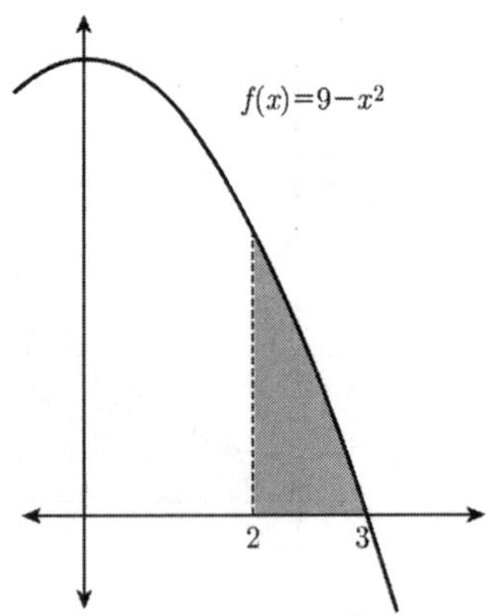

5.7 Definite Integrals

Let f be continuous on the interval $[a, b]$. The definite integral of f from a to b is

$$\int_a^b f(x)dx = \lim_{n\to\infty}\sum_{k=1}^{n} f(x_k^*)\triangle x$$

$f(x)$ is called the integrand and a and b are called the lower limit and the upper limit, respectively.

• Rules satisfied by definite integrals

1. Order of Integration: $\int_b^a f(x)dx = -\int_a^b f(x)dx$

2. Zero Width Interval: $\int_a^a f(x)dx = 0$

3. Constant Multiple: $\int_a^b kf(x)dx = k\int_a^b f(x)dx$

4. Sum and Difference: $\int_a^b (f(x)\pm g(x))dx = \int_a^b f(x)dx \pm \int_a^b g(x)dx$

5. Additivity: $\int_a^b f(x)dx + \int_b^c f(x)dx = \int_a^c f(x)dx$

6. Max-Min Inequality: If f has maximum value max f and minimum value min f on [a, b], then

$$\min f \cdot (b-a) \le \int_a^b f(x)dx \le \max f \cdot (b-a).$$

7. Domination: $f(x) \ge g(x)\, on\, [a,b] \Rightarrow \int_a^b f(x)dx \ge \int_a^b g(x)dx$

$$f(x) \ge 0\; on\, [a,b] \Rightarrow \int_a^b f(x)dx \ge 0$$

EX 1 Suppose that f and h are integrable and that

$$\int_1^9 f(x)dx = -1,\ \int_7^9 f(x)dx = 5,\ \int_7^9 h(x)dx = +4.$$

find each of the following.

(1) $\int_1^9 -2f(x)dx$

(2) $\int_7^9 [f(x)+h(x)]dx$

(3) $\int_7^9 [2f(x)-3h(x)]dx$

(4) $\int_9^1 f(x)dx$

(5) $\int_1^7 f(x)dx$

(6) $\int_9^7 [h(x)-f(x)]dx$

EX 2 Given that

$$\int_{-1}^2 f(x)dx = 2,\ \int_2^4 f(z)dz = 3,\ \int_{-1}^2 g(s)ds = -1, \text{ and } \int_4^2 g(x)dx = 4,$$

find each of the following.

(1) $\int_{-1}^2 f(t)dt$

(2) $\int_4^2 f(x)dx$

(3) $\int_4^{-1} f(y)dy$

(4) $\int_3^3 [3f(x)+2g(x)]dx$

(5) $\int_{-1}^2 [3f(x)+2g(x)]dx$

(6) $\int_{-1}^4 [2f(x)-3g(x)]dx$

EX 3 Given that

$$\int_a^b e^x \cos x\, dx = r,\ \int_b^c e^x \cos x\, dx = s, \text{ and } \int_c^a x^2 \ln(x+1)dx = p,$$

find each of the following.

(1) $\int_b^b e^x \cos x\, dx$

(2) $\int_c^a e^y \cos y\, dy$

(3) $\int_a^c t^2 \ln(t+1)dt$

(4) $\int_a^c [2e^x \cos x - 3x^2 \ln(x+1)]dx$

EX 4

(1) Given that $\int_{-2}^4 f(x)dx = 3$, $\int_2^4 f(x)dx = 2$, and $\int_2^5 f(x)dx = 6$, evaluate $\int_{-2}^5 f(x)dx$.

(2) Given that $\int_6^{-3} f(x)dx = 4$, $\int_6^{-1} f(x)dx = -1$, and $\int_3^{-3} f(x)dx = -2$, evaluate $\int_{-1}^3 f(x)dx$.

(3) Given that $\int_3^{-2} f(x)dx = 2$, $\int_5^{-2} f(x)dx = -1$, and $\int_3^{-3} f(x)dx = -4$, evaluate $\int_0^2 f(x+3)dx$.

(4) Given that $\int_1^2 f(x)dx = 3$, $\int_0^4 f(x)dx = 7$, and $\int_4^1 f(x)dx = -5$, evaluate $\int_1^2 f(2x)dx$.

5.8 Fundamental Theorem of Calculus, Part I

If $f(x)$ is continuous on the closed interval $[a,b]$ and $F(x)$ is an antiderivative of $f(x)$, then

$$\int_a^b f(x)dx = F(b) - F(a).$$

Proof Let $A(x)$ be the area under the curve $y = f(x)$ from a to x.

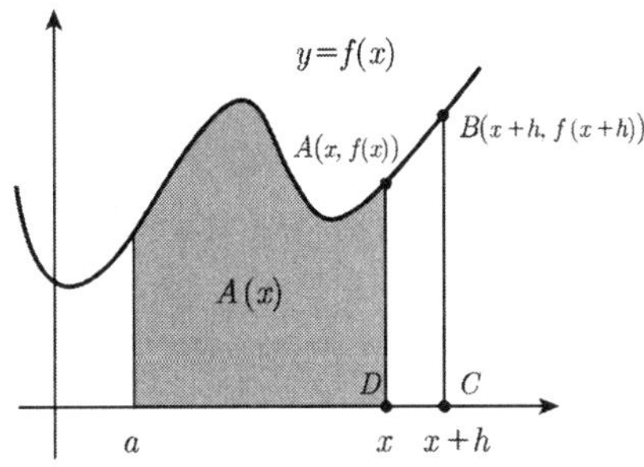

$$A(x) = \int_a^x f(x)dx \Rightarrow A(a) = 0 \text{ and } A(b) = \int_a^b f(x)dx \quad \cdots\cdots ①$$

$$A'(x) = \lim_{h\to 0}\frac{A(x+h) - A(x)}{h}$$

$$= \lim_{h\to 0}\frac{\Delta A}{h} = \lim_{h\to 0}\frac{Area\ of\ \square ABCD}{h}$$

$$= \lim_{h\to 0}\frac{\frac{1}{2}[f(x)+f(x+h)]h}{h} = \lim_{h\to 0}\frac{1}{2}[f(x)+f(x+h)]$$

$$= \frac{1}{2}[f(x)+f(x)] = f(x)$$

$$\therefore A'(x) = f(x)$$

$$A(x) = F(x) + C$$

When $x = a,\ A(a) = F(a) + C = 0$

$$\therefore C = -F(a)$$

$$\therefore A(x) = F(x) - F(a)$$

When $x = b,\ A(b) = F(b) - F(a)$ ...... ②

By ① and ②

$$\int_a^b f(x)dx = F(b) - F(a)$$

Example 1 Evaluate the following definite integrals.

(1) $\int_{-1}^{2}(3x^2 - 2x)dx$

(2) $\int_{1}^{2}\frac{x^2 + x - 2}{2x^2}dx$

Solution

(1) $\int_{-1}^{2}(3x^2 - 2x)dx$

$= \left[x^3 - x^2\right]_{-1}^{2}$

$= (8-4) - (-1-1)$

$= 6$

(2) $\int_{1}^{2}\frac{x^2 + x - 2}{2x^2}dx$

$= \frac{1}{2}\int_{1}^{2}\left(1 + \frac{1}{x} - \frac{2}{x^2}\right)dx$

$= \frac{1}{2}\left[x + \ln x + \frac{2}{x}\right]_1^2$

$= \frac{1}{2}\ln 2$

EX 1 Evaluate the following definite integrals.

(1) $\int_{-1}^{1}(x^2 - x - 1)dx$

(2) $\int_{0}^{2}(4 + 8x - 6x^2)dx$

(3) $\int_{4}^{9}\frac{2+x}{2\sqrt{x}}dx$

(4) $\int_{-1}^{8}\frac{1}{\sqrt[3]{x}}dx$

(5) $\int_{1}^{2}\frac{3x-1}{3x}dx$

(6) $\int_{2}^{1}\left(3x + \frac{2}{x}\right)^2 dx$

(7) $\int_{-\frac{\pi}{4}}^{\frac{\pi}{4}}\cos 2x\, dx$

(8) $\int_{0}^{\frac{\pi}{3}}[\tan\theta - \sec\theta\tan\theta]\, d\theta$

EX 2 Evaluate the following definite integrals.

(1) $\int_0^{\ln 2}(e^{2x}+e^{-x})dx$ (2) $\int_0^1 \frac{4x}{x^2+1}dx$

(3) $\int_{-\frac{1}{2}}^{1}\frac{4\,dx}{\sqrt{1-x^2}}$ (4) $\int_{-1}^{\sqrt{3}}\frac{3\,dx}{x^2+1}$

(5) $\int_1^{\sqrt{3}}\frac{dx}{\sqrt{4-x^2}}$ (6) $\int_0^{\sqrt{3}}\frac{dx}{x^2+9}$

(7) $\int_0^{\frac{\pi}{4}}\sin^2 x\,dx$ (8) $\int_{\frac{\pi}{12}}^{\frac{\pi}{6}}\cos^2 2\theta\,d\theta$

EX 3 Use a graphing calculator to evaluate the following definite integrals.

(1) $\int_{-2}^{2}(x^2+2x)dx$ (2) $\int_1^4\sqrt{x^2-1}\,dx$

(3) $\int_0^3\frac{10x}{x^2+2}dx$ (4) $\int_{-0.5}^{5}6e^{-x^2}dx$

5.9 Fundamental Theorem of Calculus, Part II

If f is continuous on $[a, b]$, then $F(x)=\int_a^x f(t)dt$ is continuous on $[a, b]$ and differentiable on (a, b) and its derivative is $f(x)$:

$$F'(x)=\frac{d}{dx}\int_a^x f(t)dt=f(x)$$

Proof Let $\int f(x)\,dx=F(x)+C$

$$\begin{aligned}\frac{d}{dx}\int_a^x f(t)dt &= \frac{d}{dx}[F(t)]_a^x\\ &= \frac{d}{dx}[F(x)-F(a)] \quad (\textit{fundamental Theorem of Calculus, Part I})\\ &= F'(x)-0 \quad (\because F(a)\ \textit{is a constant})\\ &= f(x)\end{aligned}$$

Example 1 Find dy/dx if

(1) $y=\int_a^x (t^3+1)dt$ (2) $y=\int_x^5 3t\sin t\,dt$

Solution

(1) $\dfrac{dy}{dx}=\dfrac{d}{dx}\int_a^x (t^3+1)dt = x^3+1$

(2) $\dfrac{dy}{dx}=\dfrac{d}{dx}\int_x^5 3t\sin t\,dt=\dfrac{d}{dx}\left(-\int_5^x 3t\sin t\,dt\right)$

$=-\dfrac{d}{dx}\int_5^x 3t\sin t\,dt$

$=-3x\sin x$

EX 1 Evaluate.

(1) $\dfrac{d}{dx}\int_0^x e^{-t}\sec^2 t\,dt$ (2) $\dfrac{d}{dx}\int_2^x \sqrt{1+t^2}\,dt$

(3) $\dfrac{d}{dx}\int_3^x \left(y^4+\dfrac{3}{\sqrt{1-y^2}}\right)dy$ (4) $\dfrac{d}{dx}\int_1^x (u-3)\ln|u+1|\,du$

EX 2 Differentiate the following integrals with respect to x.

(1) $\int_2^x \sqrt{2+\cos^3 t}\,dt$ (2) $\int_e^x \dfrac{\ln(u)}{u}du$

(3) $\int_\pi^x \dfrac{6}{3+t^4}dt$ (4) $\int_0^x u\cos u\,du$

- $\dfrac{d}{dx}\int_{p(x)}^{q(x)} f(t)\,dt = f(q(x))q'(x)-f(p(x))p'(x)$

Proof Let $\int f(x)dx = F(x)+C$

$\dfrac{d}{dx}\int_{p(x)}^{q(x)} f(t)dt=\dfrac{d}{dx}\left[F(t)\right]_{p(x)}^{q(x)}$

$=\dfrac{d}{dx}\left[F(q(x))-F(p(x))\right]$

$=F'(q(x))q'(x)-F'(p(x))p'(x)=f(q(x))q'(x)-f(p(x))p'(x)$

Example 2 Given $y=\int_{2x}^{x^2}\frac{t^3}{t^2+1}dt$, find $\frac{dy}{dx}$.

Solution

$$\frac{dy}{dx}=\frac{d}{dx}\int_{2x}^{x^2}\frac{t^3}{t^2+1}dt$$

$$=\frac{(x^2)^3}{(x^2)^2+1}\cdot 2x-\frac{(2x)^3}{(2x)^2+1}\cdot 2=\frac{2x^7}{x^4+1}-\frac{16x^3}{4x^2+1}$$

EX 3 Differentiate the following integrals with respect to x.

(1) $\int_{\tan x}^{0}\frac{dt}{1+t^2}$

(2) $\int_{2x}^{x^2}f(t)dt$

(3) $\int_{0}^{\sqrt{x}}\cos t\,dt$

(4) $\int_{-x}^{x^2}\sqrt{t^2+1}\,dt$

(5) $\int_{\cos x}^{\sin x}\frac{1}{1-t^2}dt$

(6) $\int_{\frac{1}{x}}^{3x}\cos 2t\,dt$

EX 4 Find $\frac{dy}{dx}$.

(1) $y=\int_{\sqrt{x}}^{x^2}\frac{e^t}{t}dt$

(2) $y=\int_{x+1}^{\ln x}\ln t\,dt$

(3) $y-\int_{3x}^{\tan x}\frac{1}{1-t^2}dt$

(4) $y=\int_{\cos^2 x}^{\sin^2 2x}\cos^2 t\,dt$

EX 5

(1) Let $\int_{2}^{x}f(t)\,dt=x^2\sin 3x$. Find $f(\pi)$.

(2) Let $A(x)=\int_{4}^{x^2}\frac{1}{1+t^2}dt$. Find $A(2)$ and $A'(2)$.

EX 6 Find the intervals on which the function $f(x)=\int_{1}^{x}\frac{1}{t^4+1}dt$ is concave up.

5.10 Area between a curve and the x-axis

The definite integral can be used to find the area enclosed by a curve and the x-axis in the interval $[a, b]$.

- **If a curve is above the x-axis,**

$$Area = \int_a^b f(x)dx, \text{ where } a < b$$

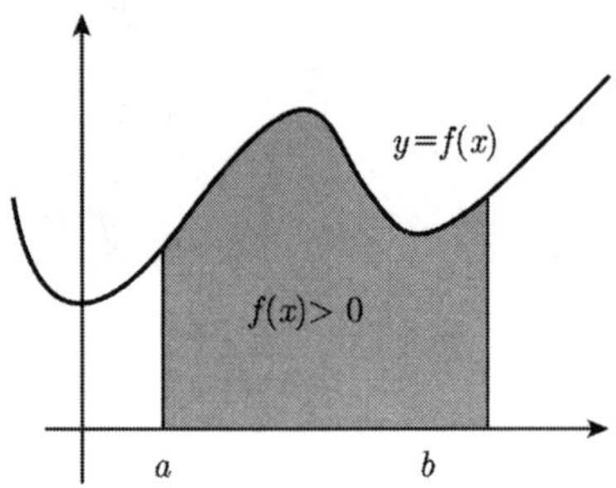

- **If a curve is below the x-axis,**

$$Area = -\int_a^b f(x)dx, \text{ where } a < b$$

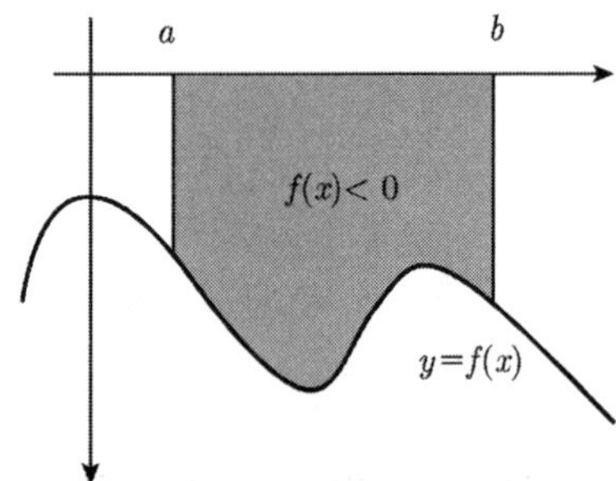

Example 1 Find the area enclosed by the curve $y = x^2 - 1$ and the x-axis in the interval [0, 2].

Solution

$$Area = -\int_0^1 (x^2-1)dx + \int_1^2 (x^2-1)dx$$
$$= -\left[\frac{1}{3}x^3 - x\right]_0^1 + \left[\frac{1}{3}x^3 - x\right]_1^2$$
$$= -\left[\left(\frac{1}{3}-1\right)-0\right] + \left[\left(\frac{8}{3}-2\right)-\left(\frac{1}{3}-1\right)\right]$$
$$= -\left(-\frac{2}{3}\right) + \left[\frac{2}{3} - \left(-\frac{2}{3}\right)\right] = \frac{2}{3} + \frac{4}{3} = 2$$

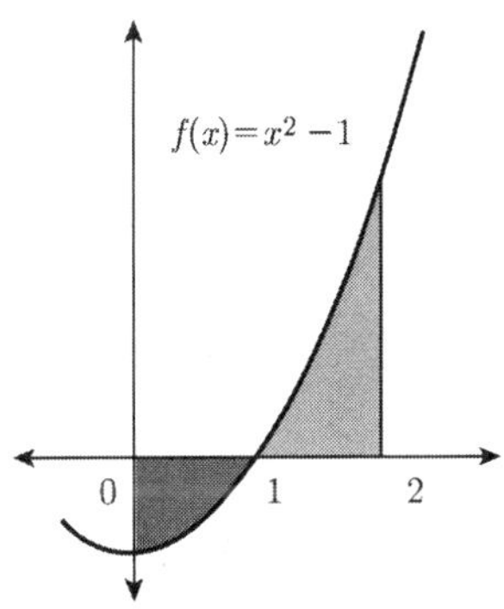

EX 1 Sketch the curve and find the area enclosed by the x-axis in the give interval.

(1) $y = x^2,\ [-1, 2]$

(2) $y = \cos x,\ \left[\frac{\pi}{2}, \pi\right]$

(3) $y = \cot x,\ \left[\frac{\pi}{4}, \frac{\pi}{2}\right]$

(4) $y = 4 - x^2,\ [1, 3]$

(5) $y = \tan x,\ \left[-\frac{\pi}{4}, \frac{\pi}{3}\right]$

(6) $y = \frac{4}{x^2+4},\ [-2, 2]$

(7) $y = 2\sin x,\ \left[\frac{\pi}{3}, \frac{3\pi}{2}\right]$

(8) $y = x^3 - 2x^2 - 3x,\ x-axis$

EX 2 Given $f(x) = \begin{cases} 2x+1, & x \le 1 \\ x^2+2, & x > 1 \end{cases}$, find the value of $\int_{-1}^{3} f(x)dx$.

EX 3 Evaluate each of the following.

(1) $\int_0^3 |x^2 - 4|dx$

(2) $\int_{-1}^3 |x^2 - 4x|dx$

(3) $\int_0^{2\pi} |\cos x|dx$

(4) $\int_0^2 |e^x - 2|dx$

EX 4 The graph of f is shown below. The areas enclosed by the curve and the x-axis in the intervals $[a, b]$, $[b, c]$ and $[c, d]$ are 10, 10, and 5, respectively. Evaluate the following definite integrals.

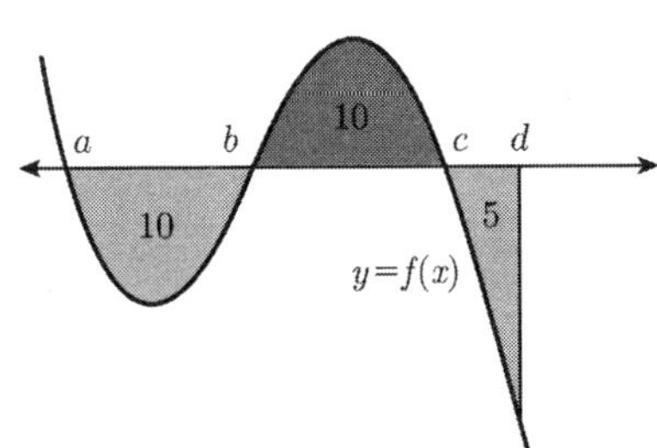

(1) $\int_a^b f(x)dx$

(2) $\int_b^c f(x)dx$

(3) $\int_a^d f(x)dx$

(4) $\int_a^d |f(x)|dx$

(5) $\int_b^a f(x)dx$

(6) $\int_d^b f(x)dx$

EX 5 Sketch a graph to evaluate the following definite integrals.

(1) $\int_0^2 \sqrt{4-x^2}\,dx$

(2) $\int_{-2}^2 \left(-\sqrt{4-x^2}\right)dx$

(3) $\int_{-3}^3 \left(5-\sqrt{9-x^2}\right)dx$

(4) $\int_0^3 |x-2|\,dx$

(5) $\int_{-2}^4 (|x+1|-1)dx$

(6) $\int_0^2 (|x-1|+3x^2)dx$

EX 6 The graph of f is shown below. Evaluate each of the following.

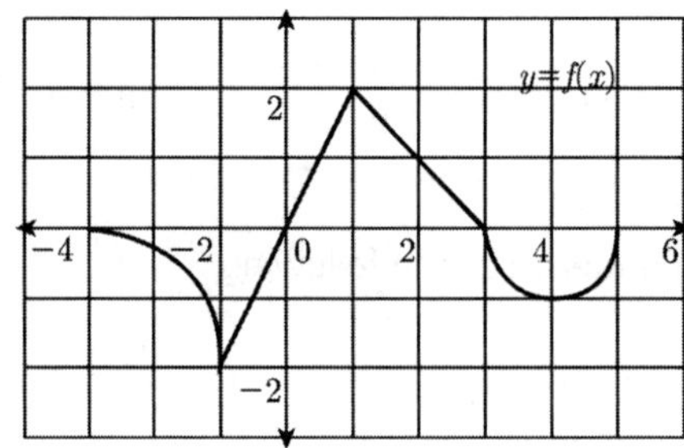

(1) $\int_{-1}^1 f(x)dx$

(2) $\int_1^5 f(x)dx$

(3) $\int_5^0 f(x)dx$

(4) $\int_{-3}^0 f(x)dx$

(5) $\int_1^5 |f(x)|dx$

(6) $\int_0^8 f(x-3)dx$

(7) $\int_0^1 [3f(x)+2x]dx$

(8) $\int_1^2 f(2x)dx$

EX 7 The graph of f is shown below. Let g be the function defined by $g(x)=\int_1^x f(t)dt$.

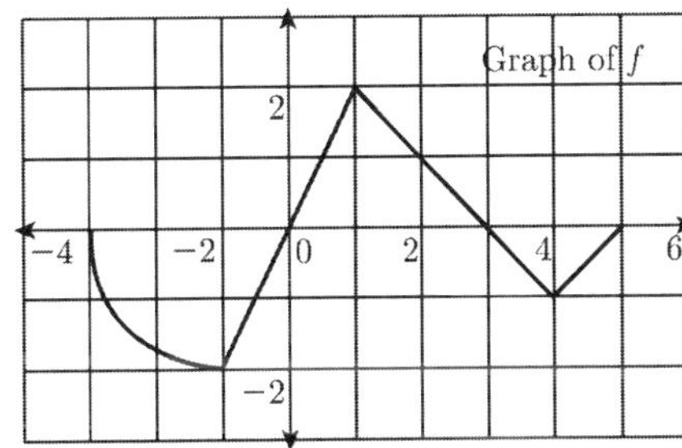

(1) Find $g(3), g(-3), g'(3)$ and $g''(3)$.

(2) Find the average rate of change of g over the interval $-3 \le x \le 3$.

(3) Find the average rate of change of g' over the interval $-1 \le x \le 3$.

(4) Find the equation of the line tangent to the graph of g at $x = -1$.

(5) Use local linearization to approximate the value of $g(-1.1)$.

(6) Find all intervals on which the graph of g is decreasing. Justify your answer.

(7) Find all intervals on which the graph of g is concave up. Justify your answer.

(8) Find values of x at which the graph of g has a relative maximum. Explain.

(9) Find all values of x at which the graph of g has a point of inflection. Explain.

(10) Find the absolute minimum value of g on the closed interval $[-3, 5]$. Show an analysis to support your answer.

5.11 Substitution for Definite Integrals

If g' is continuous on the interval $[a, b]$ and f is continuous on the range of $g(x) = u$, then

$$\int_a^b f(g(x)) \cdot g'(x)dx = \int_{g(a)}^{g(b)} f(u)du.$$

Example 1 Evaluate the following definite integral.

$$\int_0^2 x(x^2-1)^2 dx$$

Solution

Let $u = x^2 - 1$

$$\frac{du}{dx} = 2x$$

$$\frac{1}{2}du = x\,dx$$

If $x = 0,\ u = -1$

If $x = 2,\ u = 3$

$$\int_0^2 x(x^2-1)^2 dx = \int_{-1}^3 u^2 \frac{1}{2} du$$
$$= \frac{1}{6}\left[u^3\right]_{-1}^3$$
$$= \frac{1}{6}(27-(-1)) = \frac{28}{6} = \frac{14}{3}$$

EX1 Evaluate the following definite integrals.

(1) $\displaystyle\int_0^{\sqrt{3}} \frac{4x}{\sqrt{x^2+1}} dx$

(2) $\displaystyle\int_0^1 t^3(1+t^4)^3 dt$

(3) $\displaystyle\int_1^{e^{\pi/4}} \frac{4\,dt}{t(1+\ln^2 t)}$

(4) $\displaystyle\int_{-1}^1 \frac{5r}{(4+r^2)^2} dr$

(5) $\displaystyle\int_0^{\sqrt[3]{\pi^2}} \sqrt{\theta}\cos^2(\theta^{3/2}) d\theta$

(6) $\displaystyle\int_2^4 \frac{dx}{x(\ln x)^2}$

(7) $\displaystyle\int_0^{\pi} 5(5-4\cos t)^{1/4} \sin t\, dt$

(8) $\displaystyle\int_0^{\pi/4} (1+e^{\tan\theta})\sec^2\theta\, d\theta$

(9) $\displaystyle\int_0^{2\pi} \frac{\cos z}{\sqrt{4+3\sin z}} dz$

(10) $\displaystyle\int_0^{\pi/6} \cos^{-3} 2\theta \sin 2\theta\, d\theta$

(11) $\displaystyle\int_{-\pi/2}^{\pi/2} \frac{2\cos\theta\, d\theta}{1+(\sin\theta)^2}$

(12) $\displaystyle\int_{\sqrt{2}}^2 \frac{\sec^2(\sec^{-1}x)dx}{x\sqrt{x^2-1}}$

(13) $\displaystyle\int_2^{16} \frac{dx}{2x\sqrt{\ln x}}$

(14) $\displaystyle\int_0^{\pi/4} \tan x \sec^2 x\, dx$

EX2 Let $g(x)$ be a continuous function on the interval $[-4, 4]$ and $f'(x) = g(x)$. A table of values of differentiable function f is shown below. Evaluate the following definite integrals.

x	-4	-3	-2	-1	0	1	2	3	4
f	-3	1	-1	0	-2	5	4	6	2

(1) $\int_{-1}^{1}[g(x)+3x^2]dx$

(2) $\int_{-2}^{2}g(2x)dx$

(3) $\int_{-2}^{2}3g\left(\frac{x}{2}\right)dx$

(4) $\int_{-2}^{0}xg(x^2)dx$

5.12 Integration by Parts for Definite Integrals

$$\int_a^b uv'\,dx = [uv]_a^b - \int_a^b u'v\,dx$$

Example 1 Find $\int_0^4 xe^{-x}dx$.

Solution

$$\begin{aligned}\int_0^4 xe^{-x}dx &= [-xe^{-x}]_0^4 - \int_0^4(-e^{-x})dx \\ &= [-4e^{-4}-0]+\int_0^4 e^{-x}dx \\ &= -4e^{-4}-[e^{-x}]_0^4 \\ &= -4e^{-4}-e^{-4}+1 \\ &= 1-5e^{-4}\end{aligned}$$

EX 1 Evaluate each of the following definite integrals.

(1) $\int_{-1}^{1}xe^x dx$

(2) $\int_{\frac{\pi}{3}}^{\frac{\pi}{2}}x\sin x\,dx$

(3) $\int_0^{\frac{\pi}{2}}x^2\sin 2x\,dx$

(4) $\int_1^e x\ln x\,dx$

(5) $\int_0^{e-1}\ln(x+1)dx$

(6) $\int_0^1\tan^{-1}x\,dx$

(7) $\int_{\frac{1}{2}}^{1}\sin^{-1}x\,dx$

(8) $\int_0^4 e^{\sqrt{x}}dx$

(9) $\int_0^{\pi}e^x\sin x\,dx$

EX 2 Let f be the function defined for $x > 0$, with $f(e)= 2$ and f', the first derivative of f, given by $f'(x)= x^2\ln x$.

(1) Write an equation for the line tangent to the graph of f at the point $(e, 2)$.

(2) Is the graph of f concave up of concave down on the interval $1 < x < 3$? Give a reason for your answer.

(3) Use antidifferentiation to find $f(x)$.

EX 3 The derivative of a function f is given by $f'(x)= (x-3)e^x$ for $x > 0$, and $f(1)= 7$.

(1) The function f has a critical point at $x = 3$. At this point, does f have a relative minimum, a relative maximum, or neither? Justify your answer.

(2) On what intervals, if any, is the graph of f both decreasing and concave up? Explain your reasoning.

(3) Find the value of $f(3)$.

5.13 Even and Odd Functions

Let f be continuous on the symmetric interval $[-a, a]$.

(a) If f is even, then $\int_{-a}^{a} f(x)dx = 2\int_{0}^{a} f(x)dx$.

$(e.g.)\ y = x^4,\ y = x^6,\ y = \cos x\ \ etc$

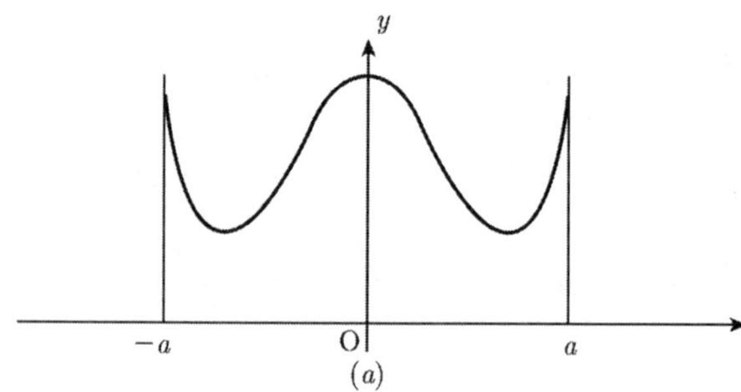

(a)

(b) If f is odd, then $\int_{-a}^{a} f(x)dx = 0$.

$(e.g.)\ y = x,\ y = x^3,\ y = \sin x,\ y = \tan x\ \ etc$

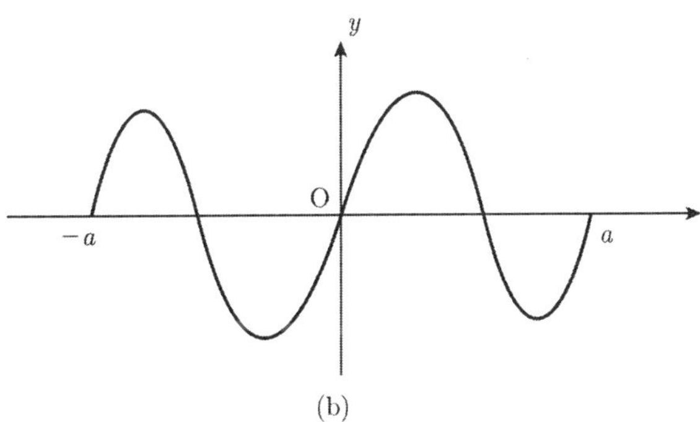

(b)

Example 1 Find $\int_{-2}^{2}(x^4-4x^2+6)dx$.

Solution

$$\int_{-2}^{2}(x^4-4x^2+6)dx = 2\int_{0}^{2}(x^4-4x^2+6)dx$$

$$= 2\left[\frac{x^5}{5}-\frac{4}{3}x^3+6x\right]_0^2$$

$$= 2\left(\frac{32}{5}-\frac{32}{3}+12\right)=\frac{232}{15}$$

EX 1 Evaluate the following definite integrals.

(1) $\int_{-1}^{1}(x^4+2x^3+3x^2+4x+5)dx$

(2) $\int_{-3}^{3}(x^5-3x^3+x-2\sin x)dx$

(3) $\int_{-2}^{2}(x^5-3x^3+3x^2+\tan x+2)dx$

(4) $\int_{-\pi}^{\pi}(x^4\sin x+x^3\cos x+3x^2\tan x+x\cos x+1)dx$

5.14 Integration by Partial Fraction

The method of partial fractions makes it possible to express a rational function $\dfrac{f(x)}{g(x)}$ as a sum of simpler fractions. Here $f(x)$ and $g(x)$ are real polynomials in x and it is assumed that $\dfrac{f(x)}{g(x)}$ is a proper fraction; that is of lower degree than $g(x)$. If not, we divide $f(x)$ by $g(x)$ to express the given rational function as the sum of a polynomial and a proper rational function. Thus,

$$\frac{x^3-x^2-2}{x(x-1)}=x-\frac{2}{x(x-1)},$$

where the fraction on the right is proper.

Theoretically, every real polynomial can be expressed as a product of (powers of) real linear factors and (powers of) real quadratic factors.

In the following, the capital letters denote constants to be determined. We consider only nonrepeating linear factors. For each distinct linear factor $(x-a)$ of $g(x)$ we set up one partial fraction of the type $\dfrac{A}{x-a}$. The techniques for determining the unknown constants are illustrated in the following examples.

Example 1 Find $\displaystyle\int \frac{x-1}{x(x-2)}dx.$

Solution

Use partial fractions; find A and B such that

$$\frac{x-1}{x(x-2)}=\frac{A}{x}+\frac{B}{x-2}.$$

Then $x-1=A(x-2)+Bx.$

Set $x=0:-1=-2A$ and $A=\dfrac{1}{2}.$

Set $x=2:1=2B$ and $B=\dfrac{1}{2}.$

So the given integral equals

$$\begin{aligned}\int\left(\frac{1}{2x}+\frac{1}{2(x-2)}\right)dx &= \frac{1}{2}\ln|x|+\frac{1}{2}\ln|x-2|+C\\ &=\frac{1}{2}\ln|x(x-2)|+C.\end{aligned}$$

EX 1 Evaluate the following indefinite integrals.

(1) $\displaystyle\int_{\frac{1}{2}}^{1} \frac{y+4}{y^2+y}dy$

(2) $\displaystyle\int \frac{dx}{x^2+2x}$

(3) $\displaystyle\int \frac{y-1}{y+1}dy$

(4) $\displaystyle\int \frac{x+4}{x^2+5x-6}dx$

(5) $\displaystyle\int \frac{dt}{t^3+t^2-2t}$

(6) $\displaystyle\int_{0}^{1} \frac{dx}{(x+1)(x^2+1)}$

(7) $\displaystyle\int \frac{4x}{(x-1)^2(x+1)}dx$

(8) $\displaystyle\int \frac{2x^2-x+4}{x(x^2+4)}dx$

Multiple Choice

01 $\int_0^1 \sqrt{x}(x+1)dx =$

(A) 0
(B) 1
(C) $\frac{16}{15}$
(D) $\frac{7}{5}$
(E) 2

02 $\int (3x+1)^5 dx =$

(A) $\frac{(3x+1)^6}{18}+C$
(B) $\frac{(3x+1)^6}{6}+C$
(C) $\frac{(3x+1)^6}{2}+C$
(D) $\frac{\left(\frac{3x^2}{2}+x\right)^6}{2}+C$
(E) $\left(\frac{3x^2}{2}+x\right)^5+C$

03 If the substitution $u=\frac{x}{2}$ is made, the integral $\int_2^4 \frac{1-\left(\frac{x}{2}\right)^2}{x}dx =$

(A) $\int_1^2 \frac{1-u^2}{u}du$
(B) $\int_2^4 \frac{1-u^2}{u}du$
(C) $\int_1^2 \frac{1-u^2}{2u}du$
(D) $\int_1^2 \frac{1-u^2}{4u}du$
(E) $\int_2^4 \frac{1-u^2}{2u}du$

04 If $F(x)=\int_1^{x^2}\sqrt{1+t^3}\,dt$, then $F'(x)=$

(A) $2x\sqrt{1+x^6}$
(B) $2x\sqrt{1+x^3}$
(C) $\sqrt{1+x^6}$
(D) $\sqrt{1+x^3}$
(E) $\int_1^{x^2}\frac{3t^2}{2\sqrt{1+t^3}}dt$

05 Which of the following is equal to $\int \frac{1}{\sqrt{25-x^2}}dx$?

(A) $\arcsin\frac{x}{5}+C$
(B) $\arcsin x+C$
(C) $\frac{1}{5}\arcsin\frac{x}{5}+C$
(D) $\sqrt{25-x^2}+C$
(E) $2\sqrt{25-x^2}+C$

06 If $\int_1^4 f(x)dx = 6$, what is the value of $\int_1^4 f(5-x)dx$?

(A) 6 (B) 3 (C) 0

(D) −1 (E) −6

07 $\int_0^1 x^3 e^{x^4} dx =$

(A) $\frac{1}{4}(e-1)$ (B) $\frac{1}{4}e$ (C) $e-1$

(D) e (E) $4(e-1)$

08 Which of the following is equal to $\int_0^\pi \sin x\, dx$?

(A) $\int_{-\frac{\pi}{2}}^{\frac{\pi}{2}} \cos x\, dx$ (B) $\int_0^\pi \cos x\, dx$ (C) $\int_{-\pi}^0 \sin x\, dx$

(D) $\int_{-\frac{\pi}{2}}^{\frac{\pi}{2}} \sin x\, dx$ (E) $\int_\pi^{2\pi} \sin x\, dx$

09 If $f(x) = \begin{cases} x & \text{for } x \le 1 \\ \frac{1}{x} & \text{for } x > 1, \end{cases}$ then $\int_0^e f(x)dx=$

(A) 0 (B) $\frac{3}{2}$ (C) 2

(D) e (E) $e+\frac{1}{2}$

10 If f is continuous on the interval $[a, b]$, then there exists c such that $a < c < b$ and $\int_a^b f(x)dx=$

(A) $\frac{f(c)}{b-a}$ (B) $\frac{f(b)-f(a)}{b-a}$ (C) $f(b)-f(a)$

(D) $f'(c)(b-a)$ (E) $f(c)(b-a)$

11 $\int x^2 \sin x\, dx =$

(A) $-x^2\cos x - 2x\sin x - 2\cos x + C$ (B) $-x^2\cos x + 2x\sin x - 2\cos x + C$

(C) $-x^2\cos x + 2x\sin x + 2\cos x + C$ (D) $-\dfrac{x^3}{3}\cos x + C$

(E) $2x\cos x + C$

12 If $0 \le x \le 4$, of the following, which is the greatest value of x such that $\int_0^x (t^2 - 2t)dt \ge \int_2^x t\,dt$?

(A) 1.35 (B) 1.38 (C) 1.41

(D) 1.48 (E) 1.59

13 $\int_1^e \left(\dfrac{x^2-1}{x}\right)dx =$

(A) $e - \dfrac{1}{e}$ (B) $e^2 - e$ (C) $\dfrac{e^2}{2} - e + \dfrac{1}{2}$

(D) $e^2 - 2$ (E) $\dfrac{e^2}{2} - \dfrac{3}{2}$

14 If f is a linear function and $0 < a < b$, then $\int_a^b f''(x)\,dx =$

(A) 0 (B) 1 (C) $\dfrac{ab}{2}$

(D) $b - a$ (E) $\dfrac{b^2 - a^2}{2}$

15 $\lim_{x \to 1} \dfrac{\int_1^x e^{t^2} dt}{x^2 - 1}$ is

(A) 0 (B) 1 (C) $\dfrac{e}{2}$

(D) e (E) nonexistent

x	2	5	7	8
$f(x)$	10	30	40	20

16 The function f is continuous on the closed interval $[2, 8]$ and has values that are given in the table above. Using the subintervals $[2, 5]$, $[5, 7]$, and $[7, 8]$, what is the trapezoidal approximation of $\int_2^8 f(x)dx$?

(A) 110 (B) 130 (C) 160

(D) 190 (E) 210

17 $\int_0^3 (x+1)^{1/2}dx =$

(A) $\frac{21}{2}$ (B) 7 (C) $\frac{16}{3}$

(D) $\frac{14}{3}$ (E) $-\frac{1}{4}$

18 $\int x\sqrt{4-x^2}\,dx =$

(A) $\frac{(4-x^2)^{3/2}}{3}+C$ (B) $-(4-x^2)^{3/2}+C$ (C) $\frac{x^2(4-x^2)^{3/2}}{3}+C$

(D) $-\frac{x^2(4-x^2)^{3/2}}{3}+C$ (E) $-\frac{(4-x^2)^{3/2}}{3}+C$

19 $\int_0^{\pi/4} \tan^2 x\,dx =$

(A) $\frac{\pi}{4}-1$ (B) $1-\frac{\pi}{4}$ (C) $\frac{1}{3}$

(D) $\sqrt{2}-1$ (E) $\frac{\pi}{4}+1$

20 $\int_1^2 \frac{x-4}{x^2}dx$

(A) $-\frac{1}{2}$ (B) $\ln 2 - 2$ (C) $\ln 2$

(D) 2 (E) $\ln 2 + 2$

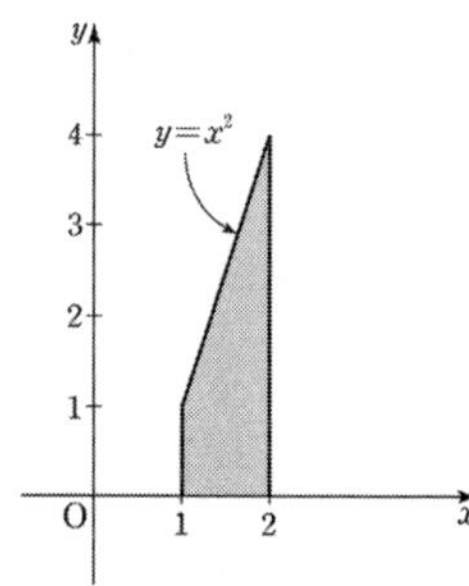

21 Calculate the approximate area of the shaded region in the figure by the trapezoidal rule, using divisions at $x = \frac{4}{3}$ and $x = \frac{5}{3}$.

(A) $\frac{50}{27}$ (B) $\frac{251}{108}$ (C) $\frac{7}{3}$

(D) $\frac{127}{54}$ (E) $\frac{77}{27}$

22 Suppose $g'(x) < 0$ for all $x \geq 0$ and $F(x) = \int_0^x t\,g'(t)dt$ for all $x \geq 0$. Which of the following statements is FALSE?

(A) F takes on negative values.

(B) F is continuous for all $x > 0$.

(C) $F(x) = x\,g(x) - \int_0^x g(t)\,dt$.

(D) $F'(x)$ exists for all $x > 0$.

(E) F is an increasing function.

23 If $\int_1^2 f(x-c)dx = 5$ where c is a constant, then $\int_{1-c}^{2-c} f(x)dx =$

(A) $5 + c$ (B) 5 (C) $5 - c$

(D) $c - 5$ (E) -5

24 If $f(x) = \ln(x + 4 + e^{-3x})$, then $f'(0)$ is

(A) $-\frac{2}{5}$ (B) $\frac{1}{5}$ (C) $\frac{1}{4}$

(D) $\frac{2}{5}$ (E) nonexistent

x	2	5	10	14
$f(x)$	12	28	34	30

25 The function f is continuous on the closed interval [2, 14] and has values as shown in the table above. Using the subintervals [2, 5], [5, 10], and [10, 14], what is the approximation of $\int_{2}^{14} f(x)\,dx$ found by using a right Riemann sum?

(A) 296 (B) 312 (C) 343

(D) 374 (E) 390

26 $\frac{d}{dx}\left(\int_{0}^{x^3} \ln(t^2+1)\,dt\right) =$

(A) $\frac{2x^3}{x^6+1}$ (B) $\frac{3x^2}{x^6+1}$ (C) $\ln(x^6+1)$

(D) $2x^3\ln(x^6+1)$ (E) $3x^2\ln(x^6+1)$

27 $\int_{0}^{1} \frac{x^2}{x^2+1}\,dx =$

(A) $\frac{4-\pi}{4}$ (B) $\ln 2$ (C) 0

(D) $\frac{1}{2}\ln 2$ (E) $\frac{4+\pi}{4}$

28 If $\int_{0}^{b} (4bx - 2x^2)\,dx = 36$, then $b =$

(A) −6 (B) −3 (C) 3

(D) 6 (E) 15

06

Applications of Definite Integrals

6.1 Average Value of a Function

If f is integrable on $[a, b]$, then its average value on $[a, b]$, also called its mean, is

$$\text{av}(f) = \frac{1}{b-a}\int_a^b f(x)dx.$$

If $f(x)$ is continuous on the interval $[a, b]$, then there must be a number c such that the area of the rectangle, with a height of f_{avg} and a base of length $b-a$, is equal to the area under the graph f over the interval $[a, b]$.

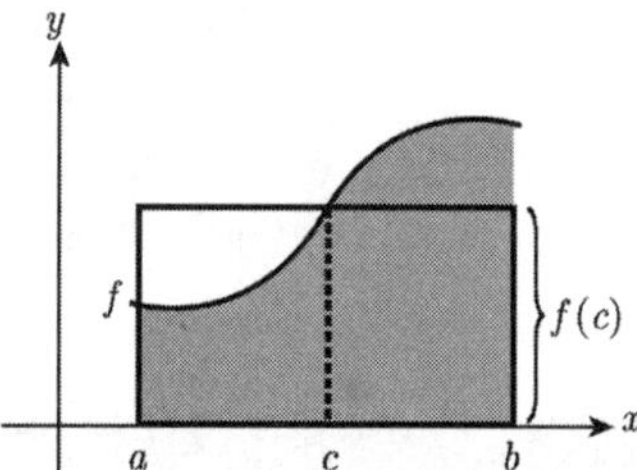

$$(b-a) \cdot f_{avg} = \int_a^b f(x)dx$$

$$\therefore\ f_{avg} = \frac{1}{b-a}\int_a^b f(x)dx$$

EX 1 Determine the average value of each of the following functions on the given interval.

(1) $y = x^2 - 1,\ [0,\ \sqrt{3}\,]$

(2) $y = (x-3)^2,\ [-1, 2]$

(3) $y = |x| - 1,\ [-1, 3]$

(4) $y = \dfrac{20}{x^2+1},\ [-1, 1]$

(5) $y = e^x - 1,\ [0, \ln 3]$

(6) $y = \sin x,\ [0, \pi]$

EX 2 The rate at which of flows out of a tank, in liters per minute, is shown in the table below. Approximate the average flow rate of oil over the time interval $0 \le t \le 18$ by using the trapezoidal rule with 3 subintervals of equal length.

t (min)	$R(t)$ (L/min)
0	8
3	11
6	13
9	16
12	12
15	13
18	8

(1) The velocity of a particle moving along a along a line is $v(t) = 5t - t^2 + 100$ at time t. Find the average velocity of the paticle over the time interval $0 \le t \le 4$.

(2) The velocity of a particle moving along a line is $v(t) = 12t^2 + 1$ at time t.

(a) Find the average velocity of the particle over the time interval $1 \le t \le 3$.

(b) Find the average acceleration of the particle over the time interval $1 \le t \le 3$.

EX 4 The average area, in square inches, of all circles with radii between 2 and 5 in. is

(a) 7π (b) 11π (c) 13π
(d) $\frac{29\pi}{2}$ (e) 17π

EX 5 Suppose the amount of a drug in a patient's bloodstream t hours after intravenous administration is $30/(t+1)^2$ mg. The average amount in the bloodstream during the first 4 hours is

(a) 6.0 mg (b) 11.0 mg (c) 16.6 mg
(d) 24.0 mg (e) none of these

EX 6 Suppose the current world population is 6 billion and the population t years from now is estimated to be $P(t) = 6e^{0.024t}$. On The basis of this supposition, the average population of the world, in billions, over the next 25 years will be approximately.

(a) 6.75 (b) 7.2 (c) 7.8
(d) 8.2 (e) 9.0

 Use a graphing calculator to solve the following.

(1) Suppose that the value of a car in dollars after t years is given by the function $V(t) = 35000e^{-0.223t}$. Determine the average value of the car over its first 10 years.

(2) Suppose that the number of bacteria in a culture is given by the function $P(t) = \dfrac{20000}{1+299e^{-0.2t}}$ where t is measured in hours. Determine the average number of bacteria in the culture during the time interval $10 \le t \le 30$.

6.2 Area Between Curves

To find the area between curves (Figure), we first find where they intersect and then write the area of a typical element for each region between the points of intersection. For the total area bounded by the curves $y = f(x)$ and $y = g(x)$ between $x = a$ and $x = e$, we see that, if they intersect at $[c, d]$, the total area is given exactly by

$$\int_a^c [f(x) - g(x)]\,dx + \int_c^e [g(x) - f(x)]\,dx.$$

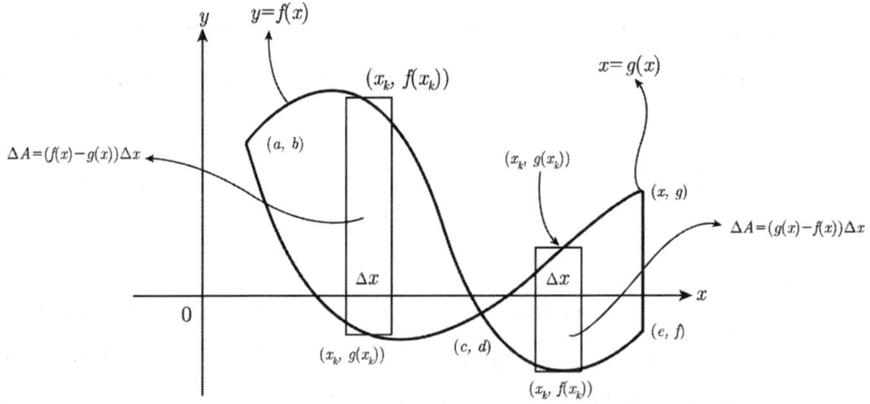

Example 1 Find the area of the region Bounded above by the curve $y = \sin x$ and below by $y = \cos x$ from $x = \dfrac{\pi}{4}$ to $x = \dfrac{5\pi}{4}$.

Solution

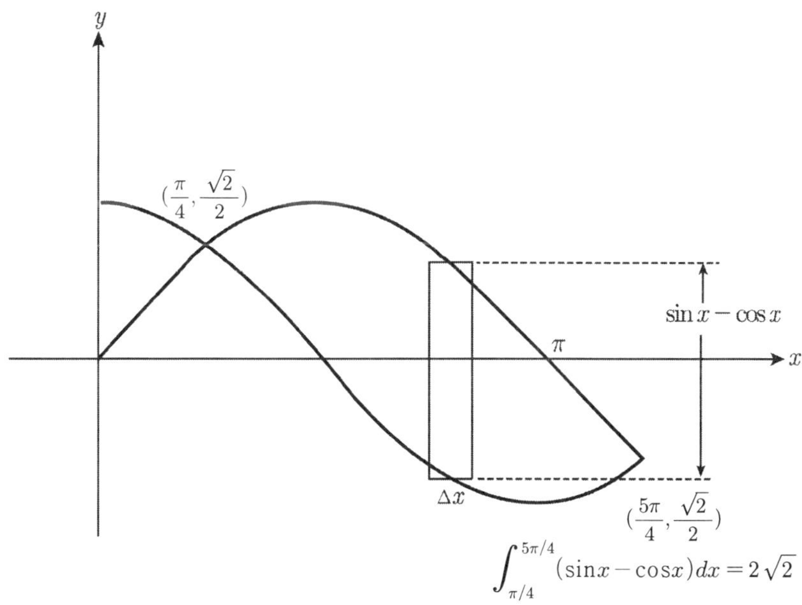

Example 2 Find the area of the region enclosed by $y = -x^2 + 3x$ and $y = x$ on the interval $[0, 3]$.

Solution

Step 1. Sketch the graphs on the same grid and find the $x-$coordinate(s) of the intersection(s).

Step 2. Determine the parts on which the graphs lie above one another.

Step 3. Calculate the separate integrals.

$y_2 = -x^2 + 3x$, $y_2 = x$

$x = -x^2 + 3x$

$x^2 - 2x = 0$

$x(x-2) = 0$

$\therefore x = 0, 2$

$$
\begin{aligned}
Area &= \int_0^2 (y_1 - y_2)dx + \int_2^3 (y_2 - y_1)\,dx \\
&= \int_0^2 (-x^2 + 3x - x)\,dx + \int_2^3 (x + x^2 - 3x)dx \\
&= \int_0^2 (-x^2 + 2x)\,dx + \int_2^3 (x^2 - 2x)dx \\
&= \left[-\frac{1}{3}x^3 + x^2\right]_0^2 + \left[\frac{1}{3}x^3 - x^2\right]_2^3 = \frac{4}{3} + \frac{4}{3} = \frac{8}{3}
\end{aligned}
$$

EX 1 Find the areas of the regions enclosed by the lines and curves is area.

(1) $y = x^4$, $y = 8x$

(2) $y = -x^2 + 3x$, $y = -x$, $x = 0$, and $x = 3$

(3) $y = 2\sin x,\ y = \sin 2x,\ x = 0$, and $x = \pi$

(4) $y = e^x,\ y = \sin x,\ x = 0$, and $x = \pi$

(5) $y = \sec^2 x,\ y = \tan^2 x,\ x = -\dfrac{\pi}{4}$, and $x = \dfrac{\pi}{4}$

(6) $y = \sin x$ and $y = \sin 2x,\ x = 0$, and $x = \pi$

EX 2 Find the areas of the regions enclosed by the lines and curves.

(1) $y = x^2 - 2x$ and $y = x$　　(2) $y = \dfrac{10}{x^2+1}$ and $y = 5$

(3) $y = \dfrac{2}{x}$ and $x + y = 3$　　(4) $y = x^3 - 2x$ and $y = x^2$

(5) $y = x^3 - 4x$ and $y = 3x - 6$　　(6) $y = |x^2 - 4|$ and $y = (x^2/2) + 4$

EX 3 Use a graphing calculator to find the area of the region enclosed by the curves.

(1) $y = x - 1$ and $y = 2\ln(x+2)$　　(2) $y = 3\sin x$ and $y = x^2 - 2$

(3) $y = e^x$ and $y = -x^2 + 20$　　(4) $y = -x^2$ and $y = 0.3x^3 - 5x$

If $f(y)$ and $g(y)$ are continuous and $f(y) \geq g(y)$ for all y in $[c, d]$, then the area of the region bounded by the graphs of $x = f(y)$, $x = g(y)$, and the horizontal lines $y = c,\ y = d$ is

$$Area = \int_c^d (x_1 - x_2)\,dy = \int_e^d [f(y) - g(y)]\,dy \text{ where } c < d$$

Proof

When $y = y_k$, the horizontal strip is a rectangle with a height of $f(y_k) - g(y_k)$ and a width of Δy.

By summing the areas of these rectangles,

$$Area = \lim_{n\to\infty} \sum_{k=1}^{n} [f(y_k) - g(y_k)]\,\Delta y$$

$$= \int_c^d [f(y) - g(y)]\,dy \text{ where } c < d$$

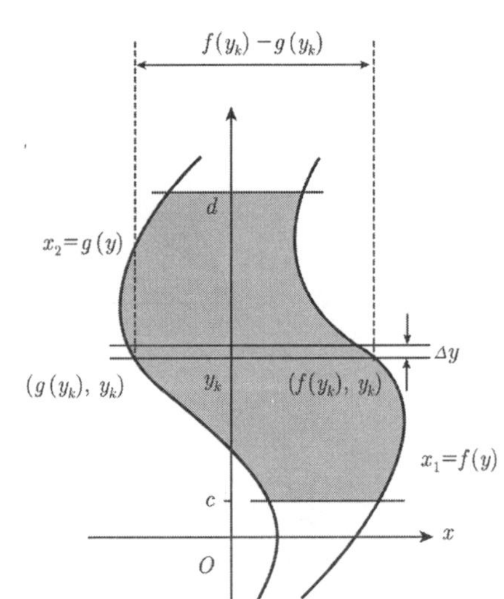

Example 3 Find the area of the region enclosed by $x=y^2$ and $x=y+2$.

Solution

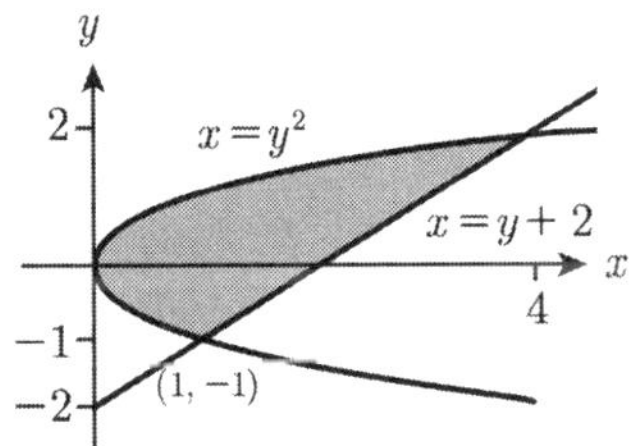

Limits of integration: $y^2=y+2 \Rightarrow (y+1)(y-2)=0$

$\Rightarrow c=-1$ and $d=2; f(y)-g(y)=(y+2)-y^2$

$\Rightarrow \mathrm{A}=\int_{-1}^{2}(y+2-y^2)\,dy=\left[\frac{y^2}{2}+2y-\frac{y^3}{3}\right]_{-1}^{2}$

$=\left(\frac{4}{2}+4-\frac{8}{3}\right)-\left(\frac{1}{2}-2+\frac{1}{3}\right)=6-\frac{8}{3}-\frac{1}{2}+2-\frac{1}{3}=\frac{9}{2}$

EX 4 Sketch the region enclosed by the curves and find its area.

(1) $x=2y^2,\ x=0$, and $y=3$

(2) $y^2=x$ and $x+y=2$

(3) $x-y^2=0$ and $x+2y^2=3$

(4) $x=y^2-4y$ and $x=-y^2+2y$

(5) $y=\sqrt{x-1},\ y=x,\ y=0$ and $y=2$

(6) $y=\ln x,\ y=\sqrt{-x},\ y=0$ and $y=1$

(7) $y=e^x,\ y=\sqrt{x},\ y=1$ and $y=e$

(8) $y=e^{-x},\ y=\frac{1}{x},\ y=1$ and $y=e$

6.3 Volume of Solids with Known Cross Sections

The slices are taken perpendicular to the x-axis so that $\triangle V=A(x)\triangle x$, where $A(x)$ is the area of a cross section and $\triangle x$ is the thickness of the slice.

$$V=\int_a^b A(x)dx \qquad \text{where } a<b$$

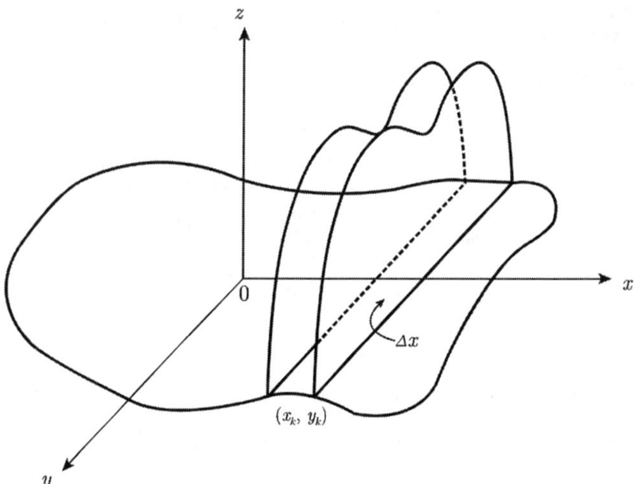

Example 1 A solid has as its base the circle $x^2 + y^2 = 9$, and all cross sections parallel to the $y-$axis and the line $y-$axis are squares. Find the volume of the solid.

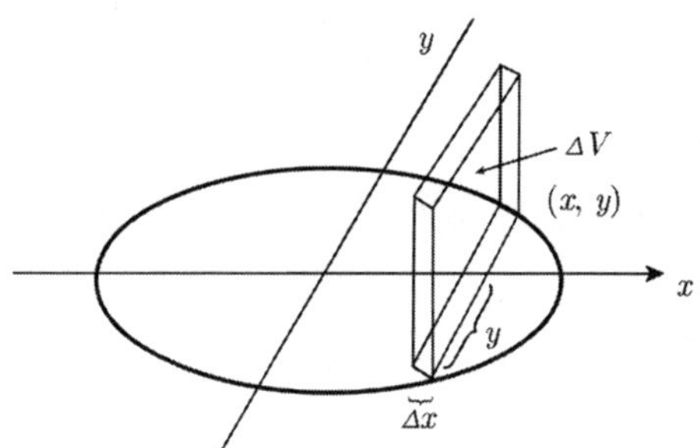

Solution

The element of volume is a square prism with sides of length $2y$ and thickness Δx, so $\Delta V = (2y)^2 \Delta x = 4y^2 \Delta x = 4(9 - x^2)\Delta x$.

Now, using symmetry across the y-axis, we find, the volume of the solid:

$$V = 2\int_0^3 4(9-x^2)\,dx = 8\int_0^3 (9-x^2)dx = 8\left(9x - \frac{x^3}{3}\right)\Bigg|_0^3 = 144.$$

EX 1 The base of a solid is the region enclosed by the parabola $x = y^2$ and the line $x = 3$. For this solid, each cross section perpendicular to the $x-$axis is a square. Find the volume of the solid.

EX 2 The solid lies between planes perpendicular to the x – axis at $x = -1$ and $x = 1$. The cross-sections perpendicular to the x – axis are circular disks whose diameters run from the parabola $y = x^2$ to the parabola $y = 2 - x^2$.

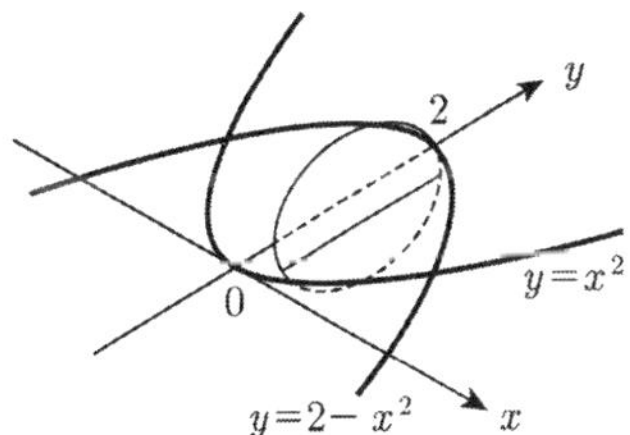

EX 3 The base of a solid is the region between the curve $y = 2\sqrt{\sin x}$ and the interval $[0, \pi]$ on the x – axis. The cross-sections perpendicular to the x – axis are equilateral triangles with bases running from the x – axis to the curve as shown in the accompanying figure.

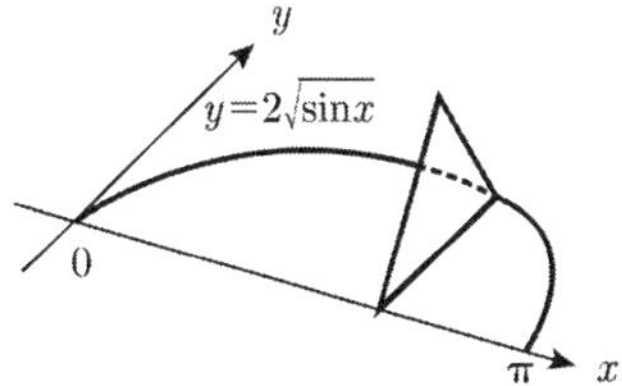

EX 4 The solid lies between planes perpendicular to the x – axis at $x = -1$ and $x = 1$. The cross-sections perpendicular to the x – axis between these planes are squares whose diagonals run from the semicircle $y = -\sqrt{1 - x^2}$ to the semicircle $y = \sqrt{1 - x^2}$.

EX 5 The functions f and g are given by $f(x) = \sqrt{x}$ and $g(x) = 6 - x$. Let R be the region bounded by the x – axis and the graphs of f and g, as shown in the figure below.

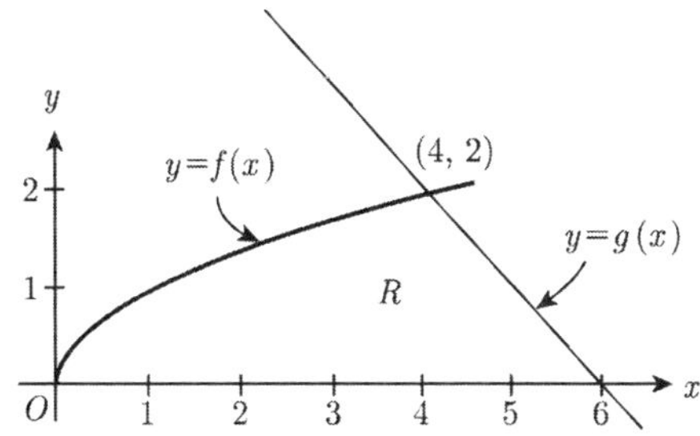

(1) Find the area of R.

(2) The region R is the base of a solid. For each y. where $0 \le y \le 2$, the cross-section of the solid taken perpendicular to the $y-$axis is a rectangle whose base lies in R and whose height is $2y$. Write, but do not evaluate, an integral expression that gives the volume of the solid.

(3) There is a point P on the graph of f at which? th line tangent to the graph of f is perpendicular to the graph of g. Find the coordinates of point P.

Let S be a solid bounded by $y = c$ and $y = d$. If the cross-sectional area of S (perpendicular to the $y-$axis) is $A(y)$, then the volume of S is

$$V = \int_c^d A(y)dy \qquad \text{where } c < d$$

EX 6 The solid lies between planes perpendicular to the $y-$axis at $y = 0$ and $y = 2$. The cross-sections perpendicular to the $y-$axis are circular disks with diameters running from the $y-$axis to the parabola $x = \sqrt{5}\,y^2$.

EX 7 The base of a solid is the region enclosed by the graphs of $y = x^2$ and $y = 1$. For this solid. each cross section perpendicular to the $y-$axis is a rectangle whose height is 4 times the length of its base. Find the volume of the solid.

EX 8 The base of the solid is the disk $x^2 + y^2 \le 1$. The cross-sections by planes perpendicular to the $y-$axis between $y = -1$ and $y = 1$ are isosceles right triangles with one leg in the disk.

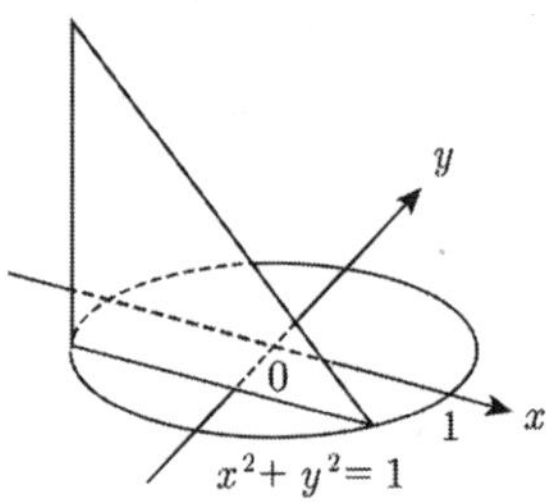

EX 9 The base of a solid is the region enclosed by the graphs of $y = \sqrt{x}$ and $y = \frac{1}{2}x$. For this solid each cross-section perpendicular to the $y-$axis is a semicircle. Find the volume of the solid.

6.4 Volume of Solids of Revolution: Disk, Washers and Cylindrical Shells

- **Disk**

① Revolution about the $x-$axis: $V=\pi\int_a^b r^2dx$

② Revolution about the $y-$aixs: $V=\pi\int_c^d r^2dy$

The region bounded by a curve and the $x-$axis is revolved around the $x-$axis, forming the slid of revolution seen in Figure. We think of the "rectangular strip" of the region at the left as generating the solid disk. $\triangle V$(an element of the volume), shown at the right

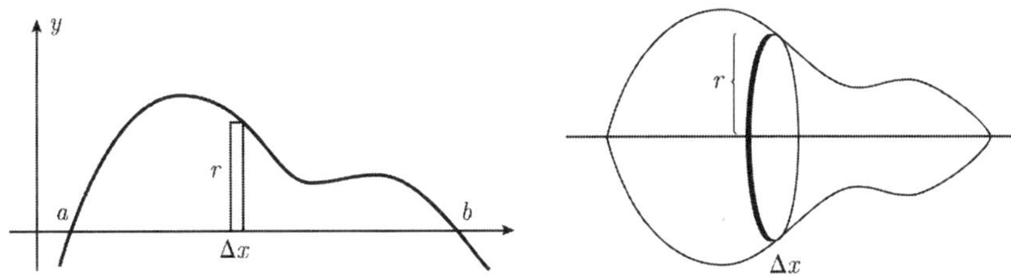

This disk is a cylinder whose radius, r, is the height of the rectangular strip, and whose height is the thickness of the strip, $\triangle x$, Thus

$$\triangle V=\pi r^2\triangle x \text{ and } V=\pi\int_a^b r^2dx.$$

Example 1 Find the volume of the solid generated by revolving the region enclosed by $y=\sqrt{r^2-x^2}$ and $y=0$ about the $x-$axis.

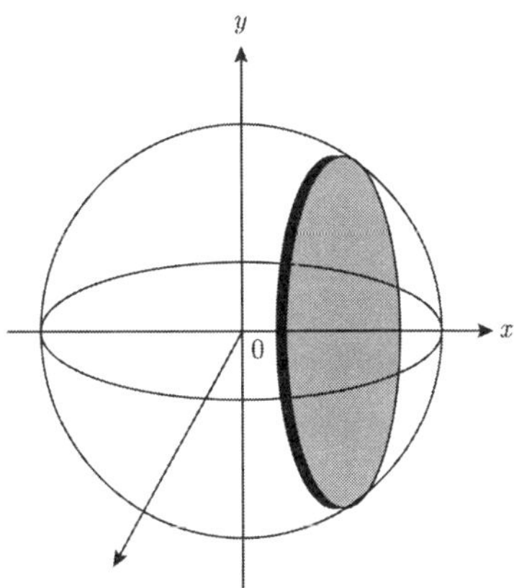

Solution

$$V=\int_{-r}^{r}\pi y^2dx=\pi\int_{-r}^{r}(r^2-x^2)dx=\pi\left[r^2x-\frac{x^3}{3}\right]\Bigg|_{-r}^{r}=\frac{4}{3}\pi r^3$$

Example 2 Find the volume of the solid generated when the region bounded by $y=x^2$, $x=2$, and $y=0$ is rotated about the line $x=2$ as shown in figure.

Solution

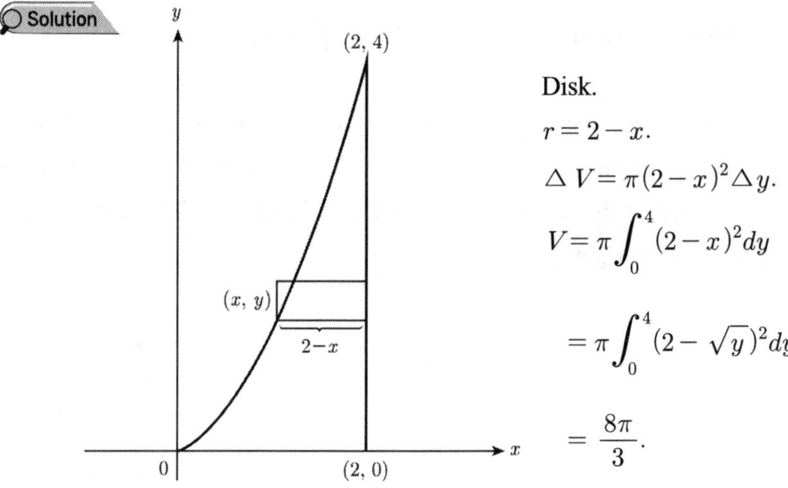

Disk.

$r=2-x.$

$\triangle V=\pi(2-x)^2\triangle y.$

$$V=\pi\int_0^4(2-x)^2dy$$

$$=\pi\int_0^4(2-\sqrt{y})^2dy$$

$$=\frac{8\pi}{3}.$$

EX 1 Find the volume of the solid generated by revolving the region enclosed by the following graphs about the x-axis.

(1) $y=x^2$, $x=2$, and $y=0$ (2) $y=\sin x$, $x=0$, $x=\pi$, and $y=0$

(3) $y=e^{-x}$, $y=0$, $x=0$, $x=1$ (4) $y=x-x^2$, $y=0$

EX 2 Find the volume of solid generated by revolving the region enclosed by the following graphs about the y-axis.

(1) $y=x^2$ and $y=2$ (2) $x=-y^2+2y$ and $x=0$

(3) $x=\sqrt{5}y^2$, $x=0$, $y=-1$, $y=1$ (4) $x=2/\sqrt{y+1}$, $x=0$, $y=0$, $y=3$

EX 3 Find the volume of the solid generated by revolving the region enclosed by $y=\sqrt{\frac{\ln x}{x}}$, $x=1$, $x=e$, and $y=0$ about the x-axis.

EX 4 Find the volume of the solid generated by revolving the region enclosed by the following graphs about the y-axis.

(1) $y = e^{-x^2}$ and $y = \frac{1}{e}$ (2) $\frac{x^2}{9} + \frac{y^2}{4} = 1$ (ellipse)

EX 5 Show that the volume of a right circular cone with radius r, height h is $\frac{1}{3}\pi r^2 h$ using the disk method. (Hint: The cone is generated by revolving the triangle about the $y-$axis.)

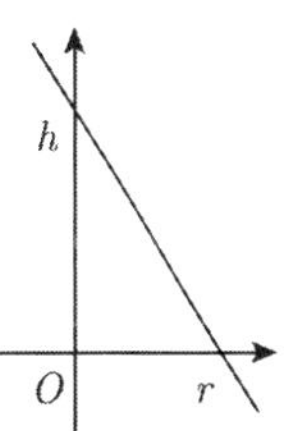

- **Washer**

① Revolution about the $x-$axis:

$$V = \pi \int_a^b (R^2 - r^2)dx = \pi \int_a^b \left[f(x)^2 - g(x)^2 \right] dx$$

② Revolution about the $y-$axis:

$$V = \pi \int_c^d (R^2 - r^2)dy = \pi \int_c^d \left[f(y)^2 - g(y)^2 \right] dy$$

The region enclosed by $y = f(x)$, $y = g(x)$ $x = a$ and $x = b$ is revolved about the $x-$axis.

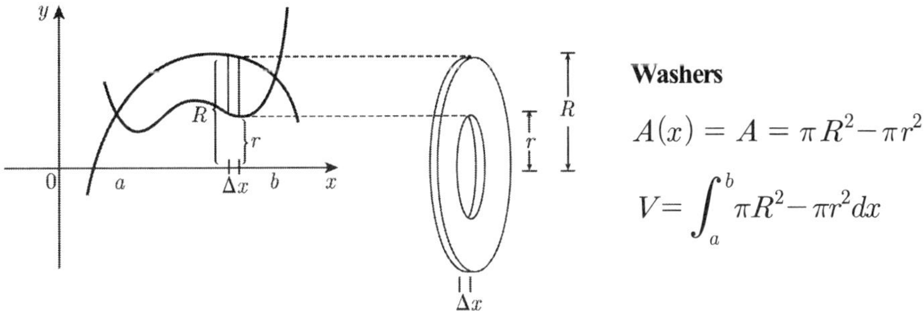

This washer's height is the thickness of the rectangular strip, Δx. The washer is a disk whose outer radius, R, is the distance to the top of the rectangular strip, with the disk of inner radius r(the distance to the bottom of the strip) removed. Thus:

$$\Delta V = \pi R^2 \Delta x \quad \pi r^2 \Delta x = \pi (R^2 - r^2) \Delta x \text{ and } V = \pi \int_a^b (R^2 - r^2)\, dx.$$

Example 3 Find the volume obtained when the region bounded by $y = x^2$ and $y = 2x$ is revolved about the $x-$axis.

Solution

Washer.

$R = y_1 = 2x$

$r = y_2 = x^2$

$\Delta V = \pi({y_1}^2 - {y_2}^2)\Delta x$

$$V = \pi\int_0^2 ({y_1}^2 - {y_2}^2)dx$$

$$= \pi\int_0^2 ((2x)^2 - (x^2)^2)dx$$

$$= \frac{64\pi}{15}.$$

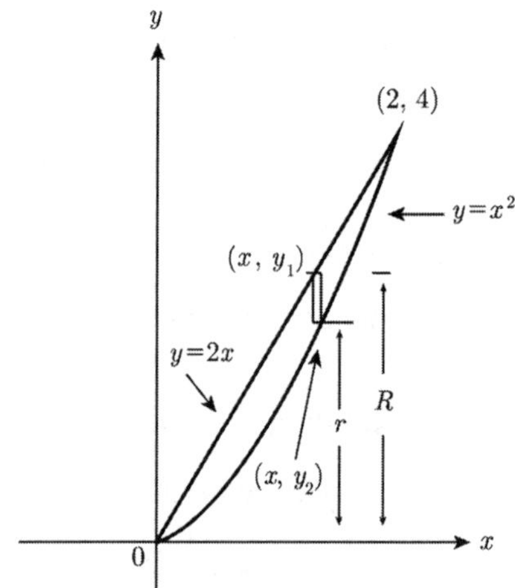

EX 6 Find the volume of the solid generated by revolving the region enclosed by the following graphs about the $x-$axis.

(1) $y = x^2 + 1,\ y = x + 3$

(2) $y = x^2,\ y^2 = 8x$

(3) $y = x^2,\ y = 4$

(4) $y = \dfrac{6}{\sqrt{x^2+1}},\ y = 3$

(5) $y = \sec x,\ y = \sqrt{2},\ -\pi/4 \le x \le \pi/4$

(6) $y = \sin x,\ y = \dfrac{1}{2},\ x = \dfrac{\pi}{6}$ and $\dfrac{\pi}{2}$

EX 7 Find the volume of the solid generated by revolving the region enclosed by the following graphs about the $y-$axis.

(1) $y = x^2, x = 2,$ and $y = 0$

(2) $y = \dfrac{1}{2}x^2,\ y = -x$

(3) $y = x^3,\ y^2 = 32x$

(4) $y = \ln x,\ y = \sqrt{x},\ y = 0$ and $y = 1$

EX 8 Find the volume of the solid generated by revolving the region bounded by the parabola $y = x^2$ and the line $y = 1$ about

(1) the line $y = 1$

(2) the line $y = 2$

(3) the line $y = -1$

EX 9 Find the volume of the solid generated by revolving the triangular region bounded by the lines $y = 2x,\ y = 0$ and $x = 1$ about

(1) the line $x = 1$

(2) the line $x = 2$

EX 10 Let R be region bounded by the graph of $y = e^{2x - x^2}$ and the horizontal line $y = 2$, Let S be region bounded by the graph of $y = e^{2x - x^2}$ and the horizontal line $y = 1$ and $y = 2$, as shown below.

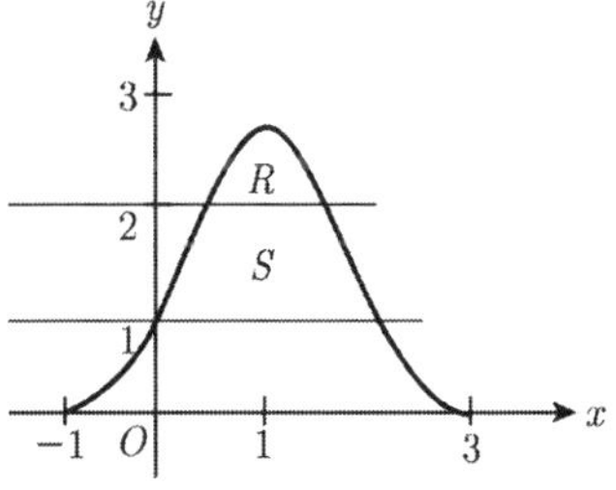

(1) Find the area of R.

(2) Find the area of S.

(3) Write, but do not evaluate, an integral expression that gives the volume of the solid generated when R is rotated about the horizontal line $y = 1$.

EX 11 Let R be the region in the first quadrant bounded by the graph of $y = 2\sqrt{x}$, the horizontal line $y = 6$, and the y−axis, as shown in the figure above.

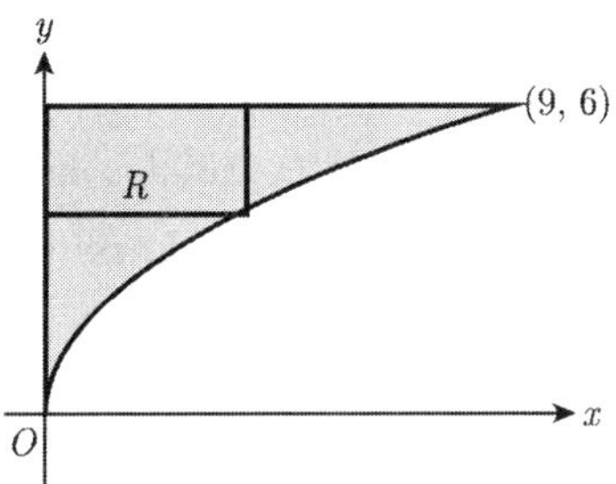

(1) Find the area of R.

(2) Write, but do not evaluate, an integral expression that gives the volume of the solid generated when R is rotated about the horizontal line $y = 7$.

(3) Region R is the base of a solid. For each y, where $0 \le y \le 6$, the cross section of the solid taken perpendicular to the y−axis is a rectangle whose height is 3 times the length of its base in region R. Write, but do evaluate, an integral expression that gives the volume of the solid.

- **Cylindrical Shell**

① Revolution about the y−axis: $V = 2\pi \int_a^b (Shell\,Radius)(Shell\,Height)dx$

② Revolution about the x−axis: $V = 2\pi \int_c^d (Shell\,Radius)(Shell\,Height)dy$

Consider the volume of solid of revolution formed when the region bounded by the two curves seen in Figure is revolved around the $y-$axis. We think of the rectangular strip of the region at the left as generating the shell, $\triangle V$(an element of the volume), shown at the below.

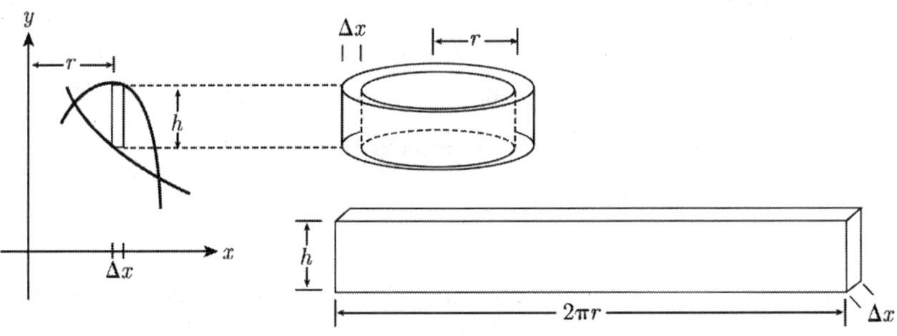

This shell's radius, r, is the distance from the axis to the rectangular strip, and its height is the rectangular strip, h. When the shell is unwound and flattened to from a rectangular solid, the length of the solid is the circumference of the cylinder, $2\pi r$, its height of the cylinder, h, and its thickness is the thickness of the rectangular strip, $\triangle x$. Thus:

$$\triangle V = 2\pi rh\triangle x \text{ and } V = 2\pi\int_a^b rh\, dx.$$

Example 3 The region enclosed by the $x-$axis and the parabola $y = f(x) = 3x - x^2$ is revolved about the vertical line $x = -1$ to generate a solid (Figure). Find the volume of the solid.

Solution

$$V = \int_0^3 2\pi(x+1)(3x - x^2)dx$$

$$= \int_0^3 2\pi(3x^2 + 3x - x^3 - x^2)dx$$

$$= 2\pi\int_0^3 (2x^2 + 3x - x^3)dx$$

$$= 2\pi\left[\frac{2}{3}x^3 + \frac{3}{2}x^2 - \frac{1}{4}x^4\right]_0^3 = \frac{45\pi}{2}.$$

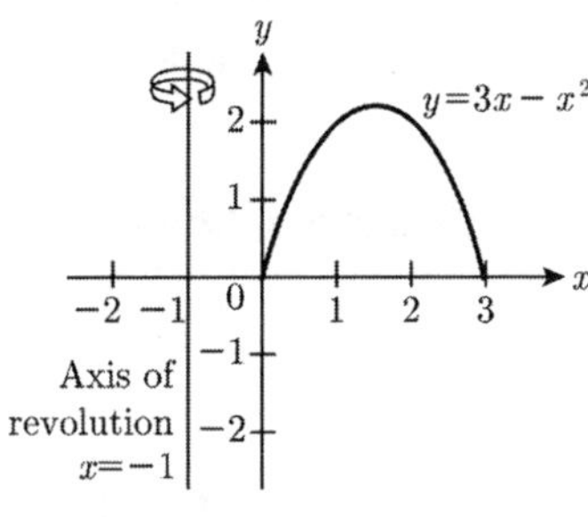

Example 4 The region bounded by the curve $y=\sqrt{x}$, the $x-$axis, and the line $x=4$ is revolved about the $x-$axis to generate a solid. Find the volume of the solid by the shell method.

Solution

$$V=\int_a^b 2\pi\left(\begin{array}{c}\text{shell}\\\text{radius}\end{array}\right)\left(\begin{array}{c}\text{shell}\\\text{height}\end{array}\right)dy$$

$$=\int_0^2 2\pi(y)(4-y^2)dy$$

$$=2\pi\int_0^2(4y-y^3)dy$$

$$=2\pi\left[2y^2-\frac{y^4}{4}\right]_0^2=8\pi.$$

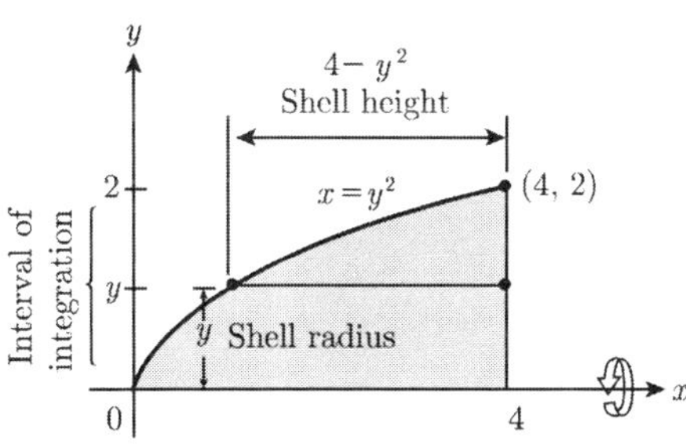

EX 12 Use the cylindrical shell method to find volume of the solid generated by revolving the region enclosed by the following graphs about the $y-$axis.

(1) $y=x^2,\ y=2-x,\ x=0,\ x=1$ (2) $y=x^2,\ x=1,\ x=2$ and $y=0$

(3) $y=2x-1,\ y=\sqrt{x},\ x=0$ (4) $y=\sin x,\ x=0,\ x=\pi$, and $y=0$

(5) $y=3x-x^2$ and $y=x$ (6) $y=\sqrt{x}$ and $y-x$

EX 13 Use the cylindrical shell method to find the volume of the solid generated by revolving the region enclosed by the following graphs about the $x-$axis.

(1) $y=\sqrt{x},\ y=0,\ y=2-x$ (2) $x=2y-y^2,\ x=y$

(3) $y=x,\ y=2x,\ y=2$ (4) $x=2y-y^2,\ x=0$

EX 14 Use the cylindrical shell method to find the volume of the solid generated by revolving the region enclosed by $y=x^3,\ y=8,\ x=0$ about the following lines.

(1) $x=3$ (2) $x=-2$

(3) $y=8$ (4) $y=-1$

EX 15 Use the cylindrical shell method to find the volume of solid generated by revolving the region enclosed by $x=\frac{y^4}{4}-\frac{y^2}{2}$ and $x=\frac{y^2}{2}$ about the line $y=5$.

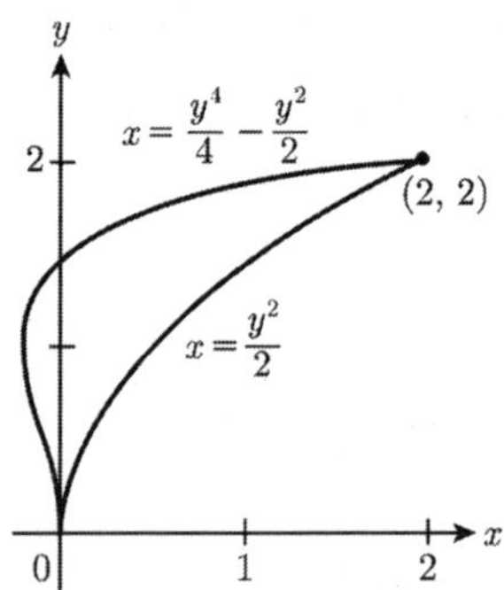

6.5 Definite Integral of a Rate is Net Change

If f is continuous and $f(t) = \frac{dF}{dt}$, then

$$\int_a^b f(t)dt = F(b) - F(a).$$

The definite integral of the rate of changes of a quantity over an interval is the net change or net accumulation of the quantity over that interval. Thus, $F(b) - F(a)$ is the net change in $F(t)$ as t varies from a to b.

Example 1 Let $G(t)$ be the rate of growth of a population at time t. Then the increase in population between times $t = a$ and $t = b$ is given by $\int_a^b G(t)\,dt$. The population may consist of people, deer, fruit files, bacteria, and so on.

Example 2 If after t minutes a chemical is decomposing at the rate of $10e^{-t}$ grams per minute, then during the first 3 minutes the amount hat has decomposed is

$$\int_0^3 10e^{-t}\,dt \simeq 9.5\,g$$

(1) A beach opens at 8 A.M. and people arrive at a rate of $R(t) = 10 + 40t$ people per hour, where t represents the number of hours the beach has been open. Assuming no one leaves before noon, at what time will there be 100 people there?

(2) Oil is leaking from a tanker at the rate of $1000e^{-0.3t}$ gal/hr, where t is given in hours. the total number of gallons of oil that will leak out during the next 8 hour is approximately.

(3) If a factory continuously dumps pollutants into a river at the rate of $\frac{\sqrt{t}}{180}$ tons per day, then the amount dumped after 7 weeks is approximately.

(4) How long will it take to release 9 tons of pollutant if the rate at which pollutant is bcing rclcased is $te^{-0.3t}$ tons per week?

(5) If you deposit \$1000 today at 8% interest compounded continuously, it will grow at the rate of $80e^{0.08t}$ dollars per year. In 6 years it will be worth (in dollars).

(6) Water is leaking from a tank at the rate of $(-0.1t^2 - 0.3t + 2)$ gal/hr. The total amount, in gallons, that will leak out in the next 3 hour is approximately.

- **New Position = Initial Position + Displacement:**

$$s(t) = s(a) + \int_a^t v(t)dt$$

- **New Velocity = Initial Velocity + Change of Velocity:**

$$v(t) = v(a) + \int_a^t a(t)dt$$

Example 3 The velocity function of a particle is given by $v(t) = 2t + 3$ and its position is 2 at time $t = 1$. Find the position function of the particle.

(Method 1) $s(t) = \int v(t)dt = \int (2t+3)dt$

$$= t^2 + 3t + C$$

$$s(1) = 2 : (1)^2 + 3(1) + C = 2,$$

$$C = -2$$

$$\therefore s(t) = t^2 + 3t - 2$$

(Method 2) $s(t) = s(1) + \int_1^t v(t)dt$

$$= 2 + \int_1^t (2t+3)dt$$

$$= 2 + \left[t^2 + 3t\right]_1^t$$

$$= 2 + (t^2 + 3t) - (1+3)$$

$$= t^2 + 3t - 2$$

EX 2 A particle moves along a straight line. Find the position of the particle at time t.

(1) $v(t) = 3t^2 + 2t,\ s(0) = 2$

(2) $v(t) = 4\cos 2t,\ s(\pi) = -3$

(3) $v(t) = \dfrac{1}{\sqrt{1+t}},\ s(3) = 1$

(4) $v(t) = \dfrac{\ln t}{t},\ s(1) = 2$

(5) $a(t) = 6t - 3,\ v(1) = 2,\ s(0) = 2$

(6) $a(t) = 2\sin 2t,\ v(0) = 3,\ s(0) = -2$

EX 3 Use a graphing calculator to evaluate the following values.

(1) Given $f'(x) = 3\sin(x^2)$ and $f(1) = 3$, find $f(3)$.

(2) Given $f'(x) = e^{-x^2}$ and $f(-1) = 2$, find $f(2)$.

(3) A particle moves along the x-axis with velocity given by $v(t) = e^{\sin t} - 1$ for $t \geq 0$. The position of the particle is 2 at time $t = 0$. Find the position of the particle at time $t = 4$.

(4) A particle moves along the x-axis with acceleration given by $a(t) = 2\cos(t + \ln t)$ for $t \geq 0$. The velocity of the particle is 3 at time $t = 1$. Find the velocity of the particle at time $t = 5$.

(5) The number of bacteria in a culture changes at a rate of $500e^{-0.02t}$ bacteria/hour at time t. Determine the increase in number of bacteria in the culture during the time interval $10 \leq t \leq 30$.

EX 4

t	0	2	4	6	8	10	12
P(t)	0	46	53	57	60	62	63

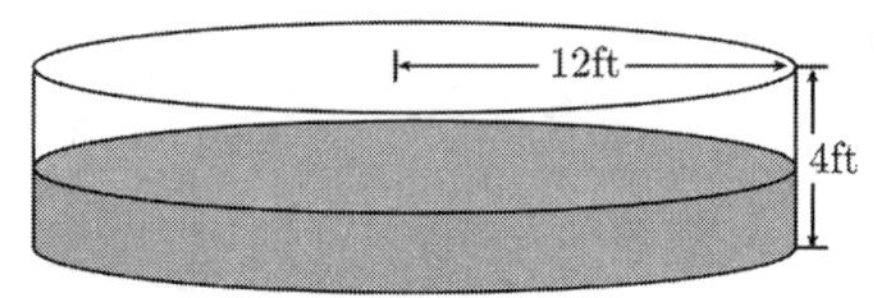

The figure above shows an aboveground swimming pool in the shape of a cylinder with a radius of 12 feet and a height of 4 feet. The pool contains 1000 cubic feet of water at time $t=0$. During the time interval $0 \le t \le 12$ hours, water is pumped into the pool at the rate $P(t)$ cubic feet per hour. The table above gives values of $P(t)$ for selected values of t. During the same time interval, water is leaking from the pool at the rate $R(t)$ cubic feet per hour, where $R(t)=25e^{-0.05t}$. (Note: The volume V of a cylinder with radius r and height h is given by $V=\pi r^2 h$.)

(1) Use a midpoint Riemann sum with three subintervals of equal length to approximate the total amount of water that was pumped into the pool during the time interval $0 \le t \le 12$ hours. Show the computations that lead to your answer.

(2) Calculate the total amount of water that leaked out of the pool during the time interval $0 \le t \le 12$ hours.

(3) Use the results from parts (1) and (2) to approximate the volume of water in the pool at time $t=12$ hours. Round your answer to the nearest cubic foot.

(4) Find the rate at which the volume of water in the pool is increasing at time $t=8$ hours. How fast is the water level in the pool rising at $t=8$ hours? Indicate units of measure in both answers.

t (minutes)	0	2	5	9	10
$H(t)$ (degrees Celsius)	66	60	52	44	43

As a pot of tea cools, the temperature of the tea is modeled by a differentiable function H for $0 \le t \le 10$, where time t is measured in minutes and temperature $H(t)$ is measured in degrees Celsius. Valucs of $H(t)$ at selected values of time t are shown in the table above.

(1) Use the date in the table to approximate the rate at which the temperature of the tea is changing at time $t=3.5$. Show the computations that lead to your answer.

(2) Using correct units, explain the meaning of $\frac{1}{10}\int_0^{10} H(t)\,dt$ in the context of this problem. Use a trapezoidal sum with the four subintervals indicated by the tables to estimate $\frac{1}{10}\int_0^{10} H(t)\,dt$.

(3) Evaluate $\int_0^{10} H'(t)dt$, Using correct units, explain the meaning of the expression in the context of this problem.

(4) At time $t=0$, biscuits with temperature $100\,^\circ\text{C}$ were removed from an oven. The temperature of the biscuits at time t is modeled by a differentiable function B for which it is known that $B'(t) = -13.84e^{-0.173t}$. Using the given models, at time $t=10$, how much cooler are the biscuits than the tea?

6.6 Motion Along a Straight Line (1-Dimension)

- **Displacement (Net Change in Position) from $t=a$ to $t=b$:**

$$\int_a^b v(t)dt = [s(t)]_a^b = s(b) - s(a)$$

- **Net Change in Velocity from $t=a$ to $t=b$:**

$$\int_a^b a(t)dt = [v(t)]_a^b = v(b) - v(a)$$

- **Distance from $t=a$ to $t=b$:**

$$\int_a^b |v(t)|\,dt$$

Example 1 The velocity function of a particle is given by $6t^2 - 18t + 12$. Find the displacement and the distance traveled by the particle during the interval $0 \le t \le 4$.

Solution

Displacement

$$= \int_0^4 v(t)dt = \int_0^4 (6t^2 - 18t + 12)dt = (2t^3 - 9t^2 + 12t)\Big|_0^4$$

$$= 128 - 144 + 48 = 32.$$

Since $v(t) = 6t^2 - 18t + 12 = 6(t-1)(t-2)$, we see that:

if $t < 1$, then $v > 0$;

if $1 < t < 2$, then $v < 0$;

if $2 < t$, then $v > 0$.

$$\text{Distance} = \int_0^4 |v(t)|\,dt$$

$$= \int_0^1 v(t)dt - \int_1^2 v(t)dt + \int_2^4 v(t)dt.$$

$$= 34$$

EX 1 A particle moves along a straight line. Find the displacement and the distance traveled by the particle during the given interval.

(1) $v(t) = 1 - \cos t,\ 0 \le t \le \dfrac{5\pi}{6}$ (2) $v(t) = 4t^3 - 4t,\ \ 0 \le t \le 2$

(3) $a(t) = 6t - 6,\ \ v(0) = 0,\ \ 0 \le t \le 3$ (4) $a(t) = e^{t-1},\ v(1) = 0,\ \ 0 \le t \le 2$

EX 2

A squirrel starts at building A at time $t = 0$ and travels along a straight wire connected to building B. For $0 \le t \le 18$, the squirrel's velocity is modeled by the piecewise-linear function defined by the graph below.

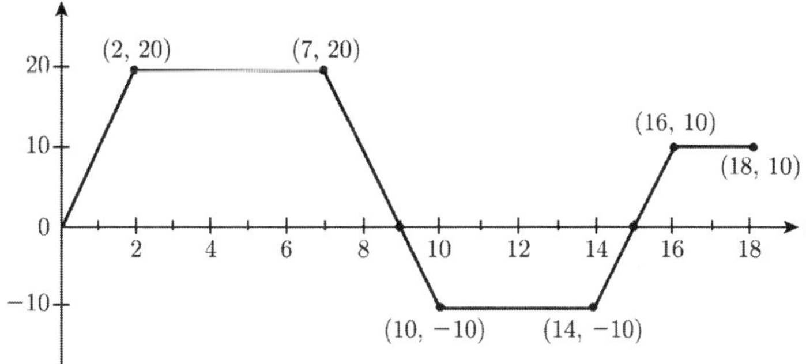

(1) At what times in the intervals $0 < t < 18$, if any, does the squirrel change direction? Give a reason for your answer.

(2) At What time in the intervals $0 \le t \le 18$ is the squirrel farthest from building A? How from building A is the squirrel at this time?

(3) Find the total distance the squirrel travels during the time interval $0 \le t \le 18$.

(4) Write expressions for the squirrel's acceleration $a(t)$. velocity $v(t)$, and distance $x(t)$ from building A that are valid for the time interval $7 < t < 10$.

EX 3

Caren rides her bicycle along a straight road from home to school, starting at home at time $t=0$ minutes and arriving at school at time $t=12$ minutes. During the time interval $0 \le t \le 12$ minutes, her velocity $v(t)$, in miles per minutes, is modeled by the piecewise-linear function whose graph is shown above.

(1) Find the acceleration of Caren's bicycle at time $t=7.5$ minutes. Indicate units of measure.

(2) Using correct units, explain the meaning of $\int_0^{12} |v(t)|\,dt$ in terms of Caren's trip. Find the value of $\int_0^{12} |v(t)|\,dt$.

(3) Shortly after leaving home, Caren realizes she left her calculus homework at home, and she returns to get it. At what time does she turn around to go back home? Give a reason for your answer.

(4) Larry also rides his bicycle along a straight road from home to school in 12 minutes. His velocity is modeled by the function w given by $w(t) = \dfrac{\pi}{15}\sin\left(\dfrac{\pi}{12}t\right)$, where $w(t)$ is in miles per minute for $0 \le t \le 12$ minutes. Who lives closer to school: Caren or Larry? Show the work that leads to your answer.

EX 4

t (seconds)	0	10	40	60
$B(t)$ (meters)	100	136	9	49
$v(t)$ (meters per second)	2.0	2.3	2.5	4.6

Ben rides a unicycle back and forth along a straight east-west track. The twice-differentiable function B models Ben's position on the track, measured in meters from the western end of the track, at time t, measured in seconds from the start of the ride. The table above gives values for $B(t)$ and Ben's velocity, $v(t)$, measured in meters per seconds, at selected times t.

(1) Use the data in the table to approximate Ben's acceleration at time $t=5$ seconds. Indicate units of measure.

(2) Using correct units, interpret the meaning of $\int_0^{60} |v(t)|\,dt$ in the context of this problem. Approximate $\int_0^{60} |v(t)|\,dt$ using a left Riemann sum with the subintervals indicated by the data in the table.

(3) For $40 \le t \le 60$, must there be a time t when Ben's velocity is 2 meters per second? Justify your answer.

(4) A light is directly above the western end of the track. Ben rides so that at time t, the distance $L(t)$ between Ben and the light satisfies $(L(t))^2 = 12^2 + (B(t))^2$. At what rate is the distance between Ben and the light changing at time $t=40$?

EX 5 There is no snow on Janet's driveway when snow begins to fall at midnight. From midnight to 9 A.M., snow accumulates on the driveway at a rate modeled by $f(t) = 7te^{\cos t}$ cubic feet per hour, where t is measured in hours since midnight, Janet starts removing snow at 6 A.M. $(t=6)$. The rate $g(t)$, in cubic feet per hour, at which Janet removes snow from the driveway at time t hours after midnight is modeled by

$$g(t) = \begin{cases} 0 & \text{for } 0 \le t < 6 \\ 125 & \text{for } 6 \le t < 7 \\ 108 & \text{for } 7 \le t \le 9 \end{cases}$$

(1) How many cubic feet of snow have accumulated on the driveway by 6 A.M.?

(2) Find the rate of change of the volume of snow on the driveway at 8 A.M..

(3) Let $h(t)$ represent the total amount of snow, in cubic feet, that Janet has removed from the driveway at time t hours after midnight. Express h as a piecewise-defined function with domain $0 \le t \le 9$.

(4) How many cubic feet of snow are on the driveway at 9 A.M.?

6.7 Motion Along a Plane Curve

If the motion of a particle P along a curve is given parametrically by the equations $x = x(t)$ and $y = y(t)$, then at time t the position vector R. The velocity vector v, and the acceleration vector a are:

$$R = x\,i + yj;$$

$$\mathrm{v} = \frac{dR}{dt} = \frac{dx}{dt}i + \frac{dy}{dt}j = v_x i + v_y j;$$

$$a = \frac{d^2R}{dt^2} = \frac{d\mathrm{v}}{dt} = \frac{\mathrm{d^2x}}{\mathrm{dt^2}}\mathrm{i} + \frac{\mathrm{d^2y}}{\mathrm{dt^2}}\mathrm{j} = \mathrm{a_x i} + \mathrm{a_y j}.$$

The components in the horizontal and vertical directions of R, v, and a are given respectively by the coefficients of i and j in the corresponding vector. The slope of v is $\frac{dy}{dx}$; its magnitude.

$$|\mathrm{v}| = \sqrt{\left(\frac{dx}{dt}\right)^2 + \left(\frac{dy}{dt}\right)^2},$$

is the speed of the particle, and the velocity vector is tangent to the path. The slope of a is $\frac{d^2y}{dt^2} / \frac{d^2x}{dt^2}$, The distance the particle travels from t_1 to t_2, it given by

$$\int_{t_1}^{t_2} |\mathrm{v}|\,dt = \int_{t_1}^{t_2} \sqrt{\left(\frac{dx}{dt}\right)^2 + \left(\frac{dy}{dt}\right)^2}\,dt.$$

Example 1 The particle $P(t) = \left((t+1)^2, \frac{t}{t+1}\right)$ in motion for 1 second, $0 \le t \le 1$. Find its position, velocity, speed, and acceleration at $t = 1$ and the distance it traveled.

Solution

From $P(t)$ we write the velocity vector:

$$\mathrm{v} = \frac{dx}{dt}i + \frac{dy}{dt}j = 2(t+1)i + \frac{1}{(t+1)^2}j.$$

Hence, at $t = 1$ the particle's velocity is $\mathrm{v} = 4i + \frac{1}{4}j$.

Speed is the magnitude of the velocity vector, so after 1 second the particle's speed is

$$|v| = \sqrt{4^2 + \left(\frac{1}{4}\right)^2} \approx 4.008 \text{ units/sec.}$$

The particle's acceleration vector at $t = 1$ is

$$a = \frac{d^2x}{dt^2}i + \frac{d^2y}{dt^2}j = 2i + \frac{-2}{(t+1)^3}j = 2i - \frac{1}{4}j.$$

On the interval $0 \le t \le 1$ the distance traveled by the particle is

$$\int_0^1 \sqrt{\left(\frac{dx}{dt}\right)^2 + \left(\frac{dy}{dt}\right)^2}\,dt = \int_0^1 \sqrt{[2(t+1)]^2 + \left[\frac{1}{(t+1)^2}\right]^2}\,dt \approx 3.057 \text{ units.}$$

EX 1 A particle $P(x,y)$ moves along a curve so that its acceleration is given by

$$a = -4\cos 2t i - 2\sin t j \left(-\frac{\pi}{2} \le t \le \frac{\pi}{2}\right);$$

when $t = 0$, the particle is at $(1,0)$ with $\frac{dx}{dt} = 0$ and $\frac{dy}{dt} = 2$. (1) Find the position vector R at any time t. (2) Find a Cartesian equation for the path of the particle, and identify the conic which P moves.

EX 2 A diver leaps from the edge of diving platform into a pool below. The figure above shows the initial position of the diver and her position at a later time. At time t seconds after she leaps, the horizontal distance from the front edge of the platform to the diver's shoulders is given by $x(t)$, and the vertical distance from the water surface to her shoulders is given by $y(t)$, where $x(t)$ and $y(t)$ are measured in meters. Suppose that the diver's shoulders are 11.4 meters above the water when she makes her leap and that

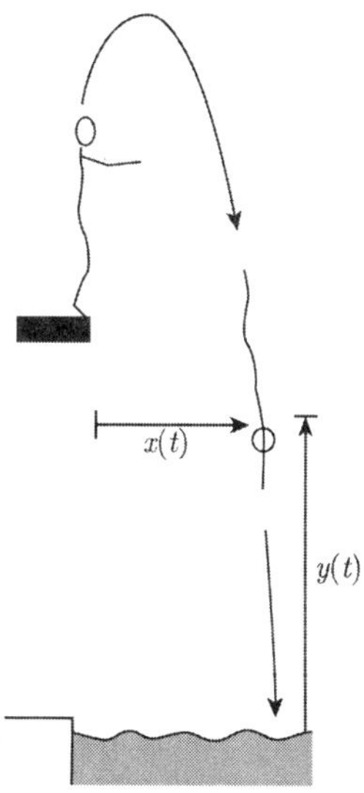

$$\frac{dx}{dt} = 0.8 \text{ and } \frac{dy}{dt} = 3.6 - 9.8t,$$

for $0 \le t \le A$, where A is the time that the diver's shoulders enter the water.

(1) Find the maximum vertical distance from the water surface to the diver's shoulders.

(2) Find A, the time that the diver's shoulders enter the water.

(3) Find the total distance traveled by the diver's shoulders from the time she leaps from the platform until the time her shoulders enter the water.

(4) Find the angle θ, $0 \le \theta \le \frac{\pi}{2}$, between the path of the diver and the water at the instant the diver's shoulders enter the water.

EX 3 A particle is moving along a curve so that its position at time t is $(x(t), y(t))$, where $x(t) = t^2 - 4t + 8$ and $y(t)$ is not explicitly given. Both x and y are measure in meters, and t is measured in seconds. It is known that $\frac{dy}{dt} = te^{t-3} - 1$.

(1) Find the speed of the particle at time $t = 3$ seconds.

(2) Find the total distance traveled by the particle for $0 \le t \le 4$ seconds.

(3) Find the time t, $0 \le t \le 4$, when the line tangent to the path of the particle is horizontal. Is the direction of motion of the particle toward the left or toward the right at that time? Given a reason for your answer.

(4) There is a point with $x-$coordinate 5 through which the particle passes twice. Find each of the following.

(i) The two values of t when that occurs.

(ii) The slopes of the lines tangent to the particle's path at that point.

(iii) The $y-$coordinate of that point, given $y(2) = 3 + \frac{1}{e}$.

EX 4 At time t, a particle moving in the $xy-$plane is at position $(x(t),\ y(t))$, where $x(t)$ and $y(t)$ are not explicitly given. For $t \ge 0$, $\frac{dx}{dt} = 4t + 1$ and $\frac{dy}{dt} = \sin(t^2)$. At time $t = 0$, $x(0) = 0$ and $y(0) = -4$.

(1) Find the speed of the particle at time $t = 3$, and find the acceleration vector of the particle at time $t = 3$.

(2) Find the slope of the line tangent to the path of the particle at time $t = 3$.

(3) Find the position of the particle at time $t = 3$.

(4) Find the total distance traveled by the particle over the time interval $0 \le t \le 3$.

6.8 Arc Length

If the derivative of function $y = f(x)$ is continuous on the interval $a \le x \le b$, then the length s of the arc of the curve of $y = f(x)$ from the point where $x = a$ to the point where $x = b$ is given by

$$s = \int_a^b \sqrt{1 + \left(\frac{dy}{dx}\right)^2}\, dx\ . \tag{1}$$

Here a small piece of the curve is equal approximately to $\sqrt{1 + (f'(x))^2}\,\Delta x$.

As $\Delta x \to 0$, the sum of these pieces approaches the definite integral above.

If the derivative of the function $x = g(y)$ is continuous on the interval $c \le y \le d$, then the length s of the arc from $y = c$ to $y = d$ is given by

$$s = \int_c^d \sqrt{1 + \left(\frac{dx}{dy}\right)^2}\, dy\ . \tag{2}$$

If a curve is defined parametrically by the equations $x = x(t)$ and $y = y(t)$, if the derivatives of the functions $x(t)$ and $y(t)$ are continuous on $[t_a, t_b]$, (and if the curve does not intersect itself, then the length of the arc from $t = t_a$ to $t = t_b$ is given by

$$s = \int_{t_a}^{t_b} \sqrt{\left(\frac{dx}{dt}\right)^2 + \left(\frac{dy}{dt}\right)^2}\, dt. \tag{3}$$

Example 1 Find the length, to three decimal places, of the arc of $y = x^{3/2}$ from $x = 1$ to $x = 8$.

Solution

Here $\dfrac{dy}{dx} = \dfrac{3}{2}x^{1/2}$, so, by (1),

$$s = \int_1^8 \sqrt{1 + \frac{9}{4}x}\, dx = 22.803.$$

EX 1

Find the length, to three decimal placves, of the curve $(x-2)^2 = 4y^3$ from $y=0$ to $y=1$.

EX 2

Find the length of the arc of the curve $y = x^{3/2}$ from $x=0$ to $x=4$.

EX 3

Find the length of the arc of $y = \ln \sec x$ from $x=0$ to $x = \frac{\pi}{3}$.

EX 4

The position (x, y) of a particle at time t is given parametrically by $x = t^2$ and $y = \frac{t^3}{3} - t$. Find the distance the particle travels between $t=1$ and $t=2$.

EX 5

Find the length one arc of the cycloid $x = t - \sin t$, $y = 1 - \cos t$.

EX 6

Find the length of $x = e^t \cos t$, $y = e^t \sin t$ from $t=2$ to $t=3$.

EX 7

Let $f(x) = e^{2x}$. Let R be the region in the first quadrant bounded by the graph of f, the coordinate axes, and the vertical line $x=k$, where $k>0$. The region R is shown in the figure below.

(1) Write, but do not evaluate, and expression involving an integral that gives the perimeter of R in terms of k.

(2) The region R is rotated about the x-axis to form a solid, Find the volume, V, of the solid in terms of k.

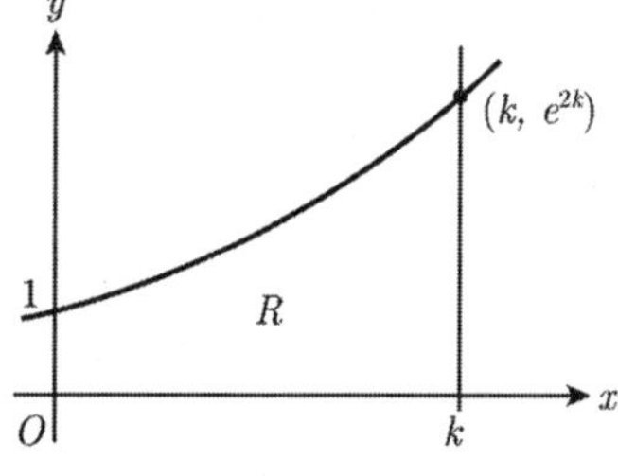

(3) The volume V, found in part (2), changes as k changes. If $\frac{dk}{dt} = \frac{1}{3}$, determine $\frac{dV}{dt}$ when $k = \frac{1}{2}$.

A baker is creating a birthday cake. The base of the cake is the region R in the first quadrant under the graph of $y=f(x)$ for $0 \le x \le 30$, where $f(x)=20\sin\left(\frac{\pi x}{30}\right)$. Both x and y are measured in centimeters. The region R is shown in the figure at the right. The derivative of f is $f'(x)=\frac{2\pi}{3}\cos\left(\frac{\pi x}{30}\right)$.

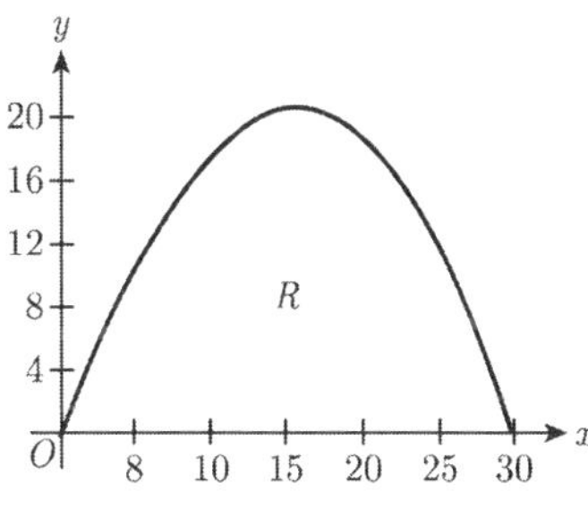

(1) The region R is cut out of a 30-centimeter-by-20-centimeter rectangular sheet of cardboard, and the remaining cardboard is discarded. Find the area of the discarded cardboard.

(2) The cake is a solid with base R. Cross sections of the cake perpendicular to the x-axis are semicircles. If the baker uses 0.05 gram of unsweetened chocolate for each cubic centimeter of cake, how many grams of unsweetened chocolate will be in the cake?

(3) Find the perimeter of the base of the cake.

6.9 Improper Integrals

There are two classes of improper integrals:

(1) those in which at least one of the limits of integration is infinite (the interval is not bounded); and

Illustrations of improper integrals of class (1) are:

$$\int_0^{\infty}\frac{dx}{\sqrt[3]{x+1}}\ ;\ \int_0^{\infty}\frac{dx}{x}\ ;\ \int_{-\infty}^{\infty}\frac{dx}{a^2+x^2}\ ;\ \int_{-\infty}^{0}e^{-x}dx\ ;$$

$$\int_{-\infty}^{-1}\frac{dx}{x^n}(n \text{ a real number})\ ;\ \int_{-\infty}^{0}\frac{dx}{e^x+e^{-x}}\ ;$$

$$\int_0^{\infty}\frac{dx}{(4+x)^2}\ ;\ \int_{-\infty}^{0}e^{-x^2}dx\ ;\ \int_1^{\infty}\frac{e^{-x^2}}{x^2}dx$$

Improper integrals of class (1), where the interval is not bounded, are handled as limits:

$$\int_a^{\infty} f(x)dx = \lim_{b\to\infty}\int_a^b f(x)dx,$$

where f is continuous on $[a, b]$. If the limit on the right exists, the improper integral on the left is said to converge to this limit; if the limit on the right fails to exist, we say that the improper integral *diverges* (or is *meaningless*).

Example 1 $\int_1^{\infty}\frac{dx}{x^2} = \lim_{b\to\infty}\int_1^b x^{2}dx = \lim_{b\to\infty} -\frac{1}{x}\Big|_1^b = \lim_{b\to\infty} -\left(\frac{1}{b}-1\right) = 1$. The given integral thus converges to 1.

Example 2 $\int_{-\infty}^{0} e^{-x}dx = \lim_{b\to-\infty} -e^{-x}\Big|_b^0 = \lim_{b\to-\infty} -(1-e^{-b}) = +\infty$. Thus, this improper integral diverges.

(2) Those of the type $\int_a^b f(x)dx$, where $f(x)$ has a point of discontinuity (becoming infinite) at $x = c$, $a \le c \le b$ (the *function* is not bounded).

The following improper integrals are of class (2):

$\int_0^1 \frac{dx}{x}$; $\int_1^2 \frac{dx}{(x-1)^n}$ (*n* a real number) ; $\int_{-1}^1 \frac{dx}{1-x^2}$; $\int_0^2 \frac{x}{\sqrt{4-x^2}}dx$; $\int_1^2 \frac{dx}{1+\sin x}$; $\int_{-1}^2 \frac{dx}{x(x-1)^2}$; $\int_0^{2\pi} \frac{\sin x\, dx}{\cos x + 1}$; $\int_a^b \frac{dx}{(x-c)^n}$ (n real; $a \le c \le b$) ; $\int_0^1 \frac{dx}{\sqrt{x+x^4}}$; $\int_{-2}^2 \sqrt{\frac{2+x}{2-x}}\,dx$.

Improper integrals of class (2), where the function has an infinite discontinuity, are handled as follows.

To investigate $\int_a^b f(x)dx$, where f becomes infinite at $x = a$, we define $\int_a^b f(x)dx$ to be $\lim_{k\to a^+}\int_k^b f(x)dx$. The given integral then converges or diverges according to whether the limit does or does not exist. If f has its discontinuity at b, we define $\int_a^b f(x)dx$ to be $\lim_{k\to b^-}\int_a^k f(x)dx$; again, the given integral converges or diverges as the limit does or does not exist. When, finally, the integrand has a discontinuity at an interior point c on the interval of integration $(a < c < b)$, we let

$$\int_a^b f(x)dx = \lim_{k \to c^-} \int_a^k f(x)dx + \lim_{m \to c^n} \int_m^b f(x)dx$$

Now the improper integral converges only if both of the limits exist. If either limit does not exist, the improper integral diverges.

Example 3 $\int_0^1 \frac{dx}{\sqrt[3]{x}} = \lim_{k \to 0^+} \int_k^1 x^{-1/3} dx = \lim_{k \to 0^+} \frac{3}{2} x^{2/3} \Big|_k^1 = \lim_{k \to 0^+} \frac{3}{2}(1 - k^{2/3}) = \frac{3}{2}$

Example 4 $\int_2^3 \frac{dt}{(3-t)^2} = \lim_{k \to 3^-} - \int_2^k (3-t)^{-2}(-dt) = \lim_{k \to 3^-} \frac{1}{3-t} \Big|_2^k = +\infty$

This integral diverges.

EX 1 Evaluate the integrals.

(1) $\int_1^\infty \frac{1}{\sqrt{x}} dx$

(2) $\int_0^\infty \frac{1}{x^2+9} dx$

(3) $\int_0^\infty \frac{dy}{e^y}$

(4) $\int_{-\infty}^0 \frac{dz}{(z-1)^2}$

(5) $\int_0^\infty \cos x \, dx$

(6) $\int_0^1 \frac{1}{x^3} dx$

(7) $\int_0^2 \frac{dy}{\sqrt{4-y^2}}$

(8) $\int_0^2 \frac{dx}{(x-1)^{2/3}}$

(9) $\int_{-2}^2 \frac{dx}{x^2}$

(10) $\int_0^{\pi/2} \frac{\sin x}{\sqrt{1-\cos x}} dx$

EX 2 Choose the alternative that gives the area, if it exists, of the region described.

(1) In the first quadrant under the curve of $y = e^{-x}$.

(a) 1 (b) e (c) $\frac{1}{e}$ (d) 2 (e) none of these

(2) In the first quadrant under the curve of $y = xe^{-x^2}$.

(a) 2 (b) $\frac{2}{e}$ (c) $\frac{1}{2}$ (d) $\frac{1}{2e}$ (e) none of these

(3) In the first quadrant above $y = 1$, between by the y-axis and the curve $xy = 1$.

(a) 1 (b) 2 (c) $\frac{1}{2}$ (d) 4 (e) none of these

(4) Between the curve $y = \dfrac{4}{1+x^2}$ and the x-axis.

(a) 2π (b) 4π (c) 8π (d) π (e) none of these

(5) Above the x-axis, between the curve $y = \dfrac{4}{\sqrt{1-x^2}}$ and its asymptotes.

(a) $\dfrac{\pi}{2}$ (b) π (c) 2π (d) 4π (e) none of these

EX 3 Choose the alternative that gives the volume, if it exists, of the solid generated.

(1) $y = \dfrac{1}{x}$, at the right by $x = 1$, and above by $y = 0$; about the x-axis.

(a) $\dfrac{\pi}{2}$ (b) π (c) 2π (d) 4π (e) none of these

(2) The first-quadrant region under $y = e^{-x}$; about the x-axis.

(a) $\dfrac{\pi}{2}$ (b) π (c) 2π (d) 4π (e) none of these

EX 4 Let f and g be the functions defined by $f(x) = \dfrac{1}{x}$ and $g(x) = \dfrac{4x}{1+4x^2}$, for all $x > 0$.

(1) Find the absolute maximum value of g on the open interval $(0, \infty)$ if the maximum exists. Find the absolute minimum value of g on the open interval $(0, \infty)$ if the minimum exists. Justify your answers.

(2) Find the area of the unbounded region in the first quadrant to the right of the vertical line $x = 1$. below the graph of f, and above the graph of g.

Multiple Choice

01 The area of the region between the graph of $y = 4x^3 + 2$ and the $x-$axis from $x = 1$ to $x = 2$ is

(A) 36　　(B) 23　　(C) 20
(D) 17　　(E) 9

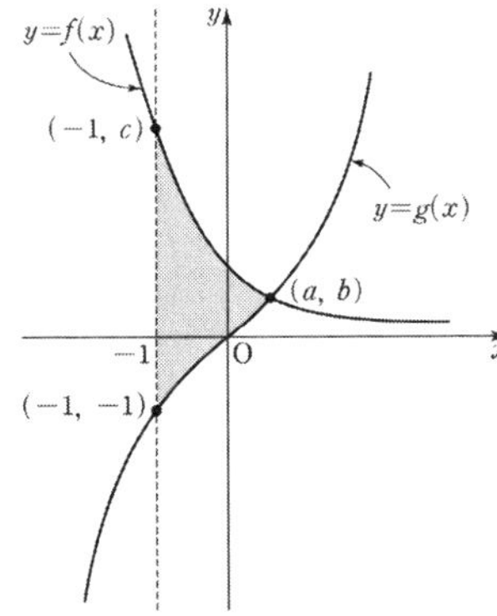

02 Let f be the function given by $f(x) = x^3 - 3x^2$. What are all values of c that satisfy the conclusion of the Mean Value Theorem of differential calculus on the closed interval $[0, 3]$?

(A) 0 only　　(B) 2 only　　(C) 3 only
(D) 0 and 3　　(E) 2 and 3

03 The region in the first quadrant between the $x-$axis and the graph of $y = 6x - x^2$ is rotated around the $y-$axis. The volume of the resulting solid of revolution is given by

(A) $\int_0^6 \pi(6x - x^2)^2 dx$　　(B) $\int_0^6 2\pi x(6x - x^2) dx$　　(C) $\int_0^6 \pi x(6x - x^2)^2 dx$
(D) $\int_0^6 \pi(3 + \sqrt{9-y})^2 dy$　　(E) $\int_0^9 \pi(3 + \sqrt{9-y})^2 dy$

04 The base of a solid is the region enclosed by the graph of $y = e^{-x}$, the coordinate axes, and the line $x = 3$. If all plane cross sections perpendicular to the $x-$axis are squares, then its volume is

(A) $\dfrac{(1-e^{-6})}{2}$ (B) $\dfrac{1}{2}e^{-6}$ (C) e^{-6}

(D) e^{-3} (E) $1-e^{-3}$

05 If n is a positive integer, then $\lim_{n\to\infty}\dfrac{1}{n}\left[\left(\dfrac{1}{n}\right)^2+\left(\dfrac{2}{n}\right)^2+\cdots+\left(\dfrac{3n}{n}\right)^2\right]$ can be expressed as

(A) $\displaystyle\int_0^1 \frac{1}{x^2}dx$ (B) $\displaystyle 3\int_0^1\left(\frac{1}{x}\right)^2 dx$ (C) $\displaystyle\int_0^3\left(\frac{1}{x}\right)^2 dx$

(D) $\displaystyle\int_0^3 x^2 dx$ (E) $\displaystyle 3\int_0^3 x^2 dx$

06 $\displaystyle\int_2^{+\infty}\frac{dx}{x^2}$ is

(A) $\dfrac{1}{2}$ (B) $\ln 2$ (C) 1

(D) 2 (E) nonexistent

07 The base of a solid is the first quadrant enclosed by the parabola $y = 4x^2$, the line $x = 1$, and the $x-$axis. Each plane section of the solid perpendicular to the $x-$axis is a square. The volume of the solid is

(A) $\dfrac{4\pi}{3}$ (B) $\dfrac{16\pi}{5}$ (C) $\dfrac{4}{3}$

(D) $\dfrac{16}{5}$ (E) $\dfrac{64}{5}$

08 The length of the curve $y = x^3$ from $x = 0$ to $x = 2$ is given by

(A) $\int_0^2 \sqrt{1+x^6}\,dx$ (B) $\int_0^2 \sqrt{1+3x^2}\,dx$ (C) $\pi\int_0^2 \sqrt{1+9x^4}\,dx$

(D) $2\pi\int_0^2 \sqrt{1+9x^4}\,dx$ (E) $\int_0^2 \sqrt{1+9x^4}\,dx$

09 Let R be the region between the graph of $y = 1$ and $y = \sin x$ from $x = 0$ to $x = \frac{\pi}{2}$. The volume of the solid obtained by revolving R about the $x-$axis is given by

(A) $2\pi\int_0^{\frac{\pi}{2}} x\sin x\,dx$ (B) $2\pi\int_0^{\frac{\pi}{2}} x\cos x\,dx$ (C) $\pi\int_0^{\frac{\pi}{2}} (1-\sin x)^2 dx$

(D) $\pi\int_0^{\frac{\pi}{2}} \sin^2 x\,dx$ (E) $\pi\int_0^{\frac{\pi}{2}} (1-\sin^2 x)dx$

10 $\int_4^0 \frac{-2x}{\sqrt[3]{9-x^2}}dx$ is

(A) $7^{\frac{2}{3}}$ (B) $\frac{3}{2}(7^{\frac{2}{3}})$ (C) $9^{\frac{2}{3}}+7^{\frac{2}{3}}$

(D) $\frac{3}{2}(9^{\frac{2}{3}}-7^{\frac{2}{3}})$ (E) nonexistent

11 The shaded region R, shown in the figure below, is rotated about the $y-$axis to form a solid whose volume is 10 cubic units. Of the following, which best approximates k?

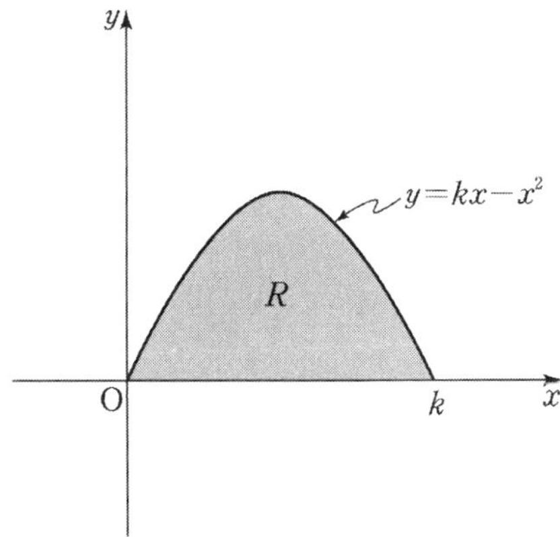

(A) 1.51 (B) 2.09 (C) 2.49

(D) 4.18 (E) 4.77

12 The length of the curve determined by the equations $x=t^2$ and $y=t$ from $t=0$ to $t=4$ is

(A) $\int_0^4 \sqrt{4t+1}\,dt$ (B) $2\int_0^4 \sqrt{t^2+1}\,dt$ (C) $\int_0^4 \sqrt{2t^2+1}\,dt$

(D) $\int_0^4 \sqrt{4t^2+1}\,dt$ (E) $2\pi\int_0^4 \sqrt{4t^2+1}\,dt$

13 A particle moves along the $x-$axis so that its acceleration at any time t is $a(t)=2t-7$. It the initial velocity of the particle is 6, at what time t during the interval $0 \le t \le 4$ is the particle farthest to the right?

(A) 0 (B) 1 (C) 2

(D) 3 (E) 4

14 The length of the path described by the parametric equations $x=\cos^3 t$ and $y=\sin^3 t$, for $0 \le t \le \frac{\pi}{2}$, is given by

(A) $\int_0^{\frac{\pi}{2}} \sqrt{3\cos^2 t+3\sin^2 t}\,dt$

(B) $\int_0^{\frac{\pi}{2}} \sqrt{-3\cos^2 t\sin t+3\sin^2 t\cos t}\,dt$

(C) $\int_0^{\frac{\pi}{2}} \sqrt{9\cos^4 t+9\sin^4 t}\,dt$

(D) $\int_0^{\frac{\pi}{2}} \sqrt{9\cos^4 t\sin^2 t+9\sin^4 t\cos^2 t}\,dt$

(E) $\int_0^{\frac{\pi}{2}} \sqrt{\cos^6 t+\sin^6 t}\,dt$

15 The base of a solid is the region in the first quadrant enclosed by the graph of $y=2-x^2$ and the coordinate axes. If every cross section of the solid perpendicular to the $y-$axis is a square, the volume of the solid is given by

(A) $\pi\int_0^2 (2-y)^2\,dy$ (B) $\int_0^2 (2-y)\,dy$ (C) $\pi\int_0^{\sqrt{2}} (2-x^2)^2\,dx$

(D) $\int_0^{\sqrt{2}} (2-x^2)^2\,dx$ (E) $\int_0^{\sqrt{2}} (2-x^2)\,dx$

16 $\int_0^\infty x^2 e^{-x^3}\,dx$ is

(A) $-\frac{1}{3}$ (B) 0 (C) $\frac{1}{3}$

(D) 1 (E) divergent

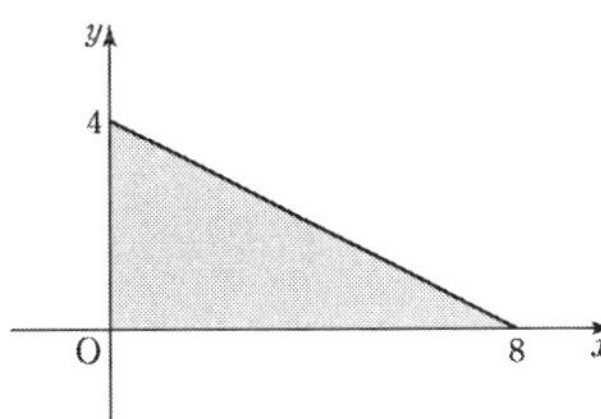

17 The base of a solid is a region in the first quadrant bounded by the x – axis, the y – axis, and the line $x+2y=8$, as shown in the figure above. If cross sections of the solid perpendicular to the x – axis are semicircles, what is the volume of the solid?

(A) 12.566 (B) 14.661 (C) 16.755

(D) 67.021 (E) 134.041

18 The acceleration α of a body moving in a straight line is given in terms of time t by $\alpha = 8 - 6t$.

If the velocity of the body is 25 at $t=1$ and if $s(t)$ is the distance of the body from the origin at time t, what is $s(4)-s(2)$?

(A) 20 (B) 24 (C) 28

(D) 32 (E) 42

19 A particle moves on the curve $y=\ln x$ so that the x – component gas velocity $x'(t)=t+1$

For $t \geq 0$. At time $t=0$, the particle is at the point $(1,0)$. At time $t=1$, the particle is at the point

(A) $(2, \ln 2)$ (B) $(e^2, 2)$ (C) $\left(\frac{5}{2}, \ln\frac{5}{2}\right)$

(D) $(3, \ln 3)$ (E) $\left(\frac{3}{2}, \ln\frac{3}{2}\right)$

20 The region bounded by the graph of $y = 2x - x^2$ and the $x-$axis is the base of a solid. For this solid, each cross section perpendicular to the $x-$axis is an equilateral triangle. What is the volume of the solid?

(A) 1.333 (B) 1.067 (C) 0.577
(D) 0.462 (E) 0.267

21 What is the average (mean) value of $3t^3 - t^2$ over the interval $-1 \le t \le 2$?

(A) $\frac{11}{4}$ (B) $\frac{7}{2}$ (C) 8
(D) $\frac{33}{4}$ (E) 16

22 Which of the following integrals gives the length of the graph of $y = \tan x$ between $x = a$ and $x = b$, where $0 < a < b < \frac{\pi}{2}$?

(A) $\int_a^b \sqrt{x^2 + \tan^2 x}\,dx$ (B) $\int_a^b \sqrt{x + \tan x}\,dx$ (C) $\int_a^b \sqrt{1 + \sec^2 x}\,dx$
(D) $\int_a^b \sqrt{1 + \tan^2 x}\,dx$ (E) $\int_a^b \sqrt{1 + \sec^4 x}\,dx$

23 $\int_1^{\infty} xe^{-x^2}dx$ is

(A) $-\frac{1}{e}$ (B) $\frac{1}{2e}$ (C) $\frac{1}{e}$
(D) $\frac{2}{e}$ (E) divergent

24 Let R be the region in the first quadrant bounded above by the graph of $y = \ln(3 - x)$, for $0 \le x \le 2$. R is the base of a solid for which each cross section perpendicular to the $x-$axis is a square. What is the volume of the solid ?

(A) 0.442 (B) 1.029 (C) 1.296
(D) 3.233 (E) 4.071

07

Differential Equations

7.1 Slope Fields (Direction Fields)

A slope field (or direction field) is a graphical solution to a first order differential equation $\dfrac{dy}{dx} = f(x,y)$.

Step 1. Select a rectangular grid of points.

Step 2. Calculate the tangent slope at each point.

Step 3. Draw a short line segment through each point with a graphical representation of its slope.

Example 1 Consider the differential equation $y' = x+1$.

(1) On a grid, like that shown in the figure, sketch a slope field for the given differential equation at the 20 points indicated.

(2) Sketch the solution curve that passes through the point $(-1,0)$.

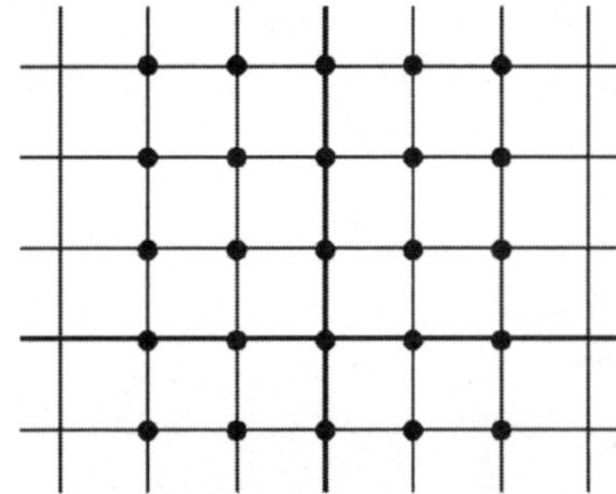

Solution

(1)

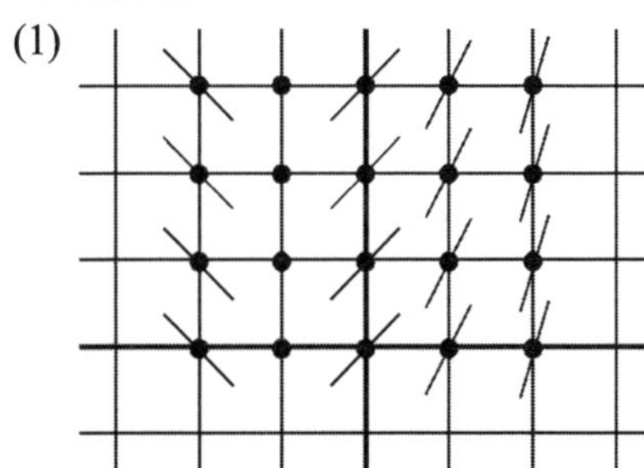

(2)

EX 1 The slope field for the differential equation $\dfrac{dy}{dx} = \dfrac{1}{x}$ is shown below.

(1) Carefully draw the solution curve that passes through the point (1, 0.5).

(2) Find the general solution for the equation.

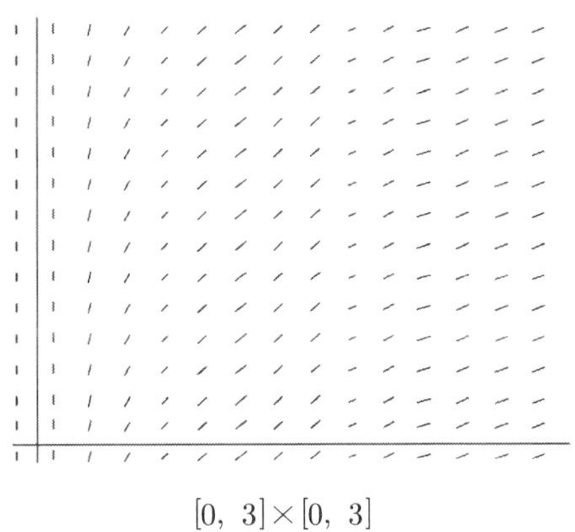

$[0,\ 3]\times[0,\ 3]$

(a) Sketch a slope field for each of the following differential equations.

(b) Sketch the solution curve that passes through the indicated point.

(1) $\dfrac{dy}{dx}=y+1,\ (0,\ 1)$

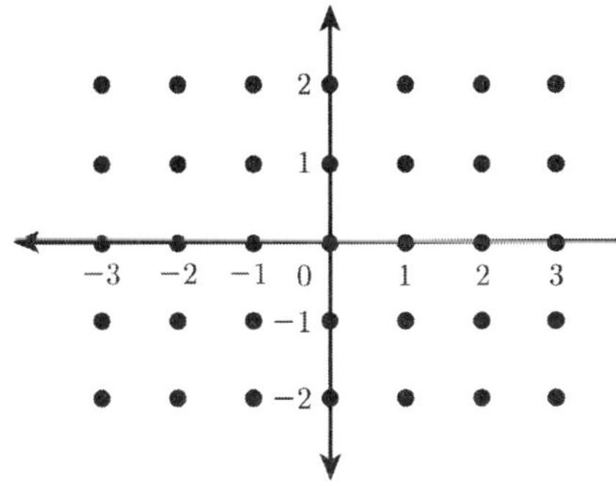

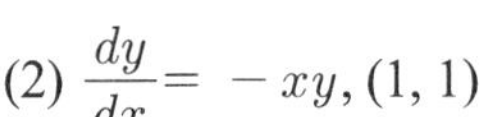

(2) $\dfrac{dy}{dx}=-xy,\ (1,\ 1)$

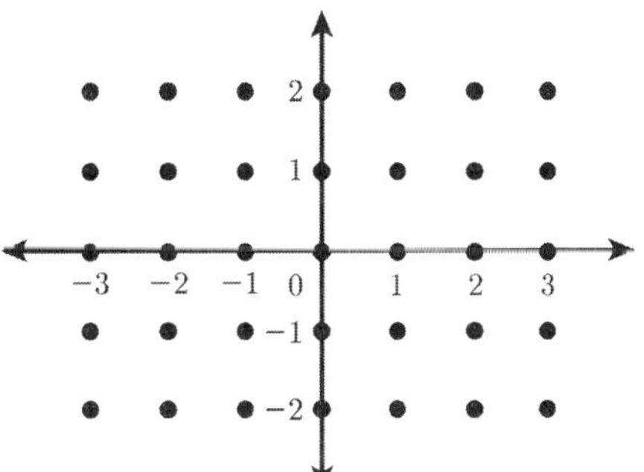

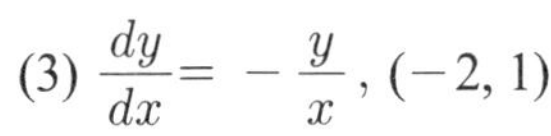

(3) $\dfrac{dy}{dx}=-\dfrac{y}{x},\ (-2,\ 1)$

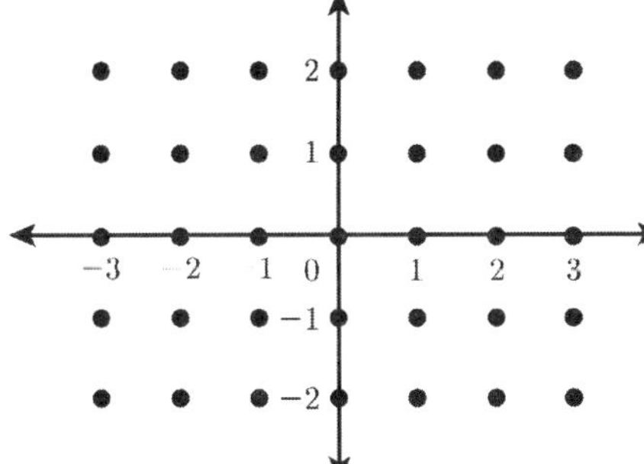

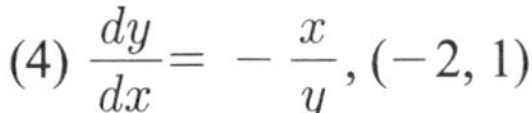

(4) $\dfrac{dy}{dx}=-\dfrac{x}{y},\ (-2,\ 1)$

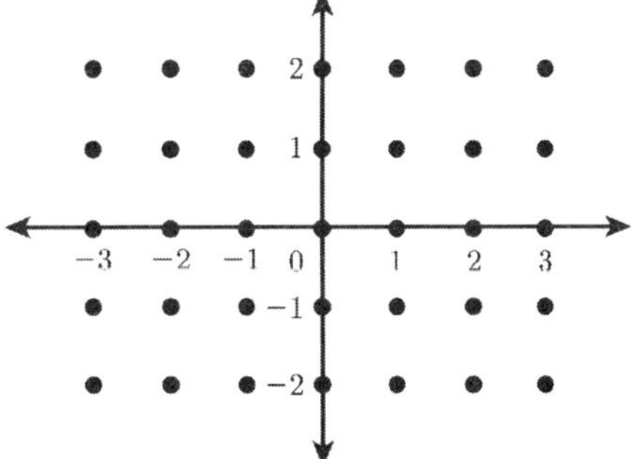

(5) $\frac{dy}{dx}=\frac{2x}{y}$, (1, 2)

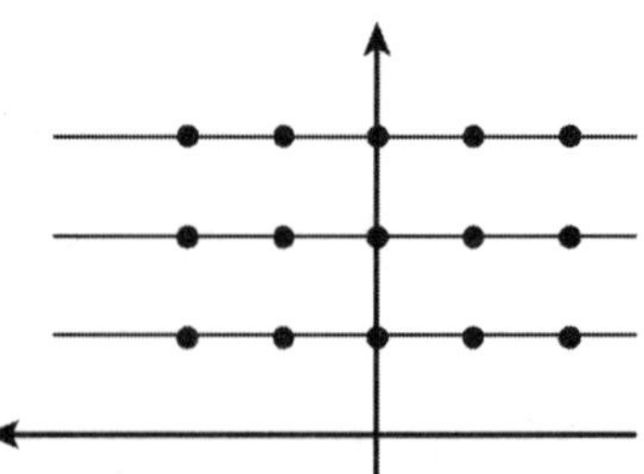

(6) $\frac{dy}{dx}=x+y$, (0, 0)

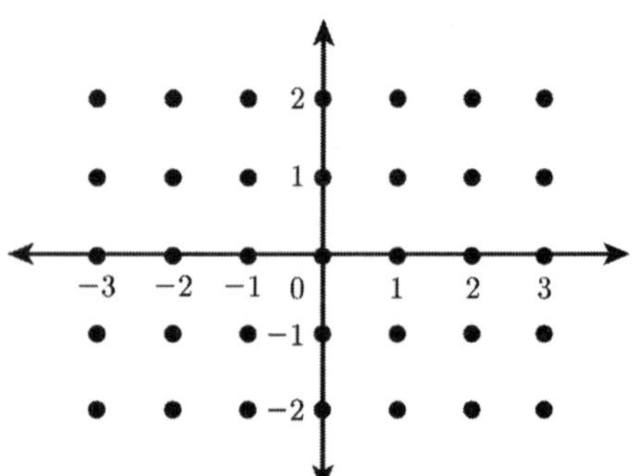

EX3 Match the slope fields with their differential equations.

(1) $y'=x+y$ (2) $y'=0.5y$ (3) $y'=0.5x-1$ (4) $y'=-\frac{x}{y}$

a)

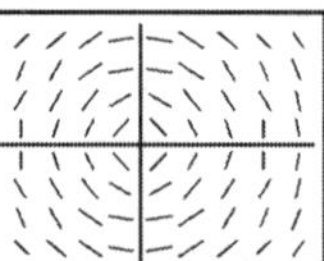

b)

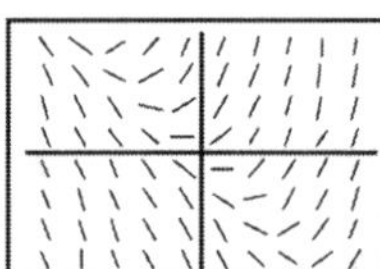

c)

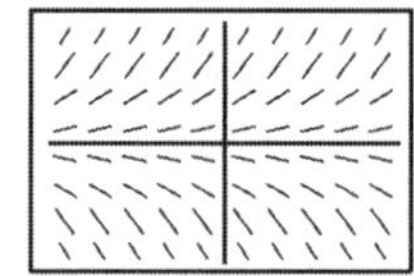

d)

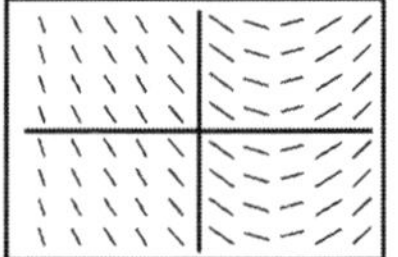

EX4 Match each differential equation with its slope field.

(1) $\frac{dy}{dx}=y(y+1)$

(2) $\frac{dy}{dx}=\sin 2x$

(3) $\frac{dy}{dx}=\cos 2x$

(4) $\frac{dy}{dx}=\frac{1}{1+x^2}$

(5) $\frac{dy}{dx}=1+y^2$

(6) $\frac{dy}{dx}=-xy$

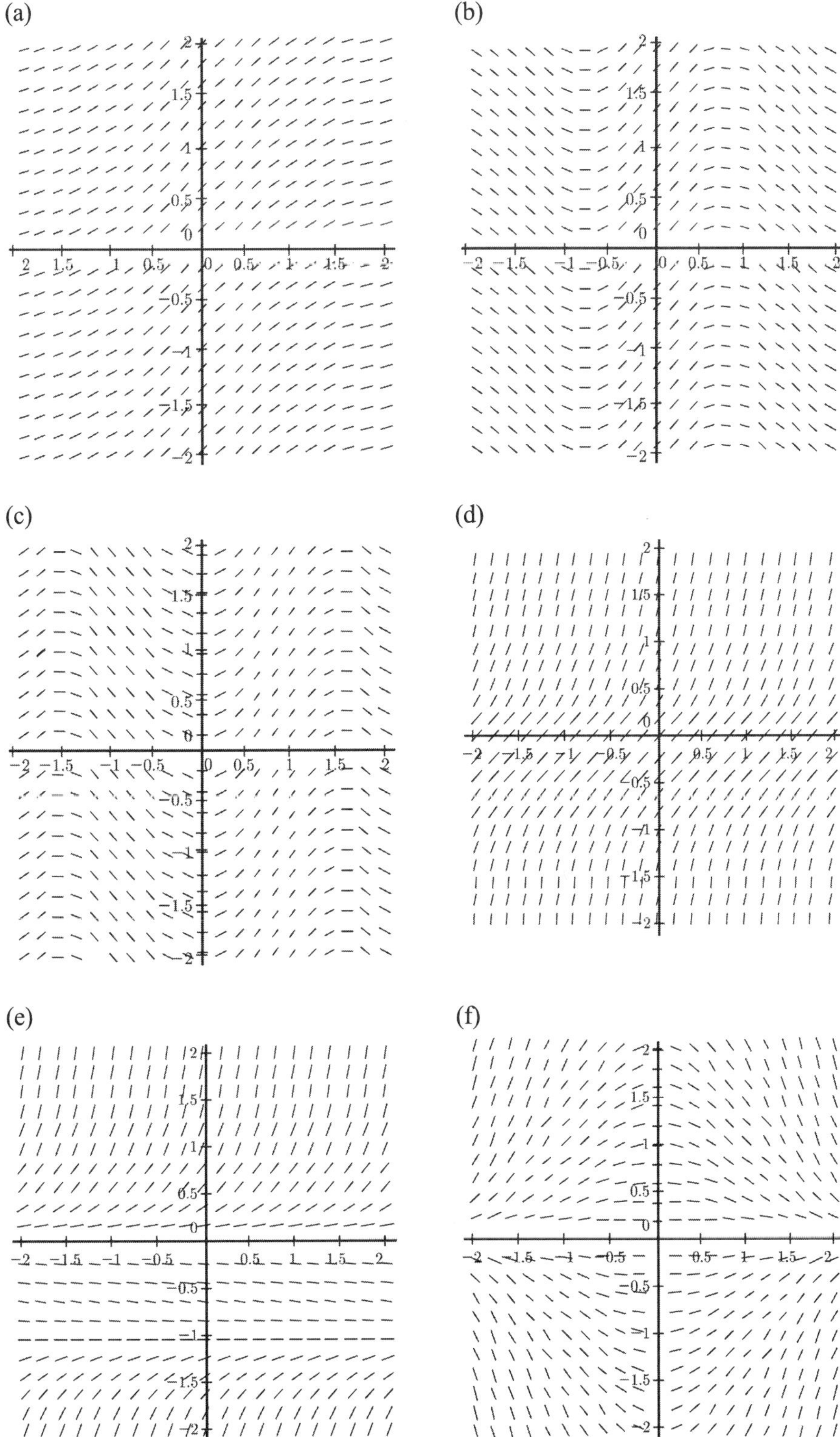

EX 5 Match each differential equation with its slope field.

(1) $\dfrac{dy}{dx} = \cos x$ (2) $\dfrac{dy}{dx} = 2x$

(3) $\dfrac{dy}{dx} = 3x^2 - 3$ (4) $\dfrac{dy}{dx} = -\dfrac{\pi}{2}$

(a)

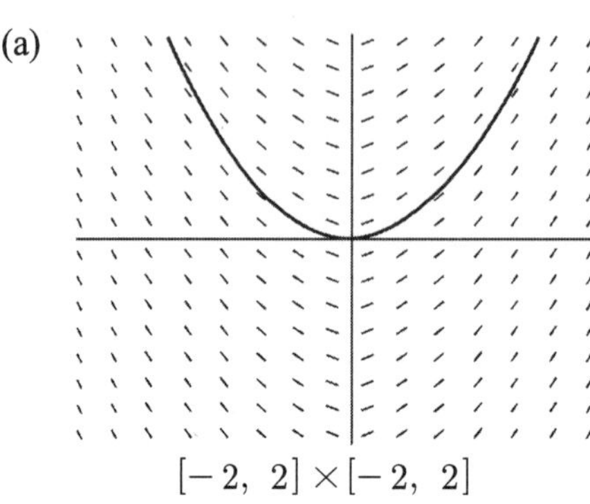

$[-2,\ 2] \times [-2,\ 2]$

(b)

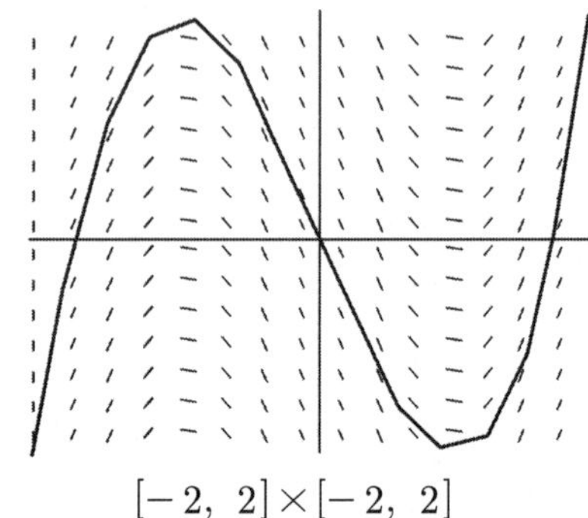

$[-2,\ 2] \times [-2,\ 2]$

(c)

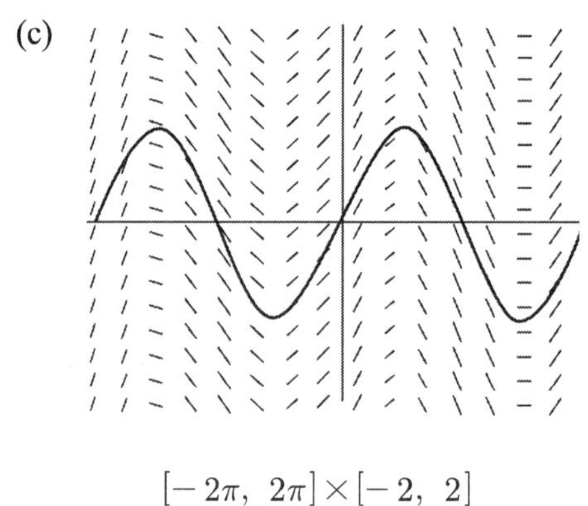

$[-2\pi,\ 2\pi] \times [-2,\ 2]$

(d)

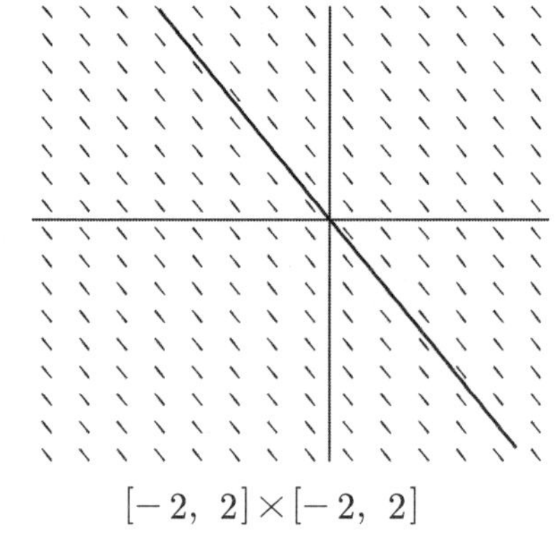

$[-2,\ 2] \times [-2,\ 2]$

7.2 First Order Separable Differential Equations

A differential equation has variables separable if it is of the form

$$\frac{dy}{dx} = \frac{f(x)}{g(y)} \quad \text{or} \quad g(y)\,dy - f(x)\,dx = 0.$$

The general solution is

$$\int g(y)\,dy - \int f(x)\,dx = C \quad \text{(C arbitrary).}$$

Example 1 Solve the d.e. $\frac{dy}{dx} = -\frac{x}{y}$, given the initial condition $y(0) = 2$.

Solution

We rewrite the equation as $y\,dy = -x\,dx$. We then integrate, getting

$$\int y\,dy = -\int x\,dx,$$

$$\frac{1}{2}y^2 = -\frac{1}{2}x^2 + k,$$

$$y^2 + x^2 = C \qquad (where\ C = 2k).$$

Since $y(0) = 2$, we get $4 + 0 = C$; the particular solution is therefore $x^2 + y^2 = 4$.

EX 1 Solve the differential equation.

(1) $\frac{dy}{dx} = \sqrt{xy},\ f(0) = 1$

(2) $\frac{dy}{dx} = -2xy^3,\ f(0) = 1$

(3) $\frac{dy}{dx} = e^{x-y},\ f(1) = 1$

(4) $2x\frac{dy}{dx} - 3y = 0,\ y(1) = 100$

EX 2 Solve the differential equation.

(1) $\frac{dy}{dx} = \frac{y}{2\sqrt{x}},\ f(4) = 1$

(2) $y' = y^2\cos 2x,\ f(\pi/4) = -1$

(3) $(\ln y)\frac{dy}{dx} = \frac{y}{x},\ f(1) = e$

(4) $y' = \frac{2+y}{x},\ y(-1) = 3$

(5) $\frac{dy}{dx} = e^y,\ f(1) = 0$

(6) $\frac{dy}{dx} = \frac{2\sqrt{y}}{x^2+1},\ y(0) = 4$

(7) $\cos^2 x\, e^y \frac{dy}{dx} = 2,\ f\left(-\frac{\pi}{4}\right) = 0$

(8) $x\frac{dy}{dx} - 4 = y^2,\ f(1) = 0$

EX 3

(1) Given $f(x)f'(x) = x$ for all real x and $f(0) = 1$, find the value of $f(x)$.

(2) Given $f'(x) = (f(x))^2 + 1$ for all real x and $f(0) = \sqrt{3}$, find the value of $f\left(\frac{\pi}{2}\right)$.

7.3 Exponential Growth and Decay

• Exponential Growth and Decay

"A positive quantity y increases (or decreases) at a rate that at any time t is proportional to the amount present." It follows that the quantity y satisfies the d.e.

$$\frac{dy}{dt}=ky$$

where $k>0$ if y is increasing and $k<0$ if y is decreasing.

Proof

$$\frac{dy}{y}=k\,dt,$$

$$\int\frac{1}{y}dy=\int k\,dt,$$

$\ln y = kt + C$ (C a constant).

$$y=e^{kt+C}=e^{kt}\cdot e^{C}$$

$$=ce^{kt}\quad(\text{where } c=e^{C}).$$

Example 1 A radioactive substance decays at a rate proportional to the amount of substance present. Radium-226 is a radioactive substance with a half-life of 1612 years. How long will it take for 200 g of this substance to decay to 150g?

Solution

$$\frac{dy}{dt}=ky$$

$$\int\frac{dy}{y}=\int k\,dt$$

$$\ln|y|=kt+C$$

$$\therefore y=Ae^{kt}$$

When $t=0,\ y=200$

$$y=200e^{kt}$$

when $t=1612,\ y=100$

$$100=200e^{k(1612)}$$

$$\frac{1}{2}=e^{1612k}$$

$$k=\frac{\ln\frac{1}{2}}{1612}=-0.00043$$

$$y=200\,e^{-0.00043t}$$

Find when $y=150$

$$150=200e^{-0.00043t}$$

$$\frac{3}{4}=e^{-0.00043t}$$

$$t=\frac{\ln\frac{3}{4}}{-0.00043}\approx 669\,yr$$

EX 1 The bacteria in a certain increase continuously at a rate proportional to the number present.

(1) If the number triples in 6 hours, how many will there be in 12 hours?

(2) In how many hours will the original number quadruple?

EX 2 It is estimated that 20% of a certain radioactive substance decays in 45 hours. What is the half-life of the substance?

EX 3 The population of a country is growing at a rate proportional to its population. If the growth rate per year is 4% of the current population, how long will it take for the population to double?

EX 4 The rate of change of y is proportional to y. If $y=4$ when $t=0$ and $y=10$ when $t=2$, What is the value of y when $t=5$?

• Restricted Growth and Decay

Suppose the temperature of an object changes at a rate proportional to the difference between its temperature(T) and that of the surrounding medium (T_S).

If $T > T_S$: Newton's Law of Cooling

$\dfrac{dT}{dt} = -k(T - T_S) \Rightarrow T = T_S + Ae^{-kt}$ where k is the positive proportional constant.

If $T < T_S$

$\dfrac{dT}{dt} = k(T_S - T) \Rightarrow T = T_S - Ae^{-kt}$ where k is the positive proportional constant.

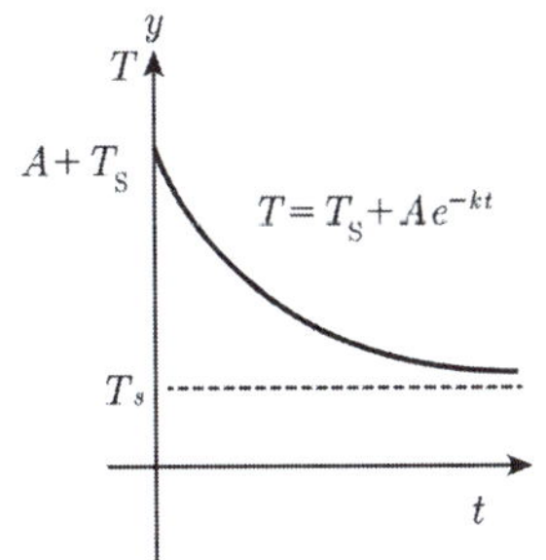

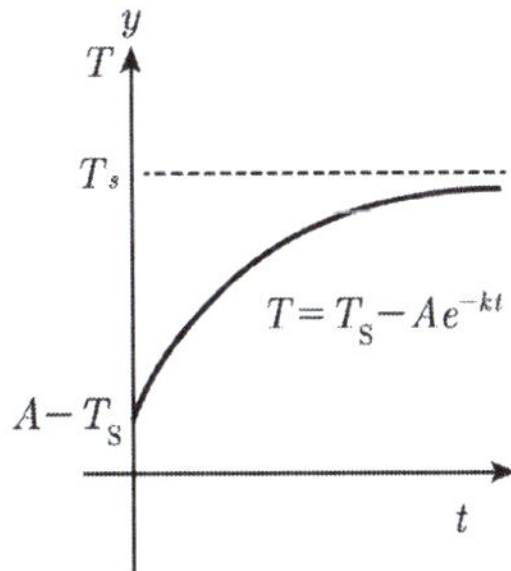

Proof

$$\frac{dT}{dt} = -k(T - T_S)$$

$$\frac{dT}{T - T_S} = -kdt$$

$$\int \frac{dT}{T - T_S} = \int -kdt$$

$$\ln|T - T_S| = -kt + C \ (where \ \ T > T_S)$$

$$|T - T_S| = e^{-kt + C}$$

$$T - T_S = \pm e^{-kt + C}$$

$$T - T_S = Ae^{-kt} \quad (where \ A = \pm e^C)$$

$$\therefore T = T_S + Ae^{-kt}$$

Example 2 A cup of coffee is initially 80℃ and is placed in a room with a temperature of 20℃. After 5 minutes, the temperature of the coffee drops to 60℃. How long will it take for the coffee to drop to 30℃ according to Newton's law of cooling?

Solution

$$\frac{dT}{dt} = -k(T - 20)$$

$$\int \frac{dT}{t - 20} = \int -kdt$$

$$\ln|T - 20| = -kt + C$$

$$\therefore \ T - 20 = Ae^{-kt}$$

When $t = 5, \ T = 60.$

$$40 = 60e^{-5k}$$

$$\frac{2}{3} = e^{-5k}$$

$$-5k = \ln\left(\frac{2}{3}\right)$$

$$k = -\frac{1}{5}\ln\left(\frac{2}{3}\right)$$

$$\therefore T - 20 = 60e^{\frac{t}{5}\ln\left(\frac{2}{3}\right)}$$

When $t = 0, \ T = 80.$

$$80 - 20 = Ae^0$$

$$A = 60$$

$$\therefore T - 20 = 60e^{-kt}$$

Find t when $T = 30.$

$$10 = 60e^{\frac{t}{5}\ln\left(\frac{2}{3}\right)}$$

$$\frac{1}{6} = e^{\frac{t}{5}\ln\left(\frac{2}{3}\right)}$$

$$\ln\left(\frac{1}{6}\right) = \frac{t}{5}\ln\left(\frac{2}{3}\right)$$

$$t = \frac{5\ln\left(\frac{1}{6}\right)}{\ln\left(\frac{2}{3}\right)}$$

$$= 22.095 \text{ min}$$

EX 5 According to Newton's law of cooling, a hot object cools at a rate proportional to the difference between its own temperature and that of its environment. If a roast at room temperature 68℉ is put into a 20℉ freezer, and if, after 2 hours, the temperature of the roast is 40℉: (1) What it its temperature after 5 hours? (2) How long will it take for the temperature of the roast to fall to 21℉?

EX 6 A bottle of water initially has a temperature of 77℉. It is left in a refrigerator set to 41℉. How long will it take for the water to reach 45℉, if it drops to 68℉ after 10 minutes?

EX 7 Advertisers generally assume that the rate at which people hear about a product is proportional to the number of people who have not yet heard about it. Suppose that the size of a community is 15,000, that to begin with no one has heard about a product, but that after 6 days 1500 people know about it. How long will it take for 2700 people to have heard of it?

EX 8 A frozen pizza with a temperature of −5℃ is put into an oven set to 250℃. How long would it take for the pizza to reach 200℃, if it is 100℃ after 7 minutes?

EX 9 A murder victim was found at midnight. A detective established the corpse's body temperature to be 25℃. Two hours later it dropped to 23℃. Suppose the temperature of a living body is 37℃ and the temperature of the room was constant at 20℃. Estimate the time at which the murder occurred.

7.4 Euler's Method

Euler's Method : Approximate value for the solution of the initial-value problem $\frac{dy}{dx} = f(x,\ y),\ y(x_0) = y_0$, with step site dx, at $x_n = x_{n-1} + dx$, are $y_n = y_{n-1} + f(x_{n-1},\ y_{n-1})dx\ \ n = 1,2,3\cdots$

that is,

$$\begin{aligned} x_1 &= x_0 + dx & y_1 &= y_0 + f(x_0, y_0)\,dx \\ x_2 &= x_1 + dx & y_2 &= y_1 + f(x_1, y_1)\,dx \\ &\vdots & &\vdots \\ x_n &= x_{n-1} + dx. & y_n &= y_{n-1} + f(x_{n-1}, y_{n-1})\,dx. \end{aligned}$$

Proof

Given a differential equation $dy/dx = f(x,y)$ and an initial condition $y(x_0) = y_0$, we can approximate the solution $y = y(x)$ by its linearization

$$L(x) = y(x_0) + y'(x_0)(x - x_0) \quad \text{or} \quad L(x) = y_0 + f(x_0, y_0)(x - x_0).$$

The function $L(x)$ gives a good approximation to the solution $y(x)$ in a short interval about x_0(Figure).

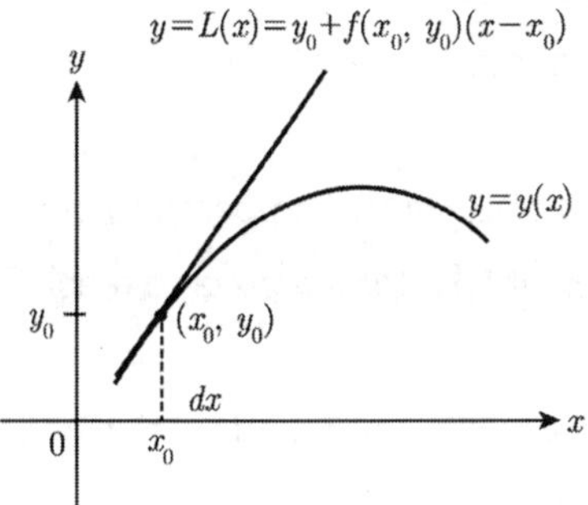

The linearization $L(x)$ of $y = y(x)$ at $x = x_0$.

We know the point $(x_0,\ y_0)$ lies on the solution curve. Suppose that we specify a new value for the independent variable to be $x_1 = x_0 \neq dx$. (Recall that $dx = \Delta x$ in the definition of differentials.) If the increment dx is small, then

$$y_1 = L(x_1) = y_0 + f(x_0, y_0)\,dx$$

is a good approximation to the exact solution value $y = y(x_1)$.

So from the point $(x_0,\ y_0)$, which lies exactly on the solution curve, we have obtained the point $(x_1,\ y_1)$, which lies very close to the point $(x_1,\ y(x_1))$ on the solution curve(Figure).

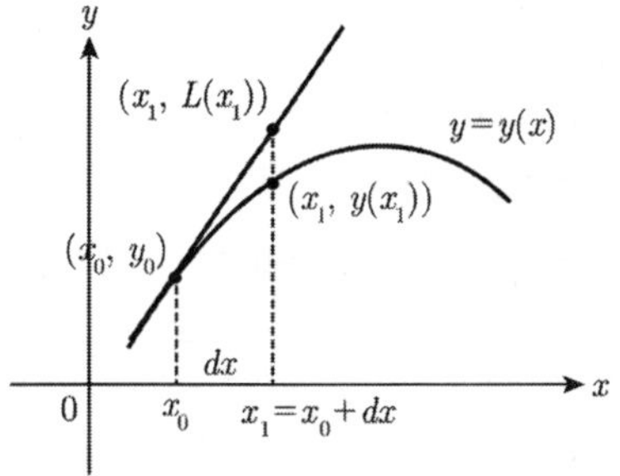

The first Euler step approximates $y(x_1)$ with $y_1 = L(x_1)$.

Using the point $(x_1,\ y_1)$ and the slope $f(x_1,\ y_1)$ of the solution curve through $(x_1,\ y_1)$, we take on second step.

Setting $x_2 = x_1 + dx$, we use the linearization of the solution curve through $(x_1,\ y_1)$ to calculate

$$y_2 = y_1 + f(x_1, y_1)\,dx.$$

This gives the next approximation $(x_2,\ y_2)$ to values along the solution curve $y = y(x)$ (Figure). Continuing in this fashion, we take a third step from the point $(x_2,\ y_2)$ with slope $f(x_2,\ y_2)$ to obtain the third approximation

$$y_3 = y_2 + f(x_2, y_2)\,dx,$$

and so on. We are literally building an approximation to one of the solutions by following the direction of the slope field of the differential equation.

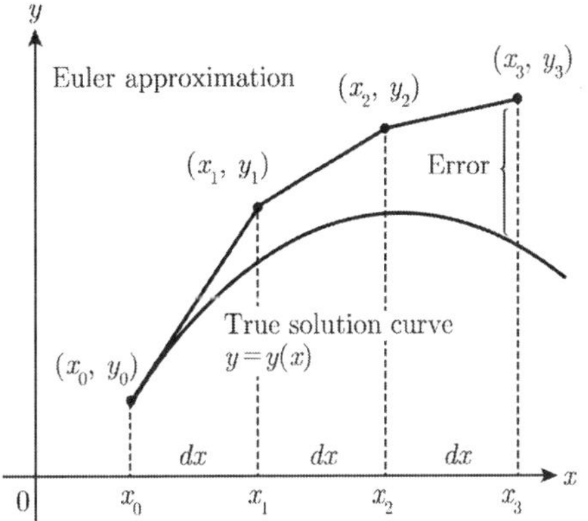

There steps in the Euler approximation
to the solution of the initial value problem $y' = f(x,y),\ y(x_0) = y_0$

Example 1 Find the first three approximations $y_1,\ y_2,\ y_3$ using Euler's method for the initial value problem.

$$y' = 1 + y, \quad y(0) = 1,$$

stating at x_0 with $dx = 0.1$.

Solution

We have $x_0 = 0,\ y_0 = 1,\ x_1 = x_0 + dx = 0.1, x_2 = x_0 + 2dx = 0.2$, and $x_3 = x_0 + 3dx = 0.3$.

$$\begin{aligned}
y_1 &= y_0 + f(x_0, y_0)\,dx \\
&= y_0 + (1 + y_0)\,dx \\
&= 1 + (1+1)(0.1) = 1.2 \\
y_2 &= y_1 + f(x_1, y_1)\,dx \\
&= y_1 + (1 + y_1)\,dx \\
&= 1.2 + (1 + 1.2)(0.1) = 1.42 \\
y_3 &= y_2 + f(x_2, y_2)\,dx \\
&= y_2 + (1 + y_2)\,dx \\
&= 1.42 + (1 + 1.42)(0.1) = 1.662
\end{aligned}$$

EX 1 Use Euler's method to calculate the first three approximations to the given initial value problem for the specified increment size.

(1) $y' = 1 - \dfrac{y}{x}$, $y(2) = -1$, $dx = 0.5$

(2) $y' = x(1-y)$, $y(1) = 0$, $dx = 0.2$

(3) $y' = 2xy + 2y$, $y(0) = 3$, $dx = 0.2$

(4) $y' = y^2(1+2x)$, $y(-1) = 1$, $dx = 0.5$

EX 2 Consider the differential equation $\dfrac{dy}{dx} = y^2(2x+2)$. ≤ t y = f(x) be the particular solution to the differential equation with initial condition $f(0) = -1$.

(1) Find $\displaystyle\lim_{x \to 0} \frac{f(x)+1}{\sin x}$. Show the work that leads to your answer.

(2) Use Euler's method, stating at $x = 0$ with two steps of equal size. to approximate $f\left(\dfrac{1}{2}\right)$.

(3) Find $y = f(x)$, the particular solution to the differential equation with initial condition $f(0) = -1$.

EX 3 Consider the differential equation $\dfrac{dy}{dx} = 1 - y$. ≤ t y = f(x) be the particular solution to this differential equation with the initial condition $f(1) = 0$. For this particular solution, $f(x) < 1$ for all values of x.

(1) Use Euler's method, starting at $x = 1$ with two steps of equal size. to approximate $f(0)$. Show the work that leads to your answer.

(2) Find $\displaystyle\lim_{x \to 1} \frac{f(x)}{x^3 - 1}$. Show the work that leads to your answer.

(3) Find the particular solution $y=f(x)$ to the differential equation $\dfrac{dy}{dx}=1-y$ with the initial condition $f(1)=0$.

EX 4 Consider the differential equation $\dfrac{dy}{dx}=3x+2y+1$.

(1) Find $\dfrac{d^2y}{dx^2}$ in terms of x and y.

(2) Find the values of the constants m, b and r for which $y=mx+b+e^{rx}$ is a solution to the differential equation.

(3) $\leq$ t $\mathrm{y}=\mathrm{f}(\mathrm{x})$ be a particular solution to the differential equation with the initial condition $f(0)=-2$. Use Euler's method, stating at $x=0$ with a step size of $\dfrac{1}{2}$, to approximate $f(1)$. Show the work that leads to your answer.

(4) $\leq$ t $\mathrm{y}=\mathrm{g}(\mathrm{x})$ be another solution to the differential equation with the initial condition $g(0)=k$, where k is a constants. Euler's method, starting at $x=0$ with a step size of 1, gives the approximation $g(1)\approx 0$. Find the value of k.

7.5 Logistic Growth

The rate of change of a quantity (for example, a population) may be proportional both to the amount (size) of the quantity and to the difference between a fixed constants A and its amount (size). If $y=f(t)$ is the amount, then

$$y'=ky(A-y), \tag{1}$$

where k and A are both positive. Equation (1) is called the logistic differential equation; it is used to model logistic growth.

The solution of the d.e. (1) is

$$y=\frac{A}{1+ce^{-Akt}} \tag{2}$$

for some positive constant c.

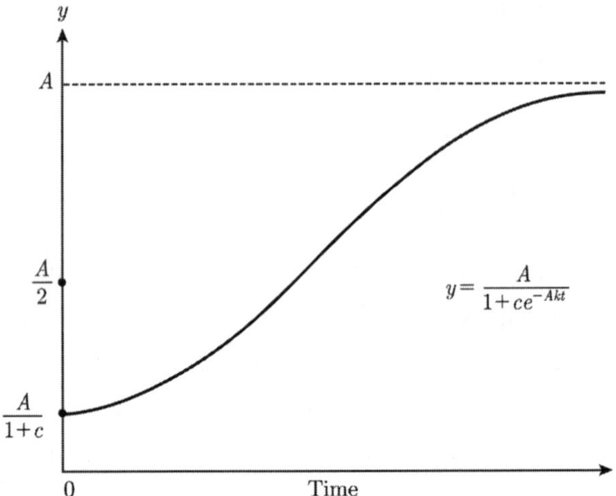

The S-shaped curve in (2) is often called a logistic curve. It shows that the rate of growth y' :

(1) Increases slowly for a while; i.e., $y'' > 0$;

(2) Attains a maximum when $y = A/2$, at the point of inflection;

(3) Then decreases $(y'' < 0)$, approaching 0 as y approaches its upper limit.

Example 1 A certain wild animal preserve can support no more than 250 lowland gorillas. twenty-eight gorillas were known to be in the preserve in 1970. Assume that the rate of growth of the population is

$$\frac{dy}{dt} = 0.0004(250 - y)y$$

where time t is in years.

Find a formula for the gorilla population in terms of t.

Solution

$k = 0.0004$ and $A = 250$, and $y = \dfrac{250}{1 + ce^{-0.1t}}$

Initial condition : $y(0) = 28$, where $t = 0$ represents the year 1970.

$28 = \dfrac{250}{1 + c} \Rightarrow 28(1 + c) = 250 \Rightarrow c = 7.93$

Formula : $P = \dfrac{250}{1 + 7.93e^{-0.1t}}$

EX 1 A 2000-gal tank can support no more than 150 guppies. Six guppies are introduced into the tank. Assume that the rate of growth of the population is

$$\frac{dy}{dt} = 0.0015(150 - y)y$$

where time t is in weeks.

(1) Find a formula for the guppy population in terms of t.

(2) How long will it take for the guppy population to be 100? 125?

EX 2 Consider the logistic differential equation $\frac{dy}{dt} = \frac{y}{8}(6 - y)$. $\leq$ t $y = f(t)$ be the particular solution to the differential equation with $f(0) = 8$

(1) A slope field for this differential equation is given below. Sketch possible solution curves through the points (3, 2) and (0, 8).
(Note: Use the axes provided in the exam booklet.)

(2) Use Euler's method, starting at $t = 0$ with two steps of equal size, to approximate $f(1)$.

(3) Write the second-degree Taylor polynomial for f about $t = 0$, and use it to approximate $f(1)$.

(4) What is the range of f for $t \geq 0$?

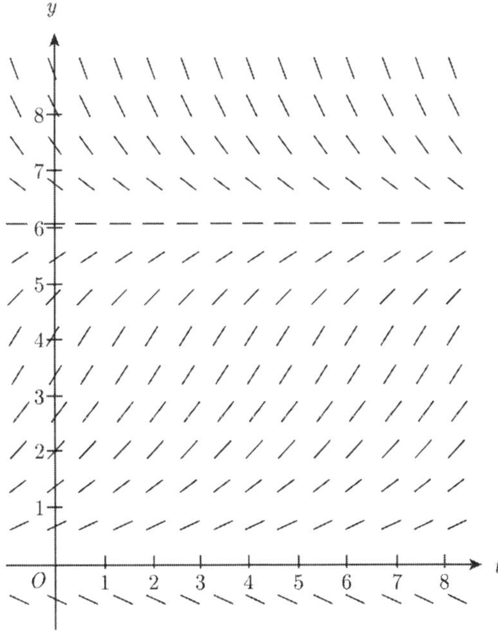

EX 3 Suppose a flu-like virus is spreading through a population of 50,000 at a rate proportional both to the number of people already infected and to the number still uninfected. If 100 people were infected yesterday and 130 are infected today:

(1) Write an expression for the number of people $N(t)$ infected after t days;

(2) Determine how many will be infected a week from today.

Multiple Choice

01 If $\frac{dy}{dt} = -2y$ and if $y = 1$ when $t = 0$, what is the value of t for which $y = \frac{1}{2}$?

(A) $-\frac{\ln 2}{2}$　(B) $-\frac{1}{4}$　(C) $\frac{\ln 2}{2}$

(D) $\frac{\sqrt{2}}{2}$　(E) $\ln 2$

02 If $\frac{dy}{dx} = y\sec^2 x$ and $y = 5$ when $x = 0$, then $y =$

(A) $e^{\tan x} + 4$　(B) $e^{\tan x} + 5$　(C) $5e^{\tan x}$

(D) $\tan x + 5$　(E) $\tan x + 5e^x$

03 If $\frac{dy}{dx} = x^2 y$, then y could be

(A) $3\ln\left(\frac{x}{3}\right)$　(B) $e^{\frac{x^3}{3}} + 7$　(C) $2e^{\frac{x^3}{3}}$

(D) $3e^{2x}$　(E) $\frac{x^3}{3} + 1$

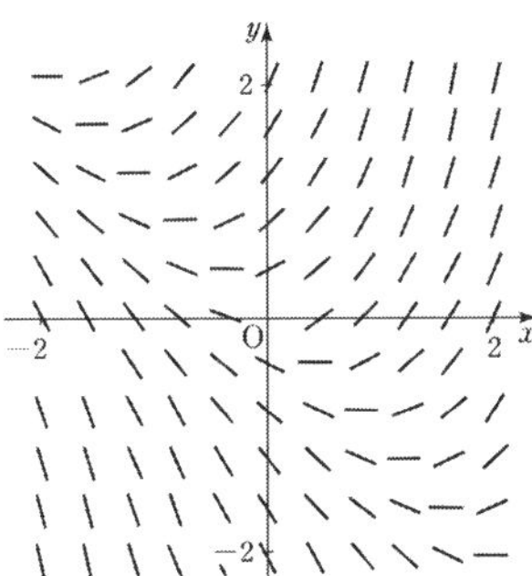

04 Shown above is a slope field for which of the following differential equations?

(A) $\frac{dy}{dx} = 1 + x$　(B) $\frac{dy}{dx} = x^2$　(C) $\frac{dy}{dx} = x + y$

(D) $\frac{dy}{dx} = \frac{x}{y}$　(E) $\frac{dy}{dx} = \ln y$

05 The number of bacteria in a culture is growing at a rate of $3{,}000e^{2t/5}$ per unit of time t. At $t=0$, the number of bacteria present was 7,500. Find the number present at $t=5$.

(A) $1{,}200e^2$ (B) $3{,}000e^2$ (C) $7{,}500e^2$

(D) $7{,}500e^5$ (E) $\frac{15{,}000}{7}e^7$

06 Let $y=f(x)$ be the solution to the differential equation $\frac{dy}{dx}=x+y$ with the intial condition $f(1)=2$. What is the approximation for $f(2)$ if Euler's method is used, starting at $x=1$ with a step size of 0.5?

(A) 3 (B) 5 (C) 6

(D) 10 (E) 12

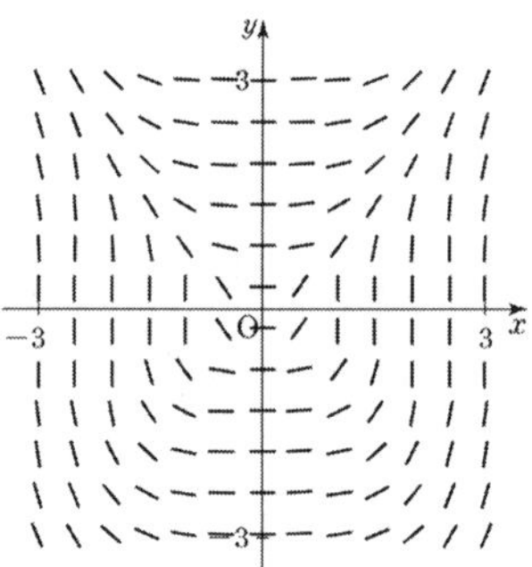

07 Shown above is a slope field for which of the following differential equations?

(A) $\frac{dy}{dx}=\frac{x}{y}$ (B) $\frac{dy}{dx}=\frac{x^2}{y^2}$ (C) $\frac{dy}{dx}=\frac{x^3}{y}$

(D) $\frac{dy}{dx}=\frac{x^2}{y}$ (E) $\frac{dy}{dx}=\frac{x^3}{y^2}$

08 Insects destroyed a crop at the rate of $\frac{100e^{-0.1t}}{2-e^{-3t}}$ tons per day, where time t is measured in days. To the nearest ton, how many tons did the insects destroy during the time interval $7 \le t \le 14$?

(A) 125 (B) 100 (C) 88

(D) 50 (E) 12

09 If $\frac{dy}{dx} = -10y$ and if $y = 50$ when $x = 0$, then $y =$

(A) $50e^{x}$ (B) $50e^{10x}$ (C) $50e^{-10x}$

(D) $50 - 10x$ (E) $50 - 5x^2$

10 If $f'(x) = -f(x)$ and $f(1) = 1$, then $f(x) =$

(A) $\frac{1}{2}e^{-2x+2}$ (B) e^{-x-1} (C) e^{1-x}

(D) e^{-x} (E) $-e^{x}$

11 Let k be a positive constant. Which of the following is a logistic differential equation?

(A) $\frac{dy}{dt} = kt$ (B) $\frac{dy}{dt} = ky$ (C) $\frac{dy}{dt} = kt(1-t)$

(D) $\frac{dy}{dt} = ky(1-t)$ (E) $\frac{dy}{dt} = ky(1-y)$

12 If $P(t)$ is the size of a population at time t. which of the following differential equations describes linear growth in the size of the population?

(A) $\frac{dP}{dt} = 200$ (B) $\frac{dP}{dt} = 200t$ (C) $\frac{dP}{dt} = 100t^2$

(D) $\frac{dP}{dt} = 200P$ (E) $\frac{dP}{dt} = 100P^2$

13-14 $a(t)$ denotes the acceleration function, $v(t)$ the velocity function, and $s(t)$ the position or height function at time t. (The acceleration due to gravity is −32 ft/sec2.)

13 If $a(t) = 4t - 1$ and $v(1) = 3$, then $v(t)$ equals

(A) $2t^2 - t$ (B) $2t^2 - t + 1$ (C) $2t^2 - t + 2$

(D) $2t^2 + 1$ (E) $2t^2 + 2$

14 If $a(t) = 20t^3 - 6t$, $s(-1) = 2$, and $s(1) = 4$, then $v(t)$ equals

(A) $t^5 - t^3$ (B) $5t^4 - 3t^2 + 1$ (C) $5t^4 - 3t^2 + 3$

(D) $t^5 - t^3 + t + 3$ (E) $t^5 - t^3 + 1$

15 A stone is thrown at a target so that its velocity after t sec is $(100 - 20t)$ft/sec. if the stone hits the target in 1 sec, then the distance from the sling to the target is

(A) 80 ft (B) 90 ft (C) 100 ft

(D) 110 ft (E) 120 ft

16 If you use Euler's method with $\Delta x = 0.1$ for the d.e. $y' = x$, with initial value $y(1) = 5$, then, when $x = 1.2$, y is approximately

(A) 5.10 (B) 5.20 (C) 5.21

(D) 6.05 (E) 7.10

17 If $\frac{ds}{dt} = \sin^2(\frac{\pi}{2}s)$ and if $s = 1$ when $t = 0$, then, when $s = \frac{3}{2}$, t is equal to

(A) $\frac{1}{2}$ (B) $\frac{\pi}{2}$ (C) 1

(D) $\frac{2}{\pi}$ (E) $-\frac{2}{\pi}$

18 The population of a city increases continuously at a rate proportional, at any time, to the population at that time. The population doubles in 50 yr. After 75 yr the ratio of the population P to the initial population P0 is

(A) $\frac{9}{4}$ (B) $\frac{5}{2}$ (C) $\frac{4}{1}$

(D) $\frac{2\sqrt{2}}{1}$ (E) $\frac{3}{2}$

19 The solution curve of $y' = y$ that passes through point $(2, 3)$ is

(A) $y = e^x + 3$ (B) $y = \sqrt{2x + 5}$ (C) $y = 0.406e^x$

(D) $y = e^x - (e^2 + 3)$ (E) $y = e^x/(0.406)$

20 A cup of coffee at temperature 180 ° F is placed on a table in a room at 68 ° F. The d.e. for its temperature at time t is $\frac{dy}{dx} = -0.11(y-68); y(0) = 180$. After 10 min the temperature (in ° F) of the coffee is

(A) 96 (B) 100 (C) 105

(D) 110 (E) 115

21 Approximately how long does it take the temperature of the coffee in Question 21 to drop to 75 ° F?

(A) 10 min (B) 15 min (C) 18 min

(D) 20 min (E) 25 min

22 Which of the following d.e.'s is not logistic?

(A) $P' = P - P^2$ (B) $\frac{dy}{dt} = 0.01y(100-y)$

(C) $\frac{dx}{dt} = 0.8x - 0.004x^2$ (D) $\frac{dR}{dt} = 0.16(350-R)$

(E) $f'(t) = kf(t) \cdot [A - f(t)]$ (where k and A are constants)

23 Suppose $P(t)$ denotes the size of an animal population at time t and its growth is described by the d.e. $\frac{dP}{dt} = 0.002P(1000-P)$. The population is growing fastest

(A) initially (B) when $P = 500$ (C) when $P = 1000$

(D) when $\frac{dP}{dt} = 0$ (E) when $\frac{d^2P}{dt^2} > 0$

24 According to Newton's law of cooling, the temperature of an object decreases at a rate proportional to the difference between its temperature an that of the surrounding air. Suppose a corpse at a temperature of 32 ° C arrives at a mortuary where the temperature is kept at 10 ° C. Then the differential equation satisfied by the temperature T of the corpse t hr later is

(A) $\frac{dT}{dt} = -k(T-10)$ (B) $\frac{dT}{dt} = k(T-32)$ (C) $\frac{dT}{dt} = 32e^{-kt}$

(D) $\frac{dT}{dt} = -kT(T-10)$ (E) $\frac{dT}{dt} = kT(T-32)$

25 If the corpse in Question 25 cools to $27\,^{\circ}C$ in 1 hr, then its temperature (in $^{\circ}C$) is given by the equation

(A) $T = 22e^{0.205t}$ (B) $T = 10e^{1.163t}$ (C) $T = 10 + 22e^{-0.258t}$

(D) $T = 32e^{-0.169t}$ (E) $T = 32 - 10e^{-0.093t}$

Solution

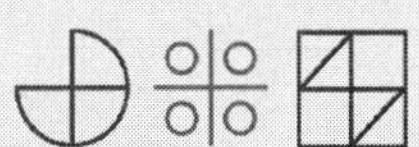

01 Functions

1.1 Function

EX1

(1) Domain: $-1 < x \le 5$

Range: $1 < y \le 3$

(2) Domain: $x \ne 2$

Range: $y \ne -1$

(3) Domain: $x \in R$

Range: $y > -2$

(4) Domain: $x \ne \pm 2$

Range: $y \le -1$ or $y > 0$

EX2

(1) $D: x \ge 2$, $R: y \ge 0$

(2) $D: x \le 4$, $R: y \ge 0$

(3) $D: x \in R$, $R: y \le \dfrac{25}{12}$

(4) $D: x \ne 2$, $R: y \ne 1$

EX3

(1) $(3)^2 - 3(3) + 4 + 1 = 9 - 9 + 5 = 5$

(2) $(2x)^2 - 3(2x) + 4 = 4x^2 - 6x + 4$

(3) $(x-2)^2 - 3(x-2) + 4$

$= x^2 - 4x + 4 - 3x + 6 + 4 = x^2 - 7x + 14$

(4) $\dfrac{(x+2)^2 - 3(x+2) + 4 - [x^2 - 3x + 4]}{2}$

$= \dfrac{x^2 + 4x + 4 - 3x - 6 + 4 - x^2 + 3x - 4}{2}$

$= \dfrac{4x-2}{2} = 2x - 1$

EX4

$\{(0,0), (1,1), (2,2), (3,3)\}$

EX5

(1) $g(f(4)) = g\left(\dfrac{4}{-3}\right) = -\dfrac{3}{4}$

(2) $f(f(-2)) = f\left(\dfrac{-2}{3}\right) = \dfrac{\left(-\dfrac{2}{3}\right) \cdot 3}{\left(1 + \dfrac{2}{3}\right) \cdot 3}$

$= \dfrac{-2}{3+2} = -\dfrac{2}{5}$

(3) $f(g(x)) = f\left(\dfrac{1}{x}\right) = \dfrac{\left(\dfrac{1}{x}\right) \cdot x}{\left(1 - \dfrac{1}{x}\right) \cdot x} = \dfrac{1}{x-1}$

(4) $g(f(x)) = g\left(\dfrac{x}{1-x}\right) = \dfrac{1}{\dfrac{x}{1-x}} = \dfrac{1-x}{x}$

EX6

(1) $y = 3x - 2, \Rightarrow x = 3y - 2$

$x + 2 = 3y,\ y = \dfrac{1}{3}x + \dfrac{2}{3}$

$\therefore f^{-1}(x) = \dfrac{1}{3}x + \dfrac{2}{3}$

(2) $y = x^3 - 5,\ \Rightarrow\ x = y^3 - 5$

$x + 5 = y^3,\ y = \sqrt[3]{x+5}$

$\therefore f^{-1}(x) = \sqrt[3]{x+5}$

(3) $y = 2(x-4)^2 + 1,$

$\Rightarrow x = 2(y-4)^2 + 1\,(x \ge 4, y \ge 1)$

$x - 1 = 2(y-4)^2,$

$\dfrac{x-1}{2} = (y-4)^2\,(x \ge 1, y \ge 4)$

$y - 4 = \pm\sqrt{\dfrac{x-1}{2}},$

$\therefore f^{-1}(x) = \sqrt{\dfrac{x-1}{2}} + 4\,(x \ge 1)$

(4) $y=\frac{x-2}{2x-3}$, $\Rightarrow x=\frac{y-2}{2y-3}$, $x(2y-3)=y-2$

$2xy-3x=y-2$, $2xy-y=3x-2$

$y(2x-1)=3x-2$, $y=\frac{3x-2}{2x-1}$

$\therefore f^{-1}(x)=\frac{3x-2}{2x-1}$

EX7

(1) $f^{-1}(5)=x$, $5=f(x)$, $5=2x^3-11$, $16=2x^3$

$8=x^3$, $\therefore x=2$

(2) $f^{-1}(2)=x$, $2=f(x)$,

$2=\sqrt[3]{x-5}+4$, $-2=\sqrt[3]{x-5}$

$-8=x-5$, $\therefore x=-3$

(3) $g(f(x))=x \Rightarrow g(x)=f^{-1}(x)$,

$g(2)=f^{-1}(2)=x$, $2=f(x)$

$2=\frac{6}{x+1}$, $1=\frac{3}{x+1}$, $x+1=3$, $\therefore x=2$

(4) $f(g(x))=x \Rightarrow g(x)=f^{-1}(x)$,

$f(x)=(x+3)^3-2$, $y=(x+3)^3-2$

$\Rightarrow x=(y+3)^3-2$, $x+2=(y+3)^3$,

$y+3=\sqrt[3]{x+2}$

$y=\sqrt[3]{x+2}-3$, $\therefore f^{-1}(x)=g(x)=\sqrt[3]{x+2}-3$

1.2 Exponents

EX1

(1) $(\sqrt[3]{64})^2=4^2=16$

(2) $-\frac{1}{(27)^{2/3}}=-\frac{1}{(\sqrt[3]{27})^2}=-\frac{1}{9}$

(3) $\left(\frac{16}{81}\right)^{\frac{3}{4}}=\left(\sqrt[4]{\frac{16}{81}}\right)^3=\left(\frac{2}{3}\right)^3=\frac{8}{27}$

(4) $-32^{-\frac{4}{5}}=-\frac{1}{(\sqrt[5]{32})^4}=-\frac{1}{16}$

EX2

(1) $a^2b^{-2}=\frac{a^2}{b^2}$

(2) $\frac{x^3}{4y^5}$

(3) $\frac{-9 \cdot 16a^{-4}b^3}{36a^{-4}b^6}=-4b^{-3}=-\frac{4}{b^3}$

(4) $\frac{1}{(3m^3n^2)^2}\left(\frac{9m^2}{n^2}\right)^2$

$=\frac{1}{9m^6n^4} \cdot \frac{81m^4}{n^4}$

$=\frac{9}{m^2n^8}$

EX3

(1) $3^2 \cdot 3^{-x}=3^{3(1+x)}$

$2-x=3+3x$

$-4x=1$

$x=-\frac{1}{4}$

(2) $2 \cdot 2^{3(x-1)}=\frac{2^{1/2}}{2^{2x}}$

$2^{1+3x-3}=2^{\frac{1}{2}-2x}$

$3x-2=\frac{1}{2}-2x$

$5x=\frac{5}{2}$, $\therefore x=\frac{1}{2}$

(3) $\left(\frac{4}{5}\right)^{-3x}=\left(\frac{5}{4}\right)^{2x-1}$

$\left(\frac{5}{4}\right)^{3x}=\left(\frac{5}{4}\right)^{2x-1}$

$3x=2x-1$

$x=-1$

(4) $2^{-4} \cdot 2^{3(x-2)}=\left(2^{\frac{2}{3}}\right)^x$

$2^{-4+3x+6}=2^{\frac{2}{3}x}$

$3x+2=\frac{2}{3}x$

$9x+6=2x$

$7x=-6$, $\therefore x=-\frac{6}{7}$

EX4

(1) $5^x=t$

$t^2=26t+25=0$

$(t-1)(t-25)=0$

$t=1, 25$

$5^x = 1, 25$

$\therefore x = 0, 2$

(2) $2^x = t$

$2t^2 - 9t + 4 = 0$

$(2t-1)(t-4) = 0$

$t = \frac{1}{2}, 4$

$2^x = \frac{1}{2}, 4$

$\therefore x = -1, 2$

(3) $2^x = t$

$t^2 + 6t - 40 = 0$

$(t-1)(t+10) = 0$

$t = 4, -10$

$2^x = 4, -10$

$2^x > 0$이므로, $2^x = 4$,

$\therefore x = 2$

(4) $3^x = t$

$3t = \frac{9}{t} + 26$

$3t^2 = 9 + 26t$

$3t^2 - 26t - 9 = 0$

$(3t+1)(t-9) = 0$

$t = -\frac{1}{3}, 9$

$3^x = -\frac{1}{3}, 9$

$3^x > 0$이므로, $3^x = 9$

$\therefore x = 2$

(1) $y = 2^x$

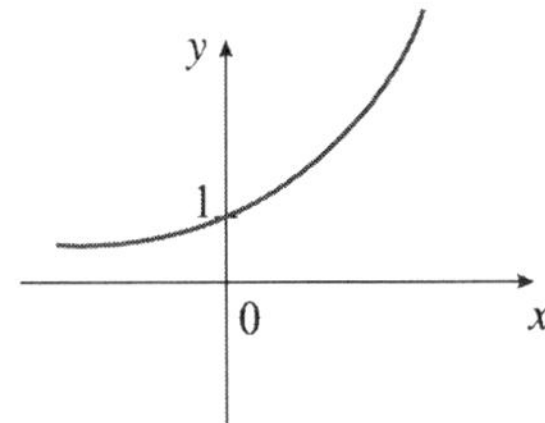

(2) $y = e^x \rightarrow y = e^x - 2$

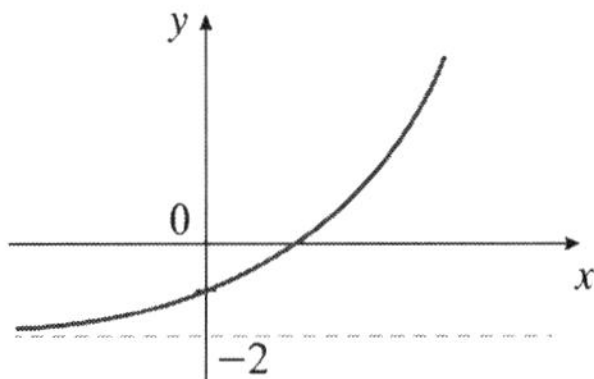

(3) $y = 3^x \rightarrow y = -3^x$

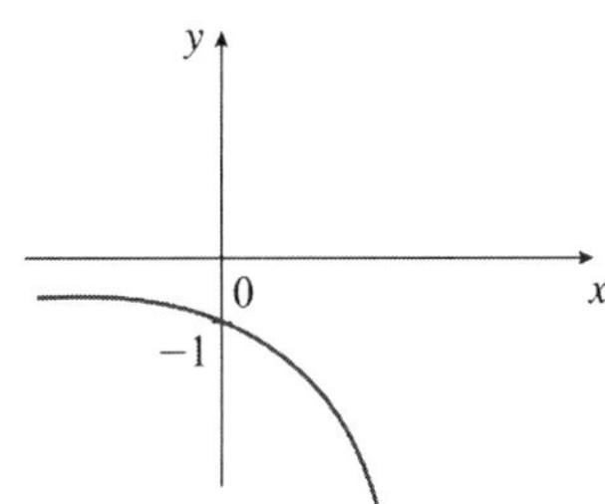

(4) $y = e^x \rightarrow y = e^{-x} \rightarrow y = e^{-x} + 2$

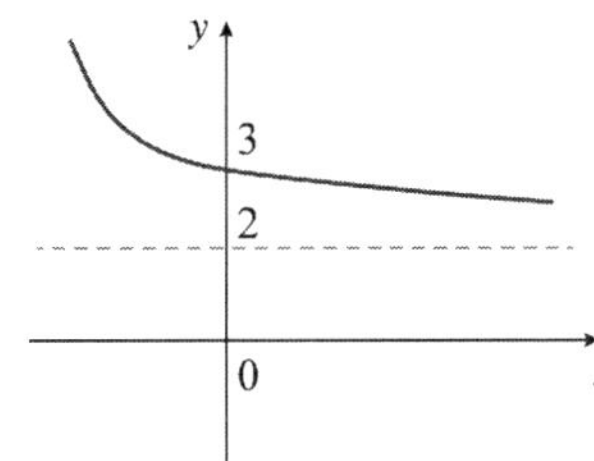

(5) $y = 3^x \rightarrow y = 2(3)^x \rightarrow y = 2(3)^{x-1}$

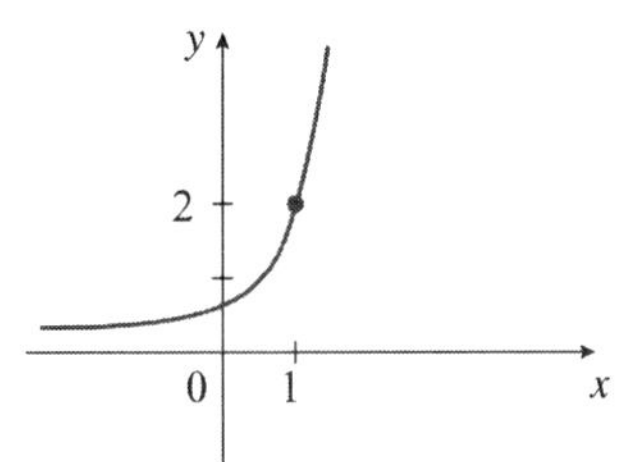

(6) $y = e^x \rightarrow y = -e^x \rightarrow y = -e^{x-3} + 2$

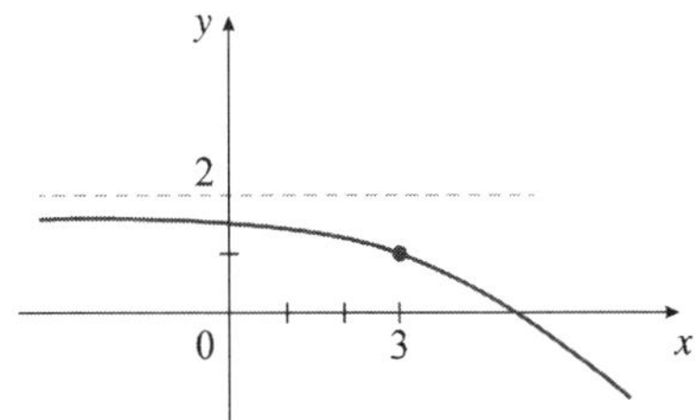

(7) $y=2^x \rightarrow y=3(2)^x \rightarrow y=3(2)^{x-1}-2$

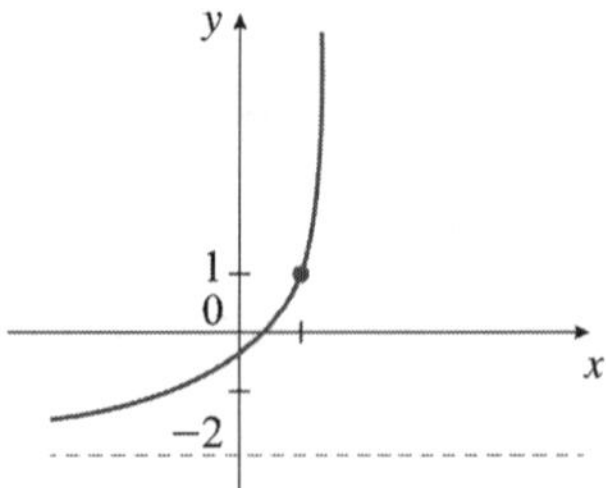

(8) $y=3^x \rightarrow y=3^{-x} \rightarrow$

$y=-3^{-x} \rightarrow y=-3^{-(x-2)}+1$

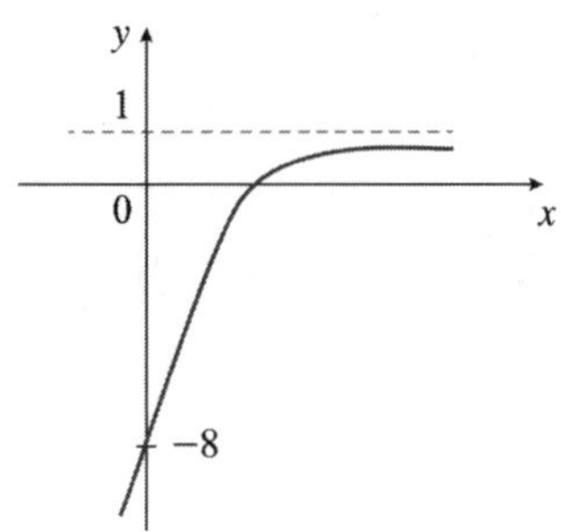

1.3 Logarithmic Functions

EX1

(1) $x=3$ (2) $x=\dfrac{1}{3}$

(3) $x=1$ (4) $x=e^3$

EX2

(1)

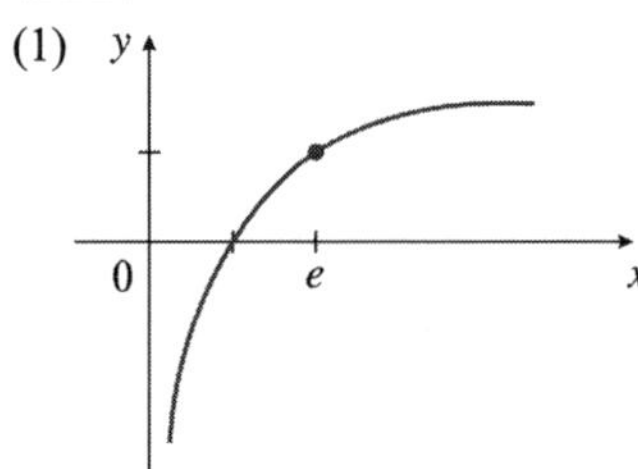

(2) $y=\ln x \rightarrow y=\ln(-x) \rightarrow y=\ln(-x)+1$

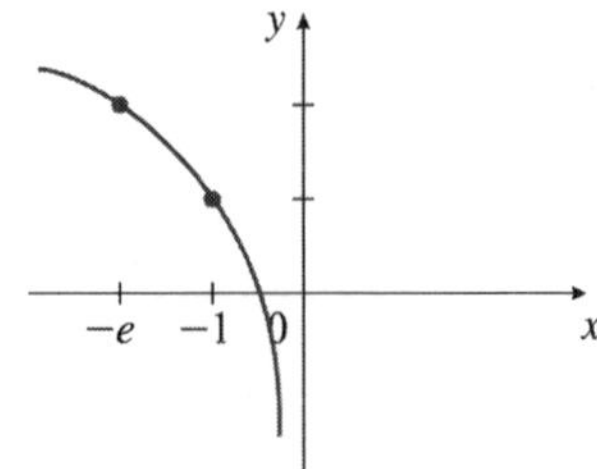

(3) $y=\log_2 x \rightarrow y=-\log_2 x$

$\rightarrow y=-\log_2(x-1)+2$

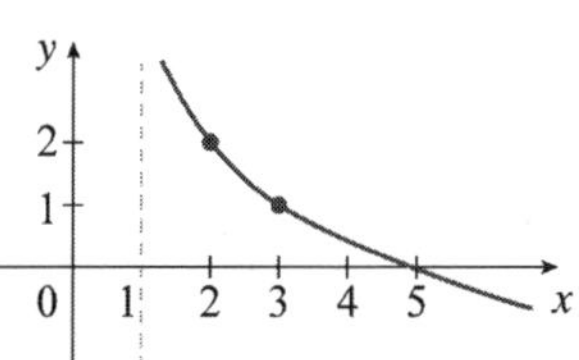

(4) $y=\log_2 x \rightarrow y=2\log x$

$\rightarrow y=2\log_2(x+4)-2$

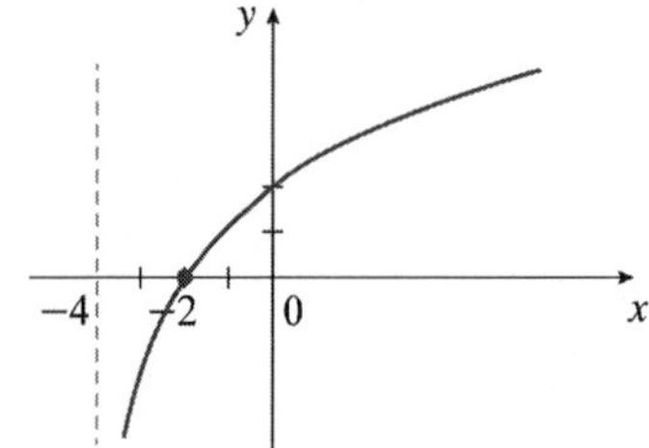

EX3

(1) $y=3(5)^x$

$\Rightarrow x=3(5)^y$

$\dfrac{x}{3}=5^y, \quad \therefore y=\log_5\left(\dfrac{x}{3}\right)$

$\therefore f^{-1}(x)=\log_5\left(\dfrac{x}{3}\right)$

(2) $y=e^{-x}$

$\Rightarrow x=e^{-y}$

$-y=\log_e x$

$y=-\ln x$

$\therefore f^{-1}(x)=-\ln x$

(3) $f^{-1}(x)=1+\log_3\left(\dfrac{x-4}{2}\right)$

(4) $y=-e^{x+3}+1$

$\Rightarrow x=-e^{y+3}+1$

$e^{y+3}=1-x$

$y+3=\log_e(1-x)$

$y=\ln(1-x)-3$

$\therefore f^{-1}(x)=\ln(1-x)-3$

EX4

(1) $\log_6 36=2$

(2) $\log\left(\frac{50 \cdot 4}{2}\right)=\log 100=2$

(3) $\log_7\left(35 \cdot \frac{14}{5} \cdot \frac{1}{2}\right)=\log_7 49=2$

(4) $\log 2+\log 2-\log 40=\log\left(\frac{2 \cdot 2}{40}\right)$

$=\log\left(\frac{1}{10}\right)=-1$

EX5

(1) $\log 4+\log a^2-\log 9-\log b^3$

$=\log 4+2\log a-\log 9-3\log b$

(2) $\log_b b^2+\log_b(x+1)^{\frac{1}{2}}-\log_b(x-1)^3$

$=2\log_b b+\frac{1}{2}\log_b(x+1)-3\log_b(x-1)$

$=2+\frac{1}{2}\log_b(x+1)-3\log_b(x-1)$

(3) $\ln(2x-1)^{\frac{1}{4}}+\ln(x-1)^{\frac{1}{2}}-\ln(x+1)^{\frac{1}{2}}$

$-\ln(5-x)^{\frac{1}{2}}$

$=\frac{1}{4}\ln(2x-1)+\frac{1}{2}\ln(x-1)-\frac{1}{2}\ln(x+1)$

$-\frac{1}{2}\ln(5-x)$

(4) $\ln e^x+\ln(\cos x)-\ln(x^2+x)^{\frac{1}{2}}$

$=x\ln e+\ln(\cos x)-\frac{1}{2}\ln(x^2+x)$

$=x+\ln(\cos x)-\frac{1}{2}\ln(x^2+x)$

EX6

(1) $\log 5^{2x}=\log 8^{x+1}$

$2x\log 5=(x+1)\log 8$

$2x\log 5=x\log 8+\log 8$

$2x\log 5-x\log 8=\log 8$

$x(2\log 5-\log 8)=\log 8$

$\therefore x=\frac{\log 8}{2\log 5-\log 8}$ $\left(\text{또는 } \frac{\log 8}{\log(25/8)}\right)$

(2) $\ln e^x=\ln 4^{2-x}$

$x\ln e=(2-x)\ln 4$

$x=2\ln 4-x\ln 4$

$x+x\ln 4=2\ln 4$

$x(1+\ln 4)=2\ln 4$

$\therefore x=\frac{2\ln 4}{1+\ln 4}$ $\left(\text{or } \frac{\ln 16}{\ln(4e)}\right)$

(3) $\log(5(4)^x)=\log 3^{2x+1}$

$\log 5+\log 4^x=(2x+1)\log 3$

$\log 5+x\log 4=2x\log 3+\log 3$

$x\log 4-2x\log 3=\log 3-\log 5$

$x(\log 4-2\log 3)=\log 3-\log 5$

$\therefore x=\frac{\log 3-\log 5}{\log 4-2\log 3}$ $\left(\text{or } \frac{\log(3/5)}{\log(4/9)}\right)$

(4) $\log(4(3)^{x-2})=\log 5^{2x}$

$\log 4+\log 3^{x-2}=2x\log 5$

$\log 4+(x-2)\log 3=2x\log 5$

$\log 4+x\log 3-2\log 3=2x\log 5$

$x\log 3-2x\log 5=2\log 3-\log 4$

$x(\log 3-2\log 5)=2\log 3-\log 4$

$\therefore x=\frac{2\log 3-\log 4}{\log 3-2\log 5}$ $\left(\text{or } \frac{\log(9/4)}{\log(3/25)}\right)$

EX7

(1) $\log_2(2x+4)-\log_2(x-1)=3$

$\log_2\left(\frac{2x+4}{x-1}\right)=\log_2 8$

$2x+4=8(x-1)$

$2x+4=8x-8$

$-6x=-12$

$\therefore x=2$

(2) $\ln\left(\frac{x}{x+1}\right)=1$

$\frac{x}{x+1}=e$

$x=e(x+1)$

$x=ex+e$

$x-ex=e$

$x(1-e)=e$

$\therefore x=\frac{e}{1-e}$

(3) $\log_3 x=t$

$t^2=2t$

$t(t-2)=0$

$t=0, 2$

$\log_3 x = 0, 2$

$x = 3^0, 3^2$

$\therefore x = 1, 9$

(4) $2(\ln x)^2 - 5\ln x + 2 = 0$

$\ln x = t$

$2t^2 - 5t + 2 = 0$

$(2t-1)(t-2) = 0$

$t = \frac{1}{2}, 2$

$\ln x = \frac{1}{2}, 2 \Rightarrow \therefore x = e^{\frac{1}{2}}, e^2$

(5) $\log(3x-2) = \log 2(x+2)$

$3x - 2 = 2(x+2)$

$3x - 2 = 2x + 4$

$\therefore x = 6$

(6) $\ln\left(\frac{3-x}{x+5}\right) = 3$

$3 - x = 3(x+5)$

$3 - x = 3x + 15$

$-4x = 12$

$\therefore x = -3$

1.4 Trigonometric Equations

EX1

(1)

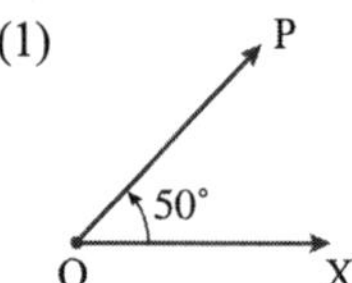

(2)

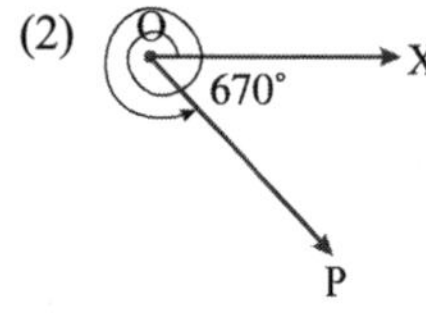

(3)

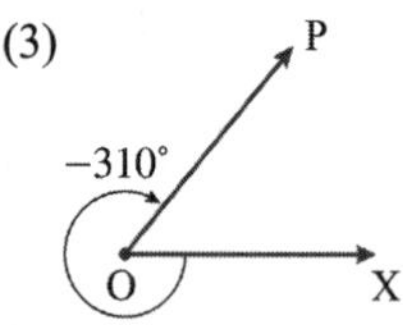

(4)

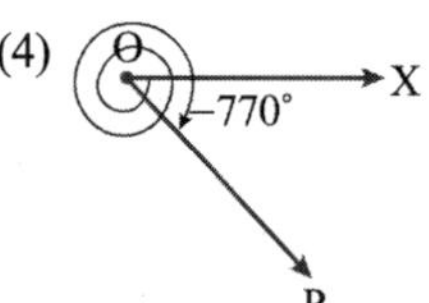

EX2

(1) π (2) $-\frac{2}{3}\pi$ (3) $\frac{5}{3}\pi$ (4) $\frac{5}{12}\pi$

EX3

(1) 120° (2) 240° (3) −330° (4) 105°

EX4

(1) $\frac{1}{2}$ (2) 0 (3) −1

(4) $-\frac{\sqrt{2}}{2}$ (5) $-\sqrt{2}$ (6) 2

(7) 0 (8) $\frac{1}{2}$

EX5

(1) $-\frac{\sqrt{3}}{2}$ (2) $-\frac{\sqrt{3}}{2}$ (3) $\frac{1}{2}$

(4) $\frac{\sqrt{3}}{2}$ (5) −1 (6) 0

EX6

(1) $\frac{\pi}{3}, \frac{2}{3}\pi$ (2) $\frac{2}{3}\pi, \frac{4}{3}\pi$ (3) $\frac{5}{4}\pi, \frac{7}{4}\pi$

(4) $\frac{3}{4}\pi, \frac{7}{4}\pi$ (5) $\frac{3}{4}\pi, \frac{5}{4}\pi$ (6) $\frac{2}{3}\pi, \frac{5}{3}\pi$

(7) 0 (8) $\frac{\pi}{2}, \frac{3}{2}\pi$

EX7

(1) 30°, 330° (2) 135°, 225°

(3) 60°, 120° (4) 225°, 315°

(5) 45°, 225° (6) 150°, 330°

EX8

(1)

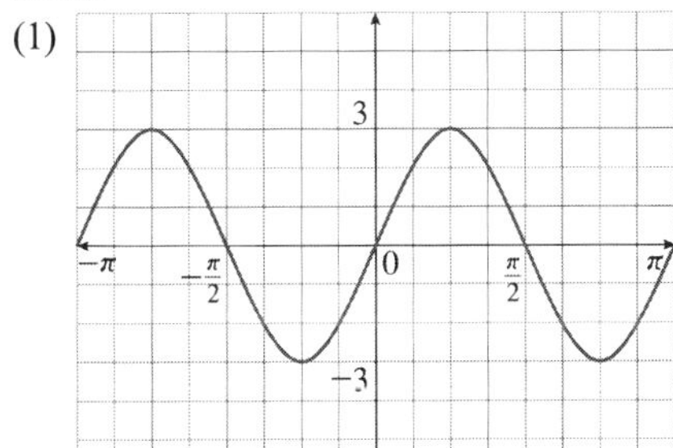

진폭$=3$, $P=\dfrac{2\pi}{2}=\pi$

(2)

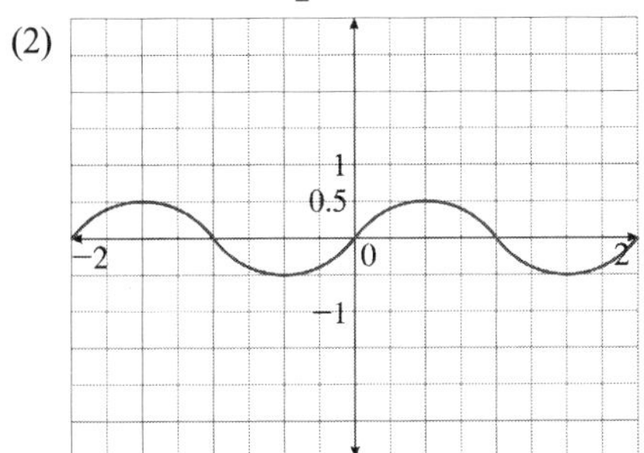

진폭$=\dfrac{1}{2}$, $P=\dfrac{2\pi}{\pi}=2$

(3)

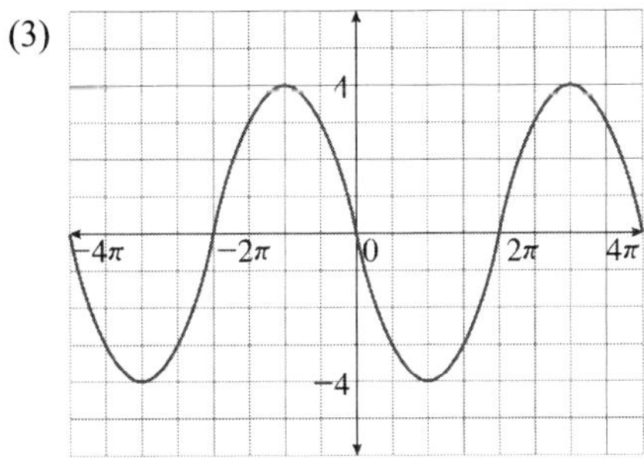

진폭$=4$, $P=\dfrac{2\pi}{1/2}=4\pi$

(4)

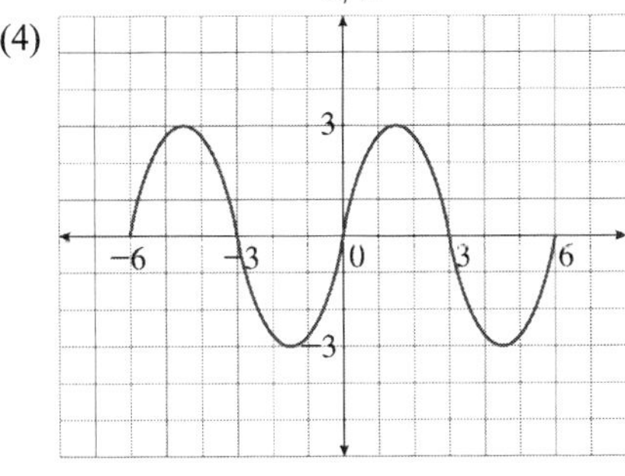

진폭$=3$, $P=\dfrac{2\pi}{\pi/3}=6$

EX9

(1)

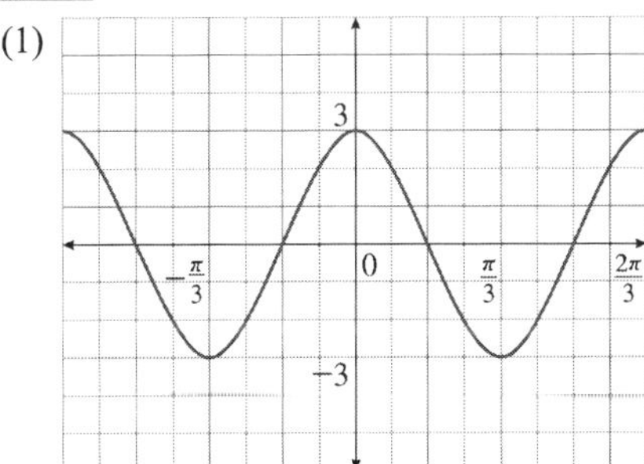

진폭$=3$, $P=\dfrac{2\pi}{3}$

(2)

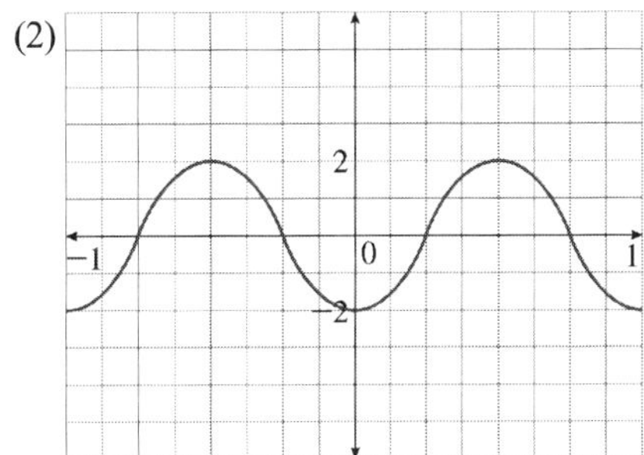

진폭$=2$, $P=\dfrac{2\pi}{2\pi}=1$

(3)

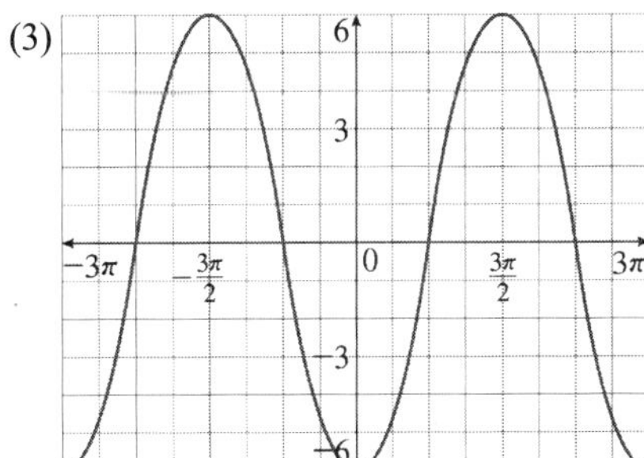

진폭$=6$, $P=\dfrac{2\pi}{2/3}=3\pi$

(4)

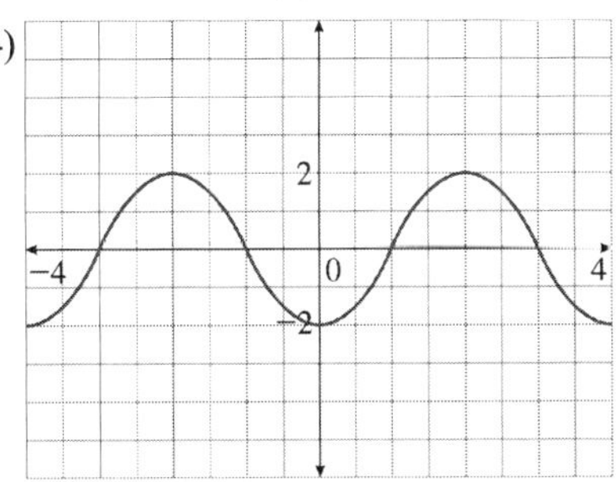

진폭$=2$, $P=\dfrac{2\pi}{\pi/2}=4$

EX10

(1)

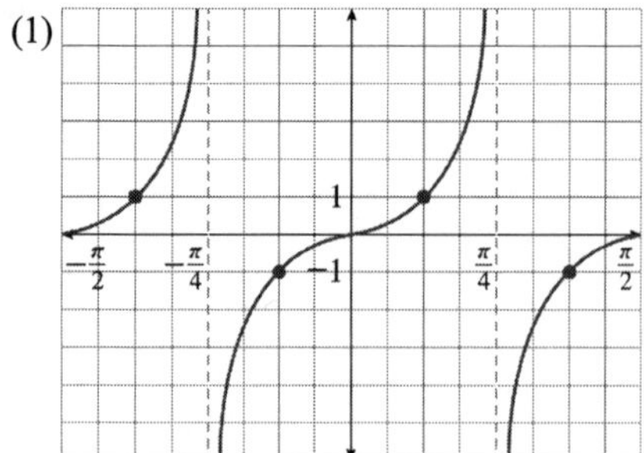

진폭 = 없음, $P=\frac{\pi}{2}$

(2)

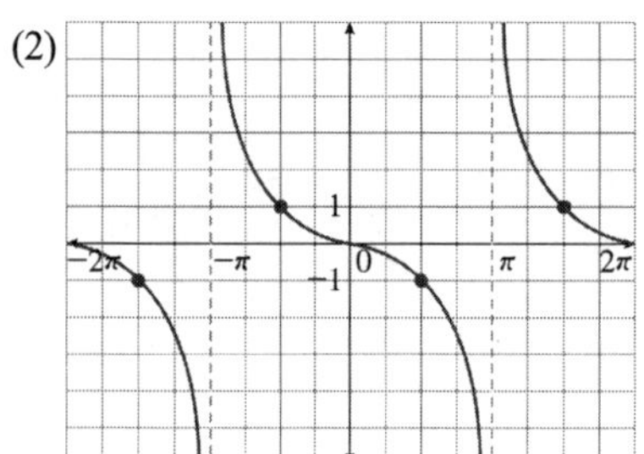

진폭 = 없음, $P=\frac{\pi}{1/2}=2\pi$

(3)

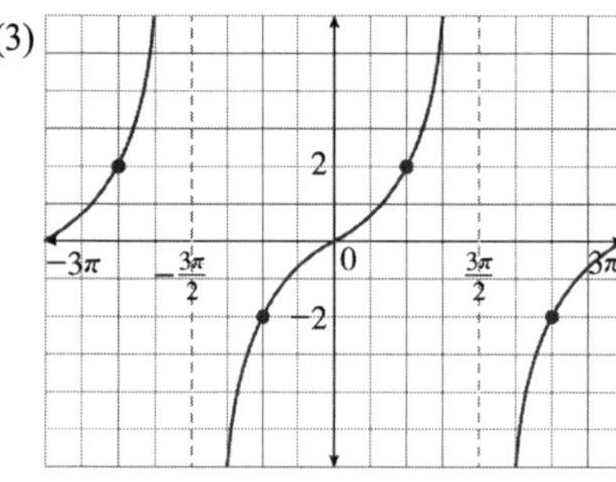

진폭 = 없음, $P=\frac{\pi}{1/3}=3\pi$

(4)

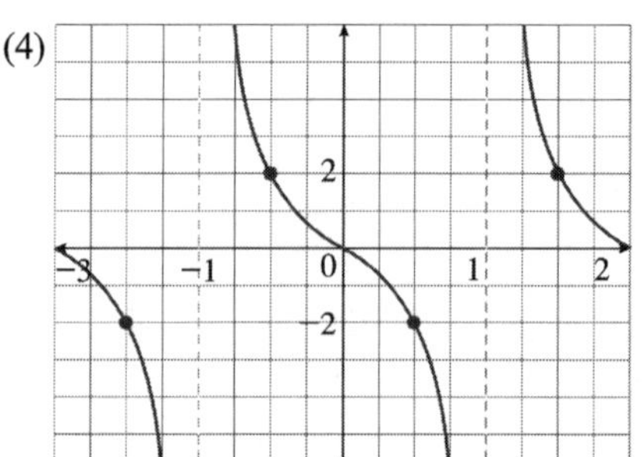

진폭 = 없음, $P=\frac{\pi}{\pi/2}=2$

EX11

(1) $-\frac{\pi}{3}$

(2) $\frac{\pi}{2}$

(3) $\sin\left(-\frac{\pi}{6}\right)$
$=-\frac{1}{2}$

(4) $\cos\left(-\frac{\pi}{2}\right)$
$=0$

EX12

(1) $\frac{5}{6}\pi$

(2) $\frac{1}{2}\pi$

(3) $\cos\left(\frac{2\pi}{3}\right)$
$=-\frac{1}{2}$

(4) $\tan(0)$
$=0$

EX13

(1) $\frac{\pi}{6}$

(2) 0

(3) $\tan\left(-\frac{\pi}{6}\right)$
$=-\frac{1}{\sqrt{3}}$

(4) $\cos\left(-\frac{\pi}{3}\right)$
$=\frac{1}{2}$

Multiple Choice

01 D 02 B 03 B 04 A 05 C 06 C 07 C 08 C 09 B 10 B 11 D 12 E 13 D 14 B 15 B
16 A 17 D 18 E 19 B 20 B 21 A 22 A 23 B 24 C 25 A 26 D 27 E 28 B 29 C 30 C

02 Limits and Continuity

2.1 Limit Existence

EX1

(1) 1 (2) 0 (3) 0 (4) 0

(5) DNE (6) 0 (7) 0 (8) DNE

(9) DNE (10) 1

EX2

(2) (a) -1 (b) 1 (c) DNE (d) DNE

(3) (a) $+\infty$ (b) 2 (c) DNE (d) -2

(4) (a) $-\infty$ (b) $-\infty$ (c) $-\infty$ (d) 1

(5) (a) 0 (b) 0 (c) 0 (d) 0

EX3

(1)

$y=3-x$, $y=\frac{x}{2}+1$

(a) 2
(b) 1
(c) DNE
(d) 3
(e) 3
(f) 3

(2)

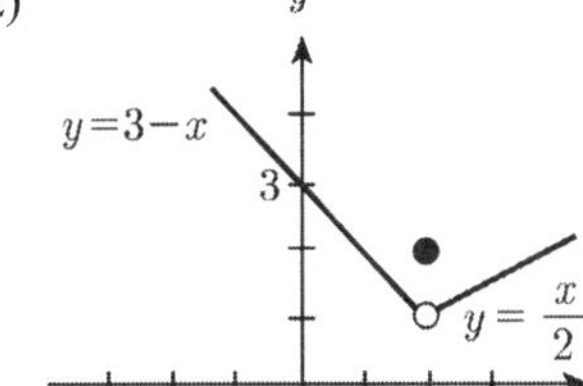

(a) 1
(b) 1
(c) 2
(d) 1
(e) 4
(f) 4
(g) 4

(3)

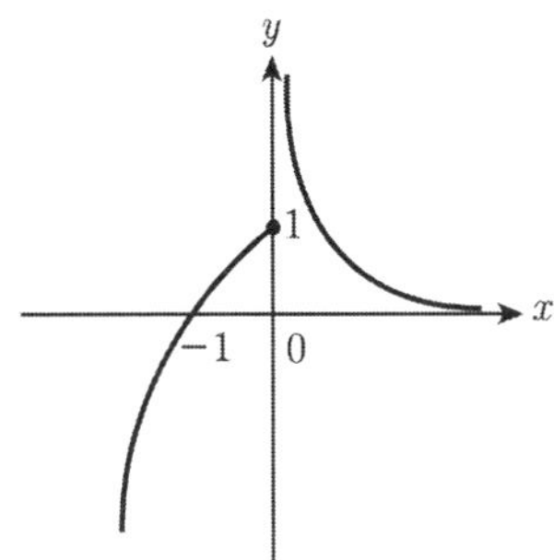

(a) 1
(b) $+\infty$
(c) DNE
(d) 1

(4)

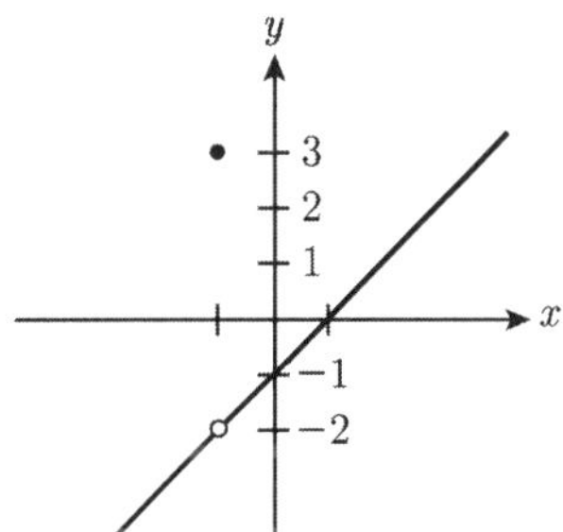

(a) -2
(b) -2
(c) -2
(d) 3

(5)

(a) -1
(b) -1
(c) -1
(d) 1
(e) 0
(f) DNE

(6)

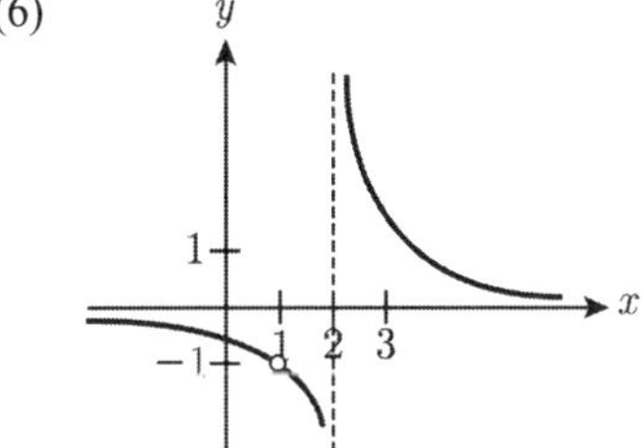

(a) -1
(b) -1
(c) -1
(d) 1

EX4

(1) c^3+4c^2-3

(2) $$\lim_{x\to c}\frac{x^4+x^2-1}{x^2+5}=\frac{\lim_{x\to c}(x^4+x^2-1)}{\lim_{x\to c}(x^2+5)}$$

$$=\frac{\lim_{x\to c}x^4+\lim_{x\to c}x^2-\lim_{x\to c}1}{\lim_{x\to c}x^2+\lim_{x\to c}5}$$

$$=\frac{c^4+c^2-1}{c^2+5}$$

(3) $\frac{1}{2}$

(4) $\lim_{x\to-2}\sqrt{4x^2-3}=\sqrt{\lim_{x\to-2}(4x^2-3)}$

$=\sqrt{\lim_{x\to-2}4x^2-\lim_{x\to-2}3}$

$=\sqrt{4(-2)^2-3}$

$=\sqrt{16-3}$

$=\sqrt{13}$

EX5

(1) -10

(2) $\lim_{x\to c}2f(x)g(x)=2\left[\lim_{x\to c}f(x)\right]\left[\lim_{x\to c}g(x)\right]$

$=2(5)(-2)=-20$

(3) -1

(4) $\lim_{x\to c}\dfrac{f(x)}{f(x)-g(x)}=\dfrac{\lim_{x\to c}f(x)}{\lim_{x\to c}f(x)-\lim_{x\to c}g(x)}$

$=\dfrac{5}{5-(-2)}=\dfrac{5}{7}$

EX6

(1) $=\dfrac{3}{2}$

(2) $=\lim_{x\to4}\dfrac{x(x-4)}{(x-1)(x-4)}=\lim_{x\to4}\dfrac{x}{(x-1)}=\dfrac{4}{3}$

(3) $=\dfrac{3}{8}$

(4) $=\lim_{x\to2}\dfrac{(x-2)(x^2+2x+4)}{(x-2)(x-1)}=\lim_{x\to2}\dfrac{x^2+2x+4}{x-1}$

$=\dfrac{4+4+4}{1}=12$

(5) $=-\dfrac{1}{2}$

(6) $=\lim_{x\to2}\dfrac{(x-2)(x^2-4)}{x^2(x-2)}$

$=\lim_{x\to2}\dfrac{(x^2-4)}{x^2}$

$=\dfrac{0}{4}=0$

(7) $=4$

(8) $=\lim_{x\to1}\dfrac{x^3+3x^2+3x+1-8}{x-1}$

$=\lim_{x\to1}\dfrac{x^3+3x^2+3x-7}{x-1}$

$=\lim_{x\to1}\dfrac{(x-1)(x^2+4x+7)}{x-1}$

$=\lim_{x\to1}(x^2+4x+7)$

$=1+4+7=12$

EX7

(1) $=\dfrac{1}{2}$

(2) $=\lim_{x\to4}\dfrac{(x-4)(2+\sqrt{x})}{(2-\sqrt{x})(2+\sqrt{x})}$

$=\lim_{x\to4}\dfrac{(x-4)(2+\sqrt{x})}{4-x}$

$=\lim_{x\to4}\dfrac{(2+\sqrt{x})}{-1}$

$=-4$

(3) $=4$

(4) $=\lim_{x\to3}\dfrac{(\sqrt{x+1}-2)(\sqrt{x+1}+2)(\sqrt{x-2}+1)}{(\sqrt{x-2}-1)(\sqrt{x+1}+2)(\sqrt{x-2}+1)}$

$=\lim_{x\to3}\dfrac{(x+1-4)(\sqrt{x-2}+1)}{(x-2-1)(\sqrt{x+1}+2)}$

$=\lim_{x\to3}\dfrac{(x-3)(\sqrt{x-2}+1)}{(x-3)(\sqrt{x+1}+2)}$

$=\lim_{x\to3}\dfrac{(\sqrt{x-2}+1)}{(\sqrt{x+1}+2)}=\dfrac{1+1}{2+2}=\dfrac{1}{2}$

(5) $=-3$

EX8

(1) $=-2$

(2) $=\lim_{h\to0}\dfrac{\left[\dfrac{1}{(h+3)^2}-\dfrac{1}{9}\right]\cdot9(h+3)^2}{h\cdot9(h+3)^2}$

$=\lim_{h\to0}\dfrac{9-(h+3)^2}{h\cdot9(h+3)^2}$

$=\lim_{h\to0}\dfrac{9-h^2-6h-9}{h\cdot9(h+3)^2}=\lim_{h\to0}\dfrac{-h^2-6h}{9h(h+3)^2}$

$=\lim_{h\to0}\dfrac{-h-6}{9(h+3)^2}=-\dfrac{6}{81}=-\dfrac{2}{27}$

(3) $=-\dfrac{1}{54}$

(4) $=\lim_{x\to2}\dfrac{\left(\dfrac{1}{\sqrt{x+2}}-\dfrac{1}{2}\right)\cdot2\sqrt{x+2}}{(x^2-4)\cdot2\sqrt{x+2}}$

$=\lim_{x\to2}\dfrac{2-\sqrt{x+2}}{(x^2-4)\cdot2\sqrt{x+2}}$

$=\lim_{x\to2}\dfrac{(2-\sqrt{x+2})\cdot(2+\sqrt{x+2})}{(x^2-4)\cdot2\sqrt{x+2}\cdot(2+\sqrt{x+2})}$

$=\lim_{x\to2}\dfrac{4-(x+2)}{(x^2-4)\cdot2\sqrt{x+2}\cdot(2+\sqrt{x+2})}$

$=\lim_{x\to2}\dfrac{2-x}{(x+2)(x-2)\cdot2\sqrt{x+2}\cdot(2+\sqrt{x+2})}$

$=\lim_{x\to2}\dfrac{-1}{(x+2)2\sqrt{x+2}(2+\sqrt{x+2})}$

$=\dfrac{-1}{(4)(2)(2)(2+2)}=-\dfrac{1}{64}$

EX9

(1) (a) $\lim_{x\to-2^+}(x+3)\dfrac{|x+2|}{x+2}$

$=\lim_{x\to-2^+}(x+3)\dfrac{(x+2)}{(x+2)}$

$(|x+2|=x+2 \text{ for } x>-2)$

$=\lim_{x\to-2^+}(x+3)=(-2)+3=1$

(b) $\lim_{x\to-2^-}(x+3)\dfrac{|x+2|}{x+2}$

$=\lim_{x\to-2^-}(x+3)\left[\dfrac{-(x+2)}{(x+2)}\right]$

$(|x+2|=-(x+2) \text{ for } x<-2)$

$=\lim_{x\to-2^-}(x+3)(-1)=-(-2+3)=-1$

(2) (a) $\lim_{x\to1^+}\dfrac{\sqrt{2}(x+1)}{|x+1|}=\lim_{x\to1^+}\dfrac{\sqrt{2}(x+1)}{(x+1)}$

$(|x-1|=x-1 \text{ for } x>1)$

$=\lim_{x\to1^+}\sqrt{2}=\sqrt{2}$

(b) $\lim_{x\to1^-}\dfrac{\sqrt{2}(x-1)}{|x-1|}=\lim_{x\to1^-}\dfrac{\sqrt{2}(x-1)}{-(x-1)}$

$(|x-1|=-(x-1) \text{ for } x<1)$

$=\lim_{x\to1^-}-\sqrt{2}=-\sqrt{2}$

(3) (a) $\lim_{\theta\to3^+}\dfrac{\lfloor\theta\rfloor}{\theta}=\dfrac{3}{3}=1$

(b) $\lim_{\theta\to3^-}\dfrac{\lfloor\theta\rfloor}{\theta}=\dfrac{2}{3}$

(4) (a) $\lim_{t\to4^+}(t-\lfloor t\rfloor)=4-4=0$

(b) $\lim_{t\to4^-}(t-\lfloor t\rfloor)=4-3=1$

EX10

(1) $=\dfrac{3}{0^-}=-\infty$

(2) $=\dfrac{-1}{0^-}=+\infty$

(3) $=\dfrac{-16}{0^+}=-\infty$

(4) DNE $\left(\lim_{x\to2^-}\dfrac{1}{x-2}=-\infty,\ \lim_{x\to2^+}\dfrac{1}{x-2}=+\infty\right)$

(5) $=\infty$

(6) $=\dfrac{-1}{0^+}=-\infty$

(7) $=+\infty$

(8) $=\dfrac{1}{0^+}=+\infty$

(9) $=+\infty$

(10) $=-\infty$

(11) $=+\infty$

(12) $=-\infty$

EX11

(1) $\lim_{x\to-1^-}f(x)=\lim_{x\to-1^-}(x+2)=1,\ \lim_{x\to-1^+}f(x)$

$=\lim_{x\to-1^+}(x^2+1)=2$

$\lim_{x\to-1^-}f(x)\neq\lim_{x\to-1^+}f(x)$

$\therefore\lim_{x\to-1}f(x)=DNE$

(2) $\lim_{x\to3^-}f(x)=\lim_{x\to3^-}4\cos\left(\dfrac{\pi}{9}x\right)=4\cos\left(\dfrac{\pi}{3}\right)=4\left(\dfrac{1}{2}\right)=2,$

$\lim_{x\to3^+}f(x)=\lim_{x\to3^+}\sqrt{x^2-5}=\sqrt{9-5}=2$

$\lim_{x\to3^-}f(x)=\lim_{x\to3^+}f(x)=2$

$\therefore\lim_{x\to3}f(x)=2$

(3) $\lim_{x\to\ln3}f(x)=6$

EX12

(1) $a=4, b=0$

(2) $\lim_{x\to1}\frac{x^2-a}{x-1}$ must me the indeterminate form $\left(\frac{0}{0}\right)$ because the limit exists.

$\lim_{x\to1}(x^2-a)=0$ $\quad\lim_{x\to1}\frac{x^2-1}{x-1}=b$

$1^2-a=0,$ $\quad\lim_{x\to1}(x+1)=b$

$\therefore a=1$ $\quad\therefore b=2$

(3) $a=4, b=4\sqrt{2}$

(4) $\lim_{x\to2}\frac{2-x}{ax^2+x-6}$ must be the indeterminate form $\left(\frac{0}{0}\right)$ because the limit exists and is not equal to zero.

$\lim_{x\to2}(ax^2+x-6)=0$ $\quad\lim_{x\to2}\frac{2-x}{x^2+x-6}=b$

$a(2)^2+(2)-6=0$ $\quad\lim_{x\to2}\frac{2-x}{(x-2)(x+3)}=b$

$4a=4$ $\quad\lim_{x\to2}\frac{-1}{x+3}=b$

$\therefore a=1$ $\quad\therefore b=-\frac{1}{5}$

2.2 Squeezing Theorem(Sandwich Theorem)

EX1

(1) $-1\le\cos\left(\frac{1}{x}\right)\le1$

$-x^2\le x^2\cos\left(\frac{1}{x}\right)\le x^2$

$\lim_{x\to0}(-x^2)=\lim_{x\to0}x^2=0.$

By the Squeezing Theorem, $\lim_{x\to0}x^2\cos\left(\frac{1}{x}\right)=0.$

(2) $-1\le\sin\left(\frac{1}{x^2}\right)\le1$

$-x\le x sin\left(\frac{1}{x^2}\right)\le x$

$\lim_{x\to0}(-x)=\lim_{x\to0}(x)=0$

By the Squeezing Theorem, $\lim_{x\to0}x\sin\left(\frac{1}{x^2}\right)=0.$

(3) $-1\le\sin x\le1$

$-\frac{1}{x}\le\frac{\sin x}{x}\le\frac{1}{x}$

$\lim_{x\to\infty}\left(-\frac{1}{x}\right)=\lim_{x\to\infty}\left(\frac{1}{x}\right)=0$

By the Squeezing Theorem, $\lim_{x\to\infty}\frac{\sin x}{x}=0.$

(4) $-1\le\cos2x\le1$

$-\frac{1}{x^2}\le\frac{\cos x}{x^2}\le\frac{1}{x^2}$

$\lim_{x\to\infty}\left(-\frac{1}{x^2}\right)=\lim_{x\to\infty}\left(\frac{1}{x^2}\right)=0$

By the Squeezing Theorem, $\lim_{x\to\infty}\frac{\cos2x}{x^2}=0.$

EX2

(1) 1

(2) $\lim_{x\to-1}(-6x-4)=\lim_{x\to-1}(x^4+x^2)=2$

By the Squeezing Theorem $\lim_{x\to-1}f(x)=2$

(3) $\sqrt{5}$

2.3 Limit of Trigonometric Functions

EX1

(1) 1

(2) $=\lim_{x\to0}\frac{\sin2x}{2x}\cdot\frac{2}{3}$

$=1\cdot\frac{2}{3}=\frac{2}{3}$

(3) $\frac{3}{4}$

(4) $=\lim_{x\to0}\frac{\frac{\sin5x}{5x}\cdot5x}{\frac{\sin2x}{2x}\cdot2x}$

$=\lim_{x\to0}\frac{5x}{2x}=\frac{5}{2}$

(5) $=\left(\lim_{x\to0}\frac{\sin2x}{2x}\right)^3$

$=\left(\lim_{x\to0}\frac{\sin2x}{2x}\bullet2\right)^3$

$=(1\cdot2)^3=8$

(6) $=\lim_{x\to 0}\left(\frac{\sin 3x}{\sin 2x}\right)^2$

$$=\lim_{x\to 0}\left(\frac{\frac{\sin 3x}{3x}\cdot 3x}{\frac{\sin 2x}{2x}\cdot 2x}\right)^2$$

$$=\left(\lim_{x\to 0}\frac{3x}{2x}\right)^2=\left(\frac{3}{2}\right)^2=\frac{9}{4}$$

(7) 2

(8) $=\lim_{x\to 0}\frac{\tan 3x}{3x}\cdot\frac{3}{2}$

$$1\cdot\frac{3}{2}=\frac{3}{2}$$

(9) $\frac{3}{8}$

(10) $=\left(\lim_{x\to 0}\frac{\tan 3x}{2x}\right)^3$

$$=\left(\lim_{x\to 0}\frac{\tan 3x}{3x}\cdot\frac{3}{2}\right)^3=\left(\frac{3}{2}\right)^3$$

$$=\frac{27}{8}$$

(11) $\frac{2}{3}$

(12) $=\lim_{x\to 0}\frac{\sin^2 x}{x^2}\cdot\sin x$

$$=1\cdot 0=0$$

(13) 0

(14) $=\lim_{x\to 0^+}\frac{\sin x\sqrt{x}\sqrt{x}}{\sin\sqrt{x}\cdot x}$

$$=\lim_{x\to 0^+}\sqrt{x}$$

$$=0$$

(15) 1

(16) 0

(17) $=\lim_{x\to 0}\frac{\sin(1-\cos x)}{(1-\cos x)}\cdot\frac{(1-\cos x)}{x^2}$

$$=\lim_{x\to 0}\frac{(1-\cos x)(1+\cos x)}{x^2(1+\cos x)}$$

$$=\lim_{x\to 0}\frac{\sin^2 x}{x^2(1+\cos x)}$$

$$=\lim_{x\to 0}\left(\frac{\sin x}{x}\right)^2\cdot\frac{1}{1+\cos x}$$

$$=1\cdot\frac{1}{1+1}=\frac{1}{2}$$

2.4 Limit of Exponential and Logarithmic Functions

EX1

(1) e^3

(2) $=\left[\lim_{x\to\infty}\left(1+\frac{3}{x}\right)^{\frac{x}{3}}\right]^3=e^3$

(3) e^3

(4) $=\left[\lim_{x\to-\infty}\left(1+\frac{2}{3x}\right)^{\frac{3x}{2}}\right]^{\frac{2}{3}}=e^{\frac{2}{3}}$

(5) $e^{\frac{3}{2}}$

(6) $=\lim_{n\to\infty}P\left[\left(1+\frac{r}{n}\right)^{\frac{r}{n}}\right]^{rt}=Pe^{rt}$

(7) e^3

(8) $=\left[\lim_{x\to 0}\left(1+\frac{x}{2}\right)^{\frac{x}{2}}\right]^{\frac{1}{4}}=e^{\frac{1}{4}}$

(9) e^3

(10) $=\left[\lim_{x\to 0}\left(1+\frac{x}{-2}\right)^{\frac{-2}{x}}\right]^{-\frac{1}{2}}=e^{-\frac{1}{2}}$

(11) e^6

(12) $=\lim_{x\to\infty}\left(2+\frac{1}{x}\right)^{3x}=(2+0)^{\infty}=\infty$

EX2

(1) $=\lim_{x\to 0}\frac{1}{x}\cdot\ln(x+1)$

$=\lim_{x\to 0}\ln(x+1)^{\frac{1}{x}}$

$=\ln e=1$

(2) Let $e^x-1=t$

$e^x=1+t$

$x=\ln(1+t)$

As $x\to 0, t\to 0$

$=\lim_{x\to 0}\frac{e^x-1}{x}=\lim_{t\to 0}\frac{t}{\ln(1+t)}$

$=\lim_{t\to 0}\frac{1}{\frac{1}{t}\ln(1+t)}$

$=\lim_{t\to 0}\frac{1}{\ln(1+t)^{\frac{1}{t}}}=\frac{1}{\ln e}=1$

2.5 Limits Involving Infinity

EX1

(1) ∞

(2) $=\lim_{x\to\infty}\frac{\frac{3x^3+4}{x^3}}{\frac{2x^3-3x+5}{x^3}}=\lim_{x\to\infty}\frac{3+\frac{4}{x^3}}{2-\frac{3}{x^2}+\frac{5}{x^3}}$

$=\frac{3+0}{2-0+0}=\frac{3}{2}$

(3) 0

(4) $=\lim_{x\to\infty}\frac{\frac{3x^2+1}{x^2}}{\frac{5x^2-2}{x^2}}=\lim_{x\to\infty}\frac{3+\frac{1}{x^2}}{5-\frac{2}{x^2}}$

$=\frac{3+0}{5-0}=\frac{3}{5}$

(5) 1

(6) $=\lim_{x\to-\infty}\frac{\frac{2x^4-4x}{x^5}}{\frac{x^5+9x-5}{x^5}}=\lim_{x\to-\infty}\frac{\frac{2}{x}-\frac{4}{x^4}}{1+\frac{9}{x^4}-\frac{5}{x^5}}$

$=\frac{0-0}{1+0-0}=0$

(7) ∞

(8) 1

(9) -1

(10) $\frac{1}{2}$

(11) 3

EX2

(1) $y=\frac{2}{5}$

(2) $y=-\frac{1}{2}$

(3) no horizontal asymptote

(4) $y=0$

(5) $y=0$

(6) $y=-2$

(7) $\lim_{x\to\pm\infty}\left(\frac{7}{x+4}+5\right)=(0+5)=5$

$\therefore y=5$

(8) $\lim_{x\to\pm\infty}\left(\frac{2x}{x^2+4}+2\right)=(0+2)=2$

$\therefore y=2$

(9) $\lim_{x\to\pm\infty}\left(-\frac{4x}{2x-3}+5\right)=(-2+5)=3$

$\therefore y=3$

EX3

(1) (a) $x=3$

(b) $\lim_{x\to 3^-}y=-\infty$

$\lim_{x\to 3^+}y=+\infty$

(2) (a) $x=-2$

(b) $\lim_{x\to -2^-}y=\lim_{x\to -2^-}\left(\frac{x^2}{x+2}\right)=\frac{4}{0^-}=-\infty$

$\lim_{x\to -2^+}y=\lim_{x\to -2^+}\left(\frac{x^2}{x+2}\right)=\frac{4}{0^+}=+\infty$

(3) (a) $x=0$

(b) $\lim_{x\to 0^-}y=+\infty$

$\lim_{x\to 0^+}y=+\infty$

(4) (a) $x=\pm 1$

(b) $\lim_{x\to -1^-}\left(\frac{x^2}{1-x^2}\right)=\frac{1}{0^-}=-\infty$

$\lim_{x\to -1^+}\left(\frac{x^2}{1-x^2}\right)=\frac{1}{0^+}=+\infty$

$\lim_{x\to 1^-}\left(\frac{x^2}{1-x^2}\right)=\frac{1}{0^+}=+\infty$

$\lim_{x\to 1^+}\left(\frac{x^2}{1-x^2}\right)=\frac{1}{0^-}=-\infty$

(5) (a) $x=\pm 2$

(b) $\lim_{x\to -2^-} y=-\infty$

$\lim_{x\to -2^+} y=+\infty$

$\lim_{x\to 2^-} y=-\infty$

$\lim_{x\to 2^+} y=+\infty$

(6) (a) $x=2$

(b) Domain : $x<2$

$\lim_{x\to 2^-}\left(\frac{x}{\sqrt{2-x}}+3\right)$

$=\frac{2}{0^+}+3=+\infty$

EX4

(1) (a) $\therefore y=2$(to the right),

$y=-2$(to the left).

(b) $\therefore |x|-1=0 \Rightarrow \therefore x=\pm 1$

(2) $|x|=\begin{cases} x \ , \text{if} \, x \ge 0 \\ -x \, , \text{if} \, x \le 0 \end{cases}$

(a) $\lim_{x\to\infty} y=\lim_{x\to\infty}\left(\frac{x-1}{2|x|-1}\right)=\lim_{x\to\infty}\frac{x-1}{2x-1}=\frac{1}{2}$

$\lim_{x\to -\infty} y=\lim_{x\to -\infty}\left(\frac{x-1}{2|x|-1}\right)=\lim_{x\to -\infty}\frac{x-1}{-2x-1}$

$=-\frac{1}{2}$

$\therefore y=\frac{1}{2}$ (to the right),

$y=-\frac{1}{2}$ (to the left)

(b) no vertical asymptote

(3) (a) $\therefore y=5$ (to the right),

$y=-2$ (to the left)

(b) $4^x-1=0$

$4^x=1$

$\therefore x=0$

(4) (a) $\lim_{x\to\infty} y=\lim_{x\to\infty}\left(\frac{3(2)^x+1}{(2)^x-3}\right)=\lim_{x\to\infty}\left(\frac{3+\frac{1}{2^x}}{1-\frac{3}{2^x}}\right)$

$=\frac{3+0}{1-0}=3$

$\lim_{x\to -\infty} y=\lim_{x\to -\infty}\left(\frac{3(2)^x+1}{(2)^x-3}\right)=\frac{0+1}{0-3}$

$=-\frac{1}{3}$

$\therefore y=3$ (to the right),

$y=-\frac{1}{3}$ (to the left)

(b) $2^x-3=0$

$2^x=3$

$\therefore x=\log_2 3$

EX5

(1) $\lim_{x\to\left(\frac{\pi}{2}\right)^-}\tan x=+\infty$ (2) $\lim_{x\to\left(\frac{\pi}{2}\right)^+}\sec x=-\infty$

(3) $\lim_{x\to 0^-}\csc x=-\infty$ (4) $\lim_{x\to 0^+}\cot x=+\infty$

(5) $\lim_{x\to 0^+}\ln x=-\infty$ (6) $\lim_{x\to 2^-}\log_3(2-x)=-\infty$

EX6

(1) 0 (2) $+\infty$

(3) $=e^{-\infty}=0$ (4) $=(0.9)^{+\infty}=0$

(5) $=e^{+\infty}=+\infty$ (6) $=\frac{1}{\ln(0^+)}=\frac{1}{-\infty}=0$

(7) $=\left(\frac{3^{-\infty}+1}{3^{-\infty}-1}\right)=\frac{0+1}{0-1}=-1$ (8) $-\frac{\pi}{2}$

EX7

(1) DNE

(2) $-1\le \sin x\le 1 \Rightarrow -\frac{1}{x}\le\frac{\sin x}{x}\le\frac{1}{x}$

$\Rightarrow \lim_{x\to\infty}\left(-\frac{1}{x}\right)=\lim_{x\to\infty}\left(\frac{1}{x}\right)=0$

$\Rightarrow$By the Squeezing Theorem, $\lim_{x\to\infty}\frac{\sin x}{x}=0.$

$-1\le \sin\left(\frac{1}{x}\right)\le 1\Rightarrow -x\le x\sin\left(\frac{1}{x}\right)\le x$

$\Rightarrow\lim_{x\to\infty}(-x)=\lim_{x\to\infty}(x)=0\Rightarrow$By the Squeezing Theorem, $\lim_{x\to\infty}x\sin\left(\frac{1}{x}\right)=0.$

(3) DNE

(4) $\lim_{x\to\infty}x\sin\frac{1}{x}=\lim_{x\to\infty}\frac{\sin\left(\frac{1}{x}\right)}{\frac{1}{x}}=1$

$(\because As\ x\to\infty,\ \frac{1}{x}\to 0)$

EX8

(1) $\frac{3}{\sqrt{2}}$

(2) $=\lim_{x\to\infty}\frac{\frac{x+3}{\sqrt{x^2}}}{\frac{\sqrt{x^2+1}+2}{\sqrt{x^2}}}$

$=\lim_{x\to\infty}\frac{\frac{x}{|x|}+\frac{3}{|x|}}{\sqrt{1+\frac{1}{x}}+\frac{2}{|x|}}$

$=\lim_{x\to\infty}\frac{\frac{x}{x}+0}{\sqrt{1+0}+0}=\frac{1}{1}=1$

(3) 1

(4) $=\lim_{x\to\infty}\frac{\frac{\sqrt{x^2-1}+3}{\sqrt{x^2}}}{\frac{2x}{\sqrt{x^2}}}$

$=\lim_{x\to\infty}\frac{\sqrt{1-\frac{1}{x^2}}+\frac{3}{|x|}}{\frac{2x}{|x|}}$

$=\frac{\sqrt{1-0}+0}{\lim_{x\to\infty}\frac{2x}{x}}=\frac{1}{2}$

EX9

(1) $\frac{5}{2}$

(2) $=\lim_{x\to-\infty}\frac{(\sqrt{x^2+x}+x)(\sqrt{x^2+x}-x)}{(\sqrt{x^2+x}-x)}$

$=\lim_{x\to-\infty}\frac{x^2+x-x^2}{\sqrt{x^2+x}-x}=\lim_{x\to-\infty}\frac{x}{\sqrt{x^2+x}-x}$

$=\frac{1}{-\sqrt{1+0}-1}=-\frac{1}{2}$

(3) $-\frac{3}{4}$

(4) $=\lim_{x\to\infty}\frac{(2x-\sqrt{4x^2+x})(2x+\sqrt{4x^2+x})}{(2x+\sqrt{4x^2+x})}$

$=\lim_{x\to\infty}\frac{4x^2-(4x^2+x)}{2x+\sqrt{4x^2+x}}=\lim_{x\to\infty}\frac{-x}{2x+\sqrt{4x^2+x}}$

$=\lim_{x\to\infty}\frac{\frac{-x}{\sqrt{x^2}}}{\frac{2x}{\sqrt{x^2}}+\frac{\sqrt{4x^2+x}}{\sqrt{x^2}}}=\lim_{x\to\infty}\frac{\frac{-x}{|x|}}{\frac{2x}{|x|}+\sqrt{4+\frac{1}{x}}}$

$=\frac{-1}{2+\sqrt{4+0}}=-\frac{1}{4}$

EX10

(1) 2	(2) -3	(3) 1
(4) DNE	(5) $+\infty$	(6) $+\infty$
(7) $+\infty$	(8) $+\infty$	(9) $-\infty$
(10) DNE	(11) 0	(12) -1

EX11

(1)

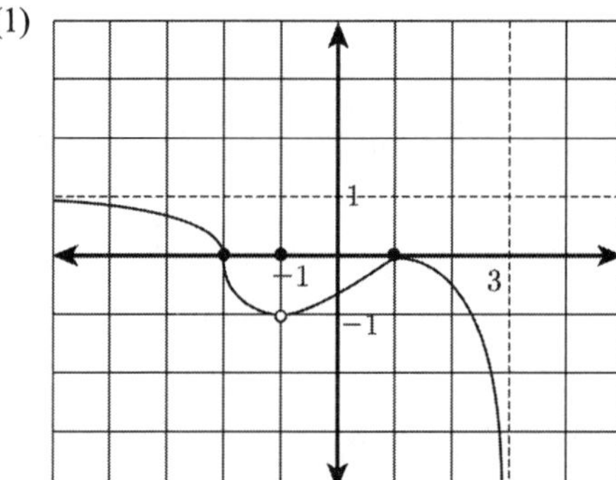

(2)

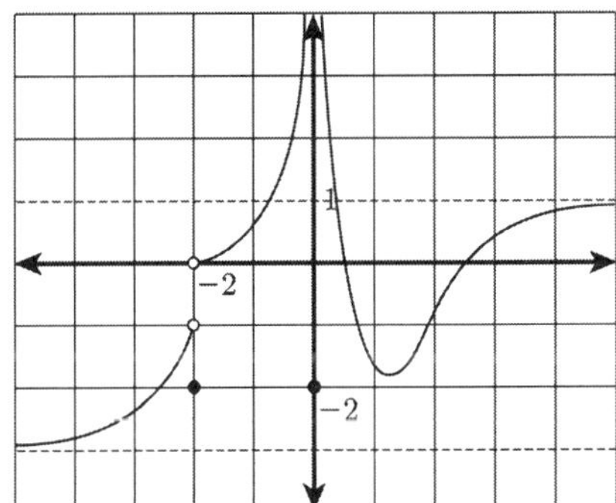

2.6 Continuity

EX1

(1) (a) $\lim_{x\to -3} f(x) \neq f(-3)$

(b) removable discontinuity

(2) (a) $\lim_{x\to -1^-} f(x) \neq \lim_{x\to -1^+} f(x)$

(b) jump discontinuity

(3) (a) $f(1) = DNE$

(b) removable discontinuity

(4) (a) $\lim_{x\to 3^-} f(x) \neq \lim_{x\to 3^+} f(x)$

(b) jump discontinuity

(5) (a) $\lim_{x\to 5} f(x) = DNE$, $f(5) = DNE$

(b) infinite discontinuity

EX2

(1)

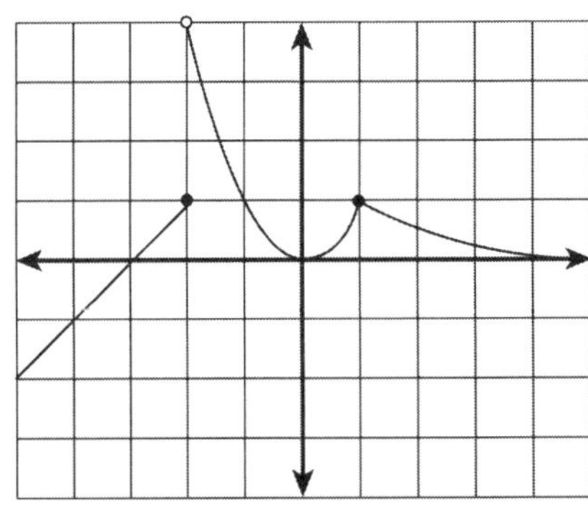

discontinuous at $x = -2$

(2)

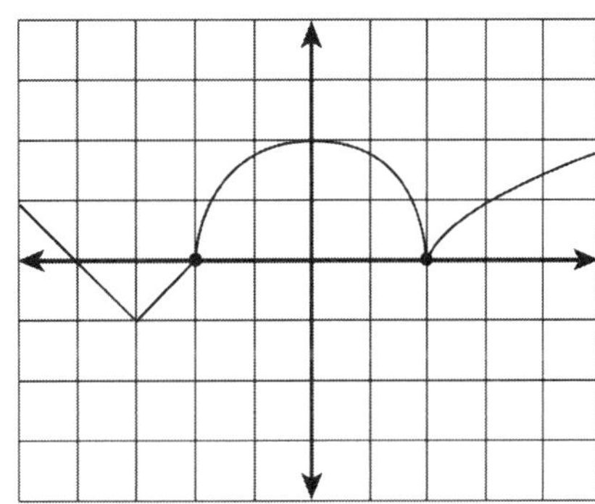

continuous everywhere

(3)

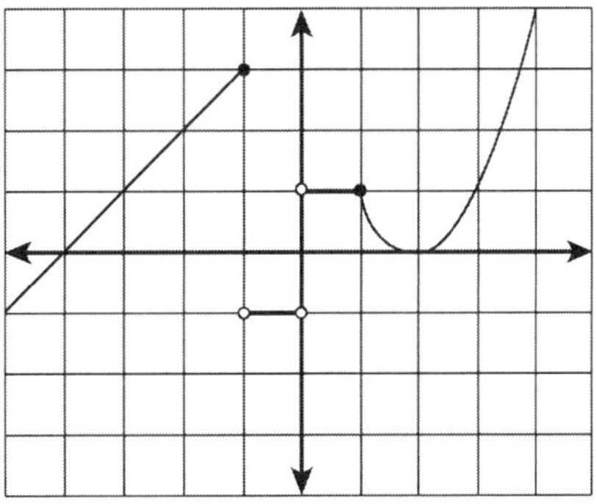

discontinuous at $x = -1, 0$

(4)

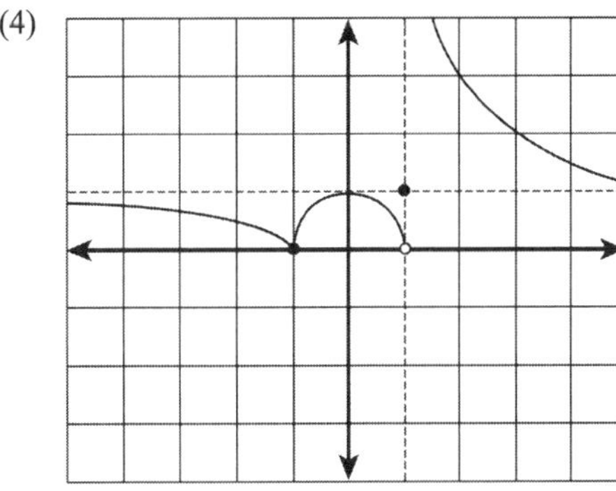

discontinuous at $x = 1$

EX3

(1) 6

(2) 7

(3) $\frac{3}{2}$

(4) $\frac{8}{5}$

EX4

(1) 1. $\lim_{x\to-1^-} f(x)=\lim_{x\to-1^-}(x+3)=2,$

$\lim_{x\to-1^+} f(x)=\lim_{x\to-1^+}\sqrt{x+5}=2$

$\therefore \lim_{x\to-1} f(x)=2$

2. $f(-1)=-1+3=2$

3. $\lim_{x\to-1} f(x)=f(-1)$

$\therefore f$ is continuous at $x=-1$

(2) 1. $\lim_{x\to1^-} f(x)=\lim_{x\to1^-}(3-\ln x)=3-\ln 1=3,$

$\lim_{x\to1^+} f(x)=\lim_{x\to1^+}(2e^{\sin(\pi x)}+1)=2e^0+1=3$

$\therefore \lim_{x\to1} f(x)=3$

2. $f(1)=2e^{\sin(\pi)}+1=2e^0+1=3$

3. $\lim_{x\to1} f(x)=f(1)$

$\therefore f$ is continuous at $x=1$

EX5

(1) $k=4$

(2) $\lim_{x\to0} f(x)=f(0)$

$\lim_{x\to0}\frac{x^2-2x}{x}=k$

$\lim_{x\to0}(x-2)=k$

$\therefore k=-2$

(3) $k=0$

(4) $\lim_{x\to0} f(x)=f(0)$

$\lim_{x\to0}\frac{\sin x}{x}=k$

$\therefore k=1$

EX6

(1) 3 or −2

(2) −2 or 0

(3) $a=\frac{5}{2}, b=-\frac{1}{2}$

(4) $a=-\frac{3}{2}, b=-\frac{3}{2}$

2.7 Average Rate of Change and Instantaneous Rate of Change

EX1

(1) a^2+ab+b^2

(2) $\frac{f(4)-f(2)}{4-2}=\frac{24-8}{2}=8$

(3) $-\frac{4}{\pi}$

(4) $\frac{f(5)-f(1)}{5-1}=\frac{3-1}{4}=\frac{1}{2}$

(5) $\frac{-1}{a(a+h)}$

(6) $\frac{f\left(\frac{\pi}{3}\right)-f\left(-\frac{\pi}{6}\right)}{\frac{\pi}{3}-\left(-\frac{\pi}{6}\right)}=\frac{\left(\frac{\sqrt{3}}{2}+\frac{1}{2}\right)-\left(-\frac{1}{2}+\frac{\sqrt{3}}{2}\right)}{\frac{\pi}{2}}$

$=\frac{1}{\frac{\pi}{2}}=\frac{2}{\pi}$

EX2

(1) (a) $f'(1)=4$

(b) $y-3=4(x-1)$

(2) (Method 1)

(a) $f'(2)=\lim_{x\to2}\frac{f(x)-f(2)}{x-2}$

$=\lim_{x\to2}\frac{x^2-6x+8}{x-2}$

$=\lim_{x\to2}\frac{(x-2)(x-4)}{(x-2)}$

$=-2$

(b) $y+8=-2(x-2)$

(Method 2)

$f'(2)=\lim_{h\to0}\frac{f(2+h)-f(2)}{h}$

$=\lim_{h\to0}\frac{(2+h)^2-6(2+h)-(-8)}{h}$

$=\lim_{h\to0}\frac{4+4h+h^2-12-6h+8}{h}$

$=\lim_{h\to0}\frac{h^2-2h}{h}=\lim_{h\to0}(h-2)=-2$

(3) (a) $f'(-1)=3$

(b) $y+1=3(x+1)$

(4) (Method 1)

(a) $f'(1)=\lim\limits_{x\to1}\dfrac{f(x)-f(1)}{x-1}$

$$=\lim_{x\to1}\frac{\left(\dfrac{1}{x+1}-\dfrac{1}{2}\right)\cdot 2(x+1)}{(x-1)\cdot 2(x+1)}$$

$$=\lim_{x\to1}\frac{2-(x+1)}{x-1\cdot 2(x+1)}$$

$$=\lim_{x\to1}\frac{1-x}{(x-1)\cdot 2(x+1)}$$

$$=\lim_{x\to1}\frac{-1}{2(x+1)}=-\frac{1}{4}$$

(b) $y-\dfrac{1}{2}=-\dfrac{1}{4}(x-1)$

(Method 2)

$$f'(1)=\lim_{h\to0}\frac{f(1+h)-f(1)}{h}$$

$$=\lim_{h\to0}\frac{\left(\dfrac{1}{h+2}-\dfrac{1}{2}\right)\cdot 2(h+2)}{h\cdot 2(h+2)}$$

$$=\lim_{h\to0}\frac{2-(h+2)}{h\cdot 2(h+2)}$$

$$=\lim_{h\to0}\frac{-h}{h\cdot 2(h+2)}=-\frac{1}{4}$$

(5) (a) $f'(6)=\dfrac{1}{4}$

(b) $y-2=\dfrac{1}{4}(x-6)$

(6) (Method 1)

(a) $f(3)=\dfrac{1}{\sqrt{4}}=\dfrac{1}{2}$, $\therefore (3,\dfrac{1}{2})$

$$f'(3)=\lim_{x\to3}\frac{f(x)-f(3)}{x-3}$$

$$=\lim_{x\to3}\frac{\left(\dfrac{1}{\sqrt{x+1}}-\dfrac{1}{2}\right)\cdot 2\sqrt{x+1}}{(x-3)\cdot 2\sqrt{x+1}}$$

$$=\lim_{x\to3}\frac{(2-\sqrt{x+1})\cdot(2+\sqrt{x+1})}{(x-3)\cdot 2\sqrt{x+1}\cdot(2+\sqrt{x+1})}$$

$$=\lim_{x\to3}\frac{4-(x+1)}{(x-3)\cdot 2\sqrt{x+1}\cdot(2+\sqrt{x+1})}$$

$$=\lim_{x\to3}\frac{(3-x)}{(x-3)\cdot 2\sqrt{x+1}\cdot(2+\sqrt{x+1})}$$

$$=\lim_{x\to3}\frac{-1}{2\sqrt{x+1}\cdot(2+\sqrt{x+1})}$$

$$=\frac{-1}{2(2)(2+2)}=-\frac{1}{16}$$

(b) $y-\dfrac{1}{2}=-\dfrac{1}{16}(x-3)$

(Method 2)

$$f'(3)=\lim_{h\to0}\frac{f(3+h)-f(3)}{h}$$

$$=\lim_{h\to0}\frac{\left(\dfrac{1}{\sqrt{4+h}}-\dfrac{1}{2}\right)\cdot 2\sqrt{4+h}}{h\cdot 2\sqrt{4+h}}$$

$$=\lim_{h\to0}\frac{(2-\sqrt{4+h})\cdot(2+\sqrt{4+h})}{h\cdot 2\sqrt{4+h}\cdot(2+\sqrt{4+h})}$$

$$=\lim_{h\to0}\frac{4-(4+h)}{h\cdot 2\sqrt{4+h}\cdot(2+\sqrt{4+h})}$$

$$=\lim_{h\to0}\frac{-h}{h\cdot 2\sqrt{4+h}\cdot(2+\sqrt{4+h})}$$

$$=\lim_{h\to0}\frac{-1}{2\sqrt{4+h}\cdot(2+\sqrt{4+h})}$$

$$=\frac{-1}{2(2)(2+2)}=-\frac{1}{16}$$

2.8 Intermediate Value Theorem(I.V.T.)

EX1

(1) $x^3-3x-8=0$

Let $f(x)=x^3-3x-8$

Since f is a polynomial function, it is continuous everywhere.

$f(2)=8-6-8=-6<0$

$f(3)=27-9-8=10>0$

$\therefore f(2)<0<f(3)$

By I.V.T. there must be at least one solution c, $2<c<3$, such that $f(c)=0$.

(2) $x^3+x^2-2x-6=0$

Let $f(x)=x^3+x^2-2x-6=0$

Since f is a polynomial function, it is continuous everywhere.

$f(1)=1+1-2-6=-6<0$

$f(2)=8+4-4-6=2>0$

$\therefore f(1)<0<f(2)$

By I.V.T. , there must be at least one solution c, $1<c<2$, such that $f(c)=0$.

EX2

(1) $v(5)=-1<0,\ v(10)=5>0$

$v(5)<0<v(10)$

$v(t)$ is continuous on the interval $[5, 10]$.

By I.V.T. there must be a value t for $5<t<10$, such that $v(t)=0$.

(2) $a(10)=2<4$, $a(20)=5>4$

$a(10)<4<a(20)$

$a(t)$ is continuous on the interval $[10, 20]$.

By I.V.T. , there must be a value t for $10<t<20$, such that $a(t)=4$.

EX3

(1) $M>0$

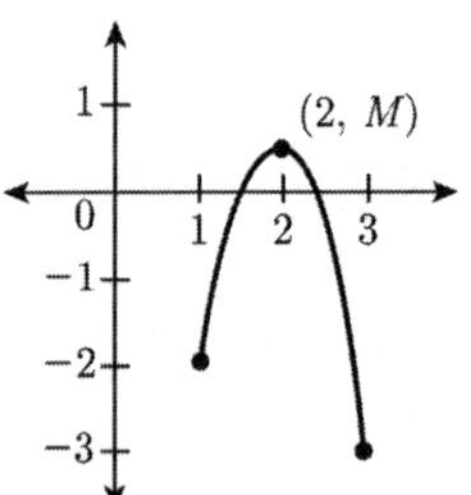

$M<0$

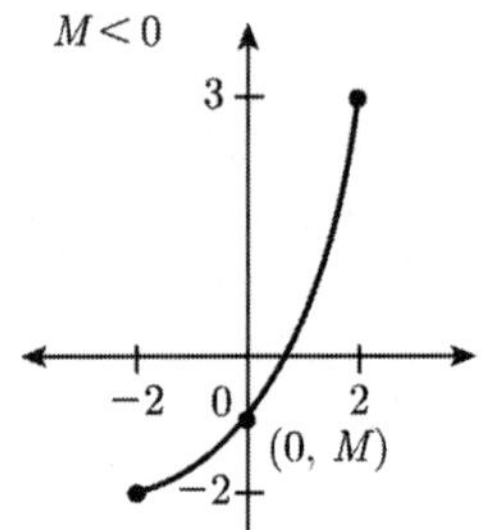

Multiple Choice

01 C 02 E 03 C 04 E 05 A 06 A 07 A 08 E 09 E 10 B 11 E 12 C 13 B 14 E 15 D
16 B 17 B 18 E 19 B 20 C 21 D 22 E 23 E

03 Derivatives

3.1 Definition for the Derivative

EX1

(1) $f'(x)=\dfrac{-3}{2(3x-2)\sqrt{3x-2}}$

(2) $f'(x)=\lim\limits_{h\to 0}\dfrac{f(x+h)-f(x)}{h}$

$=\lim\limits_{h\to 0}\dfrac{(x+h)^3-2(x+h)-(x^3-2x)}{h}$

$=\lim\limits_{h\to 0}\dfrac{x^3+3x^2h+3xh^2+h^3-2x-2h-x^3+2x}{h}$

$=\lim\limits_{h\to 0}(3x^2+3xh+h^2-2)$

$=3x^2-2$

(3) $f'(x)=1+\dfrac{1}{x^2}$

(4) $f'(x)=\lim\limits_{h\to 0}\dfrac{f(x+h)-f(x)}{h}$

$=\lim\limits_{h\to 0}\dfrac{\left[\dfrac{1}{(x+h)^2}-\dfrac{1}{x^2}\right]\cdot x^2(x+h)^2}{h\cdot x^2(x+h)^2}$

$=\lim\limits_{h\to 0}\dfrac{x^2-(x+h)^2}{h\cdot x^2(x+h)^2}$

$=\lim\limits_{h\to 0}\dfrac{-2xh-h^2}{h\cdot x^2(x+h)^2}$

$=\lim\limits_{h\to}\dfrac{-2x-h}{x^2(x+h)^2}$

$=\dfrac{-2x}{x^2(x^2)}=\dfrac{-2}{x^3}$

(5) $f'(x)=\dfrac{3}{(x+1)^2}$

(6) $f'(x)=\lim\limits_{h\to 0}\dfrac{f(x+h)-f(x)}{h}$

$=\lim\limits_{h\to 0}\dfrac{(\sqrt{(x+h)^2+1}-\sqrt{x^2+1})\cdot(\sqrt{(x+h)^2+1}+\sqrt{x^2+1})}{h\cdot(\sqrt{(x+h)^2+1}+\sqrt{x^2+1})}$

$=\lim\limits_{h\to 0}\dfrac{(x+h)^2+1-(x^2+1)}{h\cdot(\sqrt{(x+h)^2+1}+\sqrt{x^2+1})}$

$=\lim\limits_{h\to 0}\dfrac{2x+h}{2\sqrt{x^2+1}}$

$=\dfrac{2x}{2\sqrt{x^2+1}}=\dfrac{x}{\sqrt{x^2+1}}$

3.2 Differentiability

EX1

(1) (a) The function is differentiable on its domain $-3\le x\le 2$ (it is smooth)

(b) none

(c) none

(2) (a) The function is differentiable on its domain $-2\le x\le 3$ (it is smooth)

(b) none

(c) none

(3) (a) The function is differentiable on $-3\le x\le 0$ and $0<x\le 3$

(b) none

(c) The function is neither continuous nor differentiable at $x=0$ since $\lim\limits_{x\to 0^-}f(x)\ne\lim\limits_{x\to 0^+}f(x)$

(4) (a) f is differentiable on $-2\le x<1$, $-1<x<0, 0<x<2$, and $2<x\le 3$.

(b) f is continuous but not differentiable at $x=-1$; $\lim\limits_{x\to -1}f(x)=0$ exists but there is a corner at $x=-1$ since

$\lim\limits_{h\to 0^-}\dfrac{f(-1+h)-f(-1)}{h}=-3$ and

$\lim\limits_{h\to 0+}\dfrac{f(-1+h)-f(-1)}{h}=3\Rightarrow f'(-1)$

does not exist.

(c) f is neither continuous nor differentiable at $x=0$ and $x=2$: at $x=0$, $\lim_{x\to0^-}f(x)=3$ but $\lim_{x\to0^+}f(x)=0 \Rightarrow \lim_{x\to0}f(x)$ does not exist : at $x=2$, $\lim_{x\to2}f(x)$ exists but $\lim_{x\to2}f(x) \neq f(2)$

(5) (a) f is differentiable on $-1 \le x < 0$ and $0 < x \le 2$

(b) f is continuous but not differentiable at $x=0$: $\lim_{x\to0}f(x)=0$ exists but there is a cusp at $x=0$, so

$f'(0)=\lim_{h\to0}\dfrac{f(0+h)-f(0)}{h}$ does not exist

(6) (a) f is differentiable on $-3 \le x < 2$, $-2 < x < 2$, and $2 < x \le 3$

(b) f is continuous but not differentiable at $x=-2$ and $x=2$: there are corners at those points.

(c) none

EX3

(1) $y=f'(x)$

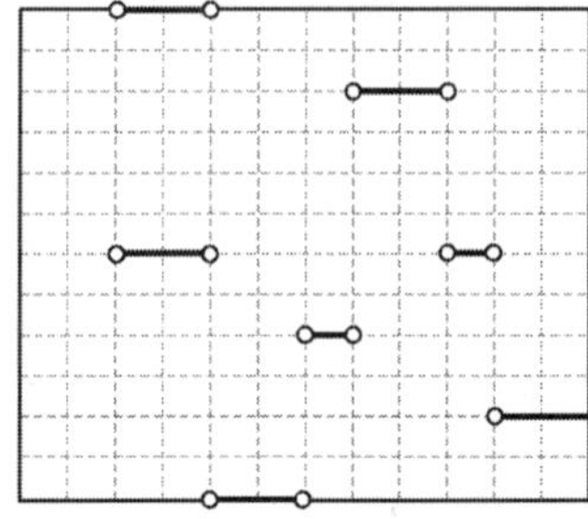

(2) $y=f'(x)$

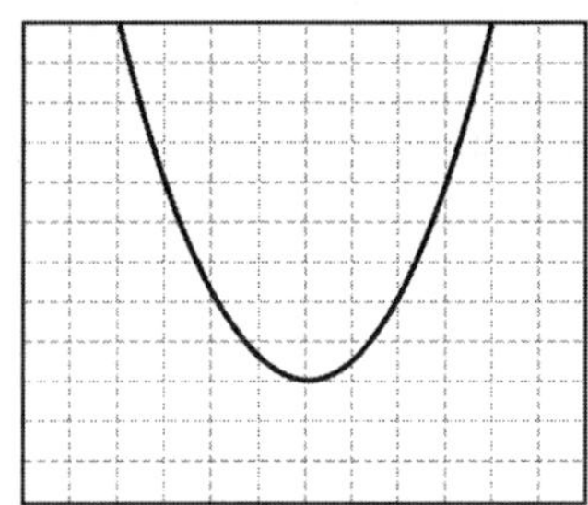

EX2

(1) T : left-hand limit = right-hand limit = 2	(2) T : f is continuous at $x=-2$	(3) F : left-hand tangent slope=1 right-hand tangent slope=-2
(4) T : left-hand limit = right-hand limit=-2	(5) F : f is discontinuous at $x=0$ $\lim_{x\to0}f(x)=-2 \neq f(0)=-1$	(6) F : left-hand tangent slope=-2 right-hand tangent slope=2
(7) T	(8) F : f is discontinuous at x=1 $\therefore f'(1)$ does not exist	(9) T : right-hand tangent slope $\lim_{x\to0^+}f'(x)=2$
(10) T : f is discontinuous at $x=1$ $\therefore f'(1)$ does not exist	(11) T : left-hand tangent slope=2 right-hand tangent slope=2	(12) T : left-hand limit=2 right-hand limit=1
(13) F : left-hand limit=2=$f(2)$	(14) T : The graph of f is smooth in the interval(2, 6) $\Rightarrow f$ is continuous at $x=3$ and left-hand slope = right-hand slope $\therefore f'(3)$ exists	(15) T : f is continuous at $x=4$
(16) F : $f'(4)=0$ (horizontal slope)	(17) T : f is continuous in the interval (2,6) $\therefore$ The limit exists for every number in the interval (2,6)	
(18) T : The graph of f is smooth in the interval (2,6) $\therefore$ left-hand slope = right-hand slope for every number in the interval (2,6)		
(19) T : tangent slope=1	(20) F : tangent slope= -2	(21) T : $f'(3)\approx-1$, $f'(4)=0$ $f'(5)\approx1$, $f'(5.5)\approx3$

(3) $y=f'(x)$

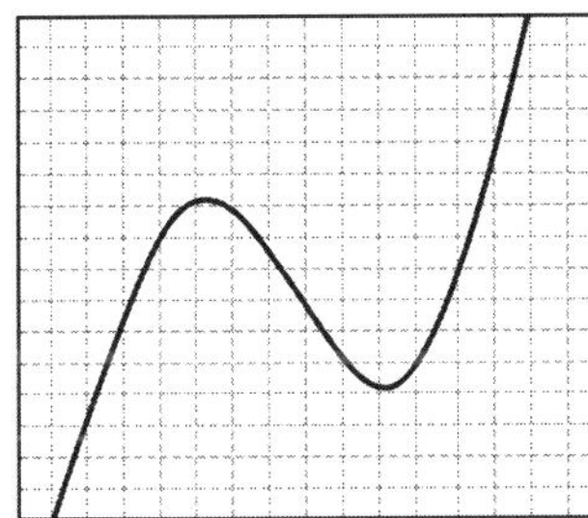

(4) $y=f'(x)$

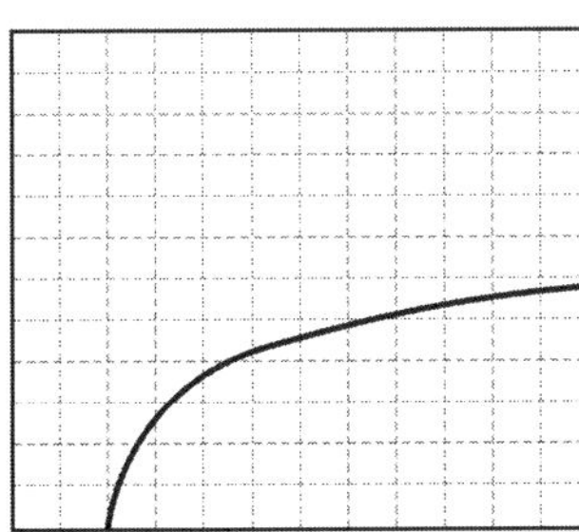

(5) $y=f'(x)$

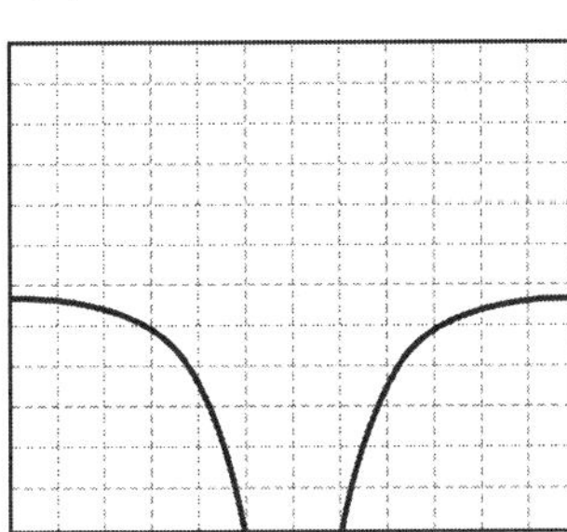

(6) $y=f'(x)$

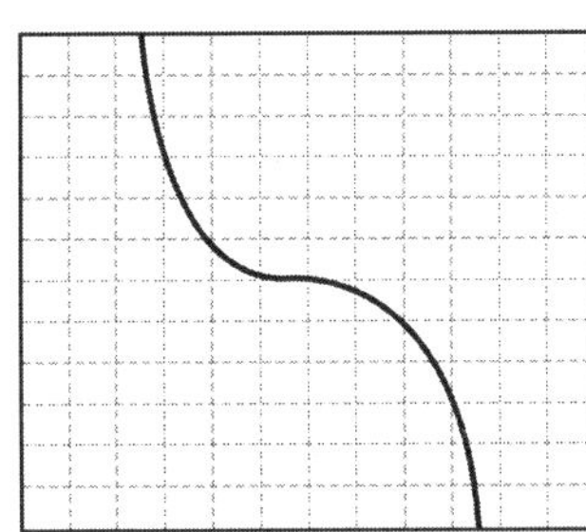

3.3 Power Rule

EX1

(1) $y'=0$

(2) $y'=0$

(3) $y'=-20x^3$

(4) $y=x^{\frac{4}{5}}$

$y'=\frac{4}{5}x^{-\frac{1}{5}}$

(5) $y'=6x^{-4}$

(6) $y=2x^{-\frac{1}{2}}$

$y'=-x^{-\frac{3}{2}}$

$=-\frac{1}{x^{3/2}}$

EX2

(1) $y'=6x-5$

(2) $y'=15x^2-\frac{1}{\sqrt{x}}$

(3) $y=1+\frac{1}{x}+2x^{-2}$

$y'=0-\frac{1}{x^2}-4x^{-3}$

$=-\frac{1}{x^2}-\frac{4}{x^3}$

(4) $y=2\sqrt{x}+\frac{2}{x}$

$y'=\frac{1}{\sqrt{x}}-\frac{2}{x^2}$

(5) $y'=2x^{-2}-8x^{-3}$

(6) $y'=-12x^{-2}+12x^{-4}-4x^{-5}$

EX3

(1) $y=1+5x^{-1}-x^{-2}$

$y'=-5x^{-2}+2x^{-3}$

(2) $y=2x^{5/2}-3x-x^{-1/2}$

$y'=5x^{3/2}-3+\frac{1}{2}x^{-3/2}$

$=5x^{3/2}-3+\frac{1}{2x^{3/2}}$

(3) $y=3x^{-1}+2x-x^2$

$y'=-3x^{-2}+2-2x$

$=-\frac{3}{x^2}+2-2x$

(4) $f(x) = 12x^{5/3} - 6x^{2/3}$

$f'(x) = 20x^{2/3} - 4x^{-1/3}$

$= 20x^{2/3} - \dfrac{4}{x^{1/3}}$

(5) $f(x) = x^{\frac{3}{2}}(x - 3x^2)$

$= x^{5/2} - 3x^{7/2}$

$f'(x) = \dfrac{5}{2}x^{3/2} - \dfrac{21}{2}x^{5/2}$

EX4

(1) $y - 3 = 2(x+2)$

(2) $y + 2 = 5(x-1)$

EX5

(1) $y = 2x + 5$

(2) $y = -2x + 9$

(3) $f(1) = 4 - 1 = 3 \Rightarrow (1,3)$

$f'(x) = 4 \cdot \dfrac{1}{2\sqrt{x}} + \dfrac{1}{x^2}$

$= \dfrac{2}{\sqrt{x}} + \dfrac{1}{x^2}$

$f'(1) = 3$

$\therefore y - 3 = 3(x-1)$

(4) $g(4) = \dfrac{8}{16} + \dfrac{2}{2} = \dfrac{1}{2} + 1 \Rightarrow (4, \dfrac{3}{2})$

$g(x) = 8x^{-2} + 2x^{-1/2}$

$g'(x) = -16x^{-3} - x^{-3/2}$

$= \dfrac{-16}{x^3} - \dfrac{1}{x\sqrt{x}}$

$g'(4) = -\dfrac{16}{64} - \dfrac{1}{4\sqrt{4}} = -\dfrac{1}{4} - \dfrac{1}{8} = -\dfrac{3}{8}$

$\therefore y - \dfrac{3}{2} = -\dfrac{3}{8}(x-4)$

EX6

(1) $y = x^3 - 3x - 2 \Rightarrow y' = 3x^2 - 3$. For the tangent to be horizontal, we need $m = y' = 0$ $\Rightarrow 0 = 3x^2 - 3 \Rightarrow 3x^2 = 3 \Rightarrow x = \pm 1$. When $x = -1, y = 0 \Rightarrow$ the tangent line has equation $y = 0$. The line perpendicular to this line at $(-1,0)$ is $x = -1$. When $x = 1, y = -4 \Rightarrow$ the tangent line has equation $y = -4$. The line perpendicular to this line at $(1,-4)$ is $x = 1$.

(2) $8x - 2y = 1 \Rightarrow y = 4x - \dfrac{1}{2} \Rightarrow m = 4$;

$g(x) = \dfrac{1}{3}x^3 - \dfrac{3}{2}x^2 + 1$

$\Rightarrow g'(x) = x^2 - 3x; x^2 - 3x = 4$

$\Rightarrow x = 4$ or $x = -1$

$\Rightarrow g(4) = \dfrac{1}{3}(4)^3 - \dfrac{3}{2}(4)^2 + 1 = -\dfrac{5}{3}$

$g(-1) = \dfrac{1}{3}(-1)^3 - \dfrac{3}{2}(-1)^2 + 1 = -\dfrac{5}{6}$

$\Rightarrow (4, -\dfrac{5}{3})$ or $(-1, -\dfrac{5}{6})$

(3) $y = 2x + 3 \Rightarrow m = 2, m_{\perp} = -\dfrac{1}{2}$;

$y = \dfrac{x}{x-2} \Rightarrow y'(x) = \dfrac{(x-2)(1) - x(1)}{(x-2)^2}$

$= \dfrac{-2}{(x-2)^2}; \dfrac{-2}{(x-2)^2}$

$= -\dfrac{1}{2} \Rightarrow 4 = (x-2)^2$

$\Rightarrow \pm 2 = x - 2 \Rightarrow x = 4$ or $x = 0$

$\Rightarrow$ if $x = 4, y = \dfrac{4}{4-2} = 2$

and if $x = 0, y = \dfrac{0}{0-2} = 0$

$\Rightarrow (4,2)$ or $(0,0)$

EX7

(1) $y = ax^2 + bx + c$ passes through $(0,0)$ $\Rightarrow 0 = a(0) + b(0) + c \Rightarrow c = 0$; $y = ax^2 + bx$ passes through $(1,2) \Rightarrow 2 = a + b$; $y' = 2ax + b$ and since the curve is tangent to $y = x$ at the origin, its slope is 1 at $x = 0 \Rightarrow y' = 1$ when $x = 0 \Rightarrow 1 = 2a(0) + b \Rightarrow b = 1$.
Then $a + b = 2 \Rightarrow a = 1$. In summary $a = b = 1$ and $c = 0$ so the curve is $y = x^2 + x$.

(2) $y=cx-x^2$ passes through $(1,0)\Rightarrow 0 = c(1)-1$ $\Rightarrow c=1\Rightarrow$ the curve is $y=x-x^2$, For this curve, $y'=1-2x$ and $x=1\Rightarrow y'=1-2=-1$. Since $y=x-x^2$ and $y=x^2+ax+b$ have common tangents at $x=0$, $y=x^2+ax+b$ must also slope −1 at $x=1$. Thus $y'=2x+a$ $\Rightarrow$ $-1=2\cdot 1+a \Rightarrow a=-3 \Rightarrow y=x^2-3x+b$. Since this last curve passes through $(1,0)$ we have $0=1-3+b\Rightarrow b=2$. In summary $a=-3$, $b=2$ and $c=1$ so the curves are $y=x^2-3x+2$ and $y=x-x^2$

EX8

(1) $y'=\dfrac{4(x^2+1)-4x(2x)}{(x^2+1)^2}=\dfrac{-4x^2+4}{(x^2+1)^2}$

$y'|_{x=1}=\dfrac{0}{4}=0\Rightarrow y-2=0(x-1)$

$\therefore\ y=2$

$y'|_{x=0}=4\Rightarrow y-0=4(x-0)$

$y=4x$

(2) $y'=\dfrac{0-8(2x)}{(x^2+4)^2}=\dfrac{-16x}{(x^2+4)^2}$

$y'|_{x=2}=\dfrac{-32}{64}=-\dfrac{1}{2}$

$y-1=-\dfrac{1}{2}(x-2)$

$\therefore y=-\dfrac{1}{2}x+2$

EX9

(1) $g'(x)=\begin{cases}2x-3 & x\ge 0\\ a & x<0\end{cases}$, since g is differentiable at $x=0\Rightarrow \lim\limits_{x\to 0^-}(2x-3)=-3$ and $\lim\limits_{x\to 0^-}a=a\Rightarrow a=-3$

(2) $f'(x)=\begin{cases}a & x\ge -1\\ 2bx & x<-1\end{cases}$, since f is differentiable at $x=-1\Rightarrow \lim\limits_{x\to -1^+}a$ and $\lim\limits_{x\to -1^-}(2bx)=-2b$ and $\Rightarrow a=-2b$, since f is continuous at $x=-1\Rightarrow \lim\limits_{x\to -1^+}(ax+b)=-a+b$ and

$\lim\limits_{x\to -1^-}(bx^2-3)=b-3 \Rightarrow -a+b=b-3$

$\Rightarrow a=3\Rightarrow 3=-2b\Rightarrow b=-\dfrac{3}{2}$

(3) $a=-\dfrac{1}{4}$, $b=-3$

(4) $a=-\dfrac{1}{2}$, $b=1$

3.4 Product Rule

EX1

(1) $y'=-5x^4+12x^2-2x-3$

(2) $y'=\dfrac{3}{4}x^{-\frac{1}{4}}+3x^{-4}+\dfrac{11}{4}x^{\frac{7}{4}}+x^{-2}$

(3) $y'=-2\sqrt{x}+(3-2x)\dfrac{1}{2\sqrt{x}}$
$=\dfrac{-4x+3-2x}{2\sqrt{x}}=\dfrac{3-6x}{2\sqrt{x}}$

EX2

(1) $f(-1)=(1-2)(1+3-1)=-3$

$\Rightarrow\ \therefore\ (-1,-3)$

$y'=(2x)(x^2-3x-1)+(x^2-2)(2x-3)$

$f'(-1)=-2(1+3-1)+(1-2)(-2-3)$

$=-2(3)+(-1)(-5)=-1$

$\therefore\ y+3=-(x+1)$

(2) $f(4)=0 \Rightarrow \therefore (4,0)$

$y'=\left(1-\dfrac{1}{\sqrt{x}}\right)(x^2-6x+2)+(x-2\sqrt{x})(2x-6)$

$f'(4)=\left(1-\dfrac{1}{2}\right)(16-24+2)+0$

$=\dfrac{1}{2}(-6)=-3$

$\therefore y=-3(x-4)$

EX3

(1) 13 (2) −7 (3) $\dfrac{7}{25}$ (4) 20

EX4

$$\frac{d}{df}(f \cdot g \cdot h) = \frac{d}{dx}[(f \cdot g) \cdot h]$$
$$= \frac{d}{dx}(f \cdot g) \cdot h + (f \cdot g) \cdot \frac{d}{dx}h$$
(by Product Rule)
$$= [f' \cdot g + f \cdot g'] \cdot h + f \cdot g \cdot h'$$
$$= f' \cdot g \cdot h + f \cdot g' \cdot h + f \cdot g \cdot h'$$

3.5 Quotient Rule

EX1

$$y' = \frac{2 \cdot (x+1) - 2x \cdot 1}{(x+1)^2} = \frac{2x+2-2x}{(x+1)^2} = \frac{2}{(x+1)^2}$$

EX2

(1) $\frac{-19}{(3x-2)^2}$

(2) $y' = \frac{-3(3x^2+x)-(4-3x)(6x+1)}{(3x^2+x)^2} = \frac{9x^2-24x-4}{(3x^2+x)^2}$

(3) $\frac{-6(x^2-2)}{(x-1)^2(x-2)^2}$

(4) $y' = \frac{\frac{1}{2\sqrt{x}}(\sqrt{x}+1) - (\sqrt{x}-1)\frac{1}{2\sqrt{x}}}{(\sqrt{x}+1)^2} = \frac{1}{\sqrt{x}(\sqrt{x}+1)^2}$

EX3

(1) $y' = \frac{1 \cdot (2x-1) - (1+x) \cdot 2}{(2x-1)^2} = \frac{-3}{(2x-1)^2}$

$f'(1) = -3$

$\therefore y-2 = -3(x-1)$

(2) $f(4) = \frac{4}{2-3} = -4 \Rightarrow (4, -4)$

$$y' = \frac{1 \cdot (\sqrt{x}-3) - x \cdot \frac{1}{2\sqrt{x}}}{(\sqrt{x}-3)^2}$$

$$f'(4) = \frac{(2-3) - 4 \cdot \frac{1}{4}}{(2-3)^2} = \frac{-1-1}{1} = -2$$

The normal line is perpendicular to the tangent line.

$\therefore\ y+4 = \frac{1}{2}(x-4)$

EX4

(1) $y' = \frac{4(x^2+1) - 4x \cdot 2x}{(x^2+1)^2} = \frac{4x^2+4-8x^2}{(x^2+1)^2} = \frac{4-4x^2}{(x^2+1)^2}$

$y' = 0 \Rightarrow 4 - 4x^2 = 0$

$x^2 = 1$

$x = \pm 1$

$f(1) = \frac{4}{1+1} = 2,$

$f(-1) = \frac{-4}{1+1} = -2$

$\therefore\ (1,2), (-1,-2)$

EX5

(1) $\frac{2}{3}$ (2) $2\pi + 5$

(3) $15 - 8\pi$ (4) $\frac{37}{6}$

EX6

$$\left(\frac{f}{g}\right)(x) = \frac{f(x)}{g(x)}$$

$$\left(\frac{f}{g}\right)'(x) = \frac{f'(x)g(x) - f(x)g'(x)}{g(x)^2}$$

$$\left(\frac{f}{g}\right)'(3) = \frac{f'(3)g(3) - f(3)g'(3)}{[g(3)^2]}$$

$$= \frac{4(-2) - 2(-5)}{(-2)^2} = \frac{2}{4} = \frac{1}{2}$$

3.6 Chain Rule

EX1

(1) $y' = -27(4-3x)^8$

(2) $y' = 4(3x^2 - \sqrt{x})^3\left(6x - \dfrac{1}{2\sqrt{x}}\right)$

(3) $y' = 4\left(\dfrac{x^2}{8} + x - \dfrac{1}{x}\right)\left(\dfrac{x}{4} + 1 + \dfrac{1}{x^2}\right)$

(4) $y' = \dfrac{2-2x}{3(2x-x^2)^{2/3}}$

EX2

(1) $y' = \dfrac{12}{(2-3x)^5}$

(2) $y' = 3(2x+1)^2(3x-4)^4(16x-3)$

(3) $y' = 2(2-5x)^3(3x^2+1)^2(-75x^2+18x-10)$

(4) $y' = -\dfrac{2(x+1)(x+2)}{(2x+1)^4}$

EX3

(1) $y' = \dfrac{x-2}{\sqrt{x^2-4x}}$

(2)
$$\begin{aligned} y' &= 6x\sqrt{x^2+1} + 3x^2 \cdot \frac{1}{2\sqrt{x^2+1}} \cdot 2x \\ &= 6x\sqrt{x^2+1} + \frac{3x^3}{\sqrt{x^2+1}} \\ &= \frac{6x(x^2+1)+3x^3}{\sqrt{x^2+1}} = \frac{9x^3+6x}{\sqrt{x^2+1}} \end{aligned}$$

(3) $y' = \dfrac{-4x-4}{(4x+1)\sqrt{4x+1}}$

(4)
$$\begin{aligned} y' &= \frac{\dfrac{1}{2\sqrt{x^2+1}} \cdot 2x \cdot (x-1) - \sqrt{x^2+1} \cdot 1}{(x-1)^2} \\ &= \frac{\left[\dfrac{x(x-1)}{\sqrt{x^2+1}} - \sqrt{x^2+1}\right] \cdot \sqrt{x^2+1}}{(x-1)^2\sqrt{(x^2+1)}} \\ &= \frac{x(x-1)-(x^2+1)}{(x-1)^2\sqrt{x^2+1}} \\ &= \frac{x^2-x-x^2-1}{(x-1)^2\sqrt{x^2+1}} \\ &= \frac{-x-1}{(x-1)^2\sqrt{x^2+1}} \end{aligned}$$

(5) $y' = \dfrac{-25(x+2)^4}{(2x-1)^6}$

(6)
$$\begin{aligned} y' &= 4\left(\frac{\sqrt{x}-1}{\sqrt{x}+1}\right)^3 \cdot \frac{\dfrac{1}{2\sqrt{x}} \cdot (\sqrt{x}+1) - (\sqrt{x}-1) \cdot \dfrac{1}{2\sqrt{x}}}{(\sqrt{x}+1)^2} \\ &= 4\left(\frac{\sqrt{x}-1}{\sqrt{x}+1}\right)^3 \cdot \frac{\left[\dfrac{\sqrt{x}+1}{2\sqrt{x}} - \dfrac{\sqrt{x}-1}{2\sqrt{x}}\right] \cdot 2\sqrt{x}}{(\sqrt{x}+1)^2 \cdot 2\sqrt{x}} \\ &= 4\left(\frac{\sqrt{x}-1}{\sqrt{x}+1}\right)^3 \cdot \frac{\sqrt{x}+1-(\sqrt{x}-1)}{2(\sqrt{x}+1)^2\sqrt{x}} \\ &= 4\left(\frac{\sqrt{x}-1}{\sqrt{x}+1}\right)^3 \cdot \frac{2}{2(\sqrt{x}+1)^2\sqrt{x}} = \frac{4(\sqrt{x}-1)^3}{(\sqrt{x}+1)^5\sqrt{x}} \end{aligned}$$

EX4

(1) $y' = \dfrac{1}{2\sqrt{x+\sqrt{x}}}\left(1 + \dfrac{1}{2\sqrt{x}}\right)$

(2) $y' = \dfrac{1}{2\sqrt{x^2+\sqrt{x^2+x}}}\left(2x + \dfrac{1}{2\sqrt{x^2+x}}(2x+1)\right)$

(3) $y' = \dfrac{1}{2\sqrt{2x-\sqrt{3x+\sqrt{x}}}}\left(2 - \dfrac{1}{2\sqrt{3x+\sqrt{x}}}\left(3 + \dfrac{1}{2\sqrt{x}}\right)\right)$

(4)
$$\begin{aligned} y' &= \frac{1}{2\sqrt{x+\sqrt{3x+\sqrt{2x+1}}}} \\ &\quad \cdot \left(1 + \frac{1}{2\sqrt{3x+\sqrt{2x+1}}}\left(3 + \frac{1}{2\sqrt{2x+1}} \cdot 2\right)\right) \\ &= 2\sqrt{x\sqrt{3x+\sqrt{2x+1}}} \\ &\quad \left(1 + \frac{1}{2\sqrt{3x+\sqrt{2x+1}}}\left(3 + \frac{1}{\sqrt{2x+1}}\right)\right) \end{aligned}$$

EX5

(1) $y = \sqrt{x^2-x+7}$ and $x = 2$

$\Rightarrow y = \sqrt{(2)^2-2+7} = \sqrt{9} = 3.$

$$\begin{aligned} y' &= \frac{1}{2}(x^2-x+7)^{-1/2}(2x-1) \\ &= \frac{2x-1}{2\sqrt{x^2-x+7}} \end{aligned}$$

$y'\big|_{x=2} = \dfrac{2(2)-1}{2\sqrt{(2)^2-(2)+7}} = \dfrac{3}{6} = \dfrac{1}{2}$

$\Rightarrow y - 3 = \dfrac{1}{2}(x-2)$

$\Rightarrow y = \dfrac{1}{2}x + 2$

(2) $y=\left(\dfrac{x-1}{x+1}\right)^2$ and $x=0 \Rightarrow y=\left(\dfrac{0-1}{0+1}\right)^2$

$=(-1)^2=1.$

$$y'=2\left(\frac{x-1}{x+1}\right)\cdot\frac{(x+1)\cdot 1-(x-1)\cdot 1}{(x+1)^2}$$

$$=2\left(\frac{x-1}{x+1}\right)\frac{2}{(x+1)^2}=\frac{4(x-1)}{(x+1)^3}$$

$$y'|_{x=0}=\frac{4(0-1)}{(0+1)^3}=\frac{-4}{1^3}=-4$$

$\Rightarrow y-1=-4(x-0)$

$\Rightarrow y=-4x+1$

EX6

(1) $y'=3[f(x)]^2\cdot f'(x)$

(2) $y'=f'(x^3)\cdot 3x^2$

$=3x^2f'(x^3)$

(3) $y'=3[f(x^3)]^2\cdot f'(x^3)\cdot 3x^2$

$=9x^2[f(x^3)]^2f'(x^3)$

(4) $y'=\dfrac{1}{2\sqrt{f(x)-3x}}(f'(x)-3)$

(5) $y=[f(x)]^{-\frac{1}{2}}$

$y'=-\dfrac{1}{2}[f(x)]^{-\frac{3}{2}}\cdot f'(x)$

(6) $y'=6x\cdot f(g(\sqrt{x}))+3x^2$

$\cdot f'(g(\sqrt{x}))\cdot g'(\sqrt{x})\cdot\dfrac{1}{2\sqrt{x}}$

$=6xf(g(\sqrt{x}))+\dfrac{3}{2}x^{3/2}f'(g(\sqrt{x}))g'(\sqrt{x})$

(7) $y'=12[f(x^2)+2x]^3\cdot[f'(x^2)\cdot 2x+2]$

$=24[f(x^2)+2x]^3[xf'(x^2)+1]$

EX7

(1) 1 (2) 6

(3) 1 (4) $-\dfrac{1}{9}$

(5) $-\dfrac{40}{3}$ (6) $-\dfrac{1}{3}$

(7) $-\dfrac{4}{9}$

3.7 Implicit Differentiation

EX1

(1) $9y^2\cdot y'$

(2) $-\dfrac{1}{2\sqrt{y}}\cdot y'$

(3) $-\dfrac{4}{y^2}\cdot y'$

(4) $=1\cdot y^2+x\cdot 2y\cdot y'$

$=y^2+2xyy'$

(5) $=\dfrac{7(1\cdot y-x\cdot y')}{y^2}$

$=7(\dfrac{y-xy'}{y^2})$

EX2

(1) $y'=-\dfrac{x^2}{y^2}$

(2) $\dfrac{d}{dx}(x^{\frac{2}{3}}+y^{\frac{2}{3}})=\dfrac{d}{dx}(1)$

$\dfrac{2}{3}x^{-\frac{1}{3}}+\dfrac{2}{3}y^{-\frac{1}{3}}\cdot y'=0$

$y'=\dfrac{-x^{-\frac{1}{3}}}{y^{-\frac{1}{3}}}=-\dfrac{y^{\frac{1}{3}}}{x^{\frac{1}{3}}}$

$\therefore \dfrac{dy}{dx}=-\dfrac{\sqrt[3]{y}}{\sqrt[3]{x}}$

(3) $\dfrac{d}{dx}(x^2-xy+y^2)=\dfrac{d}{dx}(6)$

$2x-(1\cdot y+x\cdot y')+2y\cdot y'=0$

$y'(2y-x)=y-2x$

$\therefore \dfrac{dy}{dx}=\dfrac{y-2x}{2y-x}$

(4) $y'=\dfrac{6y-x^2}{y^2-6x}$

(5) $y'=\dfrac{\sqrt{y}}{\sqrt{y}+1}$

(6) $\frac{d}{dx}(x^2y^3 - x^3y^2) = \frac{d}{dx}(-5)$

$2x \cdot y^3 + x^2 \cdot 3y^2y' - (3x^2 \cdot y^2 + x^3 \cdot 2yy')$

$= 0$

$2xy^3 + 3x^2y^2y' - 3x^2y^2 - 2x^3yy' = 0$

$y'(3x^2y^2 - 2x^3y) = 3x^2y^2 - 2xy^3$

$\therefore \frac{dy}{dx} = \frac{xy^2(3x-2y)}{x^2y(3y-2x)} = \frac{y(3x-2y)}{x(3y-2x)}$

EX3

(1) $x^2 + xy - y^2 = 1$

$\Rightarrow 2x + y + xy' - 2yy' = 0$

$\Rightarrow (x-2y)y' = -2x - y$

$\Rightarrow y' = \frac{2x+y}{2y-x} \Rightarrow y'|_{(2,3)} = \frac{7}{4}$

(2) $(x^2+y^2)^2 = (x-y)^2$ at $(1,0)$ at $(1,-1)$

$\Rightarrow 2(x^2+y^2)(2x+2y\frac{dy}{dx}) = 2(x-y)(1-\frac{dy}{dx})$

$\Rightarrow \frac{dy}{dx}[2y(x^2+y^2)+(x-y)]$

$= -2x(x^2+y^2)+(x-y)$

$\Rightarrow \frac{dy}{dx} = \frac{-2x(x^2+y^2)+(x-y)}{2y(x^2+y^2)+(x-y)}$

$\Rightarrow \frac{dy}{dx}\Big|_{(1.0)} = -1$ and $\frac{dy}{dx}\Big|_{(1,-1)} = 1$

(3) $\frac{1}{2\sqrt{4x-y^2}}(4-2yy')$

$-4 \cdot \frac{1}{2\sqrt{xy}}(y+xy') = 0$

At (2,2),

$\frac{1}{4}(4-4y') - 4 \cdot \frac{1}{4}(2+2y') = 0$

$1 - y' - 2 - 2y' = 0$

$-1 - 3y' = 0$

$y' = -\frac{1}{3} \qquad \therefore \frac{dy}{dx}\Big|_{(2,2)} = -\frac{1}{3}$

(4) $1 = 2yy'\sqrt{y-1} + y^2 \cdot \frac{1}{2\sqrt{y-1}} \cdot y'$

At (4,2),

$1 = 4y' + 4 \cdot \frac{1}{2} \cdot y'$

$1 = 6y'$

$y' = \frac{1}{6} \qquad \therefore \frac{dy}{dx}\Big|_{(4,2)} = \frac{1}{6}$

EX4

(1) slope: $y' = \frac{5}{4}$

$y - 2 = \frac{5}{4}(x-4)$

(2) slope ; $y' = 1$,
$normal; y - 1 = -1(x-1)$

$y = -x + 2$

$\rightarrow x^2 = 2x(-x+2) - 3(-x+2)^2 = 0$

$\rightarrow x^2 - 4x + 3 = 0$

$\rightarrow x = 1, x = 3$

$\rightarrow x = 3 \ \therefore (3,-1)$

EX5

(1) *Horizontal Tangent* $\Rightarrow \frac{dy}{dx} = 0$

$\frac{d}{dx}(xy^2 - x^2y) = \frac{d}{dx}(16)$

$1 \cdot y^2 + x \cdot 2y \cdot y' - (2x \cdot y + x^2 \cdot y') = 0$

$y^2 + 2xyy' - 2xy - x^2y' = 0$

$(2xy - x^2)y' = 2xy - y^2$

$\frac{dy}{dx} = \frac{y(2x-y)}{2xy-x^2}$

$\frac{dy}{dx} = 0 \Rightarrow y(2x-y) = 0$

$\therefore y = 0$ or $y = 2x$

if $y = 0$

$0 - 0 = 16$

no solution

if $y = 2x$

$x(2x)^2 - x^2(2x) = 16$

$4x^3 - 2x^3 = 16$

$2x^3 = 16$

$x^3 = 8$

$x = 2 \Rightarrow y = 2(2) = 4$

$\therefore (2,4)$

(2) $Vertical\ Tangent \Rightarrow \dfrac{dy}{dx} = undefined$

$$\frac{d}{dx}(x^2y - xy^3) = \frac{d}{dx}(6)$$
$$2xy + x^2y' - y^3 - 3xy^2y' = 0$$
$$(x^2 - 3xy^2)y' = y^3 - 2xy$$
$$\frac{dy}{dx} = \frac{y^3 - 2xy}{x^2 - 3xy^2} = \frac{y(y^2 - 2x)}{x(x - 3y^2)}$$
$$\frac{dy}{dx} = undefined \Rightarrow x(x - 3y^2) = 0$$
$$\therefore x = 0 \text{ or } x = 3y^2$$

if $x = 0$	if $x = 3y^2$
$0 - 0 = 6$	$(3y^2)^2y - 3y^2y^3 = 6$
$no\,solution$	$9y^5 - 3y^5 = 6$
	$6y^5 = 6$
	$y = 1 \Rightarrow x = 3(1)^2 = 3$
	$\therefore (3,1)$

(4) $2x - 2yy' = 0$

$$y' = \frac{x}{y}$$
$$y'' = \frac{1 \cdot y - x \cdot y'}{y^2}$$
$$= \frac{y - x\left(\dfrac{x}{y}\right)}{y^2} = \frac{\left[y - x\left(\dfrac{x}{y}\right)\right] \cdot y}{y^2 \cdot y}$$
$$= \frac{y^2 - x^2}{y^3}\left(\text{or } -\frac{4}{y^3}\right)$$

3.8 Higher Order Derivatives

EX1

(1) $y = (2x-1)^{-1}$

$$y' = -(2x-1)^{-2} \cdot 2$$
$$= -2(2x-1)^{-2}$$
$$y'' = 4(2x-1)^{-3} \cdot 2$$
$$= \frac{8}{(2x-1)^3}$$

(2) $y' = 6(x^2 - 2x)^5(2x - 2)$

$$y'' = 30(x^2-2x)^4(2x-2) \cdot (2x-2) + 6(x^2-2x)^5 \cdot 2$$
$$= 120(x^2-2x)^4(x-1)^2 + 12(x^2-2x)^5$$
$$= 12(x^2-2x)^4[10(x-1)^2 + (x^2-2x)]$$
$$= 12(x^2-2x)^4(11x^2 - 22x + 10)$$

(3) $y' = \dfrac{1}{2\sqrt{x^2+2}} \cdot 2x = \dfrac{x}{\sqrt{x^2+2}}$

$$y'' = \frac{1 \cdot \sqrt{x^2+2} - x \cdot \dfrac{1}{2\sqrt{x^2+2}} \cdot 2x}{(x^2+2)}$$
$$= \frac{\left[\sqrt{x^2+2} - \dfrac{x^2}{\sqrt{x^2+2}}\right] \cdot \sqrt{x^2+2}}{(x^2+2) \cdot \sqrt{x^2+2}}$$
$$= \frac{x^2 + 2 - x^2}{(x^2+2)\sqrt{x^2+2}} = \frac{2}{(x^2+2)\sqrt{x^2+2}}$$

EX2

(1) $y' = \dfrac{3(2x+3) - (3x-2) \cdot 2}{(2x+3)^2}$

$$= \frac{13}{(2x+3)^2} = 13(2x+3)^{-2}$$
$$y'' = -26(2x+3)^{-3} \cdot 2$$
$$= \frac{-52}{(2x+3)^3}$$
$$f''(-2) = \frac{-52}{(-1)^3} = 52$$

(2) $y = (1-2x)^{-\frac{1}{2}}$

$$y' = -\frac{1}{2}(1-2x)^{-\frac{3}{2}} \cdot (-2) = (1-2x)^{-\frac{3}{2}}$$
$$y'' = -\frac{3}{2}(1-2x)^{-\frac{5}{2}} \cdot (-2) = \frac{3}{(1-2x)^{5/2}}$$
$$f''(0) = 3$$

(3) $g'(x) = 5(3-2x^2)^4(-4x)$

$$= -20x(3-2x^2)^4$$
$$g''(x) = -20(3-2x^2)^4 - 20x \cdot 4(3-2x^2)^3(-4x)$$
$$g''(1) = -20(1) - 20 \cdot 4(1)(-4)$$
$$= 300$$

(4) Let $y = x^n - 1$

$$y' = nx^{n-1}$$
$$y'' = n(n-1)x^{n-2}$$
$$y''' = n(n-1)(n-2)x^{n-3}$$
$$f'''(1) = n(n-1)(n-2)$$

EX3

(1) $G'(x) = f(x^2) + xf'(x^2) \cdot 2x$

$= f(x^2) + 2x^2 f'(x^2)$

$G''(x) = f'(x^2) \cdot 2x + 4xf'(x^2) + 2x^2 f''(x^2) \cdot 2x$

$= 6xf'(x^2) + 4x^3 f''(x^2)$

(2) $F'(x) = 2[f(x)] \cdot f'(x)$

$F''(x) = 2f'(x)f'(x) + 2f(x)f''(x)$

$F''(2) = 2f'(2)f'(2) + 2f(2)f''(2)$

$= 2(-2)(-2) + 2(-1)(3) = 2$

(3) Let $y = [f(x^3)]$

$y' = f'(x^3) \cdot 3x^2$

$= 3x^2 f'(x^3)$

$y'' = 6xf'(x^3) + 3x^2 \cdot f''(x^3) \cdot 3x^2$

$= 6xf'(x^3) + 9x^4 \cdot f''(x^3)$

$y''(1) = 6f'(1) + 9f''(1)$

$= 6 \cdot 3 + 9 \cdot (-2) = 18 - 18$

$= 0$

EX4

(1) $2yy' + y + xy' = 0$

$y'(2y + x) = -y$

$y' = \dfrac{-y}{x+2y}, \quad \therefore \left.\dfrac{dy}{dx}\right|_{(1,1)} = \dfrac{1}{1+2} = -\dfrac{1}{3}$

$y'' = \dfrac{-y'(x+2y) - (-y)(1+2y')}{(x+2y)^2}$

$\therefore \left.\dfrac{d^2y}{dx^2}\right|_{(1,1)} = \dfrac{-\left(-\frac{1}{3}\right)(3) - (-1)\left(1 - \frac{2}{3}\right)}{(1+2)^2}$

$= \dfrac{1 + \frac{1}{3}}{9} = \dfrac{4}{27}$

(2) $2x - (y + xy') + 2yy' = 0$

$2x - y - xy' + 2yy' = 0$

$y'(-x + 2y) = -2x + y$

$y' = \dfrac{-2x+y}{-x+2y} = \dfrac{2x-y}{x-2y},$

$\therefore \left.\dfrac{dy}{dx}\right|_{(1,-1)} = \dfrac{2+1}{1+2} = 1$

$y'' = \dfrac{(2-y')(x-2y) - (2x-y)(1-2y')}{(x-2y)^2}$

$\therefore \left.\dfrac{d^2y}{dx^2}\right|_{(1,-1)} = \dfrac{(2-1)(1+2) - (2+1)(1-2)}{(1+2)^2}$

$= \dfrac{3+3}{9} = \dfrac{2}{3}$

EX5

(1) $g(1) = e^{f(1)} = e^2$

$g'(x) = e^{f(x)} f'(x), \quad g'(1) = e^{f(1)} f'(1) = -4e^2$

The tangent line is given by $y = e^2 - 4e^2(x-1)$.

(2) $g'(x) = e^{f(x)} f'(x)$

$e^{f(x)} > 0$ for all x

So, g' changes from positive to negative only when f' changes from positive to negative.

This occurs at $x = -1$ only. Thus, g has a local maximum at $x = -1$.

(3) $g''(-1) = e^{f(-1)}[(f'(-1))^2 + f''(-1)]$

$e^{f(-1)} > 0$ and $f'(-1) = 0$

Since, f' is decreasing on a neighborhood of -1, $f''(-1) < 0$. Therefore, $g''(-1) < 0$.

(4) $\dfrac{g'(3) - g'(1)}{3-1} = \dfrac{e^{f(3)} f'(3) - e^{f(1)} f'(1)}{2} = 2e^2$

3.9 Derivatives of Trigonometric Functions

EX1

(1) $y' = 3\cos(3x+1)$

(2) $y' = -\sin(-3x+2) \cdot (-3)$

$= 3\sin(-3x+2)$

(3) $y' = 2\cos\left(x^3 - \dfrac{1}{x}\right) \cdot \left(3x^2 + \dfrac{1}{x^2}\right)$

$= 2\left(3x^2 + \dfrac{1}{x^2}\right)\cos\left(x^3 - \dfrac{1}{x}\right)$

(4) $y' = -\sin(\sin x) \cdot \cos x$

(5) $y' = 10\sec^2(10x - 5)$

(6) $y' = -\csc^2(3x^2 + 2x) \cdot (6x + 2)$

$= -(6x+2)\csc^2(3x^2 + 2x)$

(7) $y' = \sec\left(\dfrac{1}{2}x\right)\tan\left(\dfrac{1}{2}x\right) \cdot \dfrac{1}{2}$

$= \dfrac{1}{2}\sec\left(\dfrac{1}{2}x\right)\tan\left(\dfrac{1}{2}x\right)$

(8) $y' = \dfrac{1}{5}\csc(5x)\cot(5x) \cdot 5$

$= \csc(5x)\cot(5x)$

(9) $y' = \sec(\tan x)\tan(\tan x) \cdot \sec^2 x$

(10) $y' = \dfrac{1}{2\sqrt{1+2\cot x}} \cdot (-2\csc^2 x)$

$= \dfrac{-\csc^2 x}{\sqrt{1+2\cot x}}$

(11) $y' = -\csc^2(\pi - \frac{1}{x}) \cdot \frac{1}{x^2}$

(12) $y = \sin[(x^3-1)^4]$

$y' = \cos[(x^3-1)^4] \cdot 4(x^3-1)^3 \cdot 3x^2$

$= 12x^2(x^3-1)^3\cos(x^3-1)^4$

EX2

(1) $y' = -4(1+\cos 2t)^{-5} \cdot (-\sin 2t) \cdot 2$

$= 8(1+\cos 2t)^{-5} \cdot \sin 2t$

(2) $y = (\sin x)^3$

$y' = 3(\sin x)^2 \cdot \cos x$

$= 3\sin^2 x cos x$

(3) $y' = -12x cos(x^2)\sin(x^2)$

(4) $y = \tan(x^3) + [\tan x]^3$

$y' = \sec^2(x^3) \cdot 3x^2 + 3[\tan x]^2 \cdot \sec^2 x$

$= 3x^2\sec^2(x^3) + 3\tan^2 x sec^2 x$

(5) $y' = 2x^{-2/3}\sec^3(x^{1/3})\tan(x^{1/3})$

(6) $y = [\sin(2x)]^3 + [\cos(2x)]^3$

$y' = 3[\sin(2x)]^2 \cdot \cos(2x)$

$\cdot 2 + 3[\cos(2x)]^2 \cdot (-\sin(2x)) \cdot 2$

$= 6\sin^2(2x)\cos(2x) - 6\cos^2(2x)\sin(2x)$

(7) $y = \tan^2(\sin^3 t)$

$\Rightarrow \dfrac{dy}{dt} = 2\tan(\sin^3 t) \cdot \sec^2(\sin^3 t)$

$\cdot (3\sin^2 t \cdot (\cos t))$

$= 6\tan(\sin^3 t)\sec^2(\sin^3 t)\sin^2 t\cos t$

EX3

(1) $y' = \sin x + x\cos x$

(2) $y' = 2x cos\left(\dfrac{1}{x}\right) + x^2\left(\sin\left(\dfrac{1}{x}\right)\right) \cdot \left(-\dfrac{1}{x^2}\right)$

$= 2x\cos\left(\dfrac{1}{x}\right) - \sin\left(\dfrac{1}{x}\right)$

(3) $y' = \dfrac{6\csc x - 3x\csc x\cot x}{2\sqrt{\csc x}}$

(4) $y' = \dfrac{\cos x(1+\cos x) - \sin x(-\sin x)}{(1+\cos x)^2}$

$= \dfrac{\cos x + \cos^2 x + \sin^2 x}{(1+\cos x)^2}$

$= \dfrac{\cos x + 1}{(1+\cos x)^2} = \dfrac{1}{1+\cos x}$

(5) $y' = 4\cos 4x cos^2 x - 2\sin 4x\cos x\sin x$

(6) $y' = (\sec\sqrt{x})\left[\dfrac{\tan\sqrt{x}\tan(\frac{1}{x})}{2\sqrt{x}} - \dfrac{\sec^2(\frac{1}{x})}{x^2}\right]$

(7) $y = \dfrac{(\cos x)^2}{\sin x}$

$y' = \dfrac{2(\cos x)(-\sin x)(\sin x) - (\cos x)^2 \cdot \cos x}{\sin^2 x}$

$= \dfrac{-2\cos x \cdot \sin^2 x - \cos^3 x}{\sin^2 x}$

$= \dfrac{-\cos x(2\sin^2 x + \cos^2 x)}{\sin^2 x}$

EX4

(1) $y' = \dfrac{\cos(2x+3y) - 2x\sin(2x+3y) - y cos x}{\sin x + 3x\sin(2x+3y)}$

(2) $-\sin(x+y) \cdot (1+y') + \cos(x-y) \cdot (1-y') = 0$

$-\sin(x+y) - y'\sin(x+y) + \cos(x-y) - y'\cos(x-y) = 0$

$-\sin(x+y) + \cos(x-y) = y'\sin(x+y) + y'\cos(x-y)$

$-\sin(x+y) + \cos(x-y) = y'[\sin(x+y) + \cos(x-y)]$

$\therefore \dfrac{dy}{dx} = \dfrac{-\sin(x+y) + \cos(x-y)}{\sin(x+y) + \cos(x-y)}$

(3) $y' = \dfrac{-1 - y sec^2(xy)}{x sec^2(xy)}$

(4) $2x + 2y \cdot y' = \sec(xy)\tan(xy) \cdot (y + xy')$

$2x + 2yy' = y sec(xy)\tan(xy) + xy'\sec(xy)\tan(xy)$

$y'[2y - x sec(xy)\tan(xy)] = y sec(xy)\tan(xy) - 2x$

$\therefore \dfrac{dy}{dx} = \dfrac{y sec(xy)\tan(xy) - 2x}{2y - x sec(xy)\tan(xy)}$

EX5

(1) $y = -\dfrac{\pi}{2}x + \pi$

(2) $f(\pi) = \pi^2\cos\pi = -\pi^2 \Rightarrow \therefore (\pi, -\pi^2)$

$y' = 2x\cos x - x^2\sin x$

$f'(\pi) = 2\pi\cos\pi - \pi^2\sin\pi$

$= -2\pi$

$\therefore y + \pi^2 = -2\pi(x - \pi)$

(3) $y-\frac{1}{4}=-\frac{\sqrt{3}}{4}\left(x-\frac{\pi}{6}\right)$

(4) $f\left(\frac{\pi}{4}\right)=2\left(\cot\left(\frac{\pi}{4}\right)\right)^2-\left(\cos\frac{\pi}{4}\right)^2$

$=2(1)^2-\left(\frac{1}{\sqrt{2}}\right)^2=\frac{3}{2}$

$\Rightarrow \therefore \left(\frac{\pi}{4},\frac{3}{2}\right)$

$y=2(\cot x)^2-(\cos x)^2$

$y'=4(\cot x)\cdot(-\csc^2 x)-2(\cos x)(-\sin x)$

$=-4\cot x\cdot\csc^2 x+2\cos x\sin x$

$f'\left(\frac{\pi}{4}\right)=-4\cot\frac{\pi}{4}\csc^2\frac{\pi}{4}+2\cos\frac{\pi}{4}\cdot\sin\frac{\pi}{4}$

$=-4(1)\cdot(\sqrt{2})^2+2\cdot\frac{1}{\sqrt{2}}\cdot\frac{1}{\sqrt{2}}$

$=-8+1=-7$

$\therefore y-\frac{3}{2}=-7\left(x-\frac{\pi}{4}\right)$

(5) $y=2x$

EX6

(1) $y=[\sin(2x)]^3$

$y'=3[\sin(2x)]^2\cdot\cos(2x)\cdot 2$

$=6[\sin(2x)]^2\cdot\cos(2x)$

$y''=12[\sin(2x)]\cdot\cos(2x)\cdot 2\cdot\cos 2x$

$+6[\sin(2x)]^2\cdot(-\sin(2x))\cdot 2$

$=24\sin 2x\cdot\cos^2 2x-12\sin^3 2x$

(2)

$y=[\cos(3x)]^3$

$y'=3[\cos(3x)]^2\cdot(-\sin(3x))\cdot 3$

$=-9[\cos(3x)]^2\cdot\sin(3x)$

$y''=-18[\cos(3x)]\cdot(-\sin(3x))\cdot 3$

$\cdot\sin(3x)-9[\cos(3x)]^2\cdot\cos(3x)\cdot 3$

$=54\cos 3x\cdot\sin^2 3x-27\cos^3 3x$

3.10 Derivatives of Exponential Functions

EX1

(1) $y'=e^{5-7x}\cdot(-7)$

$=-7e^{5-7x}$

(2) $y'=e^{\sec x}\cdot\sec x\tan x$

(3) $y'=2xe^{-x^2}\sin(e^{-x^2})$

(4) $y'=3^{-x^2}\cdot(-2x)\cdot\ln 3$

$=-2x(3)^{-x^2}\ln 3$

(5) $y'=(1-x)e^{-x}+3e^{3x}$

(6) $y'=a^{\cos x}(-\sin x)\cdot\ln a$

$=-a^{\cos x}\sin x\ln a$

(7) $y'=3x^2\cdot e^{2x}+x^3\cdot e^{2x}\cdot 2$

$=x^2e^{2x}(3+2x)$

(8) $y'=(27x^4-18x^3+6x^2+18x-6)e^{x^3}$

(9) $y'=\frac{e^{-2x}\cdot(-2)\cdot(x)-e^{-2x}\cdot 1}{x^2}$

$=\frac{-e^{-2x}(2x+1)}{x^2}$

(10) $y'=\frac{3}{2}\cos\left(\frac{x}{2}\right)e^{3\sin\left(\frac{x}{2}\right)}$

(11) $y'=\pi^{x\cos x}\cdot(\cos x+x\cdot(-\sin x))\ln\pi$

$=\pi^{x\cos x}(\cos x-x\sin x)\ln\pi$

(12) $y'=\frac{-e^{-x}\cdot(e^x-1)-(e^{-x}+1)\cdot e^x}{(e^x-1)^2}$

$=\frac{-1+e^{-x}-1-e^x}{(e^x-1)^2}$

$=\frac{e^{-x}-2-e^x}{(e^x-1)^2}$

(13) $e^{x^2y}=2x+2y$

$\Rightarrow e^{x^2y}(x^2y'+2xy)=2+2y'$

$\Rightarrow x^2e^{x^2y}y'+2xye^{x^2y}=2+2y'$

$\Rightarrow x^2e^{x^2y}y'-2y'=2-2xye^{x^2y}$

$\Rightarrow y'=\frac{2-2xye^{x^2y}}{x^2e^{x^2y}-2}$

(14) $y'=(x^2+2x)e^x\cos(x^2e^x)$

(15) $y'=2e^{-x}(4e^{-x}+2)^{-\frac{3}{2}}$

(16) $e^{x+y}\cdot(1+y')=6x\cdot y+3x^2\cdot y'$

$e^{x+y}+y'e^{x+y}=6xy+3x^2y'$

$y'(e^{x+y}-3x^2)=6xy-e^{x+y}$

$\therefore \frac{dy}{dx}=\frac{6xy-e^{x+y}}{(e^{x+y}-3x^2)}$

(17) $e^{2x} = \sin(x+3y)$

$\Rightarrow 2e^{2x} = (1+3y')\cos(x+3y)$

$\Rightarrow 1+3y' = \dfrac{2e^{2x}}{\cos(x+3y)}$

$\Rightarrow 3y' = \dfrac{2e^{2x}}{\cos(x+3y)} - 1$

$\Rightarrow y' = \dfrac{2e^{2x} - \cos(x+3y)}{3\cos(x+3y)}$

EX2

(1) $f(1) = \dfrac{e^{-1}}{\sqrt{1}} = \dfrac{1}{e} \Rightarrow \therefore \left(1, \dfrac{1}{e}\right)$

$$y' = \frac{(-e^{-x})\sqrt{x} - e^{-x} \cdot \dfrac{1}{2\sqrt{x}}}{x}$$

$$f'(1) = \frac{(-e^{-1}) - e^{-1} \cdot \dfrac{1}{2}}{1}$$

$$= -\frac{1}{e} - \frac{1}{2e} = -\frac{3}{2e}$$

$$\therefore y - \frac{1}{e} = -\frac{3}{2e}(x-1)$$

(2) $f(1) = e^1 = \Rightarrow \therefore (1, e)$

$$y' = 2xe^{\frac{1}{x}} + x^2 e^{\frac{1}{x}}\left(-\frac{1}{x^2}\right)$$

$$f'(1) = 2e + e^1 \cdot (-1)$$

$$= e$$

$\therefore y - e = e(x-1)$ (or $y = ex$)

EX3

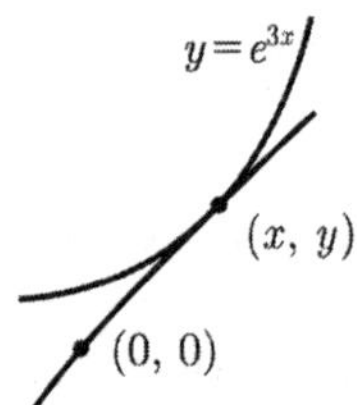

(1) $y' = e^{3x} \cdot 3 = 3e^{3x} \cdots\cdots$ ①

Let the tangent point be (x,y) tangent slope at (x,y) = secant slope between $(0,0)$ and (x,y)

$$y' = \frac{y-0}{x-0}$$

$$y' = \frac{y}{x} \cdots\cdots ②$$

By ① and ②, $3e^{3x} = \dfrac{e^{3x}}{x}$

$$3 = \frac{1}{x} \Rightarrow \therefore x = \frac{1}{3}$$

$$slope = f'\left(\frac{1}{3}\right) = 3e$$

$$\therefore y = 3ex$$

(2) $y = -\dfrac{1}{e^2}(x+2)$

EX4

$$f'(x) = e^x \cos x - e^x \sin x$$

$$= e^x(\cos x - \sin x) \cdots\cdots ①$$

$$f''(x) = e^x \cdot (\cos x - \sin x) + e^x(-\sin x - \cos x)$$

$$= e^x(\cos x - \sin x - \sin x - \cos x)$$

$$= 2e^x \sin x \cdots\cdots ②$$

Plug ① and ② $\int$ *o the original equation*

$$[-2e^x \sin x] - p[e^x(\cos x - \sin x)] + q[e^x \cos x] = 0$$

$$-2e^x \sin x - pe^x \cos x + pe^x \sin x + qe^x \cos x = 0$$

$$-2\sin x - p\cos x + p\sin x + q\cos x = 0$$

$(-2+p)\sin x + (-p+q)\cos x = 0$ for *all real νmbers* x

$-2+p=0$ and $-p+q=0$

$\therefore p=2,\ q=2$

3.11 derivatives of Logarithmic Functions

EX1

(1) $y' = \dfrac{3}{3x}$

(2) $y' = \dfrac{1}{\ln x} \cdot \dfrac{1}{x}$

(3) $y' = \dfrac{\cos x}{\sin x}$

(4) $y' = \dfrac{1}{\tan x} \cdot \sec^2 x$

$$= \frac{\sec^2 x}{\tan x}$$

(5) $y' = 5(\ln x)^4 \cdot \dfrac{1}{x}$

$$= \frac{5(\ln x)^4}{x}$$

(6) $y' = \frac{1}{\sin x} \cdot \cos x \cdot \frac{1}{\ln 3}$

$= \frac{\cot x}{\ln 3}$

(7) $y' = \frac{1}{x \ln 2}$

(8) $y' = \frac{1}{x \cos x}(\cos x - x \sin x)$

$= \frac{\cos x - x\sin x}{x \cos x}$

(9) $y' = \frac{1}{x}(\log_2 3)3^{\log_2 x}$

(10) $y' = 2x\ln x + x^2 \cdot \frac{1}{x}$

$= 2x\ln x + x$

(11) $y' = (\ln x)^{1/2} + \frac{1}{2(\ln x)^{\frac{1}{2}}}$

(12) $y' = \frac{1}{\sec x + \tan x}(\sec x \tan x + \sec^2 x)$

$= \frac{\sec x(\tan x + \sec x)}{\sec x + \tan x}$

$= \sec x$

(13) $y' = \sec^2[\log(x^2+4)]$

$\cdot \frac{1}{x^2+4} \cdot 2x \cdot \frac{1}{\ln 10}$

$= \frac{2x\sec^2[\log(x^2+4)]}{(x^2+4)\ln 10}$

(14) $y' = 4x^7(\ln x)^3 + 8x^7(\ln x)^4$

(15) $y' = \frac{\frac{1}{x} \cdot x^2 - \ln x \cdot 2x}{x^4}$

$= \frac{x - 2x\ln x}{x^4} = \frac{1-2\ln x}{x^3}$

(16) $\frac{1}{xy} \cdot (y + xy') = 1 - y'$

$y + xy' = xy - xyy'$

$y'(x + xy) = xy - y$

$\therefore \frac{dy}{dx} = \frac{y(x-1)}{x(y+1)}$

(17) $y + 1 = x \cdot [\ln y]^3$

$y' = [\ln y]^3 + x \cdot 3[\ln y]^2 \cdot \frac{1}{y} \cdot y'$

$yy' = y\ln^3 y + 3x\ln^2 y \cdot y'$

$y'(y - 3x\ln^2 y) = y\ln^3 y$

$\therefore \frac{dy}{dx} = \frac{y\ln^3 y}{(y - 3x\ln^2 y)}$

(18) $y' = \frac{\tan(\ln x)}{x}$

EX2

(1) $y' = \frac{6}{2x+1} + \frac{8x}{x^2+1}$

(2) $y' = \frac{1}{x} - 1$

(3) $y' = \frac{10x}{x^2+1} + \frac{1}{2(1-x)}$

(4) $y = \ln\sqrt{\frac{(x+1)^5}{(x+2)^{20}}} = \frac{1}{2}[5\ln(x+1) - 20\ln(x+2)]$

$\Rightarrow y' = \frac{1}{2}\left(\frac{5}{x+1} - \frac{20}{x+2}\right) = \frac{5}{2}\left[\frac{(x+2) - 4(x+1)}{(x+1)(x+2)}\right]$

$= -\frac{5}{2}\left[\frac{3x+2}{(x+1)(x+2)}\right]$

(5) $y' = \frac{1}{2}\left[\cot x - \tan x - \frac{4}{x(1+2\ln x)}\right]$

(6) $y = \log_7\left(\frac{\sin\theta\cos\theta}{e^\theta 2^\theta}\right)$

$= \frac{\ln(\sin\theta) + \ln(\cos\theta) - \ln e^\theta - 2^\theta}{\ln 7}$

$= \frac{\ln(\sin\theta) + \ln(\cos\theta) - \theta - \theta\ln 2}{\ln 7}$

$\Rightarrow \frac{dy}{d\theta} = \frac{\cos\theta}{(\sin\theta)(\ln 7)} - \frac{\sin\theta}{(\cos\theta)(\ln 7)} - \frac{1}{\ln 7} - \frac{\ln 2}{\ln 7}$

$= \left(\frac{1}{\ln 7}\right)(\cot\theta - \tan\theta - 1 - \ln 2)$

EX3

(1) $\ln y = \ln x^{\ln x}$

$\ln y = \ln x \cdot \ln x$

$\frac{1}{y} \cdot y' = \frac{1}{x}\ln x + \ln x \cdot \frac{1}{x}$

$y' = \frac{2y}{x}\ln x$

(2) $y' = (\sin x)^x[\ln(\sin x) + x\cot x]$

(3) $y' = (x+1)^x[\frac{x}{x+1} + \ln(x+1)]$

(4) $\ln y = \ln(\sin x)^{\sin x}$

$\ln y = \sin x \ln(\sin x)$

$\frac{1}{y} \cdot y' = \cos x \cdot \ln(\sin x)$

$+ \sin x \cdot \frac{1}{\sin x} \cdot \cos x$

$y' = y\cos x(\ln(\sin x) + 1)$

(5) $y' = \sqrt{(x^2+1)(x-1)^2}\left(\frac{x}{x^2+1}+\frac{1}{x-1}\right)$

(6) $\ln y = \ln(\ln x)^{\ln x}$

$\ln y = \ln x \cdot \ln(\ln x)$

$\frac{1}{y}y' = \frac{1}{x}\ln(\ln x) + \ln x \cdot \frac{1}{\ln x} \cdot \frac{1}{x}$

$y' = \frac{y}{x}(\ln(\ln x)+1)$

(7) $y = \frac{x\sqrt{x^2+1}}{(x+1)^{2/3}}\left[\frac{1}{x}+\frac{x}{x^2+1}-\frac{2}{3(x+1)}\right]$

(8) $y = \frac{\theta\sin\theta}{\sqrt{\sec\theta}}$

$\Rightarrow \ln y = \ln\theta + \ln(\sin\theta) - \frac{1}{2}\ln(\sec\theta)$

$\Rightarrow \frac{1}{y}\frac{dy}{d\theta} = \left[\frac{1}{\theta}+\frac{\cos\theta}{\sin\theta}-\frac{(\sec\theta)(\tan\theta)}{2\sec\theta}\right]$

$\Rightarrow \frac{dy}{d\theta} = \frac{\theta\sin\theta}{\sqrt{\sec\theta}}\left(\frac{1}{\theta}+\cot\theta-\frac{1}{2}\tan\theta\right)$

(9) $\ln y = \ln\left(\frac{e^{-x}\cos x}{\sqrt[3]{x^2-x}}\right)$

$\ln y = \ln e^{-x} + \ln(\cos x) - \ln(\sqrt[3]{x^2-x})$

$\ln y = -x + \ln(\cos x) - \frac{1}{3}\ln(x^2-x)$

$\frac{1}{y} \cdot y' = -1 + \frac{1}{\cos x}(-\sin x)$

$-\frac{1}{3} \cdot \frac{1}{x^2-x} \cdot (2x-1)$

$y' = y(-1-\tan x - \frac{2x-1}{3(x^2-x)})$

EX4

(1) $\ln y = \ln\left[\frac{(x+1)^2}{x^3(x-2)}\right]$

$\ln y = 2\ln(x+1) - 3\ln x - \ln(x-2)$

$\frac{1}{y}y' = 2 \cdot \frac{1}{x+1} - 3 \cdot \frac{1}{x} - \frac{1}{x-2}$

$y' = y(\frac{2}{x+1}-\frac{3}{x}-\frac{1}{x-2})$

When $x=1$, $y = \frac{(1+1)^2}{1(1-2)} = -4$

$f'(1) = -4\left(\frac{2}{2}-\frac{3}{1}-\frac{1}{-1}\right) = -4(1-3+1) = 4$

(2) $\ln y = \ln\left[\frac{\sqrt[4]{2x-1} \cdot \sqrt{x^2+1}}{\sqrt{x+1} \cdot \sqrt{5-x}}\right]$

$\ln y = \frac{1}{4}\ln(2x-1) + \frac{1}{2}\ln(x^2+1)$

$-\frac{1}{2}\ln(x+1) - \frac{1}{2}(5-x)$

$\frac{1}{y} \cdot y' = \frac{1}{4} \cdot \frac{2}{2x-1} + \frac{1}{2} \cdot \frac{2x}{x^2+1} - \frac{1}{2}$

$\cdot \frac{1}{x+1} - \frac{1}{2} \cdot \frac{-1}{5-x}$

$y' = y\left(\frac{1}{2(2x-1)}+\frac{x}{x^2+1}-\frac{1}{2(x+1)}+\frac{1}{2(5-x)}\right)$

$f(1) = \frac{1 \cdot \sqrt{2}}{\sqrt{2} \cdot 2} = \frac{1}{2} \Rightarrow \therefore \left(1.\frac{1}{2}\right)$

$f'(1) = \frac{1}{2}\left(\frac{1}{2}+\frac{1}{2}-\frac{1}{4}+\frac{1}{8}\right)$

$= \frac{1}{2} \cdot \frac{7}{8} = \frac{7}{16}$

EX5

$\leq t\ y = \sqrt{2x+\ln(f(x))}$

$y' = \frac{1}{2\sqrt{2x+\ln(f(x))}} \cdot \left(2+\frac{1}{f(x)} \cdot f'(x)\right)$

$y'(2) = \frac{1}{2\sqrt{4+\ln(f(2))}} \cdot \left(2+\frac{1}{f(2)} \cdot f'(2)\right)$

$= \frac{1}{2\sqrt{4+\ln 1}} \cdot (2+1 \cdot 4) = \frac{1}{4}(2+4) = \frac{3}{2}$

3.12 Inverse Trigonometric Functions

EX1

(1) $-\frac{\pi}{6}$

(2) $-\frac{\pi}{3}$

(3) $\frac{\pi}{3}$

(4) $\frac{\pi}{2}$

EX2

(1) $= \sin^{-1}\left(-\frac{1}{2}\right)$

$= -\frac{\pi}{6}$

(2) $= \sin\left(-\frac{\pi}{6}\right)$

$= -\frac{1}{2}$

(3) $= \sin^{-1}\left(\frac{\sqrt{3}}{2}\right)$

$= \frac{\pi}{3}$

(4) $= \cos\left(-\frac{\pi}{2}\right)$

$= 0$

EX3

(1) $\frac{2}{3}\pi$ (2) $\frac{5}{6}\pi$

(3) $\frac{1}{6}\pi$ (4) $\frac{1}{2}\pi$

EX4

(1) $=\cos^{-1}\left(-\frac{\sqrt{3}}{2}\right)$ $=\frac{5\pi}{6}$

(2) $=\cos\left(\frac{2\pi}{3}\right)$ $=-\frac{1}{2}$

(3) $=\cos^{-1}\left(\frac{\sqrt{3}}{2}\right)$ $=\frac{\pi}{6}$

(4) $=\tan(0)$ $=0$

EX5

(1) $-\frac{\pi}{4}$ (2) $\frac{\pi}{6}$ (3) $\frac{\pi}{3}$ (4) 0

EX6

(1) $=\tan^{-1}(-1)$ $=-\frac{\pi}{4}$

(2) $=\tan\left(-\frac{\pi}{6}\right)$ $=-\frac{1}{\sqrt{3}}$

(3) $=\tan^{-1}\left(-\frac{1}{\sqrt{3}}\right)$ $=-\frac{\pi}{6}$

(4) $\cos\left(-\frac{\pi}{3}\right)$ $=\frac{1}{2}$

EX7

(1) $=\sin^{-1}\left(\frac{1}{2}\right)$ $=\frac{\pi}{6}$

(2) $\sin^{-1}\left(-\frac{\sqrt{3}}{2}\right)$ $=-\frac{\pi}{3}$

(3) $=\cos^{-1}\left(-\frac{1}{\sqrt{2}}\right)$ $=\frac{3\pi}{4}$

(4) $=\sin^{-1}(0)$ $=0$

(5) $=\cos^{-1}\left(\frac{1}{2}\right)$ $=\frac{\pi}{3}$

(6) $=\cos^{-1}(0)$ $=\frac{\pi}{2}$

(7) $=\frac{\pi}{2}-\tan^{-1}\left(\frac{1}{\sqrt{3}}\right)$ $=\frac{\pi}{2}-\frac{\pi}{6}=\frac{\pi}{3}$

(8) $=\frac{\pi}{2}-\tan^{-1}(-\sqrt{3})$ $=\frac{\pi}{2}-\left(-\frac{\pi}{3}\right)=\frac{5\pi}{6}$

EX8

(1) $\sin\left(\cos^{-1}\frac{\sqrt{2}}{2}\right)=\sin\left(\frac{\pi}{4}\right)=\frac{1}{\sqrt{2}}$

(2) $\sec\left(\cos^{-1}\frac{1}{2}\right)=\sec\left(\frac{\pi}{3}\right)=2$

(3) $\tan\left(\sin^{-1}\left(-\frac{1}{2}\right)\right)=\tan\left(-\frac{\pi}{6}\right)=-\frac{1}{\sqrt{3}}$

(4) $\cot\left(\sin^{-1}\left(-\frac{\sqrt{3}}{2}\right)\right)=\cot\left(-\frac{\pi}{3}\right)=-\frac{1}{\sqrt{3}}$

3.13 Derivatives of Inverse Trigonometric Functions

EX1

(1) $y'=\frac{-6}{x\sqrt{x^4-9}}$

(2) $y'=-\frac{1}{\sqrt{1-(-3x+2)^2}}\cdot(-3)$

$=\frac{3}{\sqrt{1-(-3x+2)^2}}$

(3) $y'=\frac{-1}{\sqrt{2x-x^2}}$

(4) $y'=-2\cdot\frac{1}{\sqrt{1-(\sqrt{x-1})^2}}\cdot\frac{1}{2\sqrt{x-1}}$

$=\frac{-1}{\sqrt{2-x}\sqrt{x-1}}$

(5) $y'=\frac{1}{x[1+(\ln x)^2]}$

(6) $y'=-\frac{1}{1+\left(\frac{1}{x}\right)^2}\cdot\frac{-1}{x^2}$

$=\frac{1}{x^2+1}$

(7) $y=\ln(x^2+4)-x\tan^{-1}\left(\frac{x}{2}\right)$

$\Rightarrow \frac{dy}{dx}=\frac{2x}{x+4}-\tan^{-1}\left(\frac{x}{2}\right)-x\left[\frac{\left(\frac{1}{2}\right)}{1+\left(\frac{x}{2}\right)^2}\right]$

$=\frac{2x}{x^2+4}-\tan^{-1}\left(\frac{x}{2}\right)-\frac{2x}{4+x^2}$

$=-\tan^{-1}\left(\frac{x}{2}\right)$

(8) $y' =-\dfrac{1}{1+(\csc(x^2))^2}$

$\bullet\ (-\csc(x^2)\cot(x^2))\ \bullet\ 2x$

$=\dfrac{2x\csc(x^2)\cot(x^2)}{1+\csc^2(x^2)}$

(9) $y' =\dfrac{1}{|2x+1|\sqrt{x^2+x}}$

(10) $y' =\dfrac{1}{\sqrt{1-((1-x^2)^2)^2}}\ \bullet\ 2(1-x^2)\ \bullet\ (-2x)$

$=\dfrac{-4x(1-x^2)}{\sqrt{1-(1-x^2)^4}}$

(11)

$y=\tan^{-1}\sqrt{x^2-1}+\csc^{-1}x$

$=\tan^{-1}(x^2-1)^{1/2}+\csc^{-1}x$

$\Rightarrow \dfrac{dy}{dx}=\dfrac{\left(\frac{1}{2}\right)(x^2-1)^{-1/2}(2x)}{1+\left[(x^2-1)^{1/2}\right]^2}-\dfrac{1}{|x|\sqrt{x^2-1}}$

$=\dfrac{1}{x\sqrt{x^2-1}}-\dfrac{1}{|x|\sqrt{x^2-1}}$

$=0,$ for $x>1$

(12) $y' =\dfrac{1}{2\sqrt{x^2-1}}\ \bullet\ 2x-\dfrac{1}{|x|\sqrt{x^2-1}},\ x>1$

$=\dfrac{x}{\sqrt{x^2-1}}-\dfrac{1}{x\sqrt{x^2-1}}$

$=\dfrac{x^2-1}{x\sqrt{x^2-1}}$

$=\dfrac{\sqrt{x^2-1}}{x}$

EX2

(1) $y' =4(\cos^{-1}x)^3\ \bullet\ \dfrac{-1}{\sqrt{1-x^2}}$

$y'(0)=4(\cos^{-1}(0))^3\ \bullet\ \dfrac{-1}{\sqrt{1}}$

$=4\left(\dfrac{\pi}{2}\right)^3\ \bullet\ (-1)$

$=-\dfrac{\pi^3}{2}$

(2) $y' =\tan^{-1}(x^2-1)+x\ \bullet\ \dfrac{1}{1+(x^2-1)^2}\ \bullet\ 2x$

$=\tan^{-1}(x^2-1)+\dfrac{2x^2}{1+(x^2-1)^2}$

$y'(-\sqrt{2})=\tan^{-1}(1)+\dfrac{4}{1+1}$

$=\dfrac{\pi}{4}+2$

EX3

(1) $f(4)=\sec^{-1}(2)=\cos^{-1}\left(\dfrac{1}{2}\right)=\dfrac{\pi}{3}\Rightarrow\ \therefore\ \left(4,\dfrac{\pi}{3}\right)$

$y' =\dfrac{1}{|\sqrt{x}|\sqrt{(\sqrt{x})^2-1}}\ \bullet\ \dfrac{1}{2\sqrt{x}}$

$=\dfrac{1}{\sqrt{x}\ \bullet\ \sqrt{x-1}}\ \bullet\ \dfrac{1}{2\sqrt{x}}$

$=\dfrac{1}{2x\sqrt{x-1}}$

$f'(4)=\dfrac{1}{8\sqrt{3}}=\dfrac{\sqrt{3}}{24}$

$\therefore\ y-\dfrac{\pi}{3}=\dfrac{\sqrt{3}}{24}(x-4)$

(2) $f(\sqrt{3})=(6\tan^{-1}(\sqrt{3}))^3=\left(6\ \bullet\ \dfrac{\pi}{3}\right)^3$

$=(2\pi)^3=8\pi^3$

$\Rightarrow\ \therefore\ (\sqrt{3},8\pi^3)$

$y' =3(6\tan^{-1}x)^2\ \bullet\ 6\ \bullet\ \dfrac{1}{1+x^2}$

$f'(\sqrt{3})=3(6\tan^{-1}(\sqrt{3}))^2\ \bullet\ 6\ \bullet\ \dfrac{1}{1+3}$

$=3\ \bullet\ \left(6\ \bullet\ \dfrac{\pi}{3}\right)^2\ \bullet\ 6\ \bullet\ \dfrac{1}{4}$

$=18\pi^2$

$\therefore\ y-8\pi^3=18\pi^2(x-\sqrt{3})$

3.14 Derivatives of Inverse Functions

EX1

(1) To find the slope of g at the point where $x=6$, we must look at the point on f where $y=6$, namely, $(2,6)$. Since $f'(2)=\dfrac{1}{3}$, $g'(6)=3$.

(2) Since f passes through the point $(3,8)$, f^{-1} must pass through the point $(8,3)$, Furthermore, since the graph of f has slope 5 at $(3,8)$, the graph of f^{-1} must have slope $\dfrac{1}{5}$ at $(8,3)$.

(3) Since $f'(x)=3x^2+1$, $g'(y)=\dfrac{1}{3x^2+1}$.

To find x when $y=0$, we must solve the equation $x^3+x-2=0$. Note by inspection that $x=1$, so $g'(0)=\dfrac{1}{3(1)^2+1}=\dfrac{1}{4}$.

EX2

(1) $\dfrac{1}{3}$

(2) $\dfrac{df}{dx}=3x^2-6x \Rightarrow \left.\dfrac{df^{-1}}{dx}\right|_{x=f(3)}=\left.\dfrac{1}{\dfrac{df}{dx}}\right|_{x=3}=\dfrac{1}{9}$

(3) $\dfrac{1}{6}$

(4) $\left.\dfrac{df^{-1}}{dx}\right|_{x=4}=\left.\dfrac{df^{-1}}{dx}\right|_{x=f(2)}=\left.\dfrac{1}{\dfrac{df}{dx}}\right|_{x=2}=\dfrac{1}{\left(\dfrac{1}{3}\right)}=3$

(5) $\dfrac{1}{2}$

(6) $f(g(x))=x \Rightarrow g(x)=f^{-1}(x)$

$$g'(2)=(f^{-1})'(2)=\frac{1}{f'(f^{-1}(2))}=\frac{1}{f'(0)}=\frac{1}{2}$$

$(\therefore f^{-1}(2)=x,\quad 2=f(x)\Rightarrow 2=e^x+x+1$

$e^x+x+1=2\Rightarrow \quad \therefore x=0$

$f'(x)=e^x+1\Rightarrow \quad f'(0)=2\ )$

Multiple Choice

01 B 02 D 03 B 04 C 05 D 06 E 07 A 08 D 09 B 10 C 11 E 12 C 13 A 14 A 15 E
16 D 17 B 18 E 19 A 20 E 21 C 22 C 23 C 24 D 25 A

04 Applications of Derivatives

4.1 Critical Points

EX1

(1) $x=0$ and $x=4$.

(2) $y=x^2+\frac{2}{x}$

$\Rightarrow y'=2x-\frac{2}{x^2}=\frac{2x^3-2}{x^2}=0$

$\Rightarrow 2x^3-2=0$

$\Rightarrow x=1;\frac{2x^3-2}{x^2}=undefined$

$\Rightarrow x^2=0 \Rightarrow x=0$

The domain of the function is $(-\infty,0)\cup(0,\infty)$, thus $x=0$ is not in the domain, so the only critical point is $x=1$.

(3) $x=0, x=1$, and $x=2$.

(4) $y=4x^{\frac{1}{3}}+x^{\frac{4}{3}}$

$y'=\frac{4}{3}x^{-\frac{2}{3}}+\frac{4}{3}x^{\frac{1}{3}}$

$=\frac{4}{3\cdot x^{2/3}}+\frac{4}{3}x^{1/3}$

$=\frac{4+4x}{3\cdot x^{2/3}}$

$=\frac{4(1+x)}{3x^{2/3}}$

$y'=0\,at\ x=-1$

$y'=undefined\ at\ x=0$

$\therefore x=-1,0$

(5) $x=e^{-1/2}$

(6) $y'=1+2\cos x$

$y'=0 \Rightarrow 1+2\cos x=0$

$\cos x=-\frac{1}{2}$

$\therefore x=\frac{2\pi}{3},\frac{4\pi}{3}$

4.2 Extrema: Maximum & Minimum value

EX1

(1) neither

(2) absolute minimum, relative minimum

(3) relative maximum

(4) relative minimum

(5) relative maximum

(6) neither

(7) relative minimum

(8) absolute maximum

EX2

(1) increasing on $(0,1)$ and $(2,\infty)$,decreasing on $(-\infty,0)$ and $(1,2)$.

(2) $Domain: x>0$

$h'(x)=\frac{1}{x}-\frac{1}{x^2}=\frac{x-1}{x^2}$

$h'(x)=0\ at\ x=1$

x	0		1	
y'		$-$	0	$+$

y is increasing on $x>1$

y is decreasing on $0<x<1$

(3) y is increasing on $0\le x<\frac{\pi}{6}$, $\frac{5\pi}{6}<x\le\pi$

y is decreasing on $\frac{\pi}{6}<x<\frac{5\pi}{6}$

(4) $y'=3x^2e^{-x}-x^3e^{-x}$

$=x^2e^{-x}(3-x)$

$y'=0\ at\ x=0,3$

x		0		3	
y'	$+$	0	$+$	0	$-$

y is increasing on $x<0,\ 0<x<3$

y is decreasing on $x>3$

(5) y is increasing on $-2<x<2$

y is decreasing on $x<-2, x>2$

(6) $f(x)=x^2\ln x$

$\Rightarrow f'(x)=x+2x\ln x=x(1+2\ln x)$

$\Rightarrow$ a critical point at

$x=e^{-1/2}\Rightarrow f'=[---|+++,$

0 $\quad e^{-1/2}$

increasing on $(e^{-1/2},\infty)$,

decreasing $(0,e^{-1/2})$

EX3

(1) a local maximum is $k(3)=162$ at $t=3$,

a local minimum is $k(-3)=-162$ at $t=-3$.

(2) $Vertical\ \ Asymptote\ :\ x=0$

$$y'=\frac{e^x\cdot x-e^x\cdot 1}{x^2}$$
$$=\frac{e^x(x-1)}{x^2}$$

$y'=0\ \ at\ \ x=1$

$y'=undefined\ \ at\ \ x=0$

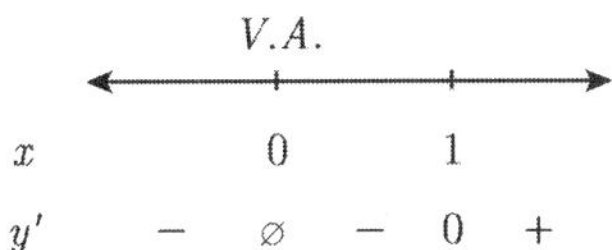

x		0		1	
y'	$-$	$\varnothing$	$-$	0	$+$

Relative Minimum $=f(1)=e$

(3) Relative Minimum $=f(0)=0$

Relative Maximum $=f(2)=\dfrac{4}{e^2}$

(4) $f(x)=x^{1/3}(x+8)=x^{4/3}+8x^{1/3}$

$\Rightarrow f'(x)=\dfrac{4}{3}x^{1/3}+\dfrac{8}{3}x^{-2/3}=\dfrac{4(x+2)}{3x^{2/3}}$

$\Rightarrow$ critical points at $x=0,-2$

$\Rightarrow f' = ---\mid+++)(+++,$
$\quad\quad -2 \quad\quad 0$

increasing on $(-2,0)\cup(0,\infty)$,

decreasing on $(-\infty,-2)$,

a local minimum is

$f(-2)=-6\sqrt[3]{2}\approx-7.56\ \ at\ x=-2.$

(5) local maximum is

$g(-2)=3\sqrt[3]{4}\approx 4.762\ \ at\ \ x=-2,$

a local minimum is

$g(0)=0$ at $x=0$.

EX4

(1)

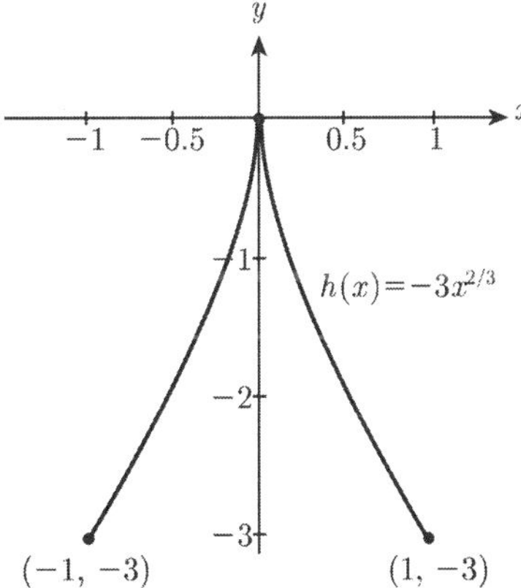

$h(x)=-3x^{2/3}\Rightarrow h'(x)=-2x^{-2/3}\Rightarrow$ a critical point at $x=0$; $h(-1)=-3$, $h(0)=0$, $h(1)=-3\Rightarrow$ the absolute maximum is $0\,at\ x=0$ and the absolute minimum is -3 at $x=1$ and at $x=-1$.

(2) Absolute Maximum $=f(2)=16$

Absolute Minimum $=f(-2)=f(4)=-16$

(3) Absolute Minimum $=f(1)=0$

Absolute Maximum $=f(e)=\dfrac{\ln e}{e}=\dfrac{1}{e}$

(4)

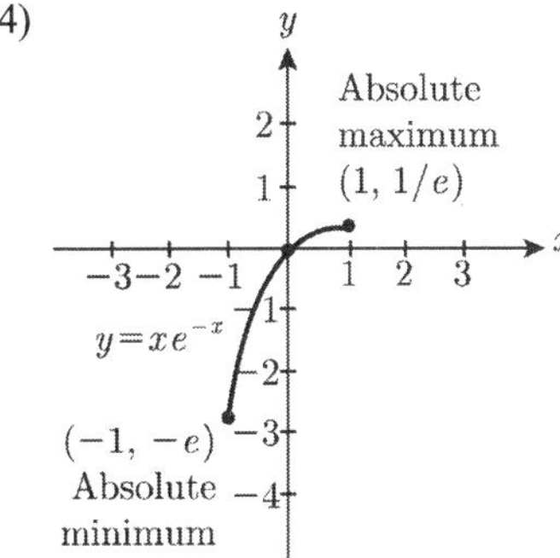

$g(x)=xe^{-x}\Rightarrow g'(x)=e^{-x}-xe^{-x}\Rightarrow$ a critical point at $x=1$; $g(-1)=-e$, and $g(1)=\dfrac{1}{e},\Rightarrow$ the absolute maximum is $\dfrac{1}{e}$ at $x=1$ and the absolute minimum is $-e$ at $x=-1$.

(5) Absolute Minimum $=f(0)=f(2)=0$

Absolute Maximum $=f(\sqrt{2})=2$

(6) $y = 4x - 6x^{\frac{2}{3}}$

$y' = 4 - 4x^{-\frac{1}{3}}$

$= 4 - \dfrac{4}{x^{1/3}} = \dfrac{4x^{1/3} - 4}{x^{1/3}} = \dfrac{4(x^{1/3} - 1)}{x^{1/3}}$

$y' = 0$ at $x = 1$, $y' = undefined$ at $x = 0$

x	-1		0		1		8
y'		$+$	$\varnothing$	$-$	0	$+$	
y	-10	↗	0	↘	-2	↗	8

Absolute Maximum $= f(8) = 8$

Absolute Minimum $= f(-1) = -10$

(7) $x = 0, \pi, 2\pi \qquad x = \dfrac{\pi}{3}, \dfrac{5\pi}{3}$

x	-1		$\pi/3$		π		$5\pi/3$		2π
y'	0	$+$	0	$-$	0	$+$	0	$-$	0
y	1	↗	$\frac{3}{2}$	↘	-3	↗	$\frac{3}{2}$	↘	1

Absolute Maximum $= f\left(\dfrac{\pi}{3}\right) = f\left(\dfrac{5\pi}{3}\right) = \dfrac{3}{2}$

Absolute Minimum $= f(\pi) = -3$

4.3 Concavity

EX1

(1)

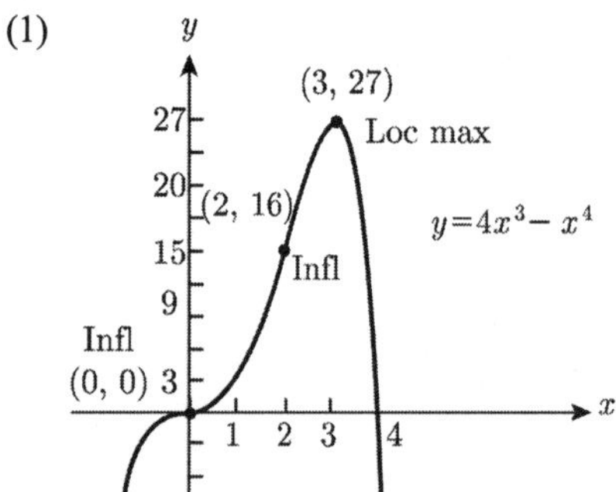

When $y = 4x^3 - x^4$,

then $y' = 12x^2 - 4x^3 = 4x^2(3 - x)$

and $y'' = 24x - 12x^2 = 12x(2 - x)$. The curve r -ise on $(-\infty, 3)$ and falls on $(3, \infty)$. At $x = 3$ there is a local maximum, but there is no loc-al minimum, The graph is concave up on $(0, 2)$, and concave down on $(-\infty, 0)$ and $(2, \infty)$. There are inflection points at $x = 0$ and $x = 2$.

(2)

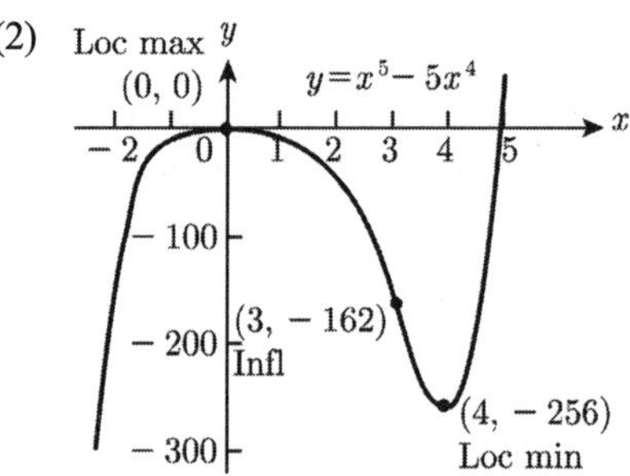

When $y = x^5 - 5x^4$, then $y' = 5x^4 - 20x^3$ $= 5x^3(x - 4)$ and $y'' = 20x^3 - 60x^2$ $= 20x^2(x - 3)$. The curve rises on $(-\infty, 0)$ and $(4, \infty)$, and falls on $(0, 4)$. There is a local maximum at $x = 0$, and a local minium at $x = 4$ the curve is concave down on $(-\infty, 3)$ and concave up on $(3, \infty)$. At $x = 3$ there is a point of inflection.

(3) y is concave up on $x > e^{-3/2}$

y is concave down on $0 < x < e^{-3/2}$

There are inflection point at $x = e^{-3/2}$

(4)

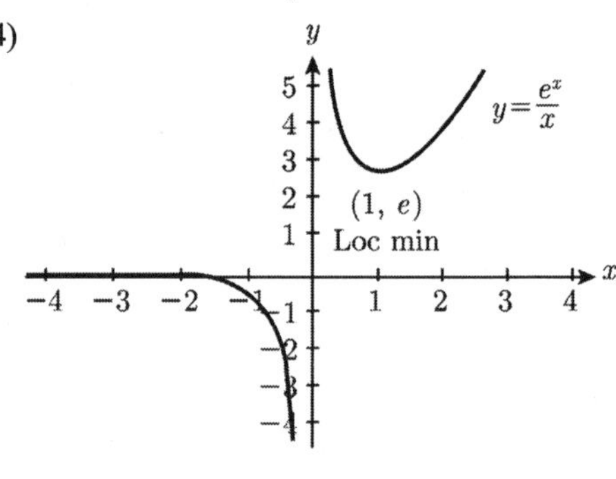

$y = \dfrac{e^x}{x}, \ \Rightarrow \ y' = \dfrac{xe^x - e^x}{x^2}$

$\Rightarrow y' = ---|---|+++$
$\qquad\qquad 0 \qquad 1$

$\Rightarrow$ the graph is rising on $(1, \infty)$, falling on $(-\infty, 0)$ and $(0, 1)$; a local minimum is e at $x = 1$;

$y'' = \dfrac{x^2(xe^x + e^x - e^x) - (xe^x - e^x)2x}{x^4}$

$= \dfrac{(x^2 - 2x + 2)e^x}{x^3}$

$\Rightarrow y'' = ---|+++$
$\qquad\qquad 0$

$\Rightarrow$ the graph is concave up on $(0, \infty)$, concave down on $(-\infty, 0)$, but has no inflection points.

EX2

(1) $a=-6,\ b=24$

(2) $f(-1)=3$

$-1+a-b+c=3$

$a-b+c=4 \cdots\cdots ①$

$f'(-1)=0$

$f'(x)=3x^2+2ax+b$

$f(-1)=3-2a+b=0 \cdots\cdots ②$

$f''(-3)=0$

$f''(x)=6x+2a$

$f''(-3)=-18+2a=0$

$\therefore a=9 \cdots\cdots\cdots\cdots ③$

② $3-2(9)+b=0$

$\therefore b=15$

① $9-15+c=4$

$\therefore\ c=10$

(3) $a=-4,\ b=-4$

(4) $f'(1)=-9$

$f'(x)=3ax^2+2bx$

$f'(x)=3a+2b=-9 \cdots\cdots ①$

$f''(2)=0$

$f''(x)=6ax+2b$

$f''(2)=12a+2b=0 \cdots\cdots ②$

$f(2)=-12$

$8a+4b+c=-12 \cdots\cdots ③$

① $3a+2b=-9$

② $12a+2b=0$

① − ② $-9a=-9$

$\therefore a=1$

① $3(1)+2b=-0$

$2b=-12$

$\therefore b=-6$

③ $8(1)+4(-6)+c=-12$

$8-24+c=-12$

$\therefore c=4$

$\therefore a=1\ b=-6\ c=4$

EX3

	$\frac{dy}{dx}$	$\frac{d^2y}{dx^2}$	Behaviour of the graph of f
A	0	+	f has a horizontal tangent slope and the graph of f is concave up. f has a relative minimum.
B	0	0	f has a horizontal tangent slope and f'' changes sign from - to +. f has a point of inflection.
C	+	0	f is increasing and f'' changes sign from + to -. f has a point of inflection
D	0	-	f has a horizontal tangent slope and the graph of f is concave down, f has a relative maximum.
E	-	-	f is decreasing and the graph of f is concave down.
F	undefined	undefined	f' changes sign from - to +. f has a relative minimum. f'' changes sign from - to +. f has a point of inflection
G	undefined	undefined	f has a vertical tangent and f'' changes sign from + to -. f has a point of inflection.

EX4

(1) $f'(x)=3x^2-6x-9$

$=3(x^2-2x-3)$

$=3(x-3)(x+1)$

$f'(x)=0\ at\ x=3,\ -1$

$f''(x)=6x-6$

$f''(3)=18-6=12>0$

$\therefore$ Relative Minimum $=f(3)=-27$

$f''(-1)=-6-6=-12<0$

$\therefore$ Relative Maximum $=f(-1)=5$

(2) $f'(x)=2e^x+2xe^x$

$=2e^x(1+x)$

$f'(x)=0\ \ at\ \ x=-1$

$f''(x)=2e^x+(2e^x+2xe^x)$

$=4e^x+2xe^x$

$=2e^x(2+x)$

$f''(-1)=2e^{-1}(2-1)=\frac{2}{e}>0$

$\therefore$ Relative Minimum $=f(-1)=-2e^{-1}$

EX5

(1) $6x-6y^2y'+12=0$

$6x+12=6y^2y'$

$x+2=y^2y'$

$y'=\dfrac{x+2}{y^2}$

$\therefore\ \dfrac{dy}{dx}=\dfrac{x+2}{y^2}$

(2) $\dfrac{x+2}{y^2}=0$

$x+2=0$

$\therefore\ x=-2$

$3(-2)^2-2y^3+12(-2)+28=0$

$12-2y^3-24+28=0$

$16-2y^3=0$

$y=2$

$\therefore\ (-2,2)$

(3) $\dfrac{d^2y}{dx^2}=\dfrac{1\cdot y^2-(x+2)\cdot 2yy'}{(y^2)^2}$

$\therefore\ \left.\dfrac{d^2y}{dx}\right|_{(-2,2)}=\dfrac{4-0}{16}=\dfrac{1}{4}$

(4) At P$(-2,2)$, $\dfrac{dy}{dx}=0$ and $\dfrac{d^2y}{dx^2}>0$.

By the second derivative test, the curve has a relative minimum at the point P.

EX6

$2x+2y\cdot y'-2y'=0$

$y'(y-1)=-x$

$\dfrac{dy}{dx}=\dfrac{-x}{y-1},\ \dfrac{dy}{dw}=0\Rightarrow x=0$

$(0)^2+y^2-2y=0$

$y(y-2)=0\Rightarrow y=0,2$

$\therefore\dfrac{dy}{dx}=0\ at\ (0,0),\ (0,2)$

$\dfrac{d^2y}{dx^2}=\dfrac{(-1)(y-1)-(-x)(y')}{(y-1)^2}$

$=\dfrac{-y+1+xy'}{(y-1)^2}$

$\left.\dfrac{d^2y}{dx^2}\right|_{(0,0)}=\dfrac{0+1+0}{(0-1)^2}=1>0$

$\therefore$ The curve has a relative minimum at $(0,0)$

$\left.\dfrac{d^2y}{dx^2}\right|_{(0,2)}=\dfrac{-2+1+0}{(2-1)^2}=-1<0$

$\therefore$ The curve has a relative maximum at $(0,2)$

4.4 Curve Sketching

EX1

(1)

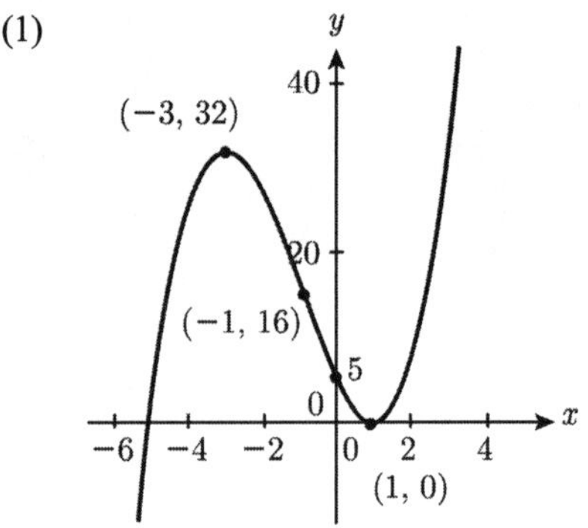

(2) $y'=-6x^2-6x+12$

$=-6(x^2+x-2)$

$=-6(x+2)(x-1)$

$y'=0\ at\ x=-2,1$

$y''=-12x-6$

$=-6(2x+1)$

$y''=0\ at\ x=-\dfrac{1}{2}$

x		-2		$-1/2$		1	
y'	$-$	0	$+$	$+$	$+$	0	$-$
y''	$+$	$+$	$+$	0	$-$	$-$	$-$
y	↘	-16	↗	$-5/2$	↗	11	↘

$f(-2)=16-12-24+4=-16$
$f(-\frac{1}{2})=\frac{1}{4}-\frac{3}{4}-6+4=-\frac{5}{2}$
$f(1)=-2-3+12+4=11$

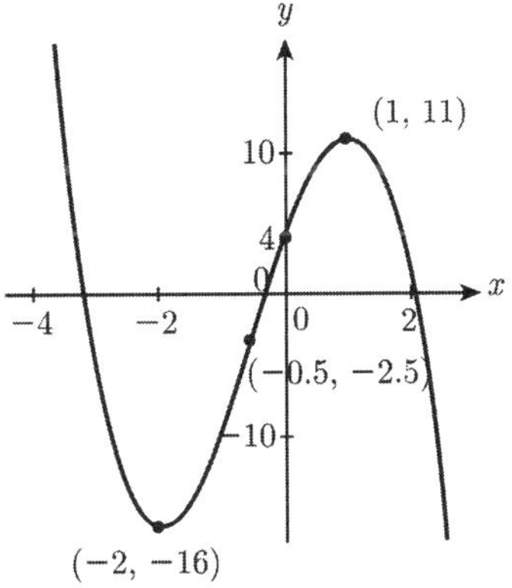

(3)

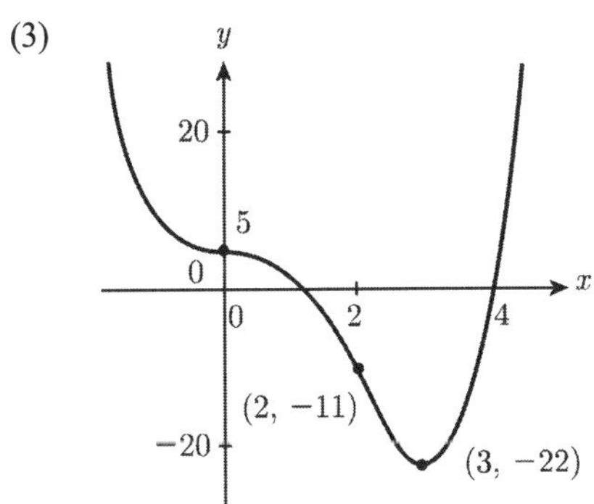

(4) $y'=-4x^3+12x$
$=-4x(x^2-3)$
$y'=0$ at $x=0,\pm\sqrt{3}$
$y''=-12x^2+12$
$=-12(x^2-1)$
$y''=0$ at $x=\pm 1$

x		$-\sqrt{3}$		-1		0		1		$\sqrt{3}$	
y'	+	0	−	−	−	0	+	+	+	0	−
y''	−	−	−	0	+	+	+	0	−	−	−
y	↗	10	↘	6	↘	1	↗	6	↗	10	↘

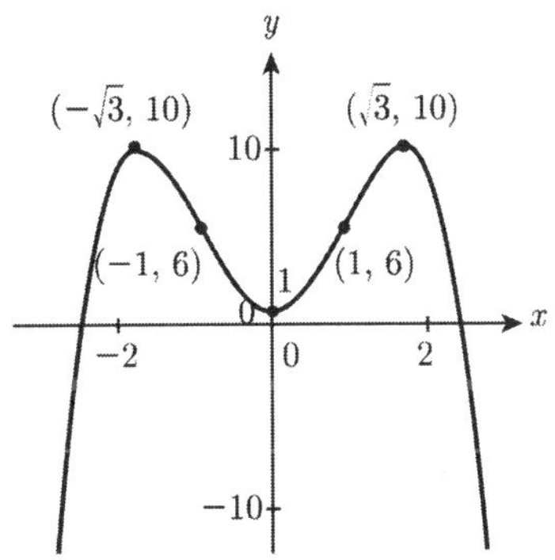

(5)

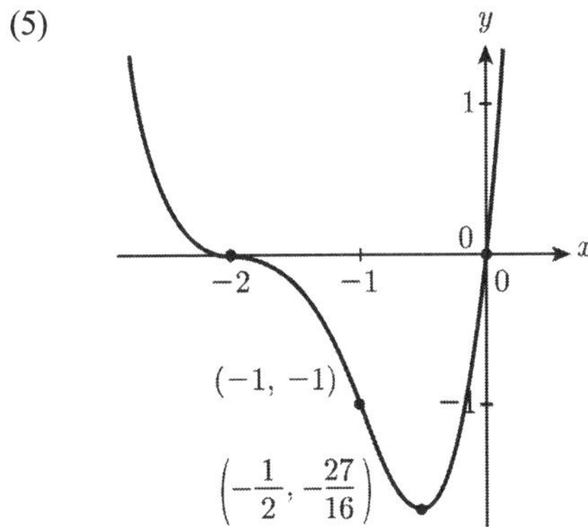

EX2

(1)

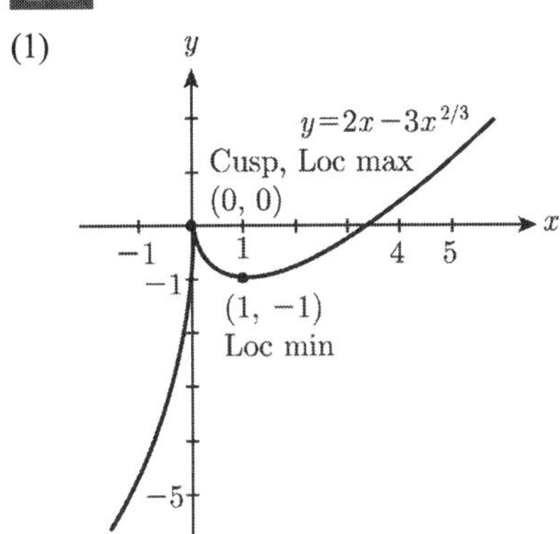

when $y=2x-3x^{2/3}$, then $y'=2-2x^{-1/3}$ and $y''=\frac{2}{3}x^{-4/3}$. The curve is rising on $(-\infty,0)$ and $(1,\infty)$, and falling on $(0,1)$. There is a local maximum at $x=0$ and a local minimum at $x=1$. The curve is concave up on $(-\infty,0)$ and $(0,\infty)$. There are no points of inflection, but a cusp exists at $x=0$.

(2)

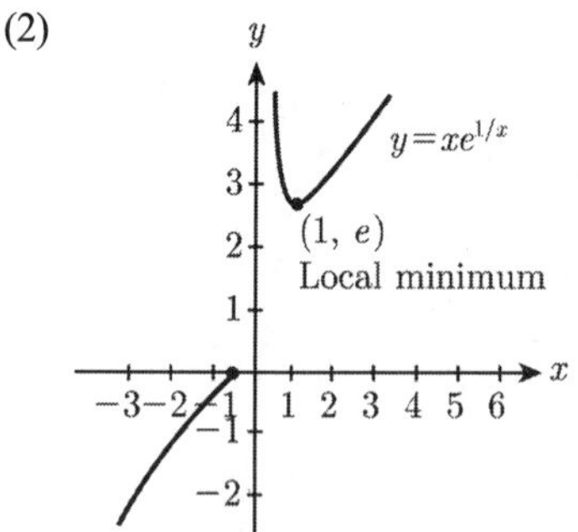

When $y=xe^{1/x}$,then

$$y'=-\frac{xe^{1/x}}{x^2}+e^{1/x}=e^{1/x}(1-\frac{1}{x})\text{ and}$$

$$y''=e^{1/x}(\frac{1}{x^2})+(1-\frac{1}{x})(-\frac{e^{1/x}}{x^2})$$
$$=\frac{e^{1/x}}{x^2}(\frac{1}{x})=\frac{e^{1/x}}{x^3}.$$

The curve is rising on $(1,\infty)$ and $(-\infty,0)$ and falling on $(0,1)$. The curve is concave down on $(-\infty,0)$ and concave up on $(0,\infty)$. There is a local minimum of e at $x=1$, but there are no inflection points.

(3)

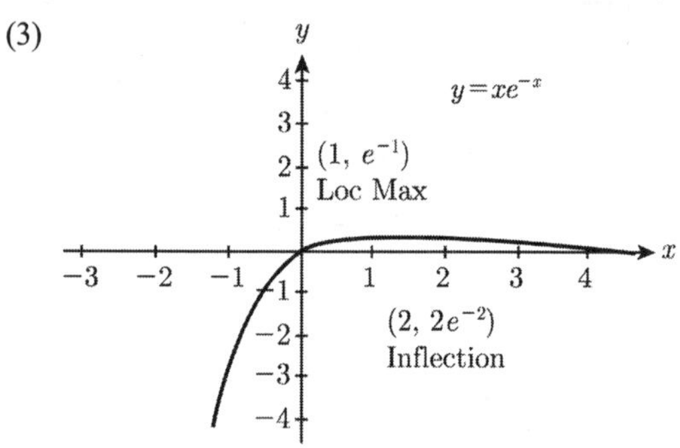

$y=xe^{-x}\Rightarrow y'=e^{-x}-xe^{-x}=(1-x)e^{-x}$

$\Rightarrow y'=+++|---$ (sign change at 1) $\Rightarrow$ the graph is increasing on $(-\infty,1)$ and decreasing on $(1,\infty)$; a local maximum is e^{-1} at $x=1$; $y''=(x-1)e^{-x}+(-1)e^{-x}=(x-2)e^{-x}$

$\Rightarrow y''=---|+++$ (sign change at 2) $\Rightarrow$ Graph is concave up on $(2,\infty)$,concave down on $(-\infty,2)$

$\Rightarrow$point of inflection at $(2,2e^{-2})$.

(4)

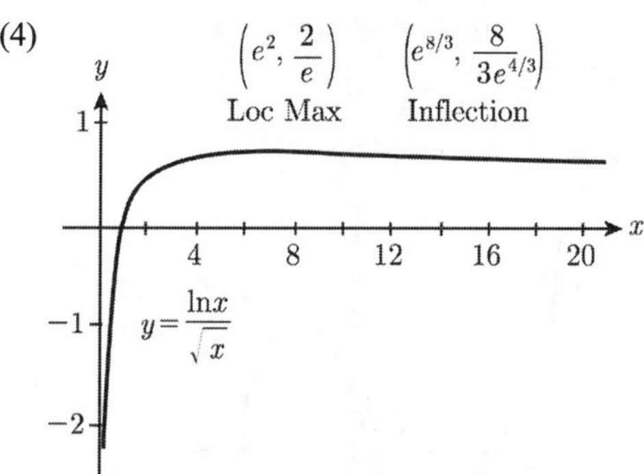

$$y=\frac{\ln x}{\sqrt{x}}$$

$$\Rightarrow y'=\frac{\sqrt{x}\cdot(\frac{1}{x})-\ln x\cdot\frac{1}{2\sqrt{x}}}{(\sqrt{x})^2}=\frac{2-\ln x}{2x^{3/2}}$$

$\Rightarrow y'=(\ +++|---$ (from 0; sign change at e^2)

$\Rightarrow$ the graph is increasing on $(0,e^2)$. decreasing on (e^2,∞); a local maximum is $\frac{2}{e}$ at $x=e^2$;

$$y''=\frac{2x^{3/2}(-\frac{1}{x})-(2-\ln x)\cdot 2\cdot\frac{2}{3}x^{1/2}}{(2x^{3/2})^2}$$
$$=\frac{3\ln x-8}{4x^{5/2}}$$

$\Rightarrow y''=(\ ---|+++$ (from 0; sign change at $e^{8/3}$) $\Rightarrow$ the graph is concave up on $(e^{8/3},\infty)$,concave down on $(0,e^{8/3})\Rightarrow$ point of inflection is $(e^{8/3},\frac{8}{3e^{4/3}})$.

(5)

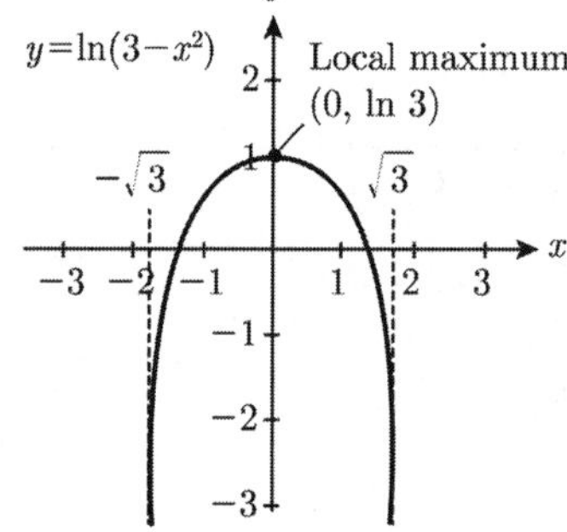

$$y=\ln(3-x^2),\Rightarrow y'=\frac{-2x}{3-x^2}=\frac{2x}{x^2-3}$$

$\Rightarrow y'=(\ +++|---\ )$ (from $-\sqrt{3}$; sign change at 0; to $\sqrt{3}$) $\Rightarrow$ the graph is rising on $(-\sqrt{3},0)$,falling on $(0,\sqrt{3})$; a local minimum is $\ln 3$ at

$x=0; y''=\dfrac{(x^2-3)(2)-(2x)(2x)}{(x^2-3)^2}$

$=\dfrac{-2(x^2+3)}{(x^2-3)^2}$

$\Rightarrow y'' = (\ ---\ )$ with sign changes at $-\sqrt{3}$ and $\sqrt{3}$

$\Rightarrow$The graph is concave down on $(-\sqrt{3},\sqrt{3})$.

(6)

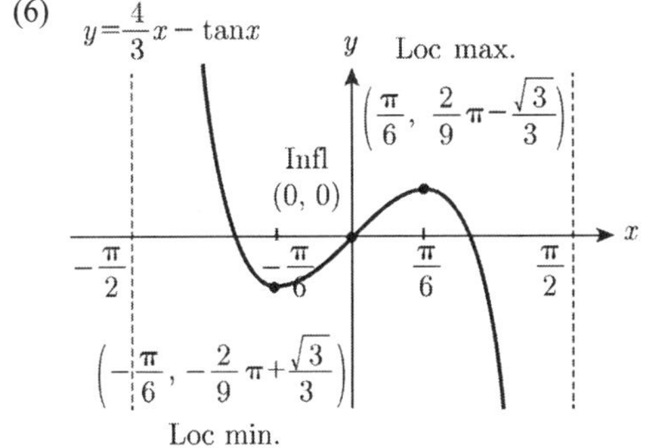

When $y=\frac{4}{3}x-\tan x$, then $y'=\frac{4}{3}-\sec^2x$ and $y''=-2\sec^2x\tan x$. The curve is increasing on $(-\frac{\pi}{6},\frac{\pi}{6})$, and decreasing on $(-\frac{\pi}{2},-\frac{\pi}{6})$ and $(\frac{\pi}{6},\frac{\pi}{2})$.

At $x=-\frac{\pi}{6}$ there is a local minimum, at $x=\frac{\pi}{6}$ there is a local maxi- mum, There are no absolute maxima or absolute minima. The curve is concave up on $(-\frac{\pi}{2},0)$, and is concave down on $(0,\frac{\pi}{2})$. At $x=0$ there is a pint of inflection.

EX3

(1)

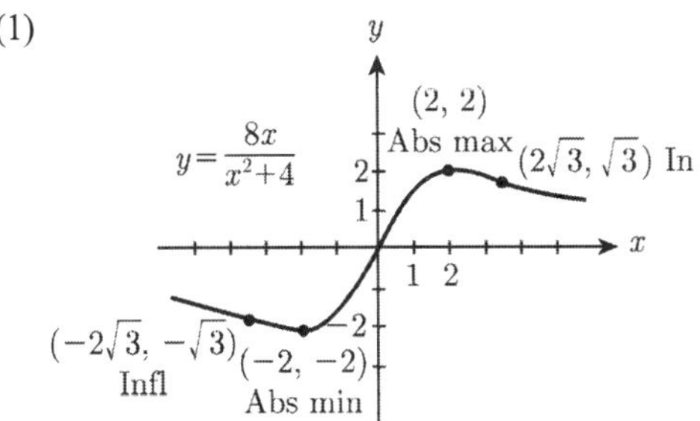

When $y=\dfrac{8x}{x^2+4}$, then $y'=\dfrac{-8(x^2-4)}{(x^2+4)^2}$ and $y''=\dfrac{16x(x^2-12)}{(x^2+4)^3}$. The curve is falling on $(-\infty,-2)$ and $(2,\infty)$,and is rising on $(-2,2)$. There is a local and absolute minimum at $x=-2$,and a local and absolute maximum at $x=2$. The curve is concave down on $(-\infty,-2\sqrt{3})$ and $(0,2\sqrt{3})$,and concave up on $(-2\sqrt{3},0)$ and $(2\sqrt{3},\infty)$.There are points of inflection at $x=-2\sqrt{3}$, $x=0$,and $x=2\sqrt{3}$. $y=0$ is a horizontal asymptote.

(2)

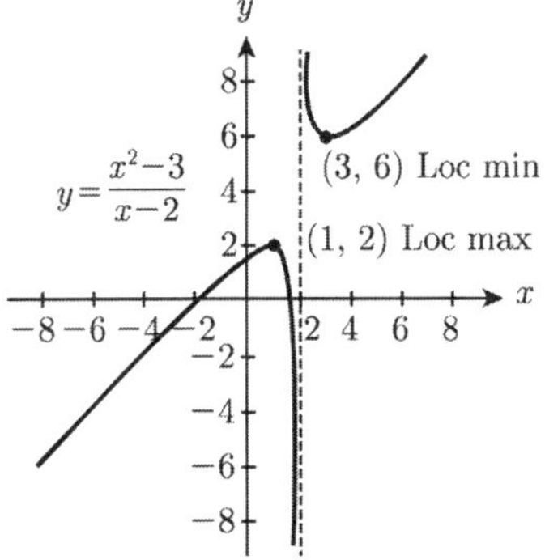

When $y=\dfrac{x^2-3}{x-2}$

then $y'=\dfrac{2x(x-2)-(x^3-3)(1)}{(x-2)^2}$

$=\dfrac{(x-3)(x-1)}{(x-2)^2}$

and

$y''=\dfrac{(2x-4)(x-2)^2-(x^2-4x+3)2(x-2)}{(x-2)^4}$

$=\dfrac{2}{(x-2)^3}$.

The curve is rising on $(-\infty,1)$ and $(3,\infty)$,and falling on $(1,2)$ and $(2,3)$. There is a local maximum at $x=1$ and a local minimum at $x=3$.The curve is concave down on $(-\infty,2)$ and concave up on $(2,\infty)$.There are no points of inflection because $x=2$ is not in the domain.

(3)

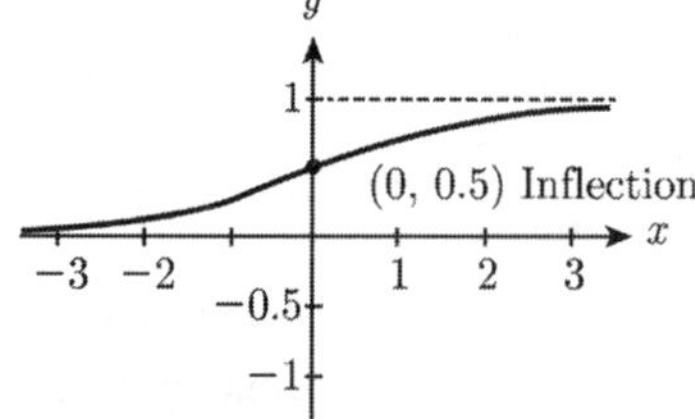

$$y=\frac{e^x}{1+e^x}$$

$$\Rightarrow y'=\frac{(e^x+1)e^x-e^x \cdot e^x}{(1+e^x)^2}=\frac{e^x}{(1+e^x)^2}$$

$\Rightarrow y' = +++$

$\Rightarrow$the graph is increasing $(-\infty,\infty)$;

$$y''=\frac{(e^x+1)^2 \cdot e^x-e^x \cdot 2(e^x+1)}{(1+e^x)^4}$$

$$=\frac{e^x(1-e^x)}{(1+e^x)^3}\Rightarrow y'' = +++\underset{0}{|}---$$

$\Rightarrow$ the graph is concave up on $(-\infty,0)$, concave down on $(0,\infty)$ $\Rightarrow$ point of inflection is $(0,\frac{1}{2})$.

(4)

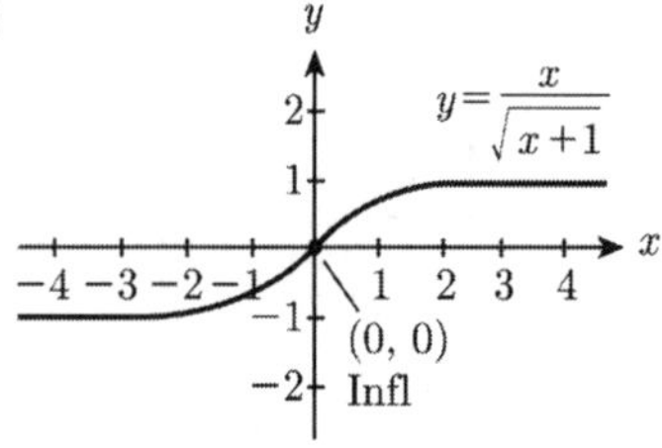

When $y=\frac{x}{\sqrt{x^2+1}}$; then $y'=\frac{1}{(x^2+1)^{3/2}}$ and $y''=\frac{-3x}{(x^2+1)^{3/2}}$. The curve is increasing on $(-\infty,\infty)$. There are no local or absolute extrema. The curve is concave up on $(-\infty,0)$ and concave down on $(0, \infty)$. At $x=0$ there is a point of inflection.

4.5 Relation of $f'(x)$, $f''(x)$ to the Graph of $f(x)$

EX1

(1)

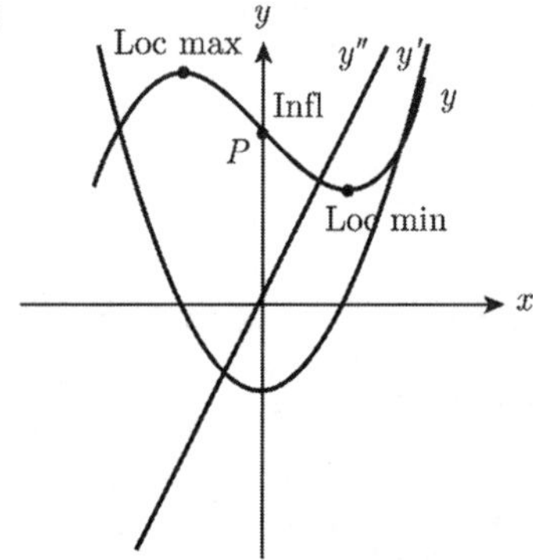

The graph of $y=f''(x)\Rightarrow$ the graph of $y=f(x)$ is concave up on $(0,\infty)$, concave down on $(-\infty,0)$ $\Rightarrow$a point of inflection at $x=0$; the graph of $y=f'(x)$

$\Rightarrow y' = +++ \mid --- \mid +++$ $\Rightarrow$the graph $y=f(x)$ has both a local maximum and a local minimum.

(2)

The graph of

$y=f''(x)\Rightarrow y'' = +++ \mid ---$

$\Rightarrow$ the graph of $y=f(x)$ has a point of inflection, the graph of

$y=f'(x)\Rightarrow y' = --- \mid +++ \mid ---$

$\Rightarrow$ the graph of $y=f(x)$ has both a local maximum and a local minimum.

(3)

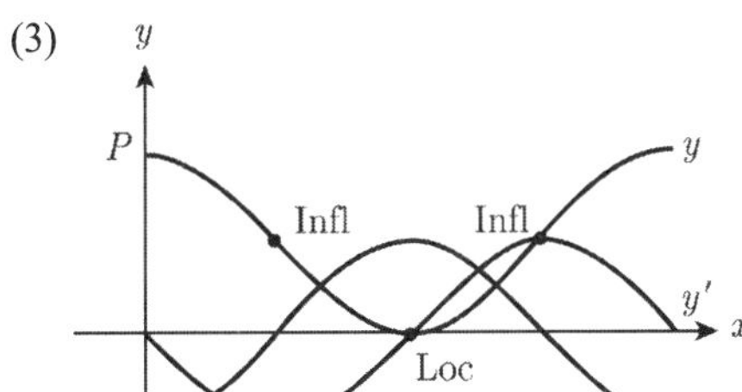

The graph of

$y = f''(x) \Rightarrow y'' = \ +++ \mid --- \Rightarrow$ the graph of $y = f(x)$ has two points of inflection, the graph of $y = f'(x) \Rightarrow y' = \ --- \mid +++ \Rightarrow$ the graph of $y = f(x)$ has both a local minimum.

(4)

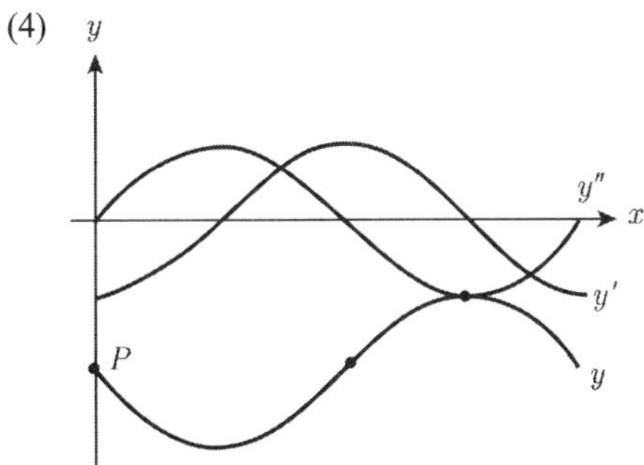

The graph of

$y = f''(x) \Rightarrow y'' = \ --- \mid +++ \mid --- \Rightarrow$

the graph of $y = f(x)$ has a point of inflection;

the graph of

$y = f'(x) \Rightarrow y' = \ --- \mid +++ \mid --- \Rightarrow$

the graph of $y = f(x)$ has both a local maximum and a local minimum.

EX2

Point	y'	y''
P	−	+
Q	+	0
R	+	−
S	0	−
T	−	−

EX3

(1)

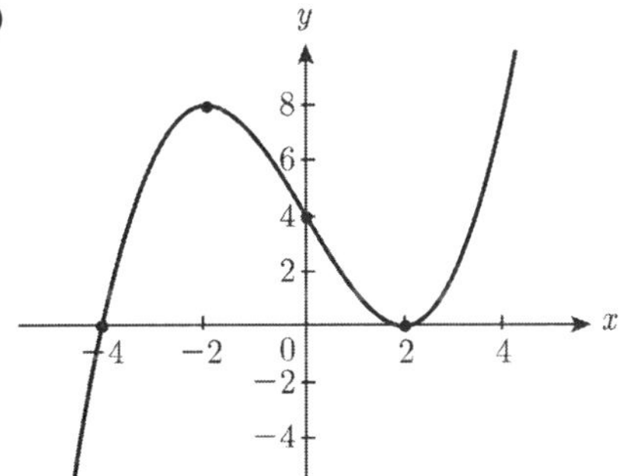

(2)

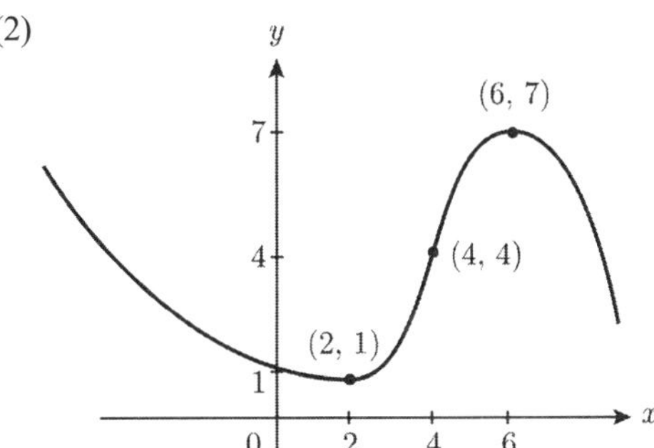

(3)

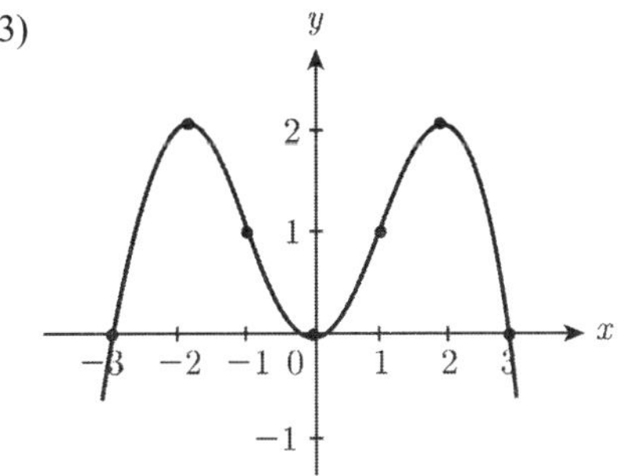

EX4

(1) $f'(-1) = 0,\ f'(1) = -2,\ f'(5) = 2,$

$f''(-1) = -1,\ f''(1) = DNE,\ f''(2) = 2,$

$f''(3) = DNE$

(2) $\dfrac{f(3) - f(-1)}{3 - (-1)} = \dfrac{2 - 0}{4} = \dfrac{1}{2}$

(3) f is increasing when $f' > 0$

$\therefore -3 \le x < -1,\ 2 < x < 4,\ 4 < x \le 5$

(4) f is concave down when $f'' < 0$ or f' is decreasing.

$\therefore -2 < x < 1,\ 3 < x < 4$

(5) f has a relative minimum when f' changes sign from - to +. This occurs at $x=2$. f has a relative maximum when f' changes sign from + to -. This occur at $x=-1$.

(6) f has a point of inflection when f'' changes sign from + to -, or vice versa. This occurs at $x=1,3,4$

EX5

(1) $c<x<e$

(2) f' is increasing when f is concave up.
$\therefore b<x<d$

(3) $x=c$

(4) $x=b,\ d$

EX6

(1) $x=6$ is the only critical point at which f' changes sign from negative to positive. Therefore, f has a local minimum at $x=6$.

(2) From part (1), the absolute minimum occurs either at $x=6$ or at an endpoint.

$$f(0)=f(8)+\int_8^0 f'(x)dx$$
$$=f(8)-\int_0^8 f'(x)dx=4-12=-8$$
$$f(6)=f(8)+\int_8^6 f'(x)dx$$
$$=f(8)-\int_6^8 f'(x)dx=4-7=-3$$
$$f(8)=4$$

The absolute minimum value of f on the closed interval $[0,8]$ is -8.

(3) The graph of f is concave down and increasing on $0<x<1$. and $3<x<4$, because f' is decreasing and positive on these intervals.

(4) $g'(x)=3[f(x)]^2 \cdot f'(x)$

$g'(3)=3[f(3)]^2 \cdot f'(3)=3(-\frac{5}{2})^2 \cdot 4=75$

EX7

(1) f has a point of inflection when f'' changes sign. This occurs at $x=b,d$.

(2) f is concave up when $f''>0$.
$\therefore b<x<d$

(3) f' is increasing when $f''>0$.
$\therefore b<x<d$

(4) f' has a relative maximum when f'' changes sign from + to - .
$\therefore x=d$.

(5) f' is concave down when f'' is decreasing.
$\therefore c<x<e,\ f<x<g$.

4.6 Motion A long a Line(1-Dimension)

EX1

(1) $v(t)=s'(t)=3t^2-12$
$a(t)=v'(t)=6t$

(2) The particle changes direction when v changes sign.
$v(t)=0 \ when\ t=2$
$\therefore t=2$

(3) v is increasing when $a=v'>0$.
$6t>0$
$\therefore t>0$

(4) The speed of the particle is decreasing when the signs of v and a are opposite.
$\therefore 0<t<2$

EX2

(1) $s(2)-s(1)=-1-1=-2m$

(2) $\frac{s(2)-s(1)}{2-1}=\frac{-2}{1}=-2m/\text{sec}$

(3) $v(t)=s'(t)=3t^2-12t+9$
$a(t)=v'(t)=6t-12$

(4) The particle is at rest when $v=0$
$$v(t)=3t^2-12t+9$$
$$=3(t^2-4t+3)$$
$$=(t-1)(t-3)$$
$\therefore t=1,3$ sec

(5) $s(4) = 1m$

$v(4) = 9m/\sec$

$speed = 9m/\sec$

$a(4) = 12m/\sec^2$

(6) The particle is moving away from the origin when he signs of $s(t)$ and $v(t)$ are the same. $s(4) = 1 > 0$ and $v(4) = 9 > 0$

$\therefore$ The particle is moving away from the origin at $t = 4$

(7) $\dfrac{v(2) - v(0)}{2-0} = \dfrac{-3-9}{2} = -6m/\sec^2$

(8) The particle is moving in the positive direction when $v > 0$.

$v(t) = 3(t-1)(t-3)$

$\therefore 0 \leq t < 1,\ t > 3$

(9) v is decreasing when $a = v' < 0$

$a(t) = 6(t-2)$

$\therefore 0 \leq t < 2$

(10) The speed of the particle is increasing when the signs of $v(t)$ and $a(t)$ are the same.

$\therefore 1 < t < 2, t > 3$

(11) $\therefore 4+4+4 = 12m$

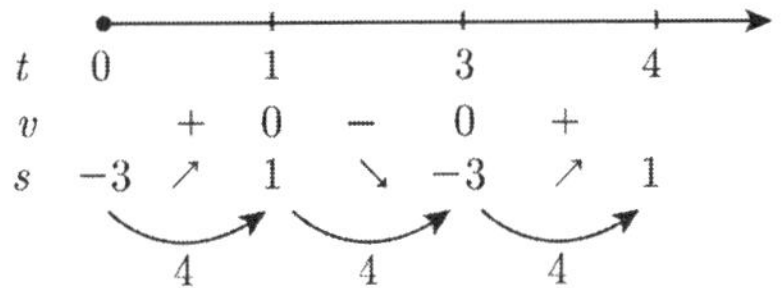

EX3

(1) The particle is at rest when $v = 0$.

$\therefore t = 0,3,5,7 \sec$

(2) The particle changes direction when v changes sign.

$\therefore t = 5,7 \sec$

(3) $speed = |v(6)| = 2m/\sec$

(4) $a(5) = v'(5) = -2m/\sec^2$

(5) $\dfrac{v(7) - v(1)}{7-1} = \dfrac{0-2}{6} = -\dfrac{1}{3}m/\sec^2$

(6) The particle is moving in the negative direction when $v < 0$.

$\therefore 5 < t < 7$

(7) The speed is decreasing when $v > 0, a < 0$ or $v < 0, a > 0$.

$\therefore 2 < t < 3,\ 4 < t < 5,\ 6 < t < 7$

(8) $4+2+2+1 = 9m$

(9) $t = 5 \sec$

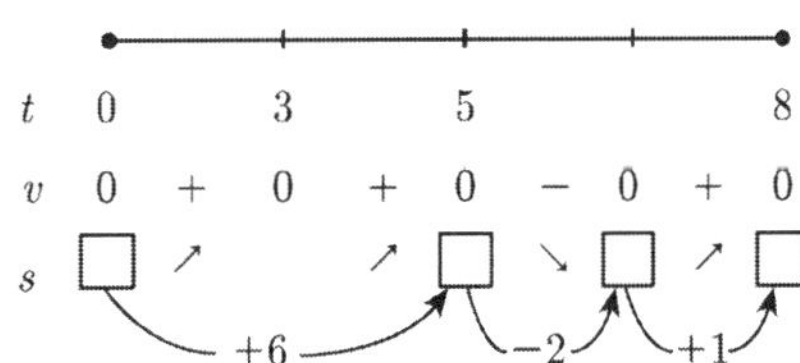

EX4

(1) Since $v(t) < 0$ for $0 < t < 3$ and $5 < t < 6$, and $v(t) > 0$ for $3 < t < 5$, we consider $t = 3$ and $t = 6$.

$x(3) = -2 + \int_0^3 v(t)dt = -2-8 = -10$

$x(6) = -2 + \int_0^6 v(t)dt = -2-8+6-2 = -9$

Therefore, the particle is farthest left at time $t = 3$ when its position is $x(3) = -10$.

(2) The particle moves continuously and monotonically from $x(0) = -2$ to $x(3) = -10$. Similarly, the particle moves continuously and monotonically from $x(3) = -10$ to $x(5) = -7$ and also from $x(5) = -7$ to $x(6) = -9$.

By the Intermediate Value Theorem, there are three values of t for which the particle is at $x(t) = -8$.

(3) The speed is decreasing on the interval $2 < t < 3$ since on this interval $v < 0$ and v is increasing.

(4) The acceleration is negative on the intervals $0 < t < 1$ and $4 < t < 6$ since velocity is decreasing on these intervals.

4.7 Motion Along a Curve: Velocity and Acceleration Vectors

EX1

(1) (E) Since $x=3\cos\frac{\pi}{3}t$ and $y=2\sin\frac{\pi}{3}t$,

we note that $\left(\frac{x}{3}\right)^2+\left(\frac{y}{2}\right)^2=1$.

(2) (A) Note that $y=-\pi\sin\frac{\pi}{3}ti+\frac{2\pi}{3}\cos\frac{\pi}{3}tj$.

At $t=3$.

$|v|=\sqrt{(-\pi\cdot 0)^2+\left(\frac{2\pi}{3}\cdot(-1)\right)^2}=\frac{2}{3}\pi$

(3) (B) $a=-\frac{\pi^2}{3}\cos\frac{\pi}{3}ti-\frac{2\pi^2}{9}\sin\frac{\pi}{3}tj$. At $t=3$,

$|a|=\sqrt{\left(-\frac{\pi^2}{3}\cdot -1\right)^2+\left(-\frac{2\pi^2}{9}\cdot 0\right)^2}=\frac{\pi^2}{3}$

(4) (D) he slope of the curve is the slope of v,

namely, $\frac{dy/dt}{dx/dt}$.

At $t=\frac{1}{2}$, the slope is equal to

$$\frac{\frac{2\pi}{3}\cdot\cos\frac{\pi}{6}}{-\pi\cdot\sin\frac{\pi}{6}}=-\frac{2}{3}\cot\frac{\pi}{6}=-\frac{2\sqrt{3}}{3}$$

EX2

(1) Eliminating t yields the equation

$y=-\frac{1}{4}x^2+2x$.

(2) $|v|=\sqrt{\left(\frac{dx}{dt}\right)^2+\left(\frac{dy}{dt}\right)^2}=\sqrt{2^2+(4-2t)^2}$.

(3) Since $|v|=2\sqrt{t^2-4t+5}$.

$\frac{d|v|}{dt}=\frac{2t-4}{\sqrt{t^2-4t+5}}=0$

if $t=2$. We note that as t increases through 2, the signs of $|v|'$ are -,0,+, assuring a minimum of $|v|$ at $t=2$. Evaluate $|v|$ at $t=2$.

(4) The direction of a is $\tan^{-1}\frac{\frac{d^2y}{dt^2}}{\frac{d^2x}{dt^2}}$. Since

$\frac{d^2x}{dt^2}=0$ and $\frac{d^2y}{dt^2}=-2$, the acceleration is always directed downward. Its magnitude, $\sqrt{0^2+(-2)^2}$, is 2 for all t.

EX3

(1) $\left.\frac{dx}{dt}\right|_{t=2}=\frac{2}{e^2}$

Because $\left.\frac{dx}{dt}\right|_{t=2}>0$, the particle is moving to the right at time $t=2$.

$\left.\frac{dy}{dx}\right|_{t=2}=\frac{dy/dt|_{t=2}}{dx/dt|_{t=2}}=3.055$(or 3.054)

(2) $x=1+\int_2^4\sqrt{(t+2)}/e^t dt$

(3) Speed $=\sqrt{(x'(4))^2+(y'(4))^2}=0.575$ (0.574)

Acceleration$=\langle x''(4),y''(4)\rangle$

$=\langle -0.041,0.989\rangle$

EX4

(1) Speed $=\sqrt{x'(4)^2+y'(4)^2}=2.912$

(2) The slope is 2, so $\frac{\frac{dy}{dt}}{\frac{dx}{dt}}=2$, or

$\ln(t^2+1)=2\arctan\left(\frac{t}{1+t}\right)$. Since $t>0$,

$t=1.35766$. At this time, the acceleration is $\langle x''(t),y''(t)\rangle|_{t=1.35766}=\langle 0.135, 0.955\rangle$.

4.8 Modeling and Optimization

EX1

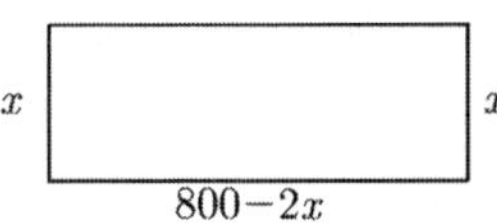

The area is $A(x)=x(800-2x)$, Where $0 \le x \le 400$. Solving $A'(x)=800-4x=0$ $\Rightarrow x=200$. With $A(0)=A(400)=0$, the maximum area is $A(200)=80{,}000m^2$. The dimensions are $200m$ by $400m$.

EX2

(1) -5 or 5

(2) Let the two numbers be x , $18-x$.

$P=x^2 \cdot (18-x)$

$\quad =18x^2-x^3$

$P'=36x-3x^2$

$\quad =3x(12-x), x>0$

$P'=0 \ at \ x=12$

$\therefore 12$ or 6

EX3

(1) $V(\frac{5}{3})=\frac{2450}{27} \approx 91in^3$ must be the maximum volume of the box with dimensions $\frac{14}{3} \times \frac{35}{3} \times \frac{5}{3}$ inches.

(2) 

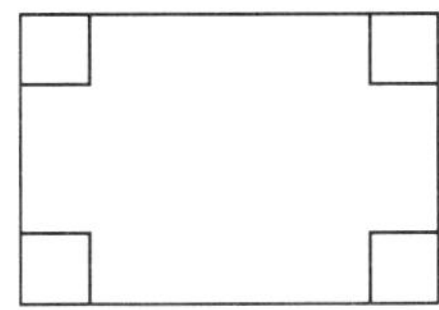

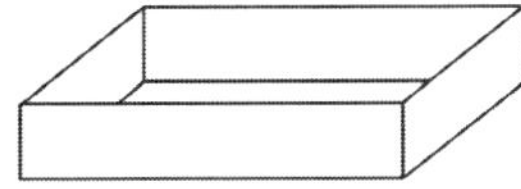

$V=x(10-2x)(16-2x)$

$\quad =4x^3+52x^2+160x$

$V'=12x^2-104x+160$

$\quad =4(3x^2-26x+40)$

$\quad =4(x-2)(3x-20), 0<x<5$

$V'=0 \ at \ x=2$

$\therefore 12cm \ by \ 6cm \ by \ 2cm$

EX4

(1) $300cm^2$

(2)

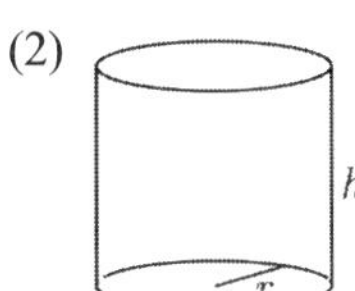

A volume $V=\pi r^2 h=1000 \Rightarrow h=\frac{1000}{\pi r^2}$.

The amount of material is the surface area given by the sides and bottom of the can $\Rightarrow S=2\pi rh+\pi r^2=\frac{2000}{r}+\pi r^2$, $0<r$. Then $\frac{dS}{dr}=-\frac{2000}{r^2}+2\pi r=0 \Rightarrow \frac{\pi r^3-1000}{r^2}=0$. The critical points are 0 and $\frac{10}{\sqrt[3]{\pi}}$,but 0 is not in the domain. Since $\frac{d^2S}{dr^2}=\frac{4000}{r^3}+2\pi>0$, we have a minimum surface area when $r=\frac{10}{\sqrt[3]{\pi}}$ cm and $h=\frac{1000}{\pi r^2}=\frac{10}{\sqrt[3]{\pi}}$ cm.

(3) $4ft \ by \ 4ft \ by \ 16ft$

(4)

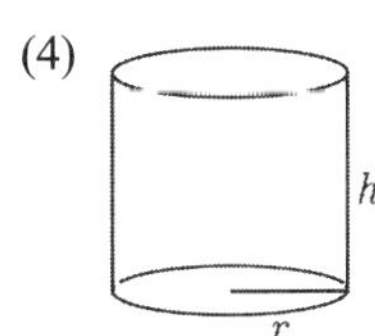

$V=\pi r^2 h=1000$

$h=\frac{1000}{\pi r^2} \cdots\cdots$ ①

$A=2\pi r^2+2\pi rh \cdots\cdots$ ②

① → ②

$A=2\pi r^2+2\pi r\left(\frac{1000}{\pi r^2}\right)$

$\quad =2\pi r^2+\frac{2000}{r}$

$A'=4\pi r-\frac{2000}{r^2}$

$\quad =\frac{4\pi r^3-2000}{r^2}$

$\quad =\frac{4(\pi r^3-500)}{r^2}$

$A' = 0 \Rightarrow \pi r^3 - 500 = 0$

$$r^3 = \frac{500}{\pi}$$

$$r = \sqrt[3]{\frac{500}{\pi}} = 5.41926$$

$$h = \frac{1000}{\pi(5.41926)^2} = 10.8385$$

$\therefore radius = 5.419cm$

$\quad height = 10.839cm$

EX5

(1) $\frac{24\pi}{4+\pi}$

(2)

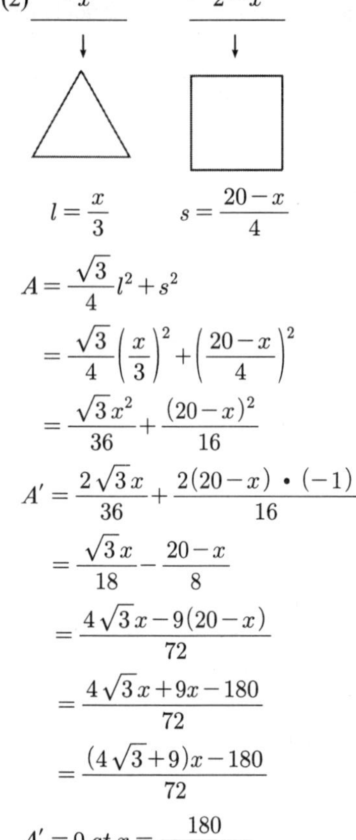

$l = \frac{x}{3} \qquad s = \frac{20-x}{4}$

$$A = \frac{\sqrt{3}}{4}l^2 + s^2 = \frac{\sqrt{3}}{4}\left(\frac{x}{3}\right)^2 + \left(\frac{20-x}{4}\right)^2 = \frac{\sqrt{3}x^2}{36} + \frac{(20-x)^2}{16}$$

$$A' = \frac{2\sqrt{3}x}{36} + \frac{2(20-x)\cdot(-1)}{16} = \frac{\sqrt{3}x}{18} - \frac{20-x}{8} = \frac{4\sqrt{3}x - 9(20-x)}{72} = \frac{4\sqrt{3}x + 9x - 180}{72} = \frac{(4\sqrt{3}+9)x - 180}{72}$$

$$A' = 0 \ at \ x = \frac{180}{4\sqrt{3}+9} = 11.301$$

EX6

(1) $10\sqrt{2}$ *by* $10\sqrt{2}$ *cm*

(2) Let $(x, y) = (x, \sqrt{9-x^2})$ be the coordinates of the corner that intersects the semicircle. Then base $= 2x$ and height $= y = \sqrt{9-x^2}$, thus the area of the inscribed rectangle is given by

$A = (2x)\sqrt{9-x^2}, 0 \le x \le 3$. Then

$$A' = 2\sqrt{9-x^2} + (2x)\frac{-x}{\sqrt{9-x^2}} = \frac{2(9-x^2) - 2x^2}{\sqrt{9-x^2}} = \frac{18-4x^2}{\sqrt{9-x^2}},$$

$A' = 0 \Rightarrow 18 - 4x^2 = 0 \Rightarrow x = \pm\frac{3\sqrt{2}}{2}$,

only $x = \frac{3\sqrt{2}}{2}$ lies in $0 \le x \le 3$. A is continuous on the closed interval $0 \le x \le 3$ $\Rightarrow$ A has an absolute maxima and absolute minima. $A(0) = 0$, $A(3) = 0$, and $A(\frac{3\sqrt{2}}{2}) = (3\sqrt{2})(\frac{3\sqrt{2}}{2}) = 9 \Rightarrow$absolute maxima. Base of rectangle is $3\sqrt{2}$ and height is $\frac{3\sqrt{2}}{2}$.

(3) $4000\pi cm^3$

(4)

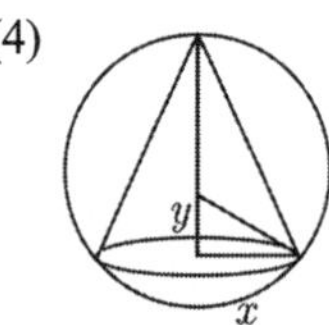

The volume of the cone is $V = \frac{1}{3}\pi r^2 h$, where $r = x = \sqrt{9-y^2}$ and $h = y + 3$

Thus,

$$V(y) = \frac{\pi}{3}(9-y^2)(y+3) = \frac{\pi}{3}(27 + 9y - 3y^2 - y^3)$$

$V'(y) = \frac{\pi}{3}(9-6y-3y^2) = \pi(1-y)(3+y).$

The critical points are -3 and 1, but -3 is not in the domain. Thus

$V''(1) = \frac{\pi}{3}(-6-6(1)) < 0 \Rightarrow$ at $y=1$ we have a maximum volume of

$V(1) = \frac{\pi}{3}(8)(4) = \frac{32\pi}{3}$ cubic units.

EX7

(1)

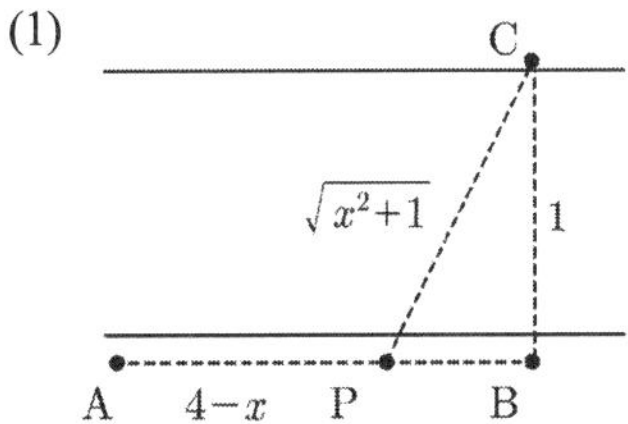

$\$50/m = \$50000/km$

$\$30/m = \$30000/km$

$$C = 50000\sqrt{x^2+1} + 30000(4-x)$$

$$C' = 50000 \cdot \frac{1}{2\sqrt{x^2+1}} \cdot 2x + 30000(-1)$$

$$= \frac{50000x - 30000\sqrt{x^2+1}}{\sqrt{x^2+1}}$$

$$= \frac{10000(5x - 3\sqrt{x^2+1})}{\sqrt{x^2+1}}$$

$$C' = 0 \Rightarrow 5x - 3\sqrt{x^2+1} = 0$$

$$5x = 3\sqrt{x^{2+1}}$$

$$25x^2 = 9(x^2+1)$$

$$25x^2 = 9x^2 + 9$$

$$16x^2 = 9$$

$$x^2 = \frac{9}{16},\ x > 0 \Rightarrow\ \therefore x = \frac{3}{4}$$

(2)

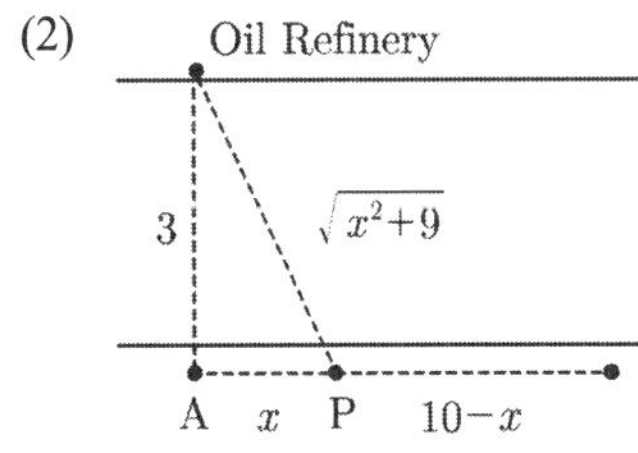

$\$20/m = \$20000/km$

$\$25/m = \$25000/km$

$$C = 25000\sqrt{x^2+9} + 20000(10-x)$$

$$C' = 25000 \cdot \frac{1}{2\sqrt{x^2+9}}(2x) + 20000(-1)$$

$$= \frac{25000x - 20000\sqrt{x^2+9}}{\sqrt{x^2+9}}$$

$$\frac{5000(5x - 4\sqrt{x^2+9})}{\sqrt{x^2+9}}$$

$$C' = 0 \Rightarrow 5x - 4\sqrt{x^2+9} = 0$$

$$5x = 4\sqrt{x^2+9}$$

$$25x^2 = 16(x^2+9)$$

$$25x^2 = 16x^2 + 144$$

$$25x^2 = 16(x^2+9)$$

$$9x^2 = 144$$

$$x^2 = 16, x > 0 \Rightarrow x = 4$$

(3)

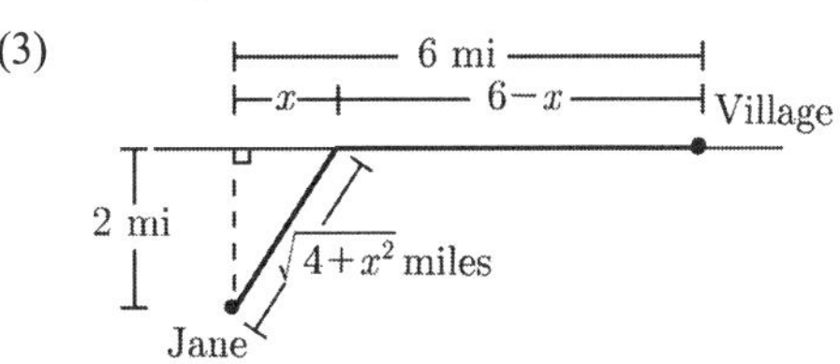

Let x be this distance from the point on the shoreline nearest Jane's boat to the point where she lands her boat. Then she needs to row $\sqrt{4+x^2}$ mi at 2 mph and walk $6-x$ at 5 mph. The total amount of time to reach the village is

$$f(x) = \frac{\sqrt{4+x^2}}{2} + \frac{6-x}{5} \text{ hours } (0 \le x \le 6).$$

Then $f'(x) = \frac{1}{2}\frac{1}{2\sqrt{4+x^2}}(2x) - \frac{1}{5}$

$$= \frac{x}{2\sqrt{4+x^2}} - \frac{1}{5}.$$

Solving $f'(x) = 0$, We have :

$$\frac{x}{\sqrt{4+x^2}} = \frac{1}{5}$$

$$\Rightarrow 5x = 2\sqrt{4+x^2}$$

$$\Rightarrow 25x^2 = 4(4+x^2)$$

$$\Rightarrow 21x^2 = 16 \Rightarrow x \pm \frac{4}{\sqrt{21}}.$$

We discard the negative value of x because it is not in the domain. Checking the endpoints and critical point, we have $f(0) = 2.2 f(\frac{4}{\sqrt{21}}) \approx 2.12$, and

$f(6) \approx 3.16$. Jane should land her boat $\frac{4}{\sqrt{21}} \approx 0.87$ miles down the shoreline from the point nearest her boat.

EX8

(1) $(4,4)$

(2)

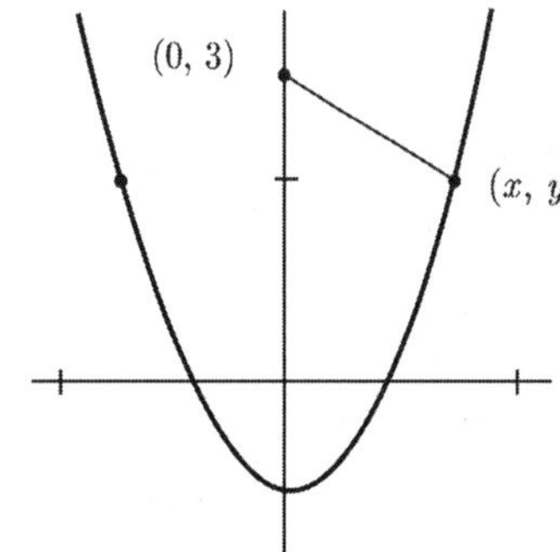

$y=\frac{1}{4}x^2-1$

$4y=x^2-4$

$x^2=4y+4 \cdots\cdots ①$

Distance Formula

$l=\sqrt{(x-0)^2+(y-3)^2}$

$=\sqrt{x^2+y^2-6y+9} \cdots\cdots ②$

① → ②

$l=\sqrt{(4y+4)+y^2-6y+9}$

$=\sqrt{y^2-2y+13}$

$l'=\frac{1}{2\sqrt{y^2-2y+13}}(2y-2)$

$l'=0 \ at \ y=1$

$\therefore (\pm 2\sqrt{2},1)$

(3) $2\sqrt{3}$

(4) $x^2-y^2=-1$

$y^2=x^2+1 \cdots\cdots ①$

$l=\sqrt{(x-4)^2+y^2} \cdots\cdots ②$

① → ②

$l=\sqrt{x^2-8x+16+x^2+1}$

$=\sqrt{2x^2-8x+17}$

$l'=\frac{1}{\sqrt{2x^2-8x+17]}}(4x-8)$

$l'=0 \ at \ x=2$

$\therefore (2, \pm\sqrt{5})$

(5) $(2,1)$

EX9

(1) $6\sqrt{3}$

(2)

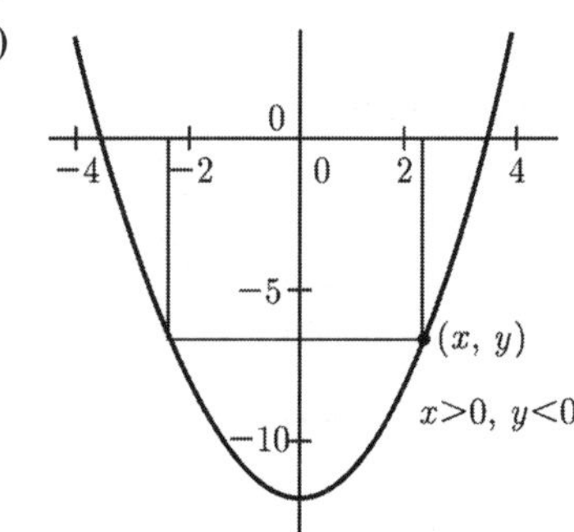

$A=2x \cdot (-y)$

$=-2xy$

$=-2x(x^2-12)$

$=-2x^3+24x$

$A'=-6x^2+24$

$=-6(x^2-4), \ x>0$

$A'=0 \ at \ x=2$

$A=32$

(3)

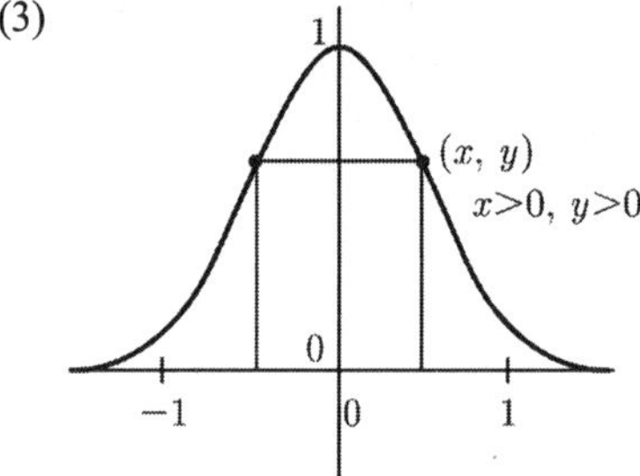

$A=2x \cdot y$

$=2xe^{-2x^2}$

$A'=2e^{-2x^2}+2xe^{-2x^2}(-4x)$

$=2e^{-2x^2}-8x^2e^{-2x^2}$

$=2e^{-2x^2}(1-4x^2), \ x>0$

$A'=0 \ at \ x=\frac{1}{2}$

$A(\frac{1}{2})=\frac{1}{\sqrt{e}}$

4.9 Related Rates

EX1

(1)

(a) $\dfrac{dV}{dt}=1\ volt/\sec$

(b) $\dfrac{dI}{dt}=-\dfrac{1}{3}\ amp/\sec$

(c) $\dfrac{dV}{dt}=R\left(\dfrac{dI}{dt}\right)+I\left(\dfrac{dR}{dt}\right) \Rightarrow \dfrac{dR}{dt}=\dfrac{1}{I}\left(\dfrac{dV}{dt}-\dfrac{V}{I}\dfrac{dI}{dt}\right)$

(d) $\dfrac{dR}{dt}=\dfrac{1}{2}\left[1-\dfrac{12}{2}\left(-\dfrac{1}{3}\right)\right]=\dfrac{3}{2}\ \Omega s/\sec,$

R is increasing

EX2

(1)

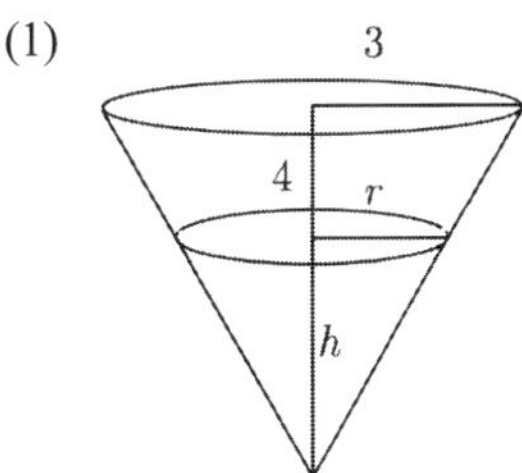

$\dfrac{dV}{dt}=18\pi\ m^3/\min$

$\left.\dfrac{dh}{dt}\right|_{h=2}=?$

$\dfrac{3}{4}=\dfrac{r}{h},\ 4r=\dfrac{3}{4}h\cdots\cdots$ ①

$V=\dfrac{1}{3}\pi r^2 h\cdots\cdots$ ②

① → ②

$V=\dfrac{\pi}{3}\left(\dfrac{3}{4}h\right)^2\cdot h$

$=\dfrac{3\pi}{16}h^3$

$\dfrac{dV}{dt}=\dfrac{9\pi}{16}h^2\cdot\dfrac{dh}{dt}$

$18\pi=\dfrac{9\pi}{16}(2)^2\cdot\dfrac{dh}{dt}$

$\therefore\ \left.\dfrac{dh}{dt}\right|_{h=2}=8\ m/\min$

(2)

$V=\dfrac{1}{3}\pi r^2 h$ and

$r=\dfrac{15h}{2} \Rightarrow V=\dfrac{1}{3}\pi\left(\dfrac{15h}{2}\right)^2 h=\dfrac{75\pi h^3}{4}$

$\Rightarrow \dfrac{dV}{dt}=\dfrac{225\pi h^2}{4}\dfrac{dh}{dt},\ -16\pi=\dfrac{225\pi}{4}h^2\dfrac{dh}{dt}$

$\Rightarrow \left.\dfrac{dh}{dt}\right|_{h=5}=\dfrac{-64}{225\times(5)^2}$

$\approx -0.0113m/\min$

$r=\dfrac{15h}{2} \Rightarrow \dfrac{dr}{dt}=\dfrac{15}{2}\dfrac{dh}{dt}$

$\Rightarrow \left.\dfrac{dr}{dt}\right|_{h=5}=\dfrac{15}{2}\times(-0.0113)\approx -0.0849\mathrm{m/sec}$

(3) If $V=\dfrac{4}{3}\pi r^3, r=5,$ and $\dfrac{dV}{dt}=100\pi\ ft^3/\min,$

then $\dfrac{dV}{dt}=4\pi r^2\dfrac{dr}{dt} \Rightarrow \dfrac{dr}{dt}=1\ ft/\min.$ Then

$S=4\pi r^2 \Rightarrow \dfrac{dS}{dt}=8\pi r\dfrac{dr}{dt}$

$=8\pi(5)(1)=40\pi\ ft^2/\min.$ the rate at which the surface area is increasing.

EX3

Given: $\dfrac{dx}{dt}=5ft/\sec,$ the ladder is 13 ft long, and $x=12, y=5$ at the instant of time

(a) Since $x^2+y^2=169$

$\Rightarrow \dfrac{dy}{dt}=-\dfrac{x}{y}\dfrac{dx}{dt}=-\left(\dfrac{12}{5}\right)(5)=-12ft/\sec,$

the ladder is sliding down the wall

(b) The area of the triangle formed by the ladder and walls is

$A=\dfrac{1}{2}xy \Rightarrow \dfrac{dA}{dt}=\left(\dfrac{1}{2}\right)\left(x\dfrac{dy}{dt}+y\dfrac{dx}{dt}\right).$

The area is changing at

$\dfrac{1}{2}[12(-12)+5(5)]=-\dfrac{119}{2}=-59.5ft^2/\sec.$

(c) $\cos\theta = \frac{x}{13} \Rightarrow -\sin\theta\frac{d\theta}{dt} = \frac{1}{13} \cdot \frac{dx}{dt}$

$\Rightarrow \frac{d\theta}{dt} = -\frac{1}{13\sin\theta} \cdot \frac{dx}{dt}$

$= -(\frac{1}{5})(5)$

$= -1\,rad/\sec$

EX4

(1) Let s represent the length of the rope and x the horizontal distance of the boat from the dock.

(a) We have

$s^2 = x^2 + 36 \Rightarrow \frac{dx}{dt} = \frac{s}{x}\frac{ds}{dx} = \frac{s}{\sqrt{s^2-36}}\frac{ds}{dt}.$

Therefore, the boat is approaching the dock at

$\left.\frac{dx}{dt}\right|_{x=10} = \frac{10}{\sqrt{10^2-36}}(-2) = -2.5ft/\sec.$

(b) $\cos\theta = \frac{6}{s} \Rightarrow -\sin\theta\frac{d\theta}{dt} = -\frac{6}{s^2}\frac{ds}{dt}$

$\Rightarrow \frac{d\theta}{dt} = \frac{6}{s^2\sin\theta}\frac{ds}{dt}.$

Thus, $s = 10, x = 8$, and

$\sin\theta = \frac{8}{10} \Rightarrow \frac{d\theta}{dt} = \frac{6}{10^2(\frac{8}{10})} \cdot (-2)$

$= -\frac{3}{20}\text{rad/sec}$

(2) Let s = distance of car from foot of perpendicular in the textbook diagram

$\Rightarrow \tan\theta = \frac{s}{132} \Rightarrow \sec^2\theta\frac{d\theta}{dt} = \frac{1}{132}\frac{ds}{dt}$

$\Rightarrow \frac{d\theta}{dt} = \frac{\cos^2\theta}{132}\frac{ds}{dt}; \frac{ds}{dt} = -264$ and

$\theta = 0 \Rightarrow \frac{d\theta}{dt} = -2rad/\sec$. A half second later the car has traveled 132 ft right of the perpendicular

$\Rightarrow |\theta| = \frac{\pi}{4}, \cos^2\theta = \frac{1}{2}$, and $\frac{ds}{dt} = 264$

(since s is increases)

$\Rightarrow \frac{d\theta}{dt} = \frac{\left(\frac{1}{2}\right)}{132}(264) = 1\,rad/\sec.$

EX5

(1) (a)

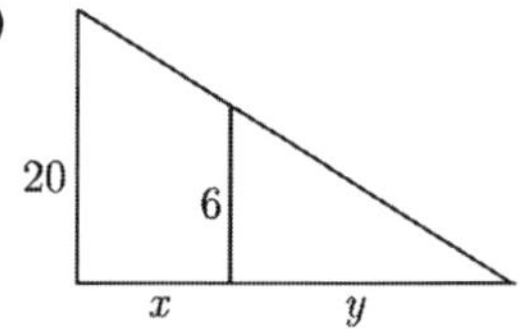

$\frac{dx}{dt} = -3ft/\sec, \frac{dy}{dt} = ?$

$\frac{6}{20} = \frac{y}{x+y}$

$6x + 6y = 20y$

$6x = 14y$

$3x = 7y$

$3\frac{dx}{dt} = 7\frac{dy}{dt}$

$\frac{dy}{dt} = \frac{3}{7}\frac{dx}{dt} = \frac{3}{7}(-3)$

$= -\frac{9}{7}ft/\sec$

(b)

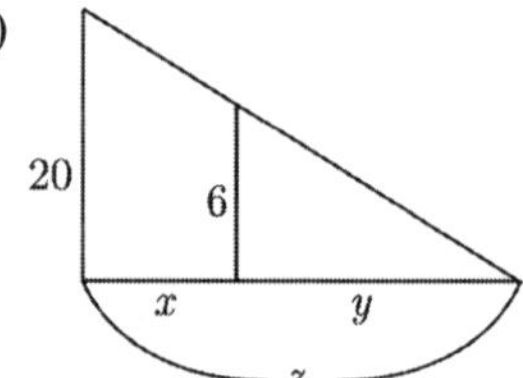

$\frac{dz}{dt} = ?$

$z = x + y$

$\frac{dz}{dt} = \frac{dx}{dt} + \frac{dy}{dt}$

$= -3 - \frac{9}{7}$

$= -\frac{30}{7}ft/\sec$

(2) Let s represent the distance between the bicycle and balloon, h the height of the balloon and x the horizontal distance between the galloon and the bicycle. The relationship between the variables is

$s^2 = h^2 + x^2$

$\Rightarrow \frac{dx}{dt} = \frac{1}{s}(h\frac{dh}{dt} + x\frac{dx}{dt})$

$\Rightarrow \frac{ds}{dt} = \frac{1}{85}[68(1) + 51(17)] = 11ft/\sec.$

(3)

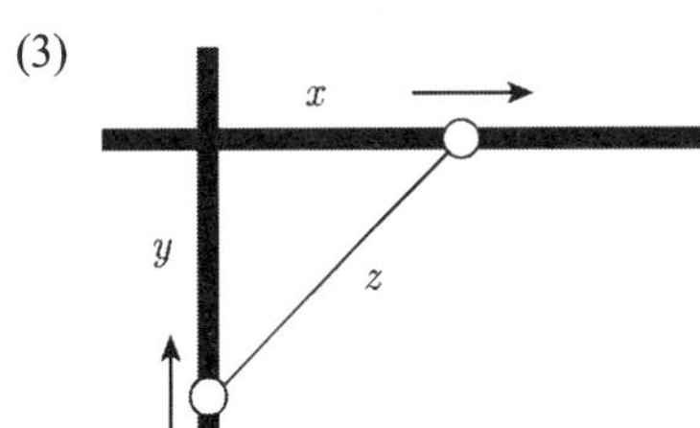

$\frac{dy}{dt}=-30mph,\ \frac{dx}{dt}=50mph$

$\left.\frac{dz}{dt}\right|_{t=1}=?$

$x^2+y^2=z^2$

$2x\frac{dx}{dt}=2y\frac{dy}{dt}=2z\frac{dz}{dt}$

$\frac{dz}{dt}=\frac{x\frac{dx}{dt}+y\frac{dy}{dt}}{z}$

When $t=1$,

$x=120,\ y=50$

$z=\sqrt{(120)^2+(50)^2}=130$

$\therefore \left.\frac{dz}{dt}\right|_{t=1}=\frac{(120)(50)+(50)(-30)}{130}$

$=\frac{450}{13}mph$

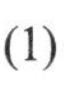

EX6

(1)

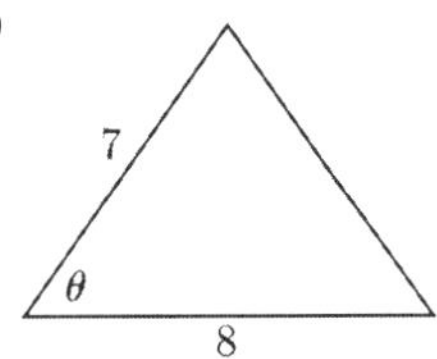

$\frac{d\theta}{dt}=-\frac{\pi}{14}\frac{rad}{\sec}$

$\left.\frac{dA}{dt}\right|_{\theta=\frac{\pi}{3}}=?$

$A=\frac{1}{2}\cdot 7\cdot 8\cdot \sin\theta$

$=28\sin\theta$

$\left.\frac{dA}{dt}\right|_{\theta=\frac{\pi}{3}}=28\cos\theta\cdot\frac{d\theta}{dt}$

$=28\cos\left(\frac{\pi}{3}\right)\cdot\left(-\frac{\pi}{14}\right)$

$=-\pi cm^2/\sec$

(2)

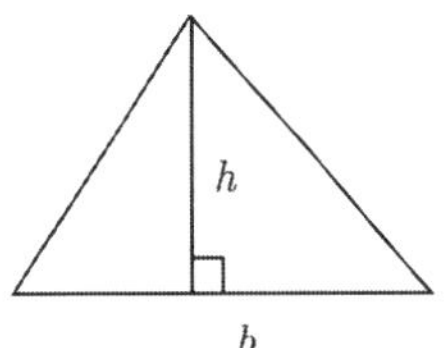

$\frac{dh}{dt}=2cm/\sec$

$\frac{db}{dt}=-3cm/\sec$

$\frac{dA}{dt}=?$

$A=\frac{1}{2}bh$

$\frac{dA}{dt}=\frac{1}{2}\frac{db}{dt}\cdot h+\frac{1}{2}b\cdot\frac{dh}{dt}$

When $b=8$ and $A=40,\ h=10$

$\frac{dA}{dt}=\frac{1}{2}(-3)(10)+\frac{1}{2}(8)(2)$

$=-7\,cm^2/\sec$

(3)

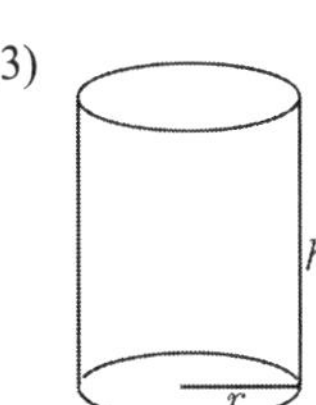

$\frac{dh}{dt}=6cm/\min,\ \frac{dV}{dt}=0,\ \left.\frac{dr}{dt}\right|_{h=15}=?$

$V=\pi r^2h$

$\frac{dV}{dt}=2\pi r\cdot\frac{dr}{dt}\cdot h+\pi r^2\cdot\frac{dh}{dt}$

When $h=15$,

$240\pi=\pi r^2(15)$

$16=r^2,\ r>0$

$\therefore\ r=4$

$0=2\pi(4)\frac{dr}{dt}(15)+\pi(4)^2\cdot(6)$

$0=120\pi\frac{dr}{dt}+96\pi$

$\therefore \left.\frac{dr}{dt}\right|_{h=15}=-\frac{4}{5}cm/\min$

EX7

(1) $y=\sqrt{x}$

$$\left.\frac{dy}{dt}\right|_{x=4}=\frac{1}{2\sqrt{x}}\bullet\frac{dx}{dt}$$
$$=\frac{1}{2\sqrt{4}}\bullet 4$$
$$=1unit/\sec$$

(2) $z^2=x^2+y^2$

$$2z\frac{dz}{dt}=2x\frac{dx}{dt}+2y\frac{dy}{dt}$$
$$\frac{dz}{dt}=\frac{x\frac{dx}{dt}+y\frac{dy}{dt}}{z}$$

$When x=4,\ y=\sqrt{4}=2$

$$\left.\frac{dz}{dt}\right|_{x=4}=\frac{(4)(4)+(2)(1)}{\sqrt{16+4}}$$
$$=\frac{18}{2\sqrt{5}}=\frac{9}{\sqrt{5}}\ unit/\sec$$

(3) $\tan\theta=\frac{y}{x}=\frac{\sqrt{x}}{x}=x^{-\frac{1}{2}}$

$$\sec^2\theta\bullet\frac{d\theta}{dt}=-\frac{1}{2}x^{-\frac{3}{2}}\bullet\frac{dx}{dt}$$

$When\ x=4 and\ y=2,\ z=2\sqrt{5}$

$$\therefore\cos\theta=\frac{4}{2\sqrt{5}}=\frac{2}{\sqrt{5}}$$
$$\left(\frac{\sqrt{5}}{2}\right)^2\bullet\frac{d\theta}{dt}=-\frac{1}{2}(4)^{-\frac{3}{2}}\bullet 4$$
$$\left.\frac{d\theta}{dt}\right|_{x=4}=-\frac{5}{4}\bullet\frac{1}{2}\bullet\frac{1}{8}\bullet 4$$
$$=-\frac{1}{5}rad/\sec$$

EX8

Let a represent the distance between point O and ship A, b the distance between point O and ship B, and D the distance between the ships, By the Law of Cosines,

$$D^2=a^2+b^2-2ab\cos120^\circ$$
$$\Rightarrow\frac{dD}{dt}=\frac{1}{2D}\left[2a\frac{da}{dt}+2b\frac{db}{dt}+a\frac{db}{dt}+b\frac{da}{dt}\right]$$

When $a=5,\frac{da}{dt}=14,b=3,$and $\frac{db}{dt}=21$, then

$\frac{dD}{dt}=\frac{413}{2D}$ where $D=7$. The ships are moving $\frac{dD}{dt}=29.5$ knots apart.

4.10 Mean Value Theorem(M.V.T)

EX1

(1) $c=\frac{1}{2}$

(2) When $f(x)=x^{2/3}$ for $0\le x\le 1$, then

$$\frac{f(1)-f(0)}{1-0}=f'(c)\Rightarrow 1=\left(\frac{2}{3}\right)c^{-1/3}\Rightarrow c=\frac{8}{27}.$$

(3) $c=1$

(4) When $f(x)=\sqrt{x-1}$ for $1\le x\le 3$, then

$$\frac{f(3)-f(1)}{3-1}=f'(c)\Rightarrow\frac{\sqrt{2}}{2}=\frac{1}{2\sqrt{c-1}}\Rightarrow c=\frac{3}{2}$$

(5) $c=\pm\sqrt{1-\frac{4}{\pi^2}}\approx\pm 0.771$

(6) When $f(x)=\ln(x-1)$for $2\le x\le 4$, then

$$f'(c)=\frac{f(4)-f(2)}{4-2}\Rightarrow\frac{1}{c-1}=\frac{\ln3-\ln1}{2}$$
$$\Rightarrow c-1=\frac{2}{\ln3}$$
$$\Rightarrow c=1+\frac{2}{\ln3}\approx 2.820$$

(7) $\frac{1+\sqrt{7}}{3}\approx 1.22$ and $\frac{1-\sqrt{7}}{3}\approx -0.549$ are both in the interval $-1\le x\le 2$.

(8) When $g(x)=\begin{cases}x^3 & -2\le x\le 0\\ x^2 & 0<x\le 2\end{cases}$, then

$\frac{g(2)-g(-2)}{2-(-2)}=g'(c)$. If $-2\le x<0$, then

$g'(x)=3x^2\Rightarrow 3=g'(c)\Rightarrow 3c^2=3\Rightarrow c=\pm1$.

Only $c=-1$ is in the interval, If $0<x\le 2$, then $g'(x)=2x\Rightarrow 3=g'(c)\Rightarrow 2c=3\Rightarrow c=\frac{3}{2}$.

EX2

(1)

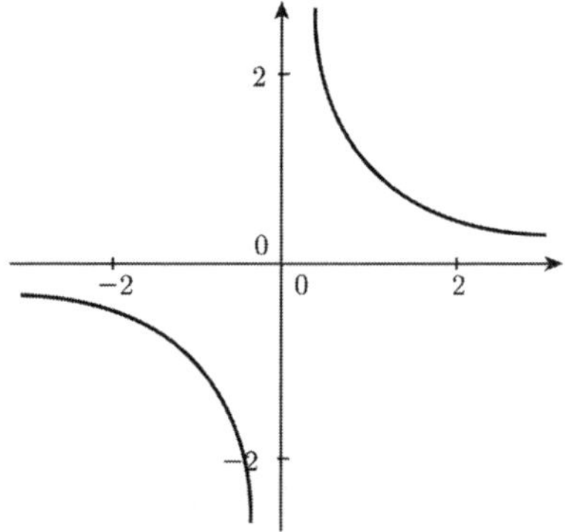

$f(x)$ is discontinuous at $x=0$.

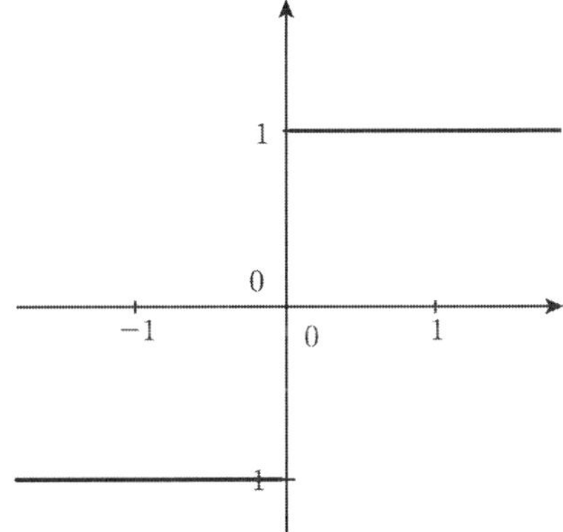

$f(x)$ is discontinuous at $x=0$.

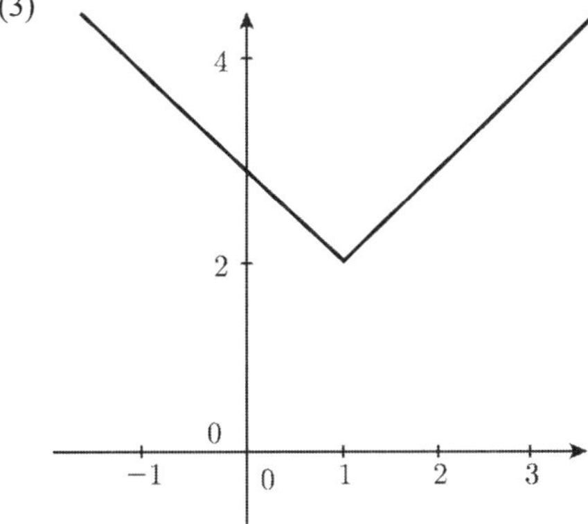

$f(x)$ is not differentiable at $x=1$

(4)

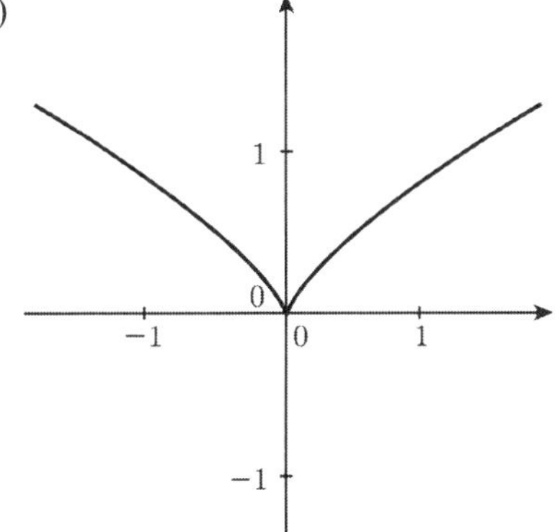

$f(x)$ is not differentiable at $x=0$.

EX3

(1) $f'(x)=2^x\ln2-\cos x$

$$f'(c)=\frac{f(\pi)-0}{\pi-0}$$

$$2^c\ln2-\cos c=\frac{(2^\pi-0)-(2^0-0)}{\pi}$$

$$2^c\ln2-\cos c=\frac{2^\pi-1}{\pi},\ 0<c<\pi$$

$\therefore c=1.743$

(2) $f'(x)=3\cos^{-1}x+3x\left(\dfrac{-1}{\sqrt{1-x^2}}\right)$

$$f'(c)=\frac{f(\frac{1}{2})-f(0)}{\frac{1}{2}-0}$$

$$3\cos^{-1}c-\frac{3c}{\sqrt{1-c^2}}=\frac{\frac{3}{2}\cdot\frac{\pi}{3}-0}{\frac{1}{2}}$$

$$3\cos^{-1}c-\frac{3c}{\sqrt{1-c^2}}=\pi,\ 0<c<\frac{1}{2}$$

$\therefore c=0.256$

(3) Note $f(0)=f(\sqrt{3})=0$ and $f'(x)$ exists on the given interval, By the MVT, there is a number, c,in the interval such that $f'(c)=0$. If $c=1$,then $6c^2-6=0$.(-1 is not in the interval.) $c=\pm1$

$\therefore c=1$(-1 is not in the interval.)

(4) Since $f(x)=3^x-x^3$then $f'(x)=3^x\ln3-3x^2$. Furthermore, f is continuous on $[0,3]$ and f' is differentiable on $(0,3)$, so MVT applies. We therefore seek c such that

$f'(c)=\dfrac{f(3)-f(0)}{3}=-\dfrac{1}{3}$.Solving

$3^c\ln3-3c^2=-\dfrac{1}{3}$ with calculator, we find that c may be either 1.244 or 2.727. These values are the x -coordinates of points on the graph of $f(x)$ at which the tangents are parallel to the secand through points (0,1) and (3,0) on the curve.

4.11 Local Linear Approximation

EX1]

(1) $f(x) = \sqrt{x}$

$f(16) = \sqrt{16} = 4 \Rightarrow (16,4)$

$f'(x) = \dfrac{1}{2\sqrt{x}}$

$f'(16) = \dfrac{1}{2\sqrt{16}} = \dfrac{1}{8}$

$y-4 = \dfrac{1}{8}(x-16)$

$y = \dfrac{1}{8}(x-16)+4$

$\sqrt{x} \approx \dfrac{1}{8}(x-16)+4 \quad at\ x=16$

$\sqrt{16.08} \approx \dfrac{1}{8}(0.08)+4 = 4.01$

(2) $f(x) = \sqrt[3]{x}$

$f(8) = \sqrt[3]{8} = 2 \Rightarrow (8,2)$

$f'(x) = \dfrac{1}{3}x^{-\frac{2}{3}}$

$f'(8) = \dfrac{1}{3}(8)^{-\frac{2}{3}} = \dfrac{1}{3} \cdot \dfrac{1}{4} = \dfrac{1}{12}$

$y = \dfrac{1}{12}(x-8)+2$

$\sqrt[3]{x} \approx \dfrac{1}{12}(x-8)+2 \ at\ x=8$

$\sqrt[3]{7.4} \approx \dfrac{1}{12}(-0.6)+2 = 1.95$

(3) $f(x) = \sqrt{x}$

$f(64) = \sqrt{64} = 8 \Rightarrow (64,8)$

$f'(x) = \dfrac{1}{2\sqrt{x}}$

$f'(64) = \dfrac{1}{16}$

$y-8 = \dfrac{1}{16}(x-64)$

$y = \dfrac{1}{16}(x-64)+8$

$\sqrt{x} \approx \dfrac{1}{16}(x-64)+8\ near\ x=64$

$\sqrt{63} \approx \dfrac{1}{16}(-1)+8 = \dfrac{127}{16}(\text{or}7.938)$

(4) $f(x) = x^{10}$

$f(1) = 1 \Rightarrow (1,1)$

$f'(x) = 10x^9$

$f'(1) = 10$

$y-1 = 10(x-1)$

$y = 10(x-1)+1$

$x^{10} \approx 10(x-1)+1\ near\ x=1$

$(1.01)^{10} \approx 10(0.01)+1 = 1.1$

(5) $f(x) = 2^x$

$f(4) = 2^4 = 16 \Rightarrow (4,16)$

$f'(x) = 2^x \cdot \ln 2$

$f'(4) = 16\ln 2$

$y-16 = 16\ln 2(x-4)$

$y = 16\ln 2(x-4)+16$

$2^x \approx 16\ln 2 \cdot (x-4)+16\ near\ x=4$

$2^{3.99} \approx 16\ln 2 \cdot (-0.01)+16$

$\approx 16 - \dfrac{4}{25}\ln 2\ (\text{or}\ 15.889)$

(6) $f(x) = \tan^{-1}x$

$f(1) = \tan^{-1}(1) = \dfrac{\pi}{4} \Rightarrow \left(1, \dfrac{\pi}{4}\right)$

$f'(x) = \dfrac{1}{1+x^2}$

$f'(1) = \dfrac{1}{2}$

$y - \dfrac{\pi}{4} = \dfrac{1}{2}(x-1)$

$y = \dfrac{1}{2}(x-1)+\dfrac{\pi}{4}$

$\tan^{-1}x \approx \dfrac{1}{2}(x-1)+\dfrac{\pi}{4}\ near\ x=1$

$\tan^{-1}(1.01) \approx \dfrac{1}{2}(0.01)+\dfrac{\pi}{4}$

$\approx \dfrac{\pi}{4}+0.005\ (\text{or}\ 0.790)$

EX2

(1) $f(x) = x^3-2x+3$

$\Rightarrow f'(x) = 3x^2-2$

$\Rightarrow L(x) = f'(2)(x-2)+f(2) = 10(x-2)+7$

$\Rightarrow L(x) = 10x-13\ at\ x=2$

(2) $f(x)=\sqrt{x^2+9}=(x^2+9)^{\frac{1}{2}}$

$\Rightarrow f'(x)=\left(\frac{1}{2}\right)(x^2+9)^{-1/2}(2x)$

$=\frac{x}{\sqrt{x^2+9}}$

$\Rightarrow L(x)=f'(-4)(x+4)+f(-4)$

$=-\frac{4}{5}(x+4)+5$

$\Rightarrow L(x)=-\frac{4}{5}x+\frac{9}{5}\ at\ x=-4$

(3) $f(x)=x+\frac{1}{x}$

$\Rightarrow f'(x)=1-x^{-2}$

$\Rightarrow L(x)=f(1)+f'(1)(x-1)$

$=2+0(x-1)=2$

(4) $f(x)=x^{1/3}$

$\Rightarrow f'(x)=\frac{1}{3x^{2/3}}$

$\Rightarrow L(x)=f'(-8)(x-(-8))+f(-8)$

$=\frac{1}{12}(x+8)-2$

$\Rightarrow L(x)=\frac{1}{12}x-\frac{4}{3}$

(5) $f(x)=\tan x$

$\Rightarrow f'(x)=\sec^2 x$

$\rightarrow L(x)=f(\pi)+f'(\pi)(x-\pi)$

$=0+1(x-\pi)=x-\pi$

EX3

(1) $r(5.4)=r(5)+r'(5)\Delta t$

$=30+2(0.4)=30.8ft$

Since the graph of r is concave down on the interval $5<t<5.4$, this estimate is greater than $r(5.4)$.

(2) $\frac{dV}{dt}=3\left(\frac{4}{3}\right)\pi r^2\frac{dr}{dt}.$

$\left.\frac{dV}{dt}\right|_{t=5}=4\pi(30)^2\,2=7200\pi\ ft^3/\text{min}$

EX4

(1) $f(2)=4\Rightarrow(2,4)$

$f'(2)=3$

$y-4=3(x-2)$

$y=3(x-2)+4$

$f(x)\approx 3(x-2)+4\ near\ x=2$

$f(2.10)\approx 3(0.1)+4=4.3$

(2) $f''>0\ on\ 0<x<3\Rightarrow$ The graph of f is concave up.

$\therefore$ The approximate value is less than the actual value.

4.12 Newton's Method

EX1

(1) $x_1=2$
$x_2=1.636363$
$x_3=1.530392$
$x_4=1.521441$
$x_5=1.521380$

(2) $x_1=1$
$x_2=1.285714$
$x_3=1.216457$
$x_4=1.208996$
$x_5=1.208918$

(3) $x_1=2$
$x_2=0.995140$
$x_3=1.030107$
$x_4=1.029867$

(4) $x_1=0$
$x_2=0.5$
$x_3=0.510958$
$x_4=0.510973$

EX2

$y=x^2+x-1\Rightarrow y'=2x+1$

$\Rightarrow x_{n+1}=x_n-\frac{x_n^2+x_n-1}{2x_n+1};\ x_0=1$

$\Rightarrow x_1=1-\frac{1+1-1}{2+1}=\frac{2}{3}$

$\Rightarrow x_2=\frac{2}{3}-\frac{\frac{4}{9}+\frac{2}{3}-1}{\frac{4}{3}+1}$

$=\frac{2}{3}-\frac{4+6-9}{12+9}$

$=\frac{2}{3}-\frac{1}{21}$

$=\frac{13}{21}\approx 0.61905\ ;\ x_0=-1$

$\Rightarrow x_1=1-\frac{1-1-1}{-2+1}=-2$

$\Rightarrow x_2=\ 2\ \ \frac{4-2-1}{-4+1}=-\frac{5}{3}\approx -1.66667$

EX3

$y = x^3 + 3x + 1 \Rightarrow y' = 3x^2 + 3$

$$\Rightarrow x_{n+1} = x_n - \frac{x_n^3 + 3x_n + 1}{3x_n^2 + 3};\ x_0 = 0$$

$$\Rightarrow x_1 = 0 - \frac{1}{3} = -\frac{1}{3}$$

$$\Rightarrow x_2 = -\frac{1}{3} - \frac{-\frac{1}{27} - 1 + 1}{\frac{1}{3} + 3}$$

$$= -\frac{1}{3} + \frac{1}{90} = -\frac{29}{90} \approx -0.32222$$

EX4

$y = x^4 + x - 3 \Rightarrow y' = 4x^3 + 1$

$$\Rightarrow x_{n+1} = x_n - \frac{x_n^4 + x_n - 3}{4x_n^3 + 1};\ x_0 = 1$$

$$\Rightarrow x_1 = 1 - \frac{1 + 1 - 3}{4 + 1} = \frac{6}{5}$$

$$\Rightarrow x_2 = \frac{6}{5} - \frac{\frac{1296}{625} + \frac{6}{5} - 3}{\frac{864}{125} + 1}$$

$$= \frac{6}{5} - \frac{1296 + 750 - 1875}{4320 + 625}$$

$$= \frac{6}{5} - \frac{171}{4945} = \frac{5763}{4945} \approx 1.16542;\ x_0 = -1$$

$$\Rightarrow x_1 = -1 - \frac{1 - 1 - 3}{-4 + 1} = -2$$

$$\Rightarrow x_2 = -2 - \frac{16 - 2 - 3}{-32 + 1} = -2 + \frac{11}{31}$$

$$= -\frac{51}{31} \approx -1.64516$$

4.13 Approximating Derivatives Numerically

EX1

The best approximation to $f'(0.10)$ is

$\frac{f(0.20) - f(0.10)}{0.20 - 0.10}$.

EX2

Use the symmetric difference quotient; then

$$f'(1.5) = \frac{f(1.6) - f(1.4)}{1.6 - 1.4} = \frac{8}{0.2}$$

EX3

$$\frac{5^{2.3} - 5^{1.97}}{2.03 - 1.97} = 40.25158$$

EX4

$$f'(2) = \frac{f(2) - f(1.98)}{2 - 1.98} = \frac{4.00 - 4.10}{0.02} = -5$$

EX5

Since an estimate of the answer for Question [EX4] is $f'(2) = -5$, them

$$(f^{-1})'(4) = \frac{1}{f'(2)} = \frac{1}{-5} = -0.2$$

EX6

(1) $\frac{f(12) - f(0)}{12 - 0} = \frac{45 - 9}{12} = \frac{36}{12} = 3$

(2) $f(0) \approx \frac{f(2) - f(0)}{2 - 10} = \frac{13 - 9}{2} = \frac{4}{2} = 2$

$f'(4) \approx \frac{f(6) - f(2)}{6 - 2} = \frac{28 - 13}{4} = \frac{15}{4}$

$f'(9) \approx \frac{f(10) - f(8)}{10 - 8} = \frac{43 - 37}{2} = 6$

$f'(12) \approx \frac{f(12) - f(10)}{12 - 10} = \frac{45 - 43}{2} = 1$

(3) $(f^{-1})'(19) = \frac{1}{f'(f^{-1}(19))} = \frac{1}{f'(4)} \approx \frac{1}{\frac{15}{4}} = \frac{4}{15}$

EX7

(1) $C'(3.5) \approx \frac{C(4) - C(3)}{4 - 3}$

$= \frac{12.8 - 11.2}{1} = 1.6$ ounces/min

(2) C is differentiable $\Rightarrow$ C is continuous(on the close interval) $\frac{C(4) - C(2)}{4 - 2} = \frac{12.8 - 8.8}{2} = 2$

Therefore, by the Mean Value Theorem. there is at least one time t, $2 < t < 4$ for which $C'(t) = 2$.

(3) $B'(t) = -16(-0.4)e^{-0.4t} = 6.4e^{-0.4t}$

$B'(5) = 6.4e^{-0.4(5)} = \frac{6.4}{e^2}$ ounces/min

4.14 Indeterminate Forms and L'Hospital's Rule

(1) $\frac{1}{4}$

(2) $\lim_{x\to\infty}\frac{5x^3-2x}{7x^3+3} = \lim_{x\to\infty}\frac{15x^2-2}{21x^2}$
$= \lim_{x\to\infty}\frac{30x}{42x}$
$= \lim_{x\to\infty}\frac{30}{42}$
$= \frac{5}{7}$

(3) -16

(4) $\lim_{\theta\to\pi/2}\frac{1-\sin\theta}{1+\cos 2\theta} = \lim_{\theta\to\pi/2}\frac{-\cos\theta}{-2\sin 2\theta}$
$= \lim_{\theta\to\pi/2}\frac{\sin\theta}{-4\cos 2\theta}$
$= \frac{1}{(-4)(-1)}$
$= \frac{1}{4}$

(5) 3

(6) $\lim_{\theta\to 0}\frac{3^{\sin\theta}-1}{\theta} = \lim_{\theta\to 0}\frac{3^{\sin\theta}((\ln 3)(\cos\theta))}{1}$
$= \frac{(3^0)(\ln 3)(1)}{1} = \ln 3$

(7) $\ln 2$

(8) $\lim_{y\to 0}\frac{\sqrt{5y+25}-5}{y} = \lim_{y\to 0}\frac{(5y+25)^{1/2}-5}{y}$
$= \lim_{y\to 0}\frac{\left(\frac{1}{2}\right)(5y+25)^{-1/2}(5)}{1}$
$= \lim_{y\to 0}\frac{5}{2\sqrt{5y+25}} = \frac{1}{2}$

(9) $\ln 2$

(10) $\lim_{x\to 1^+}\left(\frac{1}{x-1}-\frac{1}{\ln x}\right) = \lim_{x\to 1^+}\left(\frac{\ln x-(x-1)}{(x-1)(\ln x)}\right)$
$= \lim_{x\to 1^+}\left(\frac{\frac{1}{x}-1}{(\ln x)+(x-1)\left(\frac{1}{x}\right)}\right)$
$= \lim_{x\to 1^+}\left(\frac{1-x}{(x\ln x)+x-1}\right)$
$= \lim_{x\to 1^1}\left(\frac{-1}{(\ln x+1)+1}\right)$
$= \frac{-1}{(0+1)+1} = -\frac{1}{2}$

(11) 0

(12) $\lim_{\theta\to 0}\frac{\theta-\sin\theta\cos\theta}{\tan\theta-\theta} = \lim_{\theta\to 0}\frac{1+\sin^2\theta-\cos^2\theta}{\sec^2\theta-1}$
$= \lim_{\theta\to 0}\frac{2\sin^2\theta}{\tan^2\theta}$
$= \lim_{\theta\to 0}2\cos^2\theta$
$= 2$

(13) e

(14) The limit leads to the indeterminate form 1^∞

Let $f(x) = (\ln x)^{1/(x-e)} \Rightarrow \ln f(x) = \frac{\ln(\ln x)}{x-e}$

$= \lim_{x\to e^+}\ln f(x) = \lim_{x\to e^+}\frac{\ln(\ln x)}{x-e} = \lim_{x\to e^+}\frac{\left(\frac{1}{x\ln x}\right)}{1}$

$= \frac{1}{e}.$

Therefore $(\ln x)^{1/(x-e)} = \lim_{x\to e^+} f(x)$
$= \lim_{x\to e^+} e^{\ln f(x)} = e^{1/e}$

(15) e

(16) The limit lead to the indeterminate form 1^∞.

Let $f(x) = (e^x+x)^{1/x} \Rightarrow \ln f(x) = \frac{\ln(e^x+x)}{x} \Rightarrow$

$\lim_{x\to 0}\ln f(x) = \lim_{x\to 0}\frac{\ln(e^x+x)}{x}\ \lim_{x\to 0}\frac{e^x+1}{e^x+x} = 2$

Therefore

$\lim_{x\to 0}(e^x+x)^{1/x} = \lim_{x\to 0}f(x) = \lim_{x\to 0}e^{\ln f(x)} = e^2$

(17) 1

(18) $\lim_{x\to 0^+}x(\ln x)^2 = \lim_{x\to 0^+}\left(\frac{(\ln x)^2}{\frac{1}{x}}\right)$

$= \lim_{x\to 0^+}\left(\frac{2(\ln x)\frac{1}{x}}{-\frac{1}{x^2}}\right) = \lim_{x\to 0^+}\left(\frac{2\ln x}{-\frac{1}{x}}\right)$

$= \lim_{x\to 0^+}\left(\frac{\frac{2}{x}}{\frac{1}{x^2}}\right) = \lim_{x\to 0^+}\left(\frac{2x^2}{x}\right) = \lim_{x\to 0^+}(2x) = 0$

(19) 0

Multiple Choice

01 A 02 C 03 C 04 A 05 D 06 C 07 B 08 E 09 A 10 A 11 E 12 A 13 E 14 D 15 B
16 A 17 E 18 C 19 D 20 D 21 E 22 A 23 C 24 B 25 E 26 B

05 Integration

5.1 indefinite integrals

EX1

(1) $= 3x - 2x^2 + \frac{1}{2}x^4 + C$

(2) $= \int 6x^{\frac{7}{3}}dx$

$= \frac{9}{5}x^{\frac{10}{3}} + C$

(3) $= \frac{1}{5}x^5 + \frac{3}{5}x^{\frac{5}{3}} + \frac{2}{x} - \frac{1}{2}x^{\frac{2}{3}} + C$

(4) $= 2\ln|x| + \frac{1}{x} + \sqrt{x} + C$

(5) $= \int \frac{1}{\ln 2} \cdot \frac{1}{x}dx$

$= \frac{\ln|x|}{\ln 2} + C$

(6) $= \int (x^3 - 27)dx$

$= \frac{1}{4}x^4 - 27x + C$

(7) $= \int (x^2 - 1)dx = \frac{1}{3}x^3 - x + C$

(8) $= \int [4\sin^2 x + 4\sin x\cos x + \cos^2 x + \sin^2 x$

$- 4\sin x\cos x + 4\cos^2 x]dx$

$= \int (5\sin^2 x + 5\cos^2 x)dx$

$= \int 5dx$

$= 5x + C$

(9) $= \frac{1}{4}t^4 + \frac{3}{2}t^2 + \frac{2}{7}t^{\frac{7}{2}} + 2t^{\frac{3}{2}} + C$

EX2

(1) $= \frac{1}{2}x^2 - 6x + 12\ln|x| + \frac{8}{x} + C$

(2) $= \int (x^{\frac{3}{2}} - 2x^{\frac{1}{2}})dx$

$= \frac{2}{5}x^{\frac{5}{2}} - 2 \cdot \frac{2}{3}x^{\frac{3}{2}} + C$

$= \frac{2}{5}x^{\frac{5}{2}} - \frac{4}{3}x^{\frac{3}{2}} + C$

(3) $\frac{1}{2}x^2 - \ln|x| + \frac{1}{x} + C$

(4) $\frac{1}{3}x^3 - x - \frac{1}{4x} + C$

(5) $= \int (x^{\frac{1}{2}} + x)x^{\frac{2}{3}}dx$

$= \int (x^{\frac{7}{6}} + x^{\frac{5}{3}})dx$

$= \frac{6}{13}x^{\frac{13}{6}} + \frac{3}{8}x^{\frac{8}{3}} + C$

(6) $= 2y^{\frac{1}{2}} - \frac{2}{3}y^{3/2} + \frac{1}{10}y^{5/2} + C$

EX3

(1) $= e^x + 4x + C$

(2) $= \frac{5^x}{\ln 5} - \frac{7^x}{\ln 7} + C$

(3) $= \frac{10^x}{\ln 10} + 10\ln|x| + \frac{7}{2}x^{-2} + C$

EX4

(1) $= 5\sin x - \cos x + C$

(2) $= 2\ln|\sec x| + 3\ln|\sin x| + C$

(3) $= \int \csc x\, dx$

$= \ln|\csc x - \cot x| + C$

(4) $= \ln|\sec x| - 3\cos x + C$

(5) $= \sec x + 5\sin x + C$

(6) $= \tan x - 4\ln|\csc x - \cot x| + C$

(7) $-\cot x + C$

(8) $= \int \frac{1 - \cos^2 x}{1 - \cos x}dx$

$= \int \frac{(1 - \cos x)(1 + \cos x)}{1 - \cos x}dx$

$= \int (1 + \cos x)dx$

$= x + \sin x + C$

(9) $= \sec x + C$

(10) $= \int \frac{1}{\sin x} \cdot \frac{\cos x}{\sin x}dx$

$= \int \csc x \cot x\, dx$

$= -\csc x + C$

(11) $= \tan x - x + C$

(12) $= \int (\csc^2 x - 1)dx$

$= -\cot x - x + C$

(13) $= e^x - \cot x + C$

(14) $= \int (10^x - \frac{1}{\cot x})dx$

$= \int (10^x - \tan x)dx$

$= \frac{10^x}{\ln 10} - \ln|\sec x| + C$

EX5

(1) $= 5\sin^{-1}x + C$

(2) $= 3\tan^{-1}x + C$

(3) $= -7\sec^{-1}|x| + C$

(4) $= 4\tan^{-1}x + C$

EX6

(1) $= \frac{1}{8}(2x+1)^4 + C$

(2) $= \frac{1}{27}(3x-2)^9 + C$

(3) $= \frac{1}{3}(\frac{1}{2}x+3)^6 + C$

(4) $= -\frac{2}{9}(1-3x)^{\frac{3}{2}} + C$

(5) $= 7\sqrt{2x+1} + C$

(6) $= (1-x)^{-1} + C = \frac{1}{1-x} + C$

EX7

(1) $= \frac{1}{7}e^{7x} + \frac{1}{3}e^{3x} + C$

(2) $= -\frac{1}{2}e^{-2x} + \frac{1}{2}\sin 2x + C$

(3) $= -\frac{1}{2}e^{-2x} - 3e^{-\frac{1}{3}x} + C$

(4) $= -4e^{-x} + \frac{3^{x-1}}{\ln 3} + C$

(5) $= \frac{10^{5x}}{5\ln 10} - \frac{1}{2}\cos 2x + C$

(6) $= \int (4^{3x} + 4^x)dx$

$= \frac{4^{3x}}{3\ln 4} + \frac{4^x}{\ln 4} + C$

(7) $= \int (3^{x+1})^{\frac{1}{2}}dx$

$= \int 3^{\frac{1}{2}x+\frac{1}{2}}dx$

$= \frac{2(3)^{\frac{1}{2}x+\frac{1}{2}}}{\ln 3} + C$

(8) $= \frac{1}{3}\sin 3x - 2\cos(\frac{x}{2}) + C$

(9) $= \frac{1}{3}\sin(3x+5) + C$

(10) $= \frac{1}{5}\sec 5x + C$

(11) $= \frac{1}{2}e^{2x} + 2x - \frac{1}{2}e^{-2x} + C$

(12) $= \frac{1}{2}\tan 2x + \frac{1}{2}\ln|\sec 2x| + C$

EX8

(1) $= -\frac{3}{4}\cos 2x + C$

(2) $= \int \frac{1}{2}\sin\theta d\theta$

$= -\frac{1}{2}\cos\theta + C$

(3) $= -\cos 4x + C$

(4) $= \int 6\sin 6x dx$

$= -\cos 6x + C$

(5) $= \frac{1}{2}x - \frac{1}{2}\sin x + C$

(6) $= \int \frac{1-\cos 6x}{2}dx$

$= \int \left(\frac{1}{2} - \frac{1}{2}\cos 6x\right)dx$

$= \frac{1}{2}x - \frac{1}{12}\sin 6x + C$

(7) $\frac{1}{2}x + \frac{1}{4}\sin 2x + C$

(8) $= \int \frac{1+\cos 8x}{2}dx$

$= \int \left(\frac{1}{2}+\frac{1}{2}\cos 8x\right)dx$

$= \frac{1}{2}x+\frac{1}{16}\sin 8x+C$

EX9

(1) $= 5\ln|\sin x| + C$

(2) $= \ln|x^3+x| + C$

(3) $= 7\ln|x+1| + C$

(4) $= \int \frac{1}{2} \cdot \frac{2x}{x^2+1}dx$

$= \frac{1}{2}\ln|x^2+1| + C$

$= \frac{1}{2}\ln(x^2+1)+ C$

(5) $= 3\ln|x^2-4| + C$

(6) $\int \frac{2}{-3} \cdot \frac{-3}{1-3x}dx$

$= -\frac{2}{3}\ln|1-3x| + C$

(7) $= -5\ln|1+\cos x| + C$

(8) $= \int \frac{1}{4} \cdot \frac{4x}{2x^2+1}dx$

$= \frac{1}{4}\ln(2x^2+1)+C$

(9) $= \ln(e^x+e^{-x})+C$

EX10

(1) $f(x) = x+\ln x+2$

(2) $\frac{dy}{dx}=2^x-x^{\frac{5}{2}}+\sin x$

$y=\frac{2^x}{\ln 2}-\frac{2}{7}x^{\frac{7}{2}}-\cos x+C$

$y(0) = \frac{1}{\ln 2}$ $:C=1$

$\therefore y=\frac{2^x}{\ln 2}-\frac{2}{7}x^{\frac{7}{2}}-\cos x+1$

(3) $f(x) = 8\tan^{-1}x+\tan x+C$

$f(0)=1 ; C=1$

$\therefore f(x) = 8\tan^{-1}x+\tan x+1$

(4) $y=2\ln|2x-1| +\frac{1}{2} \cdot \frac{8}{4}(2x+1)^4+C$

$y=2\ln|2x-1| +(2x+1)^4+C$

$y(0)=3 :$
$C=2$

$\therefore y=2\ln|2x-1|+(2x+1)^4+2$

(5) $f(x) = \ln(x^2+4)+\frac{1}{2}x+\frac{1}{4}\sin 2x-\ln 4$

(6) $f'(x) = 2e^{\frac{1}{2}x}-6x^2+2x+C$

$f'(0) =-1 :$
$C= -3$

$\therefore f'(x) = 2e^{\frac{1}{2}x}-6x^2+2x-3$

$f(x) = 4e^{\frac{1}{2}x}-2x^3+x^2-3x+D$

$f(0)=1 :$
$D= -3$

$\therefore f(x) = 4e^{\frac{1}{2}x}-2x^3+x^2-3x-3$

(7) $y = \frac{1}{x}+2x-2$

(8) $\frac{dy}{dx}= -e^{-x}+3x^2+C$

$y= e^{-x}+x^3+Cx+D$

$y(0)=2$
$D=1$

$y(1) = \frac{1}{e}-1 :$
$C= -3$

$\therefore y=e^{-x}+x^3-3x+1$

5.2 Intergration by Substitution

EX1

(1) $=-\frac{2}{5}(1-x^2)^{5/4}+C$

(2) $u = x^2-2x+3$ $\qquad = \int u^{10} \cdot \frac{1}{2}du$

$\frac{du}{dx}=2x-2$ $\qquad = \frac{1}{2} \cdot \frac{1}{11}u^{11}+C$

$\frac{1}{2}du=(x-1)dx$ $\qquad = \frac{1}{22}(x^2-2x+3)^{11}+C$

(3) $= \dfrac{-1}{x^4+1} + C$

(4) $u = x^2 - 6x$ $\quad = \displaystyle\int \frac{1}{u^{\frac{1}{4}}} \cdot \frac{1}{2} du$

$\dfrac{du}{dx} = 2x - 6$ $\quad = \displaystyle\int \frac{1}{2} u^{-\frac{1}{4}} du$

$\dfrac{1}{2} du = (x-3)dx$ $\quad = \dfrac{1}{2} \cdot \dfrac{4}{3} u^{\frac{3}{4}} + C$

$= \dfrac{2}{3}(x^2 - 6x)^{\frac{3}{4}} + C$

(5) $= -6(1-x^3)^{1/2} + C$

(6) $u = 4x - x^2$ $\quad = \displaystyle\int \frac{1}{\sqrt{u}} \cdot -\frac{1}{2} du$

$\dfrac{du}{dx} = 4 - 2x$ $\quad = \displaystyle\int -\frac{1}{2} u^{-\frac{1}{2}} du$

$\dfrac{du}{dx} = -2(x-2)$ $\quad = -\dfrac{1}{2} \cdot 2u^{\frac{1}{2}} + C$

$-\dfrac{1}{2} du = (x-2)dx$ $\quad = -\sqrt{4x - x^2} + C$

(7) $= -\dfrac{1}{3}(7 - 3x^2)^{3/2} + C$

(8) $u = 1 + \sqrt{x}$ $\quad = \displaystyle\int u^5 \cdot 2du$

$\dfrac{du}{dx} = \dfrac{1}{2\sqrt{x}}$ $\quad = 2 \cdot \dfrac{1}{6} u^6 + C$

$2du = \dfrac{1}{\sqrt{x}} dx$ $\quad = \dfrac{1}{3}(1 + \sqrt{x})^6 + C$

(9) $= \dfrac{2^{\sqrt{x}+1}}{\ln 2} + C$

(10) $u = \sqrt{x} - 1$ $\quad = \displaystyle\int \frac{1}{u} \cdot 2du$

$\dfrac{du}{dx} = \dfrac{1}{2\sqrt{x}}$ $\quad = 2 \cdot \ln|u| + C$

$2du = \dfrac{1}{\sqrt{x}} dx$ $\quad = 2\ln|\sqrt{x} - 1| + C$

EX2

(1) $= \dfrac{1}{8}(\ln x)^4 + C$

(2) $u = \ln x$ $\quad = \displaystyle\int \frac{1}{u} \cdot du$

$\dfrac{du}{dx} = \dfrac{1}{x}$ $\quad = \ln|u| + C$

$du = \dfrac{1}{x} dx$ $\quad = \ln|\ln x| + C$

(3) $= \sqrt{\ln^2 x + 1} + C$

(4) $u = \ln(\ln x)$ $\quad = \displaystyle\int \frac{1}{u} \cdot du$

$\dfrac{du}{dx} = \dfrac{1}{\ln x} \cdot \dfrac{1}{x}$ $\quad = \ln|u| + C$

$du = \dfrac{1}{x \ln x} dx$ $\quad = \ln|\ln(\ln x)| + C$

(5) $\dfrac{1}{2}\sin^6\left(\dfrac{x}{3}\right) + C$

(6) $u = \cos 3x$ $\quad = \displaystyle\int u^3 \cdot -\frac{1}{3} du$

$\dfrac{du}{dx} = -3\sin 3x$ $\quad = -\dfrac{1}{3} \cdot \dfrac{1}{4} u^4 + C$

$-\dfrac{1}{3} du = \sin 3x dx$ $\quad = -\dfrac{1}{12}\cos^4 3x + C$

(7) $= \dfrac{e^{\sec \pi x}}{\pi} + C$

(8) $u = \tan\theta$ $\quad = \displaystyle\int u^2 \cdot du$

$\dfrac{du}{d\theta} = \sec^2\theta$ $\quad = \dfrac{1}{3} u^3 + C$

$du = \sec^2\theta d\theta$ $\quad = \dfrac{1}{3}\tan^3\theta + C$

(9) $= e^{\sin^{-1} x} + C$

(10) $= \displaystyle\int \sec^2\theta \cdot \sec^2\theta d\theta = \int (1 + \tan^2\theta) \cdot \sec^2\theta d\theta$

$\leq tu = \tan\theta$ $\quad = \displaystyle\int (1 + u^2) du$

$\dfrac{du}{dx} = \sec^2\theta$ $\quad = u + \dfrac{1}{3} u^3 + C$

$du = \sec^2\theta d\theta$ $\quad = \tan\theta + \dfrac{1}{3}\tan^3\theta + C$

(11) $= \ln|\tan^{-1} x| + C$

(12) $u = 1 + \tan 2\theta$ $\quad = \displaystyle\int \frac{1}{u} \cdot \frac{1}{2} du$

$\dfrac{du}{d\theta} = 2 \cdot \sec^2 2\theta$ $\quad = \dfrac{1}{2}\ln|u| + C$

$\dfrac{1}{2} du = \sec^2 2\theta d\theta$ $\quad = \dfrac{1}{2}\ln|1 + \tan 2\theta| + C$

(13) $= \dfrac{1}{2\cos(2x+1)} + C$

(14) $u = 1 + \sin 2x$ $\quad = \displaystyle\int \frac{1}{\sqrt{u}} \cdot \frac{1}{2} du$

$\dfrac{du}{dx} = 2\cos 2x$ $\quad = \displaystyle\int \frac{1}{2} u^{-\frac{1}{2}} du$

$\dfrac{1}{2} du = \cos 2x dx$ $\quad = \dfrac{1}{2} \cdot 2u^{\frac{1}{2}} + C$

$= \sqrt{1 + \sin 2x} + C$

(15) $= e^{\sin^2 x} + C$

(16) $u = \cos x$ $\quad = \int \frac{1}{1+u^2} \bullet -du$

$\frac{du}{dx} = -\sin x$ $\quad = -\tan^{-1}u + C$

$-du = \sin x dx$ $\quad = -\tan^{-1}(\cos x) + C$

(10) $u = e^x$ $\quad = \int \frac{e^x \bullet e^x}{e^x+1}dx$

$\frac{du}{dx} = e^x$ $\quad = \int \frac{u}{u+1}du$

$du = e^x dx$ $\quad = \int \left(1 + \frac{-1}{u+1}\right)du$

$= u - \ln|u+1| + C$

$= e^x - \ln(e^x+1) + C$

EX3

(1) $= \ln|1+e^x| + C$

(2) $\int \frac{f'(x)}{f(x)}dx = \ln|f(x)| + C$

$\int \frac{e^{2x}}{e^{2x}+1}dx = \int \frac{1}{2} \bullet \frac{2e^{2x}}{e^{2x}+1}dx$

$= \frac{1}{2}\ln(e^{2x}+1) + C$

(3) $= -e^{-x^2} + C$

(4) $u = x^2 + 2$ $\quad = \int e^{-u}\frac{1}{2}du$

$\frac{du}{dx} = 2x$ $\quad = -\frac{1}{2}e^{-u} + C$

$\frac{1}{2}du = xdx$ $\quad = -\frac{1}{2}e^{-x^2-2} + C$

(5) $= \frac{2}{3}\sqrt{(\tan^{-1}x)^3} + C$

(6) $\leq tu = \sin^{-1}(2x)$ $\quad = \int u \bullet \frac{1}{2}du$

$\frac{du}{dx} = \frac{1}{\sqrt{1-(2x)^2}} \bullet 2$ $\quad = \frac{1}{4}u^2 + C$

$= \frac{2}{\sqrt{1-4x^2}}$ $\quad = \frac{1}{4}(\sin^{-1}(2x))^2 + C$

$\frac{1}{2}du = \frac{1}{\sqrt{1-4x^2}}dx$

(7) $= \tan^{-1}(e^x) + C$

(8) $u = e^{2x}$ $\quad = \int \frac{1}{1+u^2} \bullet \frac{1}{2}du$

$\frac{du}{dx} = e^{2x} \bullet 2$ $\quad = \tan^{-1}u + C$

$\frac{1}{2}du = e^{2x}dx$ $\quad = \frac{1}{2}\tan^{-1}(e^{2x}) + C$

(9) $= \frac{1}{2}x^2 - x + 4\ln|x+1| + C$

EX4

(1) $= \frac{1}{2}\tan^{-1}(2x) + C$

(2) $= \sin^{-1}\left(\frac{x}{4}\right) + C$

(3) $= \tan^{-1}(x+1) + C$

(4) $= 3 \bullet \frac{1}{3}\tan^{-1}\left(\frac{x}{3}\right) + C$

$= \tan^{-1}\left(\frac{x}{3}\right) + C$

(5) $= \sin^{-1}\left(\frac{x}{3}\right) + C$

(6) $= \int \frac{1}{4} \bullet \frac{1}{9/4+x^2}dx$ where $a = \frac{3}{2}$

$= \frac{1}{4} \bullet \frac{1}{3/2}\tan^{-1}\left(\frac{x}{3/2}\right)$

$= \frac{1}{6}\tan^{-1}\left(\frac{2x}{3}\right) + C$

(7) $= \frac{1}{9}\int \frac{e^x dx}{1+\left(\frac{e^x}{3}\right)^2} = 3 \bullet \frac{1}{9}\int \frac{\frac{e^x}{3}dx}{1+\left(\frac{e^x}{3}\right)^2}$

$= \frac{1}{3}\tan^{-1}\frac{e^x}{3} + C$

(8) $= \ln(x^2+4) + \frac{1}{2}\tan^{-1}\frac{x}{2} + C$

(9) $u = 4 - x^2$ $\quad = \int \frac{x}{\sqrt{4-x^2}}dx + \int \frac{1}{\sqrt{4-x^2}}dx$

$\frac{du}{dx} = -2x$ $\quad = \int \frac{1}{\sqrt{u}} \bullet -\frac{1}{2}du + \sin^{-1}\left(\frac{x}{2}\right) + C$

$-\frac{1}{2}du = xdx$ $\quad = -\sqrt{u} + \sin^{-1}\left(\frac{x}{2}\right) + C$

$= -\sqrt{4-x^2} + \sin^{-1}\left(\frac{x}{2}\right) + C$

$$(10) = \int \frac{1}{x^2-2x+1+4}dx$$
$$= \int \frac{1}{(x-1)^2+4}dx$$
$$= \frac{1}{2}\tan^{-1}\left(\frac{x-1}{2}\right)+C$$

EX5

$$f(x) = \int \frac{1}{\sqrt{9-4x^2}}dx$$
$$= \int \frac{1}{2} \cdot \frac{1}{\sqrt{9/4-x^2}}dx$$
$$= \frac{1}{2}\sin^{-1}\left(\frac{x}{3/2}\right)+C$$
$$= \frac{1}{2}\sin^{-1}\left(\frac{2x}{3}\right)+C$$

$$f\left(-\frac{3}{2}\right)= \frac{\pi}{12} : \frac{\pi}{12}= \frac{1}{2}\sin^{-1}(-1)+C$$
$$\frac{\pi}{12}= \frac{1}{2} \cdot \left(-\frac{\pi}{2}\right)+C$$
$$\frac{\pi}{12}+\frac{\pi}{4} = C$$
$$C = \frac{\pi}{3}$$
$$\therefore f(x) = \frac{1}{2}\sin^{-1}\left(\frac{2x}{3}\right)+\frac{\pi}{3}$$

5.3 Integration by Parts

EX1

$(1) = x sinx + \cos x + C$

$$(2) = -2xe^{-\frac{1}{2}x} - \int -2e^{-\frac{1}{2}x}dx$$
$$= -2xe^{-\frac{1}{2}} - 4e^{-\frac{1}{2}x} + C$$

$$(3) = \frac{1}{3}x\sin 3x - \int \frac{1}{3}\sin 3x\,dx$$
$$= \frac{1}{3}x\sin 3x + \frac{1}{9}\cos 3x + C$$

$$(4) =- \frac{1}{2}x\cos 2x - \int -\frac{1}{2}\cos 2x dx$$
$$=- \frac{1}{2}x\cos 2x + \frac{1}{4}\sin 2x + C$$

$$(5) = e^x(x^2-2x+2)+C$$

$$(6) = x\ln x - \int 1\,dx$$
$$= x\ln x - x + C$$

$$(7) = \frac{x^4}{16}(4\ln x - 1)+C$$

$$(8) = -\frac{1}{x}\ln x - \int -\frac{1}{x^2}dx$$
$$= -\frac{\ln x}{x} - \frac{1}{x} + C$$

$$(9) = 2\sqrt{x}\ln x - \int \frac{1}{x} \cdot 2\sqrt{x}\,dx$$
$$= 2\sqrt{x}\ln x - \int \frac{2}{\sqrt{x}}dx$$
$$= 2\sqrt{x}\ln x - 4\sqrt{x} + C$$

$$(10) = x\tan^{-1}x - \int \frac{x}{1+x^2}dx$$
$$= x\tan^{-1}x - \frac{1}{2}\ln(1+x^2)+C$$

$$(11) = \frac{1}{2}x^2\sec^{-1}x - \int \frac{x}{2\sqrt{x^2-1}}dx$$
$$= \frac{1}{2}x^2 \cdot \sec^{-1}x - \int \frac{1}{2\sqrt{t}} \cdot \frac{1}{2}dt$$
$$= \frac{1}{2}x^2\sec^{-1}x - \frac{1}{2}\sqrt{t}+C$$
$$= \frac{1}{2}x^2\sec^{-1}x - \frac{1}{2}\sqrt{x^2-1}+C$$

EX2

$$(1) = \frac{1}{2}x^2(\ln x)^2 - \int x\ln x dx$$
$$= \frac{1}{2}x^2(\ln x)^2 - \left[\frac{1}{2}x^2\ln x - \int \frac{1}{2}x dx\right]$$
$$= \frac{1}{2}x^2(\ln x)^2 - \frac{1}{2}x^2\ln x + \frac{1}{4}x^2 + C$$

$$(2) = -\frac{1}{x}(\ln x)^2 - \int -\frac{2(\ln x)}{x^2}dx$$
$$= -\frac{1}{x}(\ln x)^2 + 2\int \frac{\ln x}{x^2}dx$$
$$= -\frac{1}{x}(\ln x)^2 + 2\left[-\frac{1}{x}\ln x - \int -\frac{1}{x^2}dx\right]$$
$$= -\frac{1}{x}(\ln x)^2 + 2\left[-\frac{1}{x}\ln x - \frac{1}{x}\right]$$
$$= -\frac{1}{x}(\ln x)^2 - \frac{2}{x}\ln x - \frac{2}{x} + C$$

EX3

$$(1) = -e^{2x}\cos x + 2\int e^{2x}\cos x dx$$
$$= -e^{2x}\cos x + 2\left[e^{2x}\sin x - \int 2e^{2x}\sin x dx\right]$$
$$= -e^{2x}\cos x + 2e^{2x}\sin x - 4\int e^{2x}\sin x dx$$
$$\int e^{2x}\sin x dx = I$$
$$I = -e^{2x}\cos x + 2e^{2x}\sin x - 4I$$
$$5I = -e^{2x}\cos x + 2e^{2x}\sin x$$
$$\therefore I = \frac{e^{2x}(-\cos x + 2\sin x)}{5} + C$$

$$(2) = \frac{1}{2}e^x\sin 2x - \frac{1}{2}\int e^x\sin 2x dx$$
$$= \frac{1}{2}e^x\sin 2x - \frac{1}{2}\left[-\frac{1}{2}e^x\cos 2x - \int -\frac{1}{2}e^x\cos 2x dx\right]$$
$$= \frac{1}{2}e^x\sin 2x + \frac{1}{4}e^x\cos 2x - \frac{1}{4}\int e^x\cos 2x dx$$
$$\int e^x\cos 2x = I$$
$$I = \frac{1}{2}e^x\sin 2x + \frac{1}{4}e^x\cos 2x - \frac{1}{4}I$$
$$4I = 2e^x\sin 2x + e^x\cos 2x - I$$
$$5I = e^x(2\sin 2x + \cos 2x)$$
$$\therefore I = \frac{e^x(2\sin 2x + \cos 2x)}{5} + C$$

$$(3) = \frac{e^{2x}}{13}(3\sin 3x + 2\cos 3x) + C$$

$$(4) = -\frac{e^{-2x}}{4}(\sin 2x + \cos 2x) + C$$

EX4

$$(1) = x^2e^x - 2xe^x + 2e^x + C$$
$$(2) = e^{-x}(-x^3 - 3x^2 - 6x - 6) + C$$
$$(3) = e^{-\frac{x}{2}}(-2x^3 - 12x^2 - 48x - 96) + C$$
$$(4) = \frac{1}{3}x^2\sin 3x + \frac{2}{9}x\cos 3x - \frac{2}{27}\sin 3x + C$$
$$(5) = -2x^3\cos\frac{1}{2}x + 12x^2\sin\frac{1}{2}x + 48x\cos\frac{1}{2}x - 96\sin\frac{1}{2}x + C$$

EX5

(1) $u = \sqrt{x}$, $u^2 = x$, $2u\frac{du}{dx} = 1$, $2udu = dx$

$$= \int e^u \cdot 2u du = \int 2ue^u du = 2u(e^u) - 2(e^u) + C = 2\sqrt{x}e^{\sqrt{x}} - 2e^{\sqrt{x}} + C$$

(2) $u = \sqrt{x}$, $u^2 = x$, $2udu = dx$

$$= \int \cos u \cdot 2u du = \int 2u \cdot \cos u du = 2u(\sin u) - 2(-\cos u) + C = 2\sqrt{x}\sin\sqrt{x} + 2\cos\sqrt{x} + C$$

(3) $u = x^2$, $du = 2xdx$, $\frac{1}{2}du = x dx$

$$= \int x^2 \cdot e^{-x^2} \cdot x dx = \int u \cdot e^{-u} \cdot \frac{1}{2}du = \int \frac{1}{2}ue^{-u}du = \frac{1}{2}u(-e^{-u}) - \frac{1}{2}(e^{-u}) + C = -\frac{1}{2}x^2e^{-x^2} - \frac{1}{2}e^{-x^2} + C$$

(4) $u = x^2$, $du = 2xdx$, $\frac{1}{2}du = x dx$

$$= \int x^2 \cdot \sin(x^2) \cdot x dx = \int u \cdot \sin u \cdot \frac{1}{2}du = \int \frac{1}{2}u\sin u du = \frac{1}{2}u(-\cos u) - \frac{1}{2}(-\sin u) + C = -\frac{1}{2}x^2\cos(x^2) + \frac{1}{2}\sin(x^2) + C$$

5.4 Rectangular Approximation Method(RAM)

EX1

(a) $L(4) = \frac{1}{2}\left[f(0)+f\left(\frac{1}{2}\right)+f(1)+f\left(\frac{3}{2}\right)\right]$

$= \frac{1}{2}\left[4+\frac{15}{4}+3+\frac{7}{4}\right]=\frac{25}{4}$ (or 6.25)

(b) $R(4) = \frac{1}{2}\left[f\left(\frac{1}{2}\right)+f(1)+f\left(\frac{3}{2}\right)+f(2)\right]$

$= \frac{1}{2}\left[\frac{15}{4}+3+\frac{7}{4}+0\right]=\frac{17}{4}$ (or 4.25)

(c) $M(4) = \frac{1}{2}\left[f\left(\frac{1}{4}\right)+f\left(\frac{3}{4}\right)+f\left(\frac{5}{4}\right)+f\left(\frac{7}{4}\right)\right]$

$= \frac{1}{2}\left[\frac{63}{16}+\frac{55}{16}+\frac{39}{16}+\frac{15}{16}\right]=\frac{43}{8}$ (or 5.375)

5.5 Trapezoidal Approximation

EX1

make a table of values for $f(x) = \frac{120}{x}$:

x	1	2	3	4	5	6
$f(x)$	120	60	40	30	24	20

Then:

$L(5) = (120+60+40+30+24) \cdot 1 = 274;$

$R(5) = (60+40+30+24+20) \cdot 1 = 174;$

$T(5) = \frac{1}{2}(120+2 \cdot 60+2 \cdot 40+2 \cdot 30+2 \cdot 24+20) = 224.$

The calculator finds that $\int_1^6 \frac{120}{x}dx$ is approximately 215.011.

EX2

$T(6) = \frac{h}{2}[y_0+2y_1+2y_2+2y_3+2y_4+2y_5+y_6\}$

$= \frac{\frac{\pi}{6}}{2}\left[0+2 \cdot \frac{1}{2}+2 \cdot \frac{\sqrt{3}}{2}+2 \cdot 1 +2 \cdot 1+2\frac{\sqrt{3}}{2}+2 \cdot \frac{1}{2}+0\right]$

$= \frac{\pi}{12}[1+\sqrt{3}+2+\sqrt{3}+1]$

$= \frac{(2+\sqrt{3})\pi}{6}$

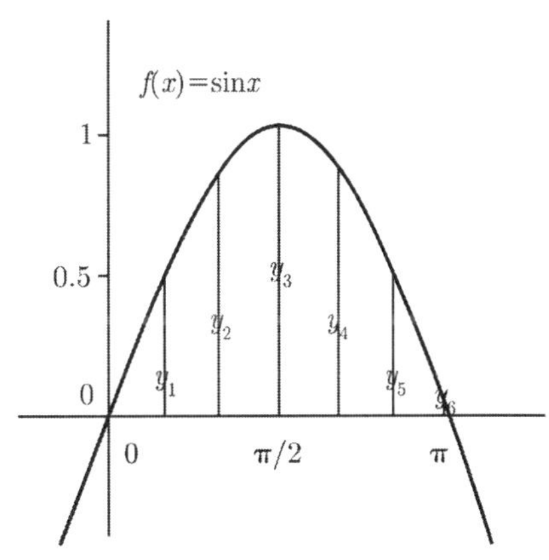

EX3

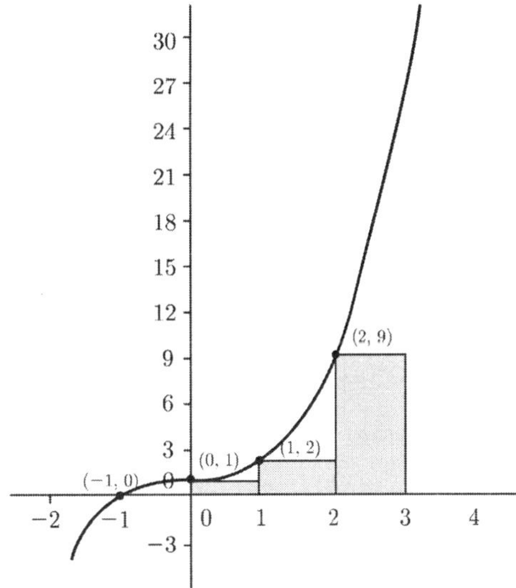

(1) $L(4) = 1 \cdot [f(-1)+f(0)+f(1)+f(2)]$

$= 0+1+2+9 = 12$

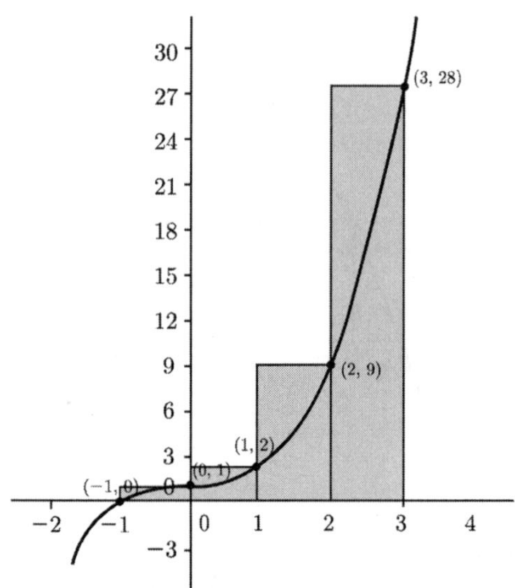

(2) $R(4) = 1 \cdot [f(0) + f(1) + f(2) + f(3)]$
$= 1 + 2 + 9 + 28 = 40$

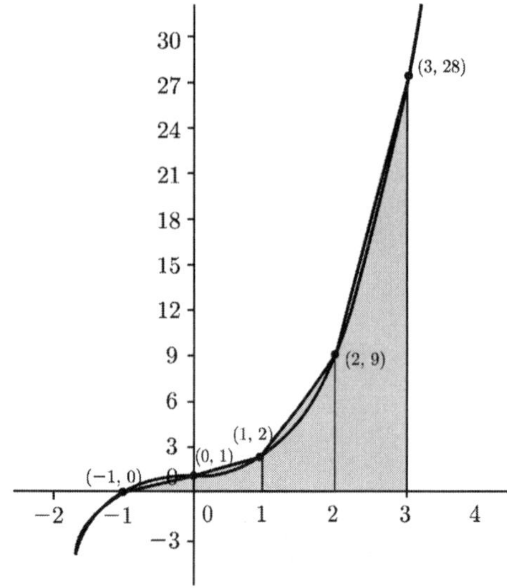

(3) $T(4)$

$= \frac{1}{2}[f(-1) + 2f(0) + 2f(1) + 2f(2) + f(3)]$

$= \frac{1}{2}[0 + 2 \cdot 1 + 2 \cdot 2 + 2 \cdot 9 + 28]$

$= 26$

EX4

(1) Any number of curves will do. The graph has only to pass through the points given in the table of speeds, as does the graph in Figure N6-12a.

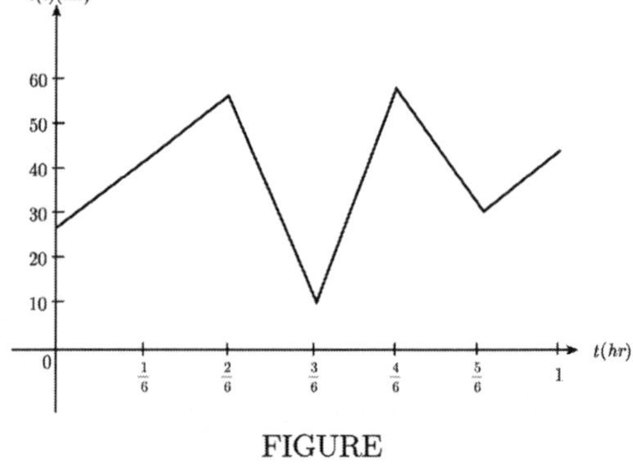

FIGURE

(2) $L(6) = (26 + 40 + 55 + 10 + 60 + 32) \cdot \frac{1}{6} = 37\frac{1}{6}\,mi;$

$R(6) = (40 + 55 + 10 + 60 + 32 + 45) \cdot \frac{1}{6} = 40\frac{1}{3}\,mi;$

$T(6) = \frac{1/6}{2}(26 + 2 \cdot 40 + 2 \cdot 55 + 2 \cdot 10 + 2 \cdot 60 + 2 \cdot 32 + 45)$

$= 38\frac{3}{4}\,mi.$

EX5

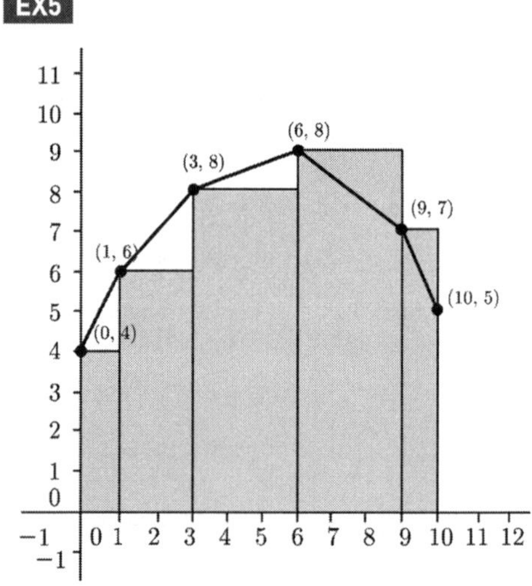

(1) $L(5) = 1 \cdot 4 + 2 \cdot 6 + 3 \cdot 8 + 3 \cdot 9 + 1 \cdot 7$
$= 74L$

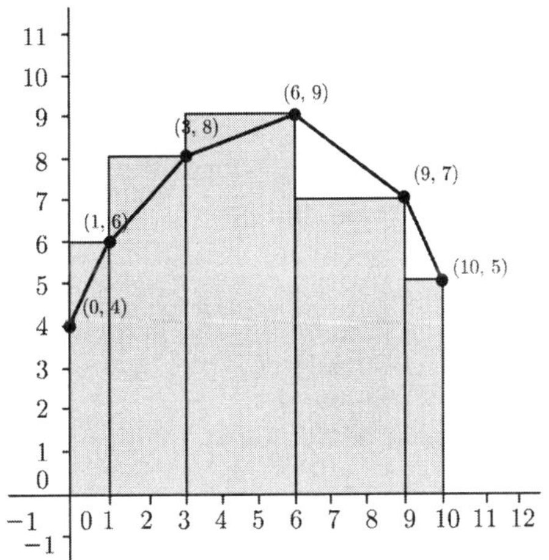

(2) $R(5) = 1 \cdot 6 + 2 \cdot 8 + 3 \cdot 9 + 3 \cdot 7 + 1 \cdot 5$
$= 75L$

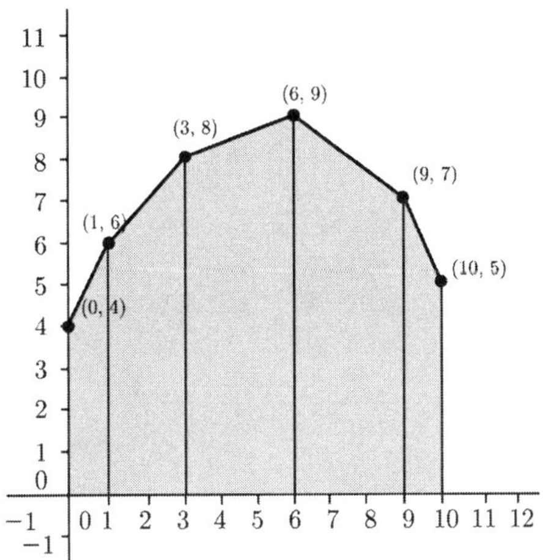

(3) $$T(5) = \frac{(4+6)\cdot 1}{2} + \frac{(6+8)\cdot 2}{2} + \frac{(8+9)\cdot 3}{2} + \frac{(9+7)\cdot 3}{2} + \frac{(7+5)\cdot 1}{2}$$

$$= 5 + 14 + \frac{51}{2} + 24 + 6$$

$$= 74.5L$$

5.6 Definition of Area as a Limit

EX1

$$y = -2x + 3,\ [0, 1]\left(note : \Delta x = \frac{1-0}{n} = \frac{1}{n}\right)$$

$$s(n) = \sum_{i=1}^{n} f\left(\frac{i}{n}\right)\left(\frac{1}{n}\right) = \sum_{i=1}^{n}\left[-2\left(\frac{i}{n}\right)+3\right]\left(\frac{1}{n}\right)$$

$$= 3 - \frac{2}{n^2}\sum_{i=1}^{n} i = 3 - \frac{2(n+1)n}{2n^2} = 3 - \frac{n+1}{n}$$

$$Area = \lim_{n\to\infty} s(n) = 2$$

EX2

$$y = x^2 + 1,\ [0,3]\left(note : \Delta x = \frac{3-0}{n} = \frac{3}{n}\right)$$

$$S(n) = \sum_{i=1}^{n} f\left(\frac{3i}{n}\right)\left(\frac{3}{n}\right) = \sum_{i=1}^{n}\left[\left(\frac{3i}{n}\right)^2 + 1\right]\left(\frac{3}{n}\right)$$

$$= \frac{27}{n^3}\sum_{i=1}^{n} i^2 + \frac{3}{n}\sum_{i=1}^{n} 1$$

$$= \frac{27}{n^3}\,\frac{n(n+1)(2n+1)}{6} + \frac{3}{n}(n) = \frac{9}{2}\,\frac{2n^2+3n+1}{n^2} + 3$$

$$Area = \mathrm{Lim}_{n\to\infty} S(n) = \frac{9}{2}(2) + 3 = 12$$

EX3

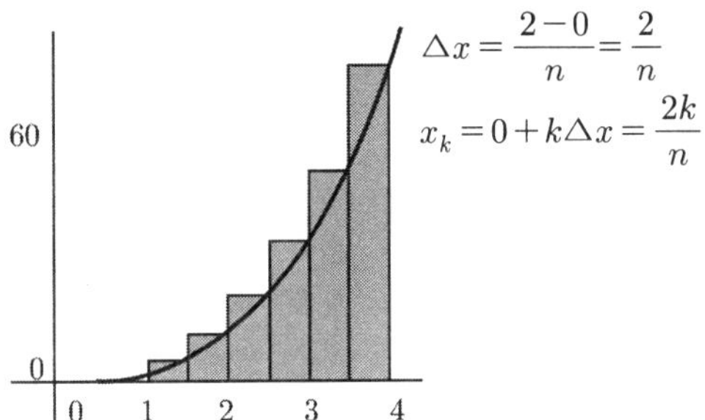

$$\Delta x = \frac{2-0}{n} = \frac{2}{n}$$

$$x_k = 0 + k\Delta x = \frac{2k}{n}$$

Area

$$= \mathrm{Lim}_{n\to\infty}\sum_{k=1}^{n} f(x_k)\Delta x$$

$$= \mathrm{Lim}_{n\to\infty}\sum_{k=1}^{n} f\left(\frac{2k}{n}\right)\frac{2}{n}$$

$$= \mathrm{Lim}_{n\to\infty}\sum_{k=1}^{n} \frac{8k^3}{n^3} \cdot \frac{2}{n}$$

$$=\lim_{n\to\infty}\sum_{k=1}^{n}\frac{16k^3}{n^4}$$
$$=\lim_{n\to\infty}\left[\frac{16}{n^4}\sum_{k=1}^{n}k^3\right]$$
$$=\lim_{n\to\infty}\frac{16}{n^4}\cdot\left[\frac{n(n+1)}{2}\right]^2$$
$$=\lim_{n\to\infty}\frac{16}{n^4}\cdot\frac{n^2(n+1)^2}{4}$$
$$=\lim_{n\to\infty}\frac{4(n+1)^2}{n^2}=4$$

EX4

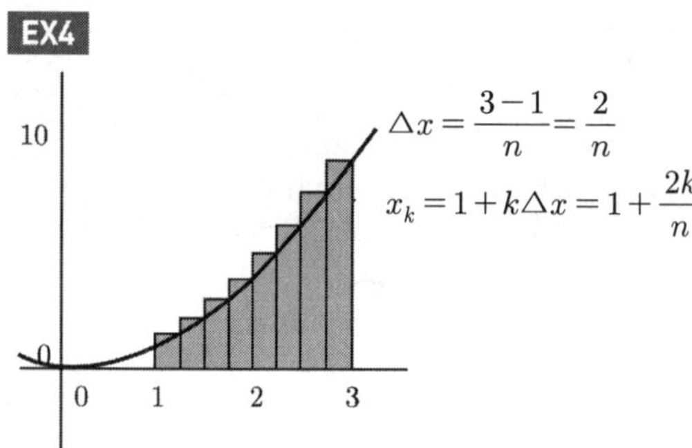

$$\Delta x=\frac{3-1}{n}=\frac{2}{n}$$
$$x_k=1+k\Delta x=1+\frac{2k}{n}$$

Area

$$=\lim_{n\to\infty}\sum_{k=1}^{n}f(x_k)\Delta x$$
$$=\lim_{n\to\infty}\sum_{k=1}^{n}f\left(1+\frac{2k}{n}\right)\frac{2}{n}$$
$$=\lim_{n\to\infty}\sum_{k=1}^{n}\left(1+\frac{2k}{n}\right)^2\frac{2}{n}$$
$$=\lim_{n\to\infty}\sum_{k=1}^{n}\left(1+\frac{4k}{n}+\frac{4k^2}{n^2}\right)\frac{2}{n}$$
$$=\lim_{n\to\infty}\sum_{k=1}^{n}\left(\frac{2}{n}+\frac{8k}{n^2}+\frac{8k^2}{n^3}\right)$$
$$=\lim_{n\to\infty}\left[\frac{2}{n}\sum_{k=1}^{n}1+\frac{8}{n^2}\sum_{k=1}^{n}k+\frac{8}{n^3}\sum_{k=1}^{n}k^2\right]$$
$$=\lim_{n\to\infty}\left[\frac{2}{n}\cdot n+\frac{8}{n^2}\cdot\frac{n(n+1)}{2}+\frac{8}{n^3}\cdot\frac{n(n+1)(2n+1)}{6}\right]$$
$$=\lim_{n\to\infty}\left[2+\frac{4(n+1)}{n}+\frac{4(n+1)(2n+1)}{3n^2}\right]$$
$$=2+4+\frac{8}{3}=\frac{26}{3}$$

EX5

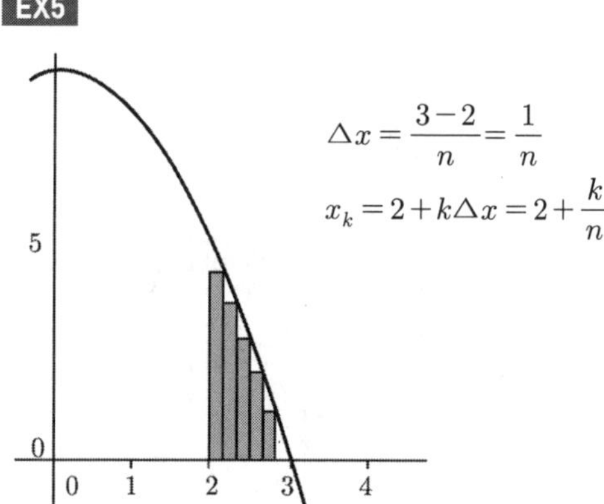

$$\Delta x=\frac{3-2}{n}=\frac{1}{n}$$
$$x_k=2+k\Delta x=2+\frac{k}{n}$$

Area

$$=\lim_{n\to\infty}\sum_{k=1}^{n}f(x_k)\Delta x$$
$$=\lim_{n\to\infty}\sum_{k=1}^{n}f\left(2+\frac{k}{n}\right)\frac{1}{n}$$
$$=\lim_{n\to\infty}\sum_{k=1}^{n}\left[9-\left(2+\frac{k}{n}\right)^2\right]\frac{1}{n}$$
$$=\lim_{n\to\infty}\sum_{k=1}^{n}\left(9-4-\frac{4k}{n}-\frac{k^2}{n^2}\right)\frac{1}{n}$$
$$=\lim_{n\to\infty}\sum_{k=1}^{n}\left(\frac{5}{n}-\frac{4k}{n^2}-\frac{k^2}{n^3}\right)$$
$$=\lim_{n\to\infty}\left[\frac{5}{n}\sum_{k=1}^{n}1-\frac{4}{n^2}\sum_{k=1}^{n}k-\frac{1}{n^3}\sum_{k=1}^{n}k^2\right]$$
$$=\lim_{n\to\infty}\left[\frac{5}{n}\cdot n-\frac{4}{n^2}\cdot\frac{n(n+1)}{2}-\frac{1}{n^3}\cdot\frac{n(n+1)(2n+1)}{6}\right]$$
$$=\lim_{n\to\infty}\left[5-\frac{2(n+1)}{n}-\frac{(n+1)(2n+1)}{6n^2}\right]$$
$$=5-2-\frac{1}{3}=\frac{8}{3}$$

5.7 Definite Integrals

EX1

(1) $\int_1^9 -2f(x)dx = -2\int_1^9 f(x)dx$

$= -2(-1) = 2$

(2) $\int_7^9 [f(x)+h(x)]dx$

$= \int_7^9 f(x)\,dx + \int_7^9 h(x)\,dx = 5+4 = 9$

(3) $\int_7^9 [2f(x)-3h(x)]dx$

$= 2\int_7^9 f(x)dx - 3\int_7^9 h(x)\,dx$

$= 2(5)-3(4) = -2$

(4) $\int_9^1 f(x)\,dx = -\int_1^9 f(x)\,dx = -(-1) = 1$

(5) $\int_1^7 f(x)\,dx = \int_1^9 f(x)\,dx - \int_7^9 f(x)\,dx$

$= -1-5 = -6$

(6) $\int_9^7 [h(x)-f(x)]\,dx = \int_7^9 [f(x)-h(x)]dx$

$= \int_7^9 f(x)\,dx - \int_7^9 h(x)\,dx = 5-4 = 1$

EX2

(1) $= \int_{-1}^2 f(x)dx$

$= 2$

(2) $= -\int_2^4 f(x)dx$

$= -3$

(3) $= -\left[\int_{-1}^2 f(x)dx + \int_2^4 f(x)dx\right]$

$= -(2+3)$

$= -5$

(4) 0

(5) $= 3\int_{-1}^2 f(x)dx + 2\int_{-1}^2 g(x)dx$

$= 3(2)+2(-1)$

$= 4$

(6) $= 2\int_{-1}^4 f(x)dx - 3\int_{-1}^4 g(x)dx$

$= 2\left[\int_{-1}^2 f(x)dx + \int_2^4 f(x)dx\right]$

$-3\left[\int_{-1}^2 g(x)dx + \int_2^4 g(x)dx\right]$

$= 2(2+3) - 3[-1+(-4)]$

$= 10+15 = 25$

EX3

(1) 0

(2) $= -\int_a^c e^x \cos x\,dx$

$= -\left[\int_a^b e^x\cos x\,dx + \int_b^c e^x \cos x\,dx\right]$

$= -(r+s) = -r-s$

(3) $= -\int_c^a x^2 \ln(x+1)dx$

$= -p$

(4) $= 2\int_a^c e^x\cos x dx - 3\int_a^c x^2\ln(x+1)dx$

$= 2(r+s) - 3(-p)$

$= 2r+2s+3p$

EX4

(1) $\int_{-2}^5 f(x)dx$

$= \int_{-2}^4 f(x)dx + \int_4^2 f(x)dx + \int_2^5 f(x)dx$

$= 3+(-2)+6$

$= 7$

(2) $\int_{-1}^3 f(x)dx$

$= \int_{-1}^6 f(x)dx + \int_6^{-3} f(x)dx + \int_{-3}^3 f(x)dx$

$= 1+4+2$

$= 7$

(3) $\leq tu = x+3$ $\quad = \int_0^2 f(x+3)dx$

$\frac{du}{dx} = 1$ $\quad = \int_3^5 f(u)du = \int_3^5 f(x)dx$

if $x = 0, u = 3$

if $x = 2, u = 5$ $\quad = \int_3^{-2} f(x)dx + \int_{-2}^5 f(x)dx$

$= 2+1$

$= 3$

(4) $\leq t\, u=2x$ $= \int_1^2 f(2x)dx = \int_2^4 f(u)\cdot\frac{1}{2}du$

$\frac{du}{dx}=2$ $= \frac{1}{2}\int_2^4 f(u)du = \frac{1}{2}\int_2^4 f(x)dx$

$\frac{1}{2}du = dx$

if $x=1, u=2$ $= \frac{1}{2}\left[\int_2^1 f(x)dx + \int_1^4 f(x)dx\right]$

if $x=2, u=4$ $= \frac{1}{2}(-3+5) = 1$

(5) $= \frac{\pi}{6}$

(6) $= \left[\frac{1}{3}\tan^{-1}\left(\frac{x}{3}\right)\right]_0^{\sqrt{3}}$

$= \frac{1}{3}\tan^{-1}\left(\frac{\sqrt{3}}{3}\right) - \frac{1}{3}\tan^{-1}(0)$

$= \frac{\pi}{18}$

(7) $= \frac{\pi}{8} - \frac{1}{4}$

(8) $= \int_{\frac{\pi}{12}}^{\frac{\pi}{6}} \frac{1+\cos4\theta}{2}d\theta$

$= \int_{\frac{\pi}{12}}^{\frac{\pi}{6}} \left(\frac{1}{2}+\frac{1}{2}\cos4\theta\right)d\theta$

$= \left[\frac{1}{2}\theta + \frac{1}{8}\sin4\theta\right]_{\frac{\pi}{12}}^{\frac{\pi}{6}}$

$= \frac{\pi}{24}$

5.8 Fundamental Theorem of Calculus, Part I

EX1

(1) $-\frac{4}{3}$

(2) $= \left[4x+4x^2-2x^3\right]_0^2$

$= 8$

(3) $= \frac{25}{3}$

(4) $= \int_{-1}^{8} x^{-\frac{1}{3}} dx$

$= \left[\frac{3}{2}x^{\frac{2}{3}}\right]_{-1}^{8} = \frac{9}{2}$

(5) $= 1 - \frac{1}{3}\ln2$

(6) $= \int_2^1 (9x^2+12+\frac{4}{x^2})dx$

$= \left[3x^3+12x-\frac{4}{x}\right]_2^1 = -35$

(7) $= 1$

(8) $= \left[\ln|\sec\theta| - \sec\theta\right]_0^{\frac{\pi}{3}}$

$= \ln2 - 1$

EX2

(1) $= 2$

(2) $= \left[2\ln(x^2+1)\right]_0^1$

$= \ln4$

(3) $= \frac{8\pi}{3}$

(4) $= \left[3\tan^{-1}x\right]_{-1}^{\sqrt{3}}$

$= 3\tan^{-1}(\sqrt{3}) - 3\tan^{-1}(-1)$

$= 3\cdot\frac{\pi}{3} - 3\cdot\left(-\frac{\pi}{4}\right)$

$= \frac{7}{4}\pi$

EX3

(1) 5.333

(2) 6.714

(3) 8.524

(4) 8.085

5.9 Fundamental Theorem of Calculus, Part II

EX1

(1) $= e^{-x}\sec^2x$

(2) $= \sqrt{1+x^2}$

(3) $= x^4 + \frac{3}{\sqrt{1-x^2}}$

(4) $= (x-3)\ln|x+1|$

EX2

(1) $\sqrt{2+\cos^3x}$

(2) $= \frac{d}{dx}\int_e^x \frac{\ln(u)}{u}du$

$= \frac{\ln x}{x}$

(3) $= \frac{6}{3+x^4}$

(4) $= \dfrac{d}{dx}\displaystyle\int_{o}^{x} u\cos u du$

$= x\cos x$

EX3

(1) $= -\dfrac{\sec^2 x}{1+\tan^2 x}$

(2) $\dfrac{d}{dx}\displaystyle\int_{2x}^{x^2} f(t)dt$

$= f(x^2) \cdot 2x - f(2x) \cdot 2$

$= 2xf(x^2) - 2f(2x)$

(3) $= \dfrac{1}{2\sqrt{x}} \cos\sqrt{x}$

(4) $\dfrac{d}{dx}\displaystyle\int_{-x}^{x^2} \sqrt{t^2+1}\,dt$

$= \sqrt{(x^2)^2+1} \cdot 2x - \sqrt{(-x)^2+1} \cdot (-1)$

$= 2x\sqrt{x^4+1} + \sqrt{x^2+1}$

(5) $= \dfrac{\cos x}{1-\sin^2 x} + \dfrac{\sin x}{1-\cos^2 x}$

(6) $\dfrac{d}{dx}\displaystyle\int_{\frac{1}{x}}^{3x} \cos(2t)dt$

$= \cos(2(3x)) \cdot 3 - \cos\left(2\left(\dfrac{1}{x}\right)\right) \cdot \left(-\dfrac{1}{x^2}\right)$

$= 3\cos 6x + \dfrac{1}{x^2}\cos\left(\dfrac{2}{x}\right)$

EX4

(1) $= \dfrac{2e^{x^2}}{x} - \dfrac{e^{\sqrt{x}}}{2x}$

(2) $\dfrac{dy}{dx} = \ln(\ln x) \cdot \dfrac{1}{x} - \ln(x+1) \cdot 1$

$= \dfrac{\ln(\ln x)}{x} - \ln(x+1)$

(3) $= \dfrac{\sec^2 x}{1-\tan^2 x} - \dfrac{3}{1-9x^2}$

(4) $\sin^2 2x \Rightarrow (\sin 2x)^2, \quad \cos^2 x \Rightarrow (\cos x)^2$

$\dfrac{dy}{dx} = \cos^2(\sin^2 2x) \cdot 2(\sin 2x) \cdot \cos 2x$

$\cdot\, 2 - \cos^2(\cos^2 x) \cdot 2(\cos x) \cdot (-\sin x)$

$= 4\cos^2(\sin^2 2x)\sin 2x \cos 2x$

$+ 2\cos^2(\cos^2 x)\cos x \sin x$

EX5

(1) Differentiate each side with respect to x.

$\dfrac{d}{dx}\displaystyle\int_{2}^{x} f(t)\,dt = \dfrac{d}{dx}(x^2\sin 3x)$

$f(x) = 2x \cdot \sin 3x + x^2 \cdot \cos 3x \cdot 3$

$f(\pi) = 2\pi \cdot \sin 3\pi + \pi^2 \cdot \cos 3\pi \cdot 3 = -3\pi^2$

(2) $A(2) = \displaystyle\int_{4}^{4} \dfrac{1}{1+t^2}dt$ Differentiate each side with

$\therefore A(2) = 0$

respect to x.

$A'(x) = \dfrac{d}{dx}\displaystyle\int_{4}^{x^2} \dfrac{1}{1+t^2}dt$

$= \dfrac{1}{1+(x^2)^2} \cdot 2x = \dfrac{2x}{1+x^4}$

$\therefore A'(2) = \dfrac{4}{17}$

EX6

$f'(x) = \dfrac{d}{dx}\displaystyle\int_{1}^{x} \dfrac{1}{t^4+1}dt = \dfrac{1}{x^4+1} = (x^4+1)^{-1}$

$f''(x) = -(x^4+1)^{-2} \cdot 4x^3$

$= -\dfrac{4x^3}{(x^4+1)^2}$

$f''(x) = 0$ at $x = 0$

x		0	
y''	+	0	−

$\therefore f$ is concave up on $x < 0$

5.10 Area between a Curve and the x-axis :

EX1

(1) $Area = 3$

(2)

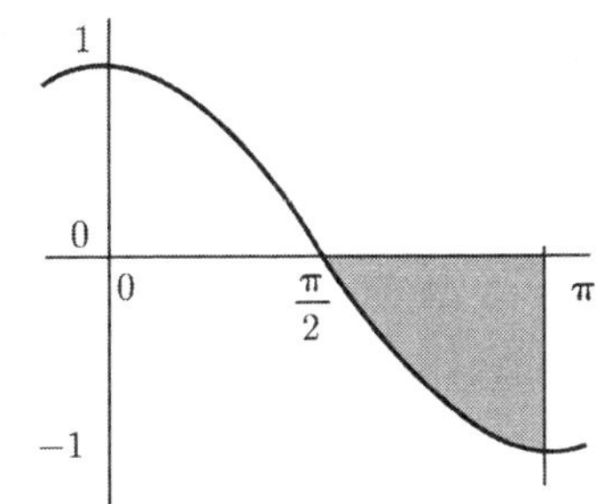

$$Area = -\int_{\frac{\pi}{2}}^{\pi} \cos x dx$$
$$= -[\sin x]_{\frac{\pi}{2}}^{\pi}$$
$$= -\left(\sin\pi - \sin\frac{\pi}{2}\right)$$
$$= -(0-1) = 1$$

(3) $Area = \frac{1}{2}\ln 2$

(4)

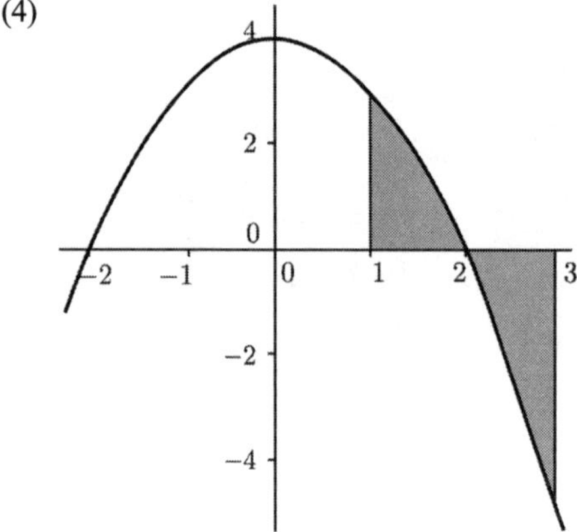

$Area$
$$= \int_1^2 (4-x^2)dx - \int_2^3 (4-x^2)dx$$
$$= \left[4x - \frac{1}{3}x^3\right]_1^2 - \left[4x - \frac{1}{3}x^3\right]_2^3$$
$$= \frac{5}{3} + \frac{7}{3} = 4$$

(5)

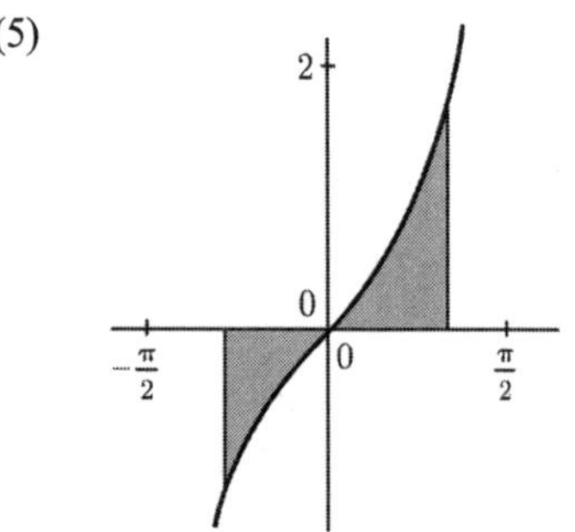

$Area$
$$= -\int_{-\frac{\pi}{4}}^{0} \tan x dx + \int_0^{\frac{\pi}{3}} \tan x dx$$
$$= -[\ln|\sec x|]_{-\frac{\pi}{4}}^{0} + [\ln|\sec x|]_0^{\frac{\pi}{3}}$$
$$= \ln\sqrt{2} + \ln 2 = \ln(2\sqrt{2})$$

(6)

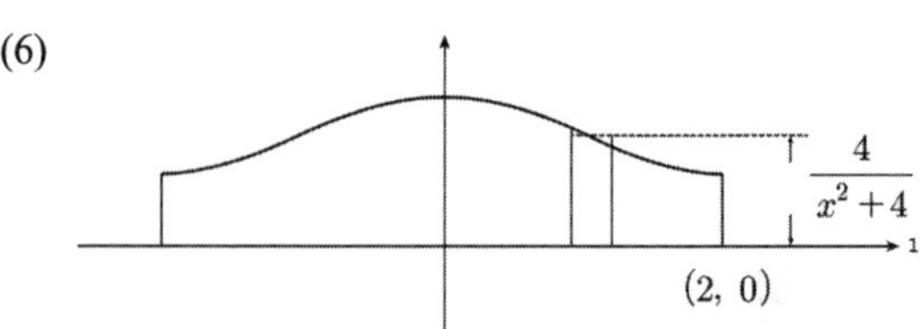

$$Area = 2\int_0^2 \frac{4}{x^2+4}dx = \pi$$

(7)

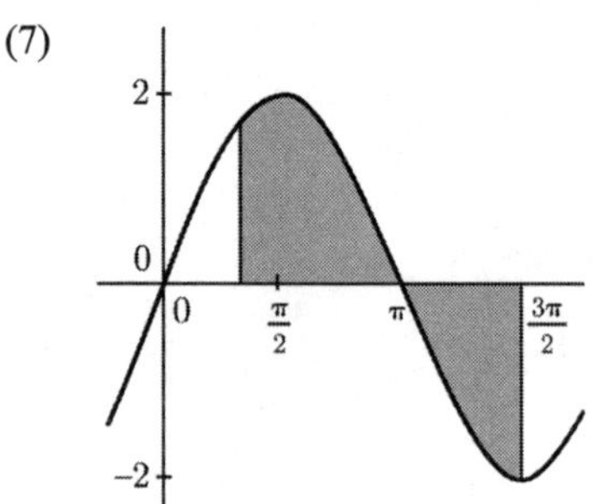

$Area$
$$= \int_{\frac{\pi}{3}}^{\pi} 2\sin x dx - \int_{\pi}^{\frac{3}{2}\pi} 2\sin x dx$$
$$= -2[\cos x]_{\frac{\pi}{3}}^{\pi} + 2[\cos x]_{\pi}^{\frac{3}{2}\pi}$$
$$= -2\left[-1-\left(\frac{1}{2}\right)\right] + 2[0-(-1)]$$
$$= 3+2 = 5$$

(8)

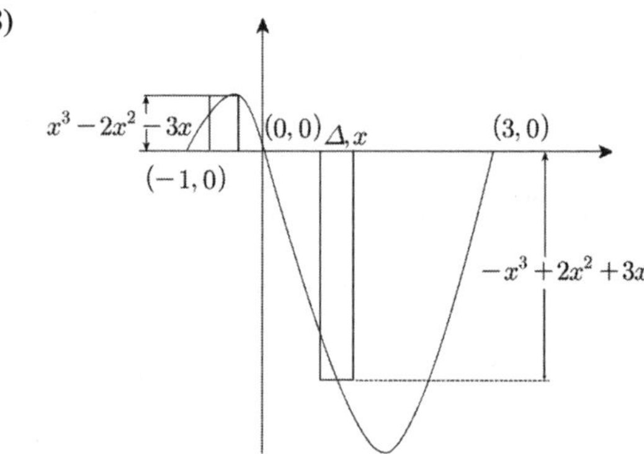

$$Area = \int_{-1}^{0} (x^3-2x^2-3x)dx$$
$$- \int_0^3 (x^3-2x^2-3x)dx$$
$$= \frac{71}{6}$$

EX2

$$= \int_{-1}^{1} f(x)dx + \int_{1}^{3} f(x)dx$$

$$= \int_{-1}^{1} (2x+1)dx + \int_{1}^{3} (x^2+2)dx$$

$$= [x^2+x]_{-1}^{1} + \left[\frac{1}{3}x^3+2x\right]_{1}^{3} = \frac{44}{3}$$

EX3

(1) $\frac{23}{3}$

(2)

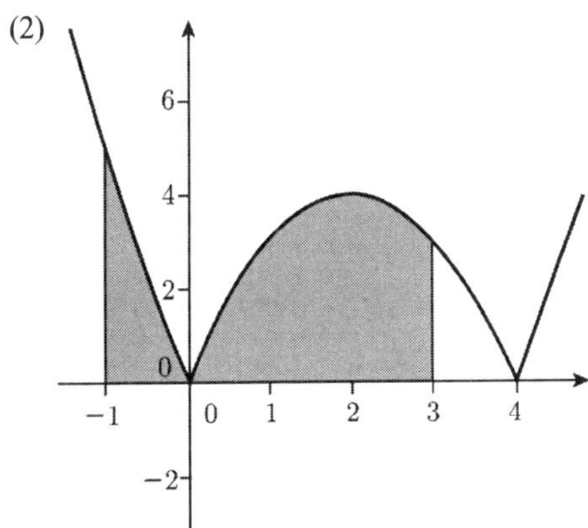

$$= \int_{-1}^{0} (x^2-4x)dx + \int_{0}^{3} (-x^2+4x)dx$$

$$= \left[\frac{1}{3}x^3-2x^2\right]_{-1}^{0} + \left[-\frac{1}{3}x^3+2x^2\right]_{0}^{3}$$

$$= \frac{7}{3}+9 = \frac{34}{3}$$

(3) 4

(4)

$e^x - 2 = 0$

$e^x = 2,\ x = \ln2$

$$= \int_{0}^{\ln2} (-e^x+2)dx + \int_{\ln2}^{2} (e^x-2)dx$$

$$= [-e^x+2x]_{0}^{\ln2} + [e^x-2x]_{\ln2}^{2}$$

$$= e^2+4\ln2-7$$

EX4

(1) $= -10$

(2) $= 10$

(3) $= -10+10-5$

$= -5$

(4) $= 10+10+5$

$= 25$

(5) $\int_{b}^{a} f(x)dx = -\int_{a}^{b} f(x)dx = 10$

(6) $\int_{d}^{b} f(x)dx = -\int_{b}^{d} f(x)dx$

$= -\left[\int_{b}^{c} f(x)dx + \int_{c}^{d} f(x)dx\right]$

$= -[10-5] = -5$

EX5

(1) π

(2)

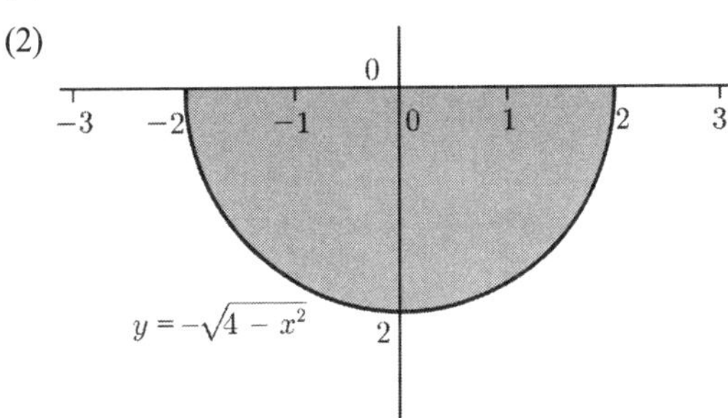

$= -\frac{1}{2}\pi(2)^2$

$= -2\pi$

(3) $30 - \frac{9}{2}\pi$

(4)

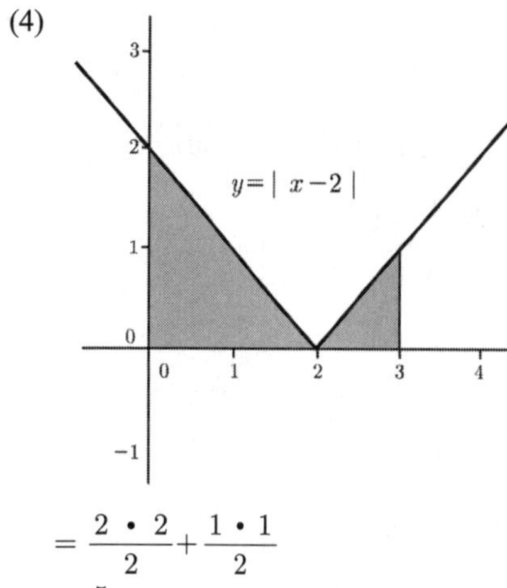

$= \frac{2 \cdot 2}{2} + \frac{1 \cdot 1}{2}$

$= \frac{5}{2}$

(5) 7

(6)

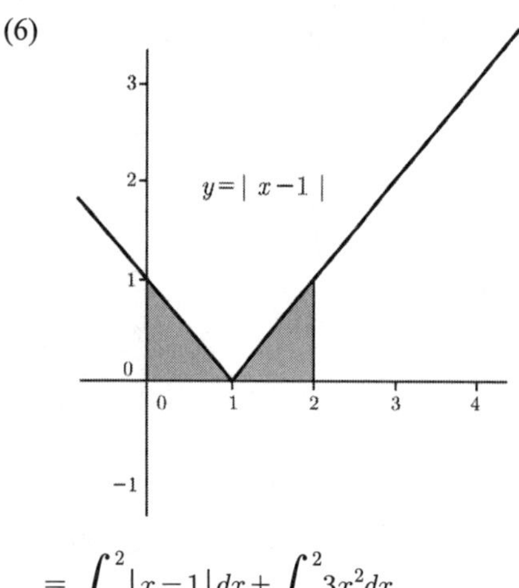

$= \int_0^2 |x-1|dx + \int_0^2 3x^2 dx$

$= 1 + [x^3]_0^2$

$= 9$

EX6

(1) 0

(2) $= \int_1^3 f(x)dx = \int_3^5 f(x)dx$

$= 2 - \frac{1}{2}\pi(1)^2 = 2 - \frac{\pi}{2}$

(3) $\frac{\pi}{2} - 3$

(4) $= \int_{-3}^{-1} f(x)dx + \int_{-1}^{0} f(x)dx$

$= -(4-\pi) + (-1) = \pi - 5$

(5) $2 + \frac{\pi}{2}$

(6) $\leq tu = x-3$

$\frac{du}{dx} = 1$

$du = dx$

$if\, x = 0,\ u = -3$

$if\, x = 8,\ u = 5$

$= \int_{-3}^{5} f(u)du = \int_{-3}^{5} f(x)dx$

$= \int_{-3}^{0} f(x)dx + \int_{0}^{5} f(x)dx$

$= \pi - 5 + \left(3 - \frac{\pi}{2}\right)$

$= \frac{\pi}{2} - 2$

(7) 4

(8) $\leq tu = 2x$

$\frac{du}{dx} = 2$

$\frac{1}{2}du = dx$

$if\, x = 1, u = 2$

$if\, x = 2, u = 4$

$= \int_2^4 f(u)\frac{1}{2}du$

$= \frac{1}{2}\int_2^4 f(x)dx$

$= \frac{1}{2}\left[\int_2^3 f(x)dx + \int_3^4 f(x)dx\right]$

$= \frac{1}{2}\left[\frac{1}{2} - \frac{\pi}{4}\right]$

$= \frac{1}{4} - \frac{\pi}{8}$

EX7

(1) $g(3) = \int_1^3 f(t)dt = 2$

$g(-3) = \int_1^{-3} f(x)dx = -\int_{-3}^{1} f(x)dx$

$= -\left[\int_{-3}^{-1} f(x)dx + \int_{-1}^{0} f(x)dx + \int_{0}^{1} f(x)dx\right]$

$= -\left[-\frac{1}{4}\pi(2)^2 - 1 + 1\right] = \pi$

$g(x) = \int_1^x f(t)dt \Rightarrow g'(x) = f(x) \Rightarrow g''(x) = f'(x)$

$g'(3) = f(3) = 0,$

$g''(3) = f'(3) = -1$

(2) $\frac{g(3) - g(-3)}{3-(-3)} = \frac{2-\pi}{6}$

(3) $\frac{g'(3) - g'(-1)}{3-(-1)} = \frac{f(3) - f(-1)}{4} = \frac{0-(-2)}{4} = \frac{1}{2}$

(4) $g(-1) = \int_{1}^{-1} f(t)dt = 0 \Rightarrow (-1, 0)$

$g'(-1) = f(-1) = -2$

$\therefore y - 0 = -2(x+1)$

(5) $g(x) \approx -2(x+1)$ *near* $x = -1$

$g(-1.1) \simeq -2(-0.1) = 0.2$

(6) *g is decreasing when* $g' = f < 0$.

$\therefore\ -4 < x < 0,\ 3 < x < 5$

(7) *g is concave up when* $g'' = f' > 0$.

$\therefore -1 < x < 1,\ 4 < x < 5$

(8) g has a relative maximum when

$g' = f$ changes sign from + to −

$\therefore x = 3$

(9) g has a point of inflection when $g'' = f'$ changes sign.

$\therefore x = -1, 1, 4$

(10)

x	-3		0		3		5
$g' = f$	0	−	0	+	0	−	0
g		↘		↗		↘	

$g(0)$ and $g(5)$ are the candidates for the absolute minimum.

$g(0) = \int_{1}^{0} f(t)dt = -\int_{0}^{1} f(t)dt = -1$

$g(5) = \int_{1}^{5} f(t)dt = \int_{1}^{3} f(t)dt + \int_{3}^{5} f(t)dt$

$= 2 - 1 = 1$

$g(0) < g(5)$

$\therefore$ absolute minimum $= g(0) = -1$

5.11 Substitution for Definite Intergrals

EX1

(1) 4

(2) $u = 1 + t^4 \Rightarrow du = 4t^3\,dt \Rightarrow \frac{1}{4}du = t^3\,dt; t = 0$

$\Rightarrow u = 1, t = 1 \Rightarrow u = 2$

$\int_0^1 t^3(1+t^4)\,dt = \int_1^2 \frac{1}{4}u^3 du$

$= \left[\frac{u^4}{16}\right]_1^2 = \frac{2^4}{16} - \frac{1^4}{16} = \frac{15}{16}$

(3) $4\tan^{-1}\frac{\pi}{4}$

(4) $u = 4 + r^2 \Rightarrow du = 2r\,dr \Rightarrow \frac{1}{2}du = r\,dr;$

$r = -1 \Rightarrow u = 5, r = 1 \Rightarrow u = 5$

$\int_{-1}^{1} \frac{5r}{(4+r^2)}dr = 5\int_5^5 \frac{1}{2}u^{-2}du = 0$

(5) $\frac{\pi}{3}$

(6) $tu = \ln x \Rightarrow du = \frac{1}{x}dx; x = 2$

$\Rightarrow u = \ln 2$ and $x = 4 \Rightarrow u = \ln 4;$

$\int_2^4 \frac{dx}{x(\ln x)^2} = \int_{\ln 2}^{\ln 4} u^{-2}du = \left[-\frac{1}{u}\right]_{\ln 2}^{\ln 4}$

$= -\frac{1}{\ln 4} + \frac{1}{\ln 2} = -\frac{1}{2\ln 2} + \frac{1}{\ln 2} = \frac{1}{2\ln 2} = \frac{1}{\ln 4}$

(7) $3^{5/2} - 1$

(8) $u = \tan\theta \Rightarrow du = \sec^2\theta\,d\theta;$

$\theta = 0 \Rightarrow u = 0, \theta = \frac{\pi}{4} \Rightarrow u = 1;$

$\int_0^{\pi/4} (1 + e^{\tan\theta})\sec^2\theta\,d\theta$

$= \int_0^{\pi/4} \sec^2\theta\,d\theta + \int_0^1 e^u du$

$= [\tan\theta]_0^{\pi/4} + [e^u]_0^1 = \left[\tan\left(\frac{\pi}{4}\right) - \tan(0)\right] + (e^1 - e^0)$

$= (1-0) + (e-1) = e$

(9) 0

(10) $u = \cos 2\theta \Rightarrow du = -2\sin 2\theta\,d\theta$

$\Rightarrow -\frac{1}{2}du = \sin 2\theta\,d\theta; \theta = 0$

$\Rightarrow u = 1, \theta = \frac{\pi}{6} \Rightarrow u = \cos 2\left(\frac{\pi}{6}\right) = \frac{1}{2}$

$\int_0^{\pi/6} \cos^{-3}2\theta\sin 2\theta\,d\theta = \int_1^{1/2} u^{-3}\left(-\frac{1}{2}du\right)$

$= -\frac{1}{2}\int_1^{1/2} u^{-3}du = \left[-\frac{1}{2}\left(\frac{u^{-2}}{-2}\right)\right]_1^{1/2}$

$= \frac{1}{4\left(\frac{1}{2}\right)^2} - \frac{1}{4(1)^2} = \frac{3}{4}$

(11) π

(12) $\int_{\sqrt{2}}^{2} \frac{\sec^2(\sec^{-1}x)}{x\sqrt{x^2-1}}dx = \int_{\pi/4}^{\pi/3} \sec^2 u\,du,$ *where* u

$= \sec^{-1}x$ and $du = \frac{dx}{x\sqrt{x^2-1}};$

$x = \sqrt{2} \Rightarrow u = \frac{\pi}{4}, x = 2 \Rightarrow u = \frac{\pi}{3}$

$$= [\tan u]_{\pi/4}^{\pi/3} = \tan\frac{\pi}{3} - \tan\frac{\pi}{4} = \sqrt{3} - 1$$

(13) $\sqrt{\ln 2}$

(14) $u = \tan x \Rightarrow du = \sec^2 x\, dx$;

$x = 0 \Rightarrow u = 0,\ x = \frac{\pi}{4} \Rightarrow u = 1$

$$\int_0^{\pi/4} \tan x \sec^2 x\, dx$$

$$= \int_0^1 u\, du = \left[\frac{u^2}{2}\right]^1 = \frac{1^2}{2} - 0 = \frac{1}{2}$$

EX2

(1) $f'(x) = g(x) \Rightarrow \int g(x)dx = f(x) + C$

$= [f(x) + x^3]_{-1}^{1}$

$= [f(1) + 1] - [f(-1) - 1]$

$= 6 - (-1) = 7$

(2) $u = 2x$

$du = 2dx$

$\frac{1}{2}du = dx$

if $x = -2, u = -4$

if $x = 2, \quad u = 4$

$$= \int_{-4}^{4} g(u) \cdot \frac{1}{2}du$$
$$= \int_{-4}^{4} \frac{1}{2} g(x)dx$$
$$= \left[\frac{1}{2}f(x)\right]_{-4}^{4}$$
$$= \frac{1}{2}(f(4) - f(-4))$$
$$= \frac{1}{2}(2 - (-3)) = \frac{5}{2}$$

(3) $u = \frac{x}{2}$

$du = \frac{1}{2}dx$

if $x = -2, u = -1$

if $x = 2, \quad u = 1$

$$= \int_{-1}^{1} 3g(u) \cdot 2du$$
$$= \int_{-1}^{1} 6g(x)dx$$
$$= [6f(x)]_{-1}^{1}$$
$$= 6[f(1) - f(-1)]$$
$$= 6(5 - 0)$$
$$= 30$$

(4) $u = x^2$

$du = 2x dx$

$\frac{1}{2}du = x\, dx$

if $x = -2, u = 4$

if $x = 0, \quad u = 0$

$$= \int_4^0 g(u) \cdot \frac{1}{2}du$$
$$= \int_4^0 \frac{1}{2}g(x)dx$$
$$= \left[\frac{1}{2}f(x)\right]_4^0$$
$$= \frac{1}{2}[f(0) - f(4)]$$
$$= \frac{1}{2}[-2 - 2] = -2$$

5.12 Integration by Parts for Definite Intergrals

EX1

(1) $= \frac{2}{e}$

(2) $= [x(-\cos x) - 1(-\sin x)]_{\frac{\pi}{3}}^{\frac{\pi}{2}}$

$= [-x\cos x + \sin x]_{\frac{\pi}{3}}^{\frac{\pi}{2}}$

$= \frac{\pi}{6} - \frac{\sqrt{3}}{2} + 1$

(3) $= \frac{\pi^2 - 4}{8}$

(4) $u = \ln x,\ v' = x$

$u' = \frac{1}{x},\ v = \frac{1}{2}x^2$

$= \left[\frac{1}{2}x^2 \ln x\right]_1^e - \int_1^e \frac{1}{2}x dx$

$= \left[\frac{1}{2}x^2 \ln x\right]_1^e - \left[\frac{1}{4}x^2\right]_1^e$

$= \frac{1}{4}e^2 + \frac{1}{4}$

(5) 1

(6) $u = \tan^{-1}x,\ v' = 1$

$u' = \frac{1}{1 + x^2},\ v = x$

$= [x\tan^{-1}x]_0^1 - \int_0^1 \frac{x}{1 + x^2}dx$

$= [x\tan^{-1}x]_0^1 - \left[\frac{1}{2}\ln(1 + x^2)\right]_0^1$

$= \frac{\pi}{4} - \ln\sqrt{2}$

(7) $= \frac{5}{12}\pi - \frac{\sqrt{3}}{2}$

(8) $u = \sqrt{x}$

$u^2 = x$

$2udu = dx$

if $x = 0, u = 0$

if $x = 4, u = 2$

$= \int_0^2 e^u \cdot 2udu$

$= \int_0^2 2ue^u du$

$= [2u(e^u) - 2(e^u)]_0^2$

$= [e^u(2u - 2)]_0^2$

$= 2e^2 + 2$

(9) $= \dfrac{e^{\pi}+1}{2}$

EX2

(1) $f'(e) = e^2$

An equation for the line tangent to the graph of f at the point (e,2) is

y-2=$e^2(x-e)$.

(2) $f''(x) = x + 2x\ln x$.

For $1 < x < 3, x > 0$ and ln $x > 0$, so $f''(x) > 0$. Thus, the graph of f is concave up on (1,3).

(3) Since $f(x) = \int (x^2 \ln x)\, dx$, we consider integration by parts.

$u = \ln x \qquad dv = x^2\, dx$

$du = \frac{1}{x} dx \qquad v = \int (x^2)\, dx = \frac{1}{3}x^3$

Therefore,

$$f(x) = \int (x^2 \ln x)\, dx$$
$$= \frac{1}{3}x^3 \ln x - \int \left(\frac{1}{3}x^3 \cdot \frac{1}{x}\right) dx$$
$$= \frac{1}{3}x^3 \ln x - \frac{1}{9}x^3 + C.$$

Since $f(e) = 2$, $2 = \frac{e^3}{3} - \frac{e^3}{9} + C$ and $C = 2 - \frac{2}{9}e^3$.

Thus, $f(x) = \frac{x^3}{3}\ln x - \frac{1}{9}x^3 + 2 - \frac{2}{9}e^3$.

EX3

(1) $f'(x) < 0$ for $0 < x < 3$

and $f'(x) > 0$ for $x > 3$

Therefore, f has a relative minimum at $x = 3$.

(2) $f''(x) = e^x + (x-3)e^2 = (x-2)e^x$

$f''(x) > 0$ for $x > 2$

$f'(x) < 0$ for $0 < x < 3$

Therefore, the graph of f is both decreasing and concave up on the interval $2 < x < 3$.

(3) $f(3) = f(1) + \int_1^3 f'(x)\, dx = 7 + \int_1^3 (x-3)e^x\, dx$

$u = x - 3 \quad dv = e^x\, dx$

$du = dx \qquad v = e^x$

$$f(3) = 7 + (x-3)e^x|_1^3 - \int_1^3 e^x\, dx$$
$$= 7 + |(x-3)e^x - e^x|_1^3$$
$$= 7 + 3e - e^3$$

5.13 Even and Odd Functions

EX1

(1) $= \int_{-1}^{1} (x^4 + 3x^2 + 5)dx$
$$= 2\int_0^1 (x^4 + 3x^2 + 5)dx$$
$$= 2\left[\frac{1}{5}x^5 + x^3 + 5x\right]_0^1$$
$$= \frac{62}{5}$$

(2) 0

(3) $= \int_{-2}^{2} (3x^2 + 2)dx$
$$= 2\int_0^2 (3x^2 + 2)dx$$
$$= 2[x^3 + 2x]_0^2$$
$$= 24$$

(4) $= \int_{-\pi}^{\pi} 1\, dx$
$$= 2\int_0^{\pi} 1\, dx$$
$$= 2[x]_0^{\pi} = 2\pi$$

5.14 Integration by Partial Fraction

(1) $\ln \frac{27}{4}$

(2) $\frac{1}{x^2 + 2x} = \frac{A}{x} + \frac{B}{x+2} \Rightarrow 1 = A(x+2) + Bx;$

$x = 0 \Rightarrow A = \frac{1}{2}; x = -2 \Rightarrow B = -\frac{1}{2};$

$$\int \frac{dx}{x^2 + 2x} = \frac{1}{2}\int \frac{dx}{x} - \frac{1}{2}\int \frac{dx}{x+2}$$
$$= \frac{1}{2}[\ln|x| - \ln|x+2|] + C$$

(3) $y - 2\ln|y+1| + C$

(4) $\dfrac{x+4}{x^2+5x-6} = \dfrac{A}{x+6} + \dfrac{B}{x-1}$

$\Rightarrow x+4 = A(x-1) + B(x+6); x = 1$

$\Rightarrow B = \dfrac{5}{7}; x = -6 \Rightarrow A = \dfrac{-2}{-7} = \dfrac{2}{7};$

$$\int \frac{x+4}{x^2+5x-6}dx = \frac{2}{7}\int \frac{dx}{x+6} + \frac{5}{7}\int \frac{dx}{x-1}$$

$$= \frac{2}{7}\ln|x+6| + \frac{5}{7}\ln|x-1| + C$$

$$= \frac{1}{7}\ln|(x+6)^2(x-1)^5| + C$$

(5) $\dfrac{1}{t^3+t^2-2t} = \dfrac{A}{t} + \dfrac{B}{t+2} + \dfrac{C}{t-1}$

$\Rightarrow 1 = A(t+2)(t-1) + Bt(t-1) + Ct(t+2);$

$t = 0 \Rightarrow A = -\dfrac{1}{2}; t = -2 \Rightarrow B = \dfrac{1}{6};$

$t = 1 \Rightarrow C = \dfrac{1}{3}; \displaystyle\int \frac{dt}{t^3+t^2-2t}$

$$= -\frac{1}{2}\int \frac{dt}{t} + \frac{1}{6}\int \frac{dt}{t+2} + \frac{1}{3}\int \frac{dt}{t-1}$$

$$= -\frac{1}{2}\ln|t| + \frac{1}{6}\ln|t+2| + \frac{1}{3}\ln|t-1| + C$$

(6) $\dfrac{1}{(x+1)(x^2+1)} = \dfrac{A}{x+1} + \dfrac{Bx+C}{x^2+1}$

$\Rightarrow 1 = A(x^2+1) + (Bx+C)(x+1);$

$x = -1 \Rightarrow A = \dfrac{1}{2};$ coefficient of x^2

$= A+B \Rightarrow A+B = 0 \Rightarrow B = -\dfrac{1}{2};$

constant $= A + C \Rightarrow A + C = 1$

$\Rightarrow C = \dfrac{1}{2}; \displaystyle\int_0^1 \frac{dx}{(x+1)(x^2+1)}$

$$\frac{1}{2}\int_0^1 \frac{dx}{x+1} + \frac{1}{2}\int_0^1 \frac{(-x+1)}{x^2+1}dx$$

$$= \left[\frac{1}{2}\ln|x+1| - \frac{1}{4}\ln(x^2+1) + \frac{1}{2}\tan^{-1}x\right]_0^1$$

$$= \left(\frac{1}{2}\ln 2 - \frac{1}{4}\ln 2 + \frac{1}{2}\tan^{-1}1\right)$$

$$- \left(\frac{1}{2}\ln 1 - \frac{1}{4}\ln 1 + \frac{1}{2}\tan^{-1}0\right)$$

$$= \frac{1}{4}\ln 2 + \frac{1}{2}\left(\frac{\pi}{4}\right) = \frac{(\pi+2\ln 2)}{8}$$

(7) $\dfrac{4x}{(x-1)^2(x+1)}$

$$= \frac{A}{x-1} + \frac{b}{(x-1)^2} + \frac{c}{x+1}$$

Multiplying by the least common denominator, $(x-1)^2(x-1)$

we get

$4x = A(x-1)(x+1) + B(x+1) + C(x-1)^2$

$= A(x+C)x^2 + (B-2C)x + (-A+B+C)$

Now we equate coefficieuts;

$A+C=0 \quad B-2C=4, \quad -A+B+C=0$

Solving, we obtain $A=1$, $B=2$, and $C=-1$, so

$$\int \frac{1}{x-1} + \frac{2}{(x-1)^2} - \frac{1}{x+1}dx$$

$$= \ln|x-1| - \frac{2}{x-1} - \ln|x+1| + c$$

(8) Since $x^3+4x = x(x^2+4)$ can't be factored further, we write

$$\frac{2x^2-x+4}{x(x^2+4)} = \frac{A}{x} + \frac{Bx+C}{x^2+4}$$

Multiplying by $x(x^2+4)$, we have

$$2x^2-x+4 = A(x^2+4) + (Bx+C)x$$

$$= (A+B)x^2 + Cx + 4A$$

Equating coefficients, we obtain

$$A+B=2 \quad C=-1 \quad 4A=4$$

Thus $A=1, B=1,$ and $C=-1$ and so

$$\int \frac{2x^2-x+4}{x^3+4x}dx = \int\left(\frac{1}{x} + \frac{x-1}{x^2+4}\right)dx$$

In order to integrate the second term we split it into two parts:

$$\int \frac{x-1}{x^2+4}dx = \int \frac{x}{x^2+4}dx - \int \frac{1}{x^2+4}dx$$

$$\int \frac{2x^2-x+4}{x(x^2+4)}dz$$

$$= \int \frac{1}{x}dx + \int \frac{x}{x^2+4}dx - \int \frac{1}{x^2+4}dz$$

$$= \ln|x| + \frac{1}{2}\ln(x^2+4) - \frac{1}{2}\tan^{-1}(x/2) + C$$

Multiple Choice

01 C 02 A 03 A 04 A 05 A 06 A 07 A 08 A 09 B 10 E 11 C 12 B 13 E 14 A 15 C
16 C 17 D 18 E 19 B 20 B 21 D 22 E 23 B 24 A 25 D 26 E 27 A 28 C

06 Applications of Definite Integrals

6.1 Average Value of a Function

EX1

(1) y_{avg}

$$=\left(\frac{1}{\sqrt{3}-0}\right)\int_0^{\sqrt{3}}(x^2-1)dx$$

$$=\frac{1}{\sqrt{3}}\int_0^{\sqrt{3}}x^2dx-\frac{1}{\sqrt{3}}\int_0^{\sqrt{3}}1dx$$

$$=\frac{1}{\sqrt{3}}\left(\frac{(\sqrt{3})^3}{3}\right)-\frac{1}{\sqrt{3}}\ (\sqrt{3}-0)$$

$$=1-1=0.$$

(2) $y_{avg}=\dfrac{1}{2-(-1)}\displaystyle\int_{-1}^{2}y\,dx$

$$=\frac{1}{3}\int_{-1}^{2}(x-3)^2\,dx$$

$$=\frac{1}{3}\left[\frac{1}{3}(x-3)^3\right]_{-1}^{2}$$

$$=\frac{1}{9}\left[(-1)^3-(-4)^3\right]=\frac{1}{9}[-1+64]=7$$

(3) y_{avg}

$$=\left(\frac{1}{3-(-1)}\right)\int_{-1}^{3}(|x|-1)dx$$

$$=\frac{1}{4}\int_{-1}^{1}(|x|-1)dx+\frac{1}{4}\int_{1}^{3}(|x|-1)dx$$

$$=\frac{1}{4}(-1+2)=\frac{1}{4}$$

(4) $y_{avg}=\dfrac{1}{2}\displaystyle\int_{-1}^{1}\frac{20}{x^2+1}dx$

$$=\frac{1}{2}\left[20\tan^{-1}x\right]_{-1}^{1}$$

$$=10\left[\tan^{-1}x\right]_{-1}^{1}$$

$$=10\left[\frac{\pi}{4}-\left(-\frac{\pi}{4}\right)\right]=5\pi$$

(5) $y_{avg}=\dfrac{1}{\ln 3}\displaystyle\int_0^{\ln3}(e^x-1)dx$

$$=\frac{1}{\ln 3}\left[e^x-x\right]_0^{\ln3}$$

$$=\frac{1}{\ln 3}[(e^{\ln3}-\ln3)-1]$$

$$=\frac{1}{\ln 3}[3-\ln3-1]=\frac{2-\ln3}{\ln3}$$

(6) $y_{avg}=\dfrac{1}{\pi}\displaystyle\int_0^{\pi}y\,dx$

$$=\frac{1}{\pi}\int_0^{\pi}\sin x\ dx$$

$$=\frac{1}{\pi}\left[-\cos x\right]_0^{\pi}$$

$$=\frac{1}{\pi}(1-(-1))=\frac{2}{\pi}$$

EX2

$$R_{avg}=\frac{1}{18-0}\int_0^{18}R(t)\,dt$$

$$=\frac{1}{18}\left[\frac{(8+13)\cdot 6}{2}+\frac{(13+12)\cdot 6}{2}+\frac{(12+8)\cdot 6}{2}\right]$$

$$=\frac{1}{18}[63+75+60]=11(L/\min)$$

EX3

(1) Average velocity

$$=\frac{1}{4-0}\int_0^4(5t-t^2+100)\,dt=104\frac{2}{3}\text{ mph.}$$

(2) (a) Average velocity

$$=v_{avg}$$

$$=\frac{1}{3-1}\int_1^3 v(t)dt$$

$$=\frac{1}{2}\int_1^3(12t^2+1)dt$$

$$=\frac{1}{2}[4t^3+t]_1^3=\frac{1}{2}[111-5]=53$$

(b) Average accerleration

$$=a_{avg}=\frac{v(3)-v(1)}{3-1}=\frac{109-13}{2}=48$$

EX4

(C) Average area

$$= \frac{1}{5-2}\int_2^5 \pi x^2 dx = \frac{\pi}{3}\left(\frac{125-8}{3}\right) = 13\pi \text{in}^2.$$

EX5

(A) Average volume

$$= \frac{1}{4-0}\int_0^4 \frac{30}{(t+1)^2}dt \simeq 6 \text{ mg}.$$

EX6

(D) $\frac{1}{25-0}\int_0^{25} P(t)dt = \frac{1}{25}\int_0^{25} 6e^{0.024t} \simeq 8.2$

(billion people).

EX7

(1) $V_{avg} = \frac{1}{10}\int_0^{10} V(t)\,dt$

$$= \frac{1}{10}\int_0^{10} 35{,}000e^{-0.223t}\,dt$$

$$= \int_0^{10} 3{,}500e^{-0.223t}dt = \$14{,}007.40$$

(2) $P_{avg} = \frac{1}{30-10}\int_{10}^{30} P(t)dt$

$$= \frac{1}{20}\int_{10}^{30} \frac{20{,}000}{1+299e^{-0.2t}}dt$$

$$= \int_{10}^{30} \frac{1{,}000}{1+299e^{-0.2t}}dt = 4148.442 \simeq 4148$$

6.2 Area Between Curves

EX1

(1) 48/5

(2)

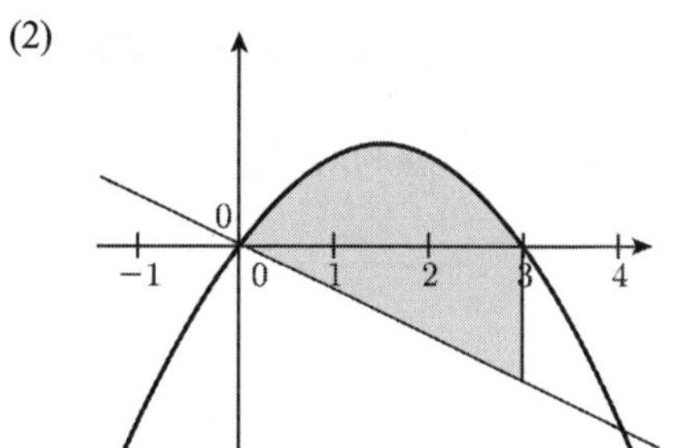

$$Area = \int_0^3 [-x^2+3x-(-x)]\,dx$$

$$= \int_0^2 (-x^2+4x)dx$$

$$= \left[-\frac{1}{3}x^3+2x^2\right]_0^3$$

$$= (-9+18) = 9$$

(3)

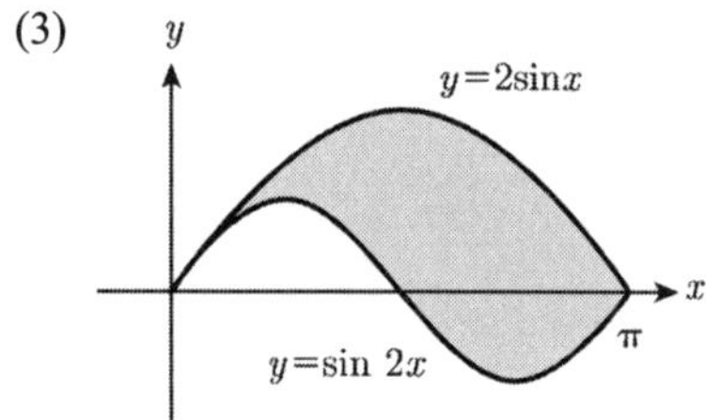

$$\Rightarrow A = \int_0^{\pi} (2\sin x - \sin 2x)\,dx$$

$$= \left[-2\cos x + \frac{\cos 2x}{2}\right]_0^{\pi}$$

$$= \left[-2(-1)+\frac{1}{2}\right] - \left(-2\cdot 1+\frac{1}{2}\right) = 4$$

(4)

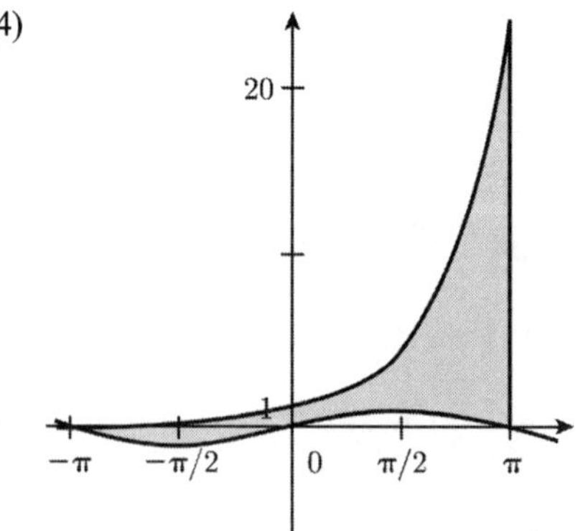

$$Area = \int_0^{\pi}(e^x - \sin x)dx$$

$$= \left[e^x + \cos x\right]_0^{\pi}$$

$$= (e^{\pi} - 1) - (1 + 1)$$

$$= e^{\pi} - 3$$

(5) $\pi/2$

(6)

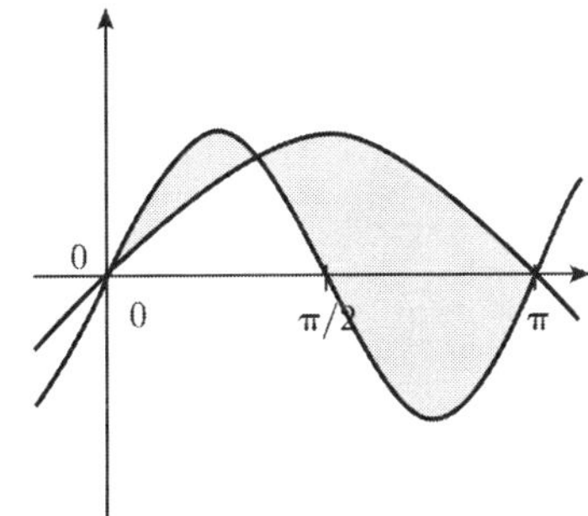

$$Area = \int_0^{\frac{\pi}{3}}(\sin 2x - \sin x)dx$$

$$+ \int_{\frac{\pi}{3}}^{\pi}(\sin x - \sin 2x)dx$$

$$= \left[-\frac{1}{2}\cos 2x + \cos x\right]_0^{\frac{\pi}{3}} + \left[-\cos x + \frac{1}{2}\cos 2x\right]_{\frac{\pi}{3}}^{\pi}$$

$$= \left(\frac{1}{4} + \frac{1}{2}\right) - \left(-\frac{1}{2} + 1\right) + \left(1 + \frac{1}{2}\right) - \left(-\frac{1}{2} - \frac{1}{4}\right)$$

$$= \frac{1}{4} + \frac{9}{4} = \frac{5}{2}$$

EX2

(1) $9/2$

(2)

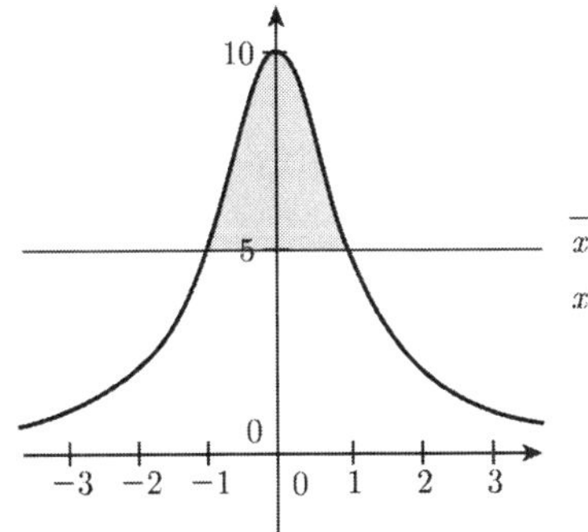

$\frac{10}{x^2+3} = 5$

$x^2 + 1 = 2$

$\therefore x = \pm 1$

$$Area = \int_{-1}^{1}\left(\frac{10}{x^2+1} - 5\right)dx$$

$$= 2\int_0^1\left(\frac{10}{x^2+1} - 5\right)dx$$

$$= 2\left[10\tan^{-1}x - 5x\right]_0^1$$

$$= 2\left[\left(\frac{10\pi}{4} - 5\right) - 0\right] = 5\pi - 10$$

(3) $\frac{3}{2} - \ln 4$

(4)

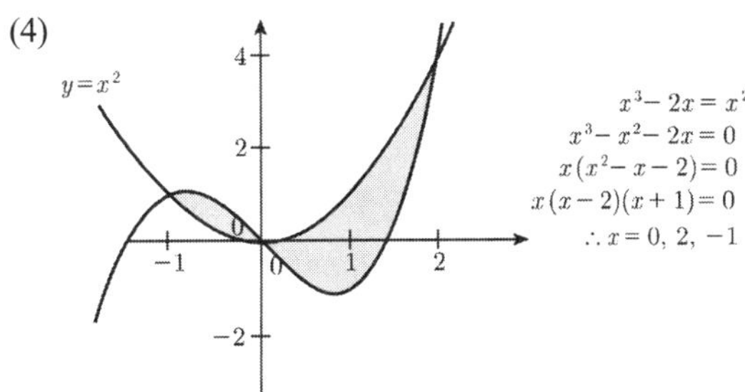

$$Area = \int_{-1}^{0}(x^3 - 2x - x^2)dx + \int_0^2(x^2 - x^3 + 2x)\,dx$$

$$= \left[\frac{1}{4}x^4 - x^2 - \frac{1}{3}x^3\right]_{-1}^{0} + \left[\frac{1}{3}x^3 - \frac{1}{4}x^4 + x^2\right]_0^2$$

$$= 0 - \left(\frac{1}{4} - 1 + \frac{1}{3}\right) + \left(\frac{8}{3} - 4 + 4\right) - 0$$

$$= \frac{5}{12} + \frac{8}{3} = \frac{37}{12}$$

(5) $131/4$

(6) $y = |x^2 - 4| = \begin{cases} x^2 - 4, & x \le -2 \text{ or } x \ge 2 \\ 4 - x^2, & -2 \le x \le 2 \end{cases}$

for $x \le -2$ and $x \ge 2$: $x^2 - 4 = \frac{x^2}{2} + 4$

$\Rightarrow 2x^2 - 8 = x^2 + 8 \;\Rightarrow x^2 = 16 \;\Rightarrow x = \pm 4$;

for $-2 \le x \le 2$: $4 - x^2 = \frac{x^2}{2} + 4$

$\Rightarrow 8 - 2x^2 = x^2 + 8$

$\Rightarrow x^2 = 0 \Rightarrow x = 0$; by symmetry of the graph,

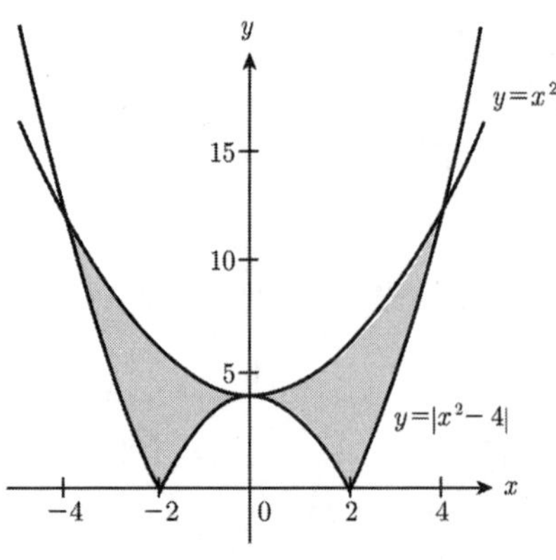

$$A=2\int_0^2\left[\left(\frac{x^2}{2}+4\right)-(4-x^2)\right]dx$$

$$+2\int_2^4\left[\left(\frac{x^2}{2}+4\right)-(x^2-4)\right]dx$$

$$=2\left[\frac{x^3}{2}\right]_0^2+2\left[8x-\frac{x^3}{6}\right]_2^4$$

$$=2\left(\frac{8}{2}-0\right)+2\left(32-\frac{64}{6}-16+\frac{8}{6}\right)$$

$$=\frac{64}{3}$$

EX3

(1) $Area=10.225$

$[-10, 10]\times[-10, 10]$

$x_1=-1.746749$
$x_2=4.8478775$

(2) $Area=6.2347$

$[-10, 10]\times[-10, 10]$

$x_1=-0.585$
$x_2=2.131$

(3) $Area=92.3188504\approx 92.319$

$[-10, 10]\times[-10, 30]$

$x_1=-4.470857$
$x_2=2.5879522$

(4) $Area=72.5308651\approx 72.531$

$[-10, 10]\times[-50, 10]$

$x_1=-6.076252$
$x_2=0$
$x_3=2.7429189$

EX4

(1) 18

(2)

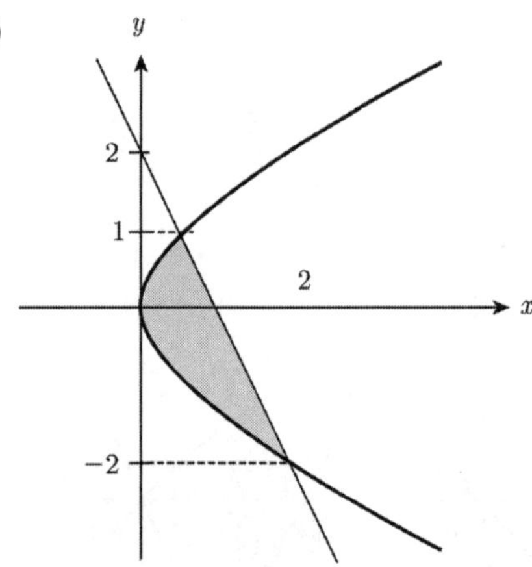

$$Area=\int_{-2}^{1}(x_1-x_2)dy$$

$$=\int_{-2}^{1}(2-y-y^2)dy$$

$$=\left[2y-\frac{1}{2}y^2-\frac{1}{3}y^3\right]_{-2}^{1}$$

$$=\left(2-\frac{1}{2}-\frac{1}{3}\right)-\left(-4-2+\frac{8}{3}\right)=\frac{9}{2}$$

(3) 4

(4)

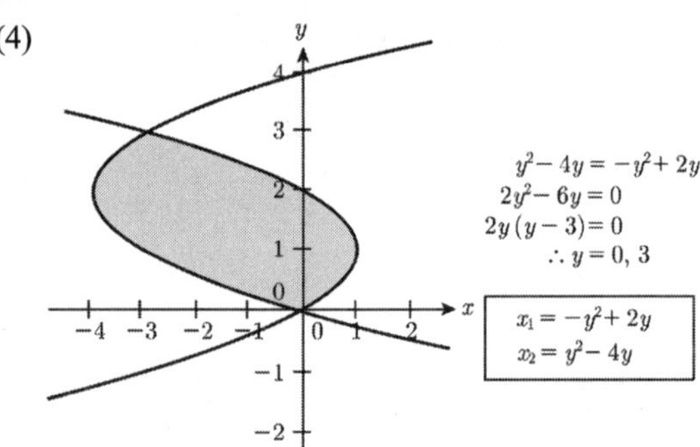

$$Area=\int_0^3(x_1-x_2)dy$$

$$=\int_0^3[(-y^2+2y-(y^2-4y)]dy$$

$$=\int_0^3(-2y^2+6y)dy$$

$$=\left[-\frac{2}{3}y^3+3y^2\right]_0^3=-18+27=9$$

(5)

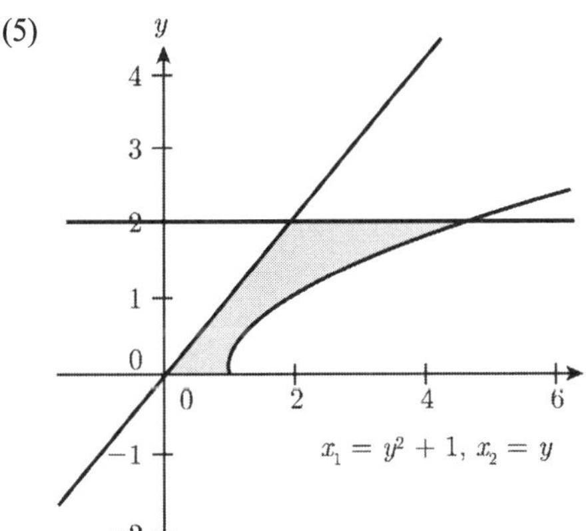

$$Area = \int_0^2 (x_1 - x_2)dy$$
$$= \int_0^2 (y^2 + 1 - y)dy$$
$$= \left[\frac{1}{3}y^3 + y - \frac{1}{2}y^2\right]_0^2$$
$$= \frac{8}{3} + 2 - 2 = \frac{8}{3}$$

(6)

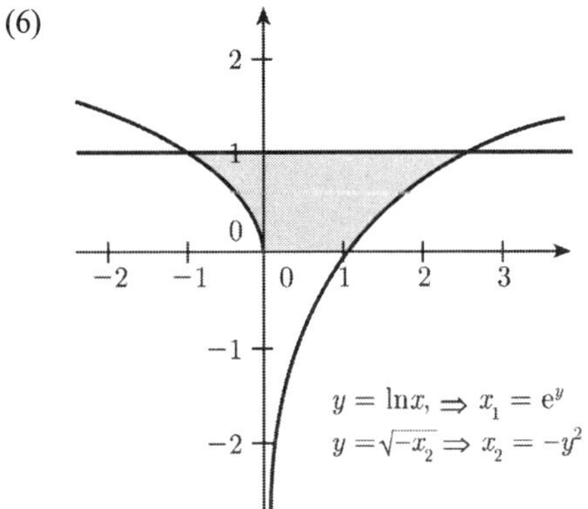

$$Area = \int_0^1 (x_1 - x_2)dy$$
$$= \int_0^1 (e^y + y^2)dy$$
$$= \left[e^y + \frac{1}{3}y^3\right]_0^1$$
$$= \left(e + \frac{1}{3}\right) - (1 - 0) = e - \frac{2}{3}$$

(7)

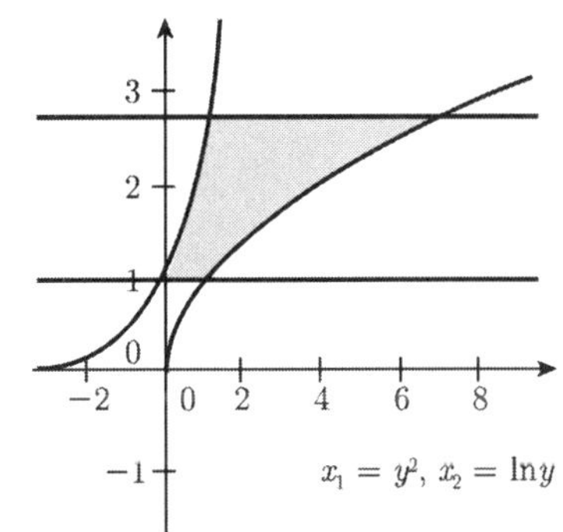

$$Area = \int_1^e (x_1 - x_2)dy$$
$$= \int_1^e (y^2 - \ln y)dy$$
$$= \left[\frac{1}{3}y^3 + (y\ln y + y)\right]_1^e = \left[\frac{1}{3}y^3 - y\ln y + y\right]_1^e$$
$$= \left(\frac{1}{3}e^3 - e\ln e + e\right) - \left(\frac{1}{3} - \ln 1 + 1\right)$$
$$= \frac{1}{3}e^3 - \frac{4}{3}$$

(8)

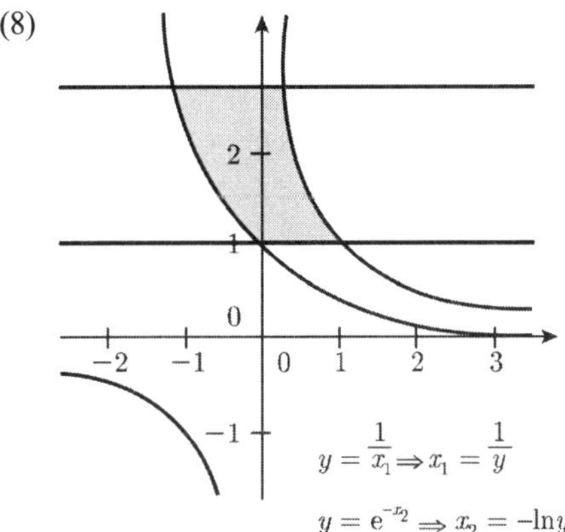

$$Area = \int_1^e (x_1 - x_2)dy$$
$$= \int_1^e \left(\frac{1}{y} + \ln y\right)dy$$
$$= [\ln|y| + y\ln y - y]_1^e$$
$$= (\ln e + e\ln e - e) - (\ln 1 + \ln 1 - 1)$$
$$= 1 - (-1) = 2$$

6.3 Volume of Solids with Known Cross Sections

EX1

$Area = (2y)^2 = 4y^2$

$$V = \int_0^3 4y^2 dx = \int_0^3 4x\,dx$$

$$= [2x^2]_0^3$$

$$= 18$$

EX2

$$A(x) = \frac{\pi\,(diameter)^2}{4}$$
$$= \frac{\pi[(2-x^2)-x^2]^2}{4} = \frac{\pi[2(1-x^2)]^2}{4}$$
$$= \pi(1-2x^2+x^4); a = -1, b = 1;$$

$$V = \int_a^b A(x)\,dx$$

$$= \int_{-1}^1 \pi(1-2x^2+x^4)dx = \pi\left[x - \frac{2}{3}x^3 + \frac{x^5}{5}\right]_{-1}^1$$

$$= 2\pi\left(1 - \frac{2}{3} + \frac{1}{5}\right) = \frac{16\pi}{15}$$

EX3

$2\sqrt{3}$

EX4

$$A(x) = \frac{(diagonal)^2}{2}$$
$$= \frac{[\sqrt{1-x^2}-(-\sqrt{1-x^2})]^2}{2} = \frac{(2\sqrt{1-x^2})^2}{2}$$
$$= 2(1-x^2);\ a = -1, b = 1;$$

$$V = \int_a^b A(x)\,dx = 2\int_{-1}^1 (1-x^2)dx$$

$$= 2\left[x - \frac{x^3}{3}\right]_{-1}^1 = 4\left(1 - \frac{1}{3}\right)$$

EX5

(1) $Area = \int_0^4 \sqrt{x}\,dx + \frac{1}{2} \cdot 2 \cdot 2$

$$= \frac{2}{3}x^{3/2}\Big|_{x=0}^{x=4} + 2 = \frac{22}{3}$$

(2) $y = \sqrt{x} \Rightarrow x = y^2$

$y = 6 - x \Rightarrow x = 6 - y$

Width $= (6-y) - y^2$

Volume $= \int_0^2 2y(6-y-y^2)\,dy$

(3) $g'(x) = -1$

Thus a line perpendicular to the graph of g has slope 1

$$f'(x) = \frac{1}{2\sqrt{x}}$$

$$\frac{1}{2\sqrt{x}} = 1 \Rightarrow x = \frac{1}{4}$$

The point P has coordinates $\left(\frac{1}{4}, \frac{1}{2}\right)$

EX6

$$A(y) = \frac{\pi}{4}(diameter)^2 = \frac{\pi}{4}(\sqrt{5}y^2 - 0)^2 = \frac{5\pi}{4}y^4;$$

$$c = 0,\ d = 2;\ V = \int_c^d \frac{5\pi}{4}y^4 dy$$

$$= \left[\left(\frac{5\pi}{4}\right)\left(\frac{y^5}{5}\right)\right]_0^2 = \frac{\pi}{4}(2^5 - 0) = 8\pi$$

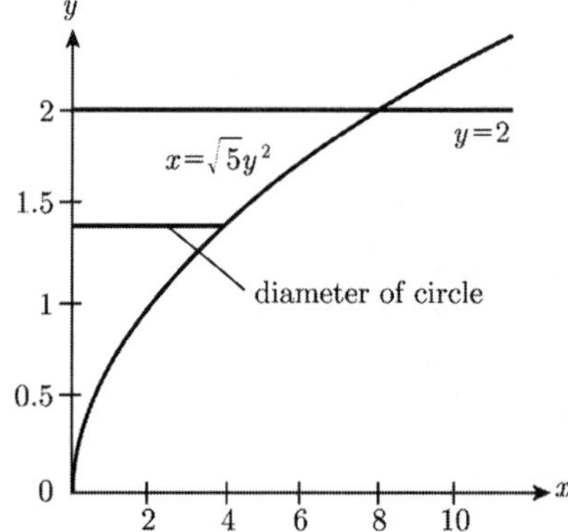

EX7

$Area=(2x)(8x)=16x^2$

$$V=\int_0^1 16x^2dy=\int_0^1 16y\,dy$$
$$=[8y^2]_0^1$$
$$=8$$

EX8

$$A(y)=\frac{1}{2}(\leq g)(\leq g)$$
$$=\frac{1}{2}[\sqrt{1-y^2}-(-\sqrt{1-y^2})]^2$$
$$=\frac{1}{2}(2\sqrt{1-y^2})^2=2(1-y^2);$$
$$c=-1, d=1;$$
$$V=\int_c^d A(y)dy=\int_{-1}^1 2(1-y^2)\,dy$$
$$=2\left[y-\frac{y^3}{3}\right]_{-1}^2=4\left(1-\frac{1}{3}\right)=\frac{8}{3}$$

EX9

$2\pi/15$

6.4 Volume of Solids of Revolution : Disks, Washers and Cylindrical Shells

EX1

(1) $32\pi/5$

(2)

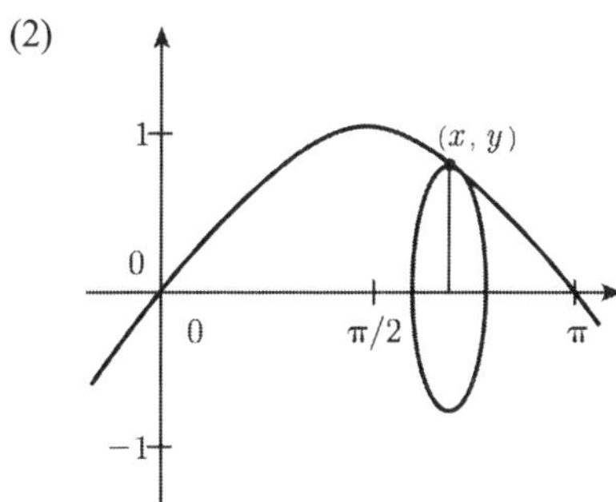

$$V=\pi\int_0^\pi y^2\,dx$$
$$=\pi\int_0^\pi \sin^2 x\,dx$$
$$=\pi\int_0^\pi\left[\frac{1}{2}-\frac{1}{2}\cos 2x\right]dx$$
$$=\pi\left[\frac{1}{2}x-\frac{1}{4}\sin 2x\right]_0^\pi$$
$$=\pi\left[\frac{\pi}{2}\right]=\frac{\pi^2}{2}$$

(3) $\frac{\pi}{2}(1-\frac{1}{e^2})$

(4) $\frac{\pi}{30}$

EX2

(1) 2π

(2)

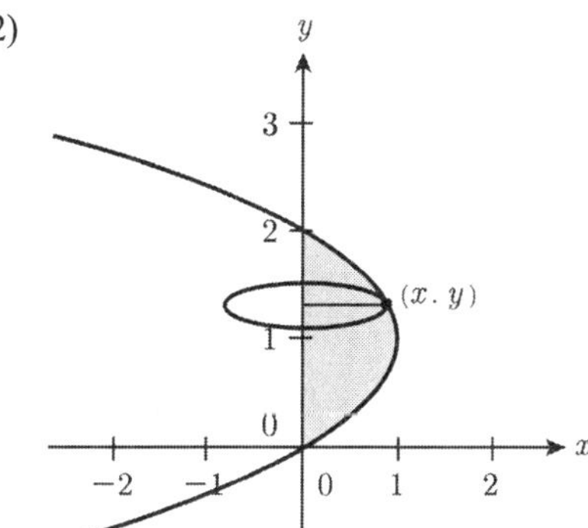

$Area=\pi x^2$

$$V=\pi\int_0^2(-y^2+2y)^2\,dy$$
$$=\pi\int_0^2\left[y^4-4y^3+4y^2\right]dy$$
$$=\pi\left[\frac{1}{5}y^5-y^4+\frac{4}{3}y^3\right]_0^2$$
$$=\pi\left[\frac{32}{5}-16+\frac{32}{5}\right]$$
$$=\frac{16\pi}{15}$$

(3) 2π

(4)

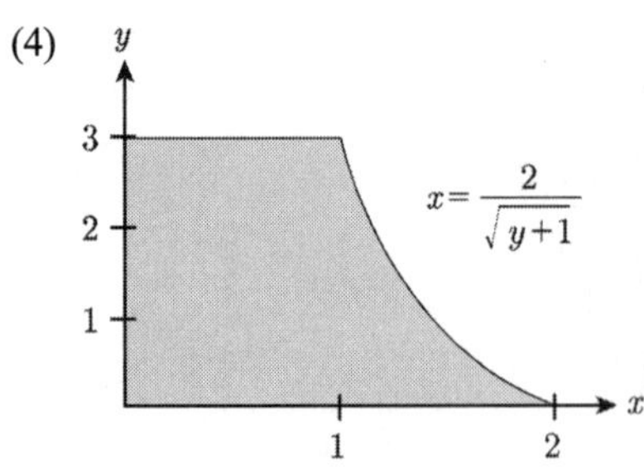

$$R(y)=\frac{2}{\sqrt{y+1}} \Rightarrow$$

$$V=\int_0^3 \pi[R(y)]^2 dy=4\pi\int_0^3 \frac{1}{y+1}dy$$

$$=4\pi[\ln|y+1|]_0^3$$

$$=4\pi[\ln 4-\ln 1]$$

$$=4\pi \ln 4$$

EX3

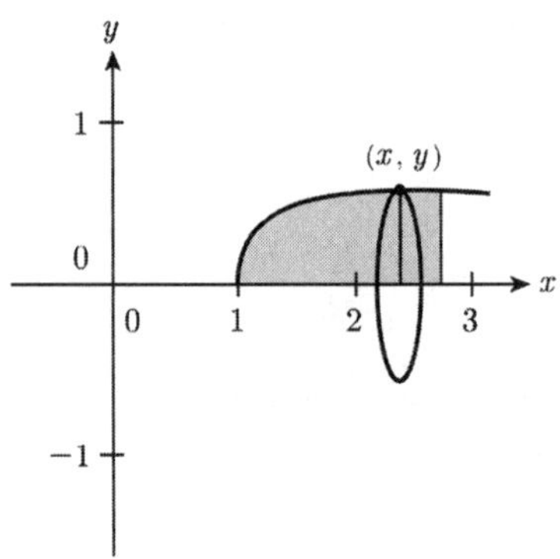

$Area=\pi y^2$

$$V=\pi\int_1^e y^2 dx \qquad u=\ln x$$

$$=\pi\int_1^e \frac{\ln x}{x}dx \qquad du=\frac{1}{x}dx$$

$$=\pi\int_0^1 u\,du \qquad if\ x=1,\ u=0$$

$$=\pi\left[\frac{1}{2}u^2\right]_0^1 \qquad if\ x=e,\ u=1$$

$$=\frac{\pi}{2}$$

EX4

(1)

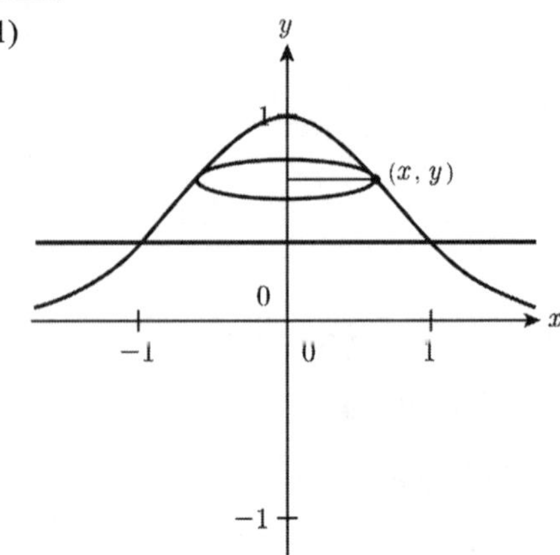

$Area=\pi x^2$

$y=e^{-x^2} \Rightarrow \ln y=-x^2$

$x^2=-\ln y$

$$V=\pi\int_{\frac{1}{e}}^1 x^2 dy$$

$$=\pi\int_{\frac{1}{e}}^1 -\ln y\, dy$$

$$=\pi[-y\ln y+y]_{\frac{1}{e}}^1$$

$$=\pi\left[(0+1)-\left(\frac{1}{e}+\frac{1}{e}\right)\right]$$

$$=\pi\left(1-\frac{2}{e}\right)$$

(2)

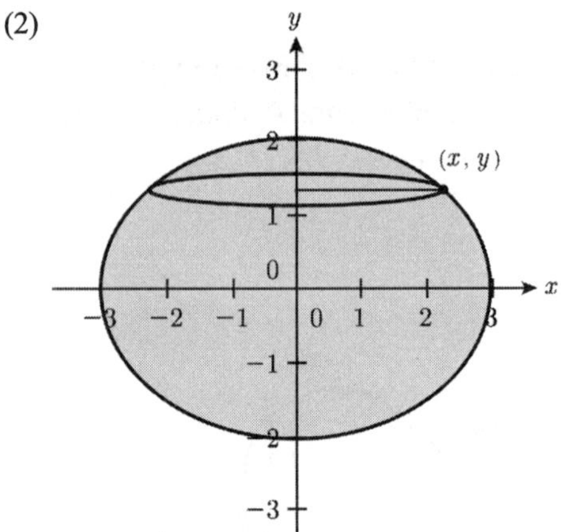

$Area=\pi x^2$

$$\frac{x^2}{9}+\frac{y^2}{4}=1 \Rightarrow x^2=9-\frac{9}{4}y^2$$

$$V=\pi\int_{-2}^2 x^2 dy$$

$$=\pi\int_{-2}^2\left(9-\frac{9}{4}y^2\right)dy$$

$$=2\pi\int_0^2\left(9-\frac{9}{4}y^2\right)dy$$

$$=2\pi\left[9y-\frac{3}{4}y^3\right]_0^2$$

$$=2\pi(18-6)=24\pi$$

EX5

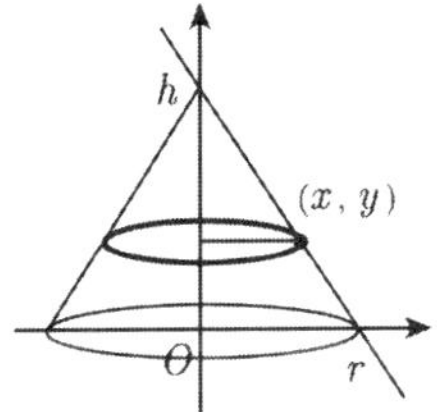

$Area=\pi x^2$

$y=-\frac{h}{r}x+h$

$\Rightarrow x=r\left(1-\frac{y}{h}\right)$

$$V=\pi\int_0^h x^2\,dy$$
$$=\pi\int_0^h r^2\left(1-\frac{y}{h}\right)^2 dy$$
$$=\pi r^2\left[y-\frac{y^2}{h}+\frac{y^3}{3h^2}\right]_0^h$$
$$=\pi r^2\left[h-h+\frac{h}{3}\right]=\frac{1}{3}\pi r^2 h$$

EX6

(1) $117\pi/5$

(2)

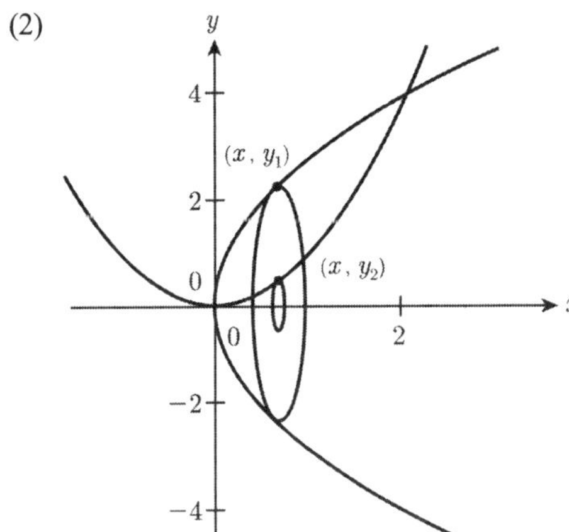

$(x^2)^2=8x$

$x^4=8x$

$x(x^3-8)=0$

$x=0,\ 2$

$y_1{}^2=8x$

$y_2=x^2$

$$V=\pi\int_0^2 (y_1{}^2-y_2{}^2)dx$$
$$=\pi\int_0^2 (8x-x^4)dx$$
$$=\pi\left[4x^2-\frac{1}{5}x^5\right]_0^2$$
$$=\pi\left(16-\frac{32}{5}\right)=\frac{48\pi}{5}$$

(3) $256\pi/5$

(4)

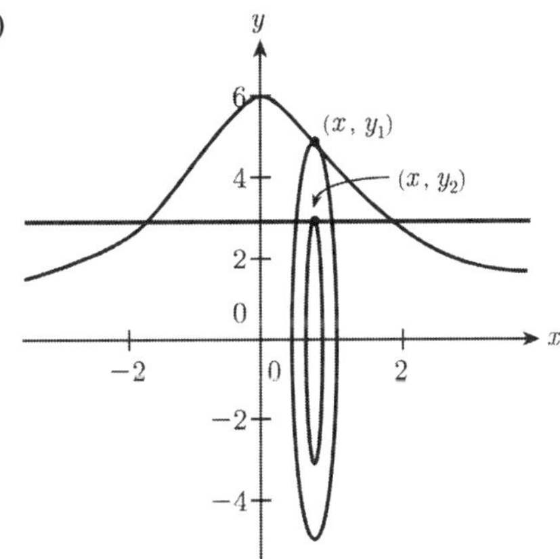

$\frac{6}{\sqrt{x^2+1}}=3$

$\sqrt{x^2+1}=2$

$x^2+1=4$

$x=\pm\sqrt{3}$

$$V=\pi\int_{-\sqrt{3}}^{\sqrt{3}} (y_1{}^2-y_2{}^2)dx$$
$$=\pi\int_{-\sqrt{3}}^{\sqrt{3}}\left(\frac{36}{x^2+1}-9\right)dx$$
$$=2\pi\int_0^{\sqrt{3}}\left(\frac{36}{x^2+1}-9\right)dx$$
$$=2\pi[36\tan^{-1}x-9x]_0^{\sqrt{3}}$$
$$=2\pi\left(36\cdot\frac{\pi}{3}-9\sqrt{3}\right)$$
$$=2\pi(12\pi-9\sqrt{3})$$
$$=6\pi(4\pi-3\sqrt{3})$$

(5) $\pi(\pi-2)$

(6)

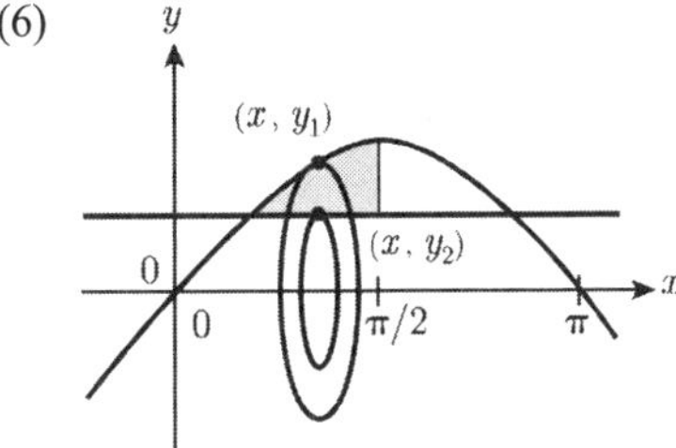

$$V=\pi\int_{\frac{\pi}{6}}^{\frac{\pi}{2}} (y_1{}^2-y_2{}^2)dx$$
$$=\pi\int_{\frac{\pi}{6}}^{\frac{\pi}{2}}\left[\sin^2 x-\frac{1}{4}\right]dx$$
$$=\pi\int_{\frac{\pi}{6}}^{\frac{\pi}{2}}\left[\frac{1}{2}-\frac{1}{2}\cos 2x-\frac{1}{4}\right]dx$$
$$=\pi\left[\frac{1}{4}x-\frac{1}{4}\sin 2x\right]_{\frac{\pi}{6}}^{\frac{\pi}{2}}$$
$$=\pi\left[\left(\frac{\pi}{8}-0\right)-\left(\frac{\pi}{24}-\frac{\sqrt{3}}{8}\right)\right]$$
$$=\pi\left(\frac{\pi}{12}+\frac{\sqrt{3}}{8}\right)$$

EX7

(1) 8π

(2)

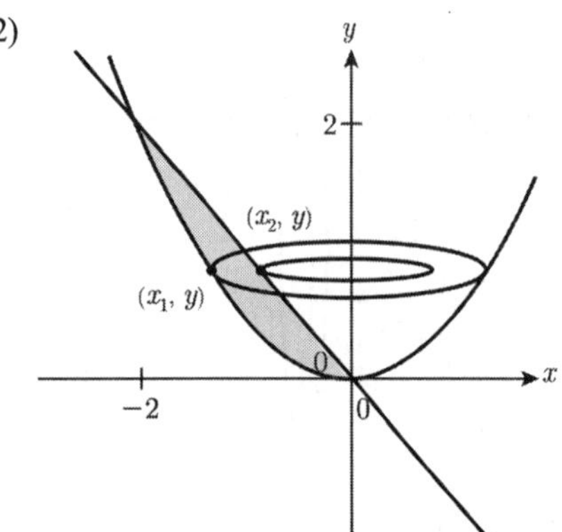

$$\frac{1}{2}x^2 = -x$$
$$x^2 = -2x$$
$$x(x+2)=0$$
$$x=0,\ -2 \Rightarrow y=0,\ 2$$
$$y=\frac{1}{2}{x_1}^2 \Rightarrow {x_1}^2=2y$$
$$y=-x_2 \Rightarrow {x_2}^2=y^2$$

$$V=\pi\int_0^2({x_1}^2-{x_2}^2)dy$$
$$=\pi\int_0^2(2y-y^2)\,dy$$
$$=\pi\left[y^2-\frac{1}{3}y^3\right]_0^2$$
$$=\pi\left(4-\frac{8}{3}\right)$$
$$=\frac{4\pi}{3}$$

(3)

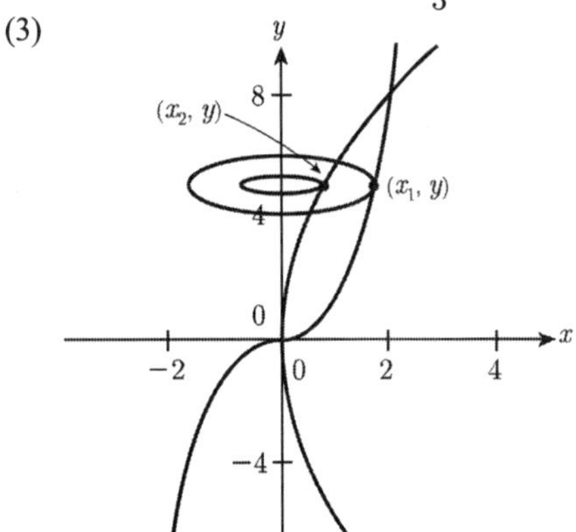

$$(x^3)^2=32x$$
$$x^6=32x$$
$$x(x^5-32)=0$$
$$x=0,2 \Rightarrow y=0,\ 8$$
$$y={x_1}^3 \Rightarrow x_1=y^{\frac{1}{3}}$$
$$y^2=32x_2 \Rightarrow x_2=\frac{1}{32}y^2$$

$$V=\pi\int_0^8({x_1}^2-{x_2}^2)dy$$
$$=\pi\int_0^8\left[y^{\frac{2}{3}}-\frac{y^4}{1024}\right]dy$$
$$=\pi\left[\frac{3}{5}y^{\frac{5}{3}}-\frac{y^5}{5\cdot1024}\right]_0^8$$
$$=\pi\left[\frac{3}{5}(8)^{\frac{5}{3}}-\frac{(8)^5}{5\cdot1024}\right]$$
$$=\pi\left(\frac{3}{5}(2)^5-\frac{2^{15}}{5\cdot1024}\right)$$
$$=\pi\left(\frac{96}{5}-\frac{2^5}{5}\right)$$
$$=\frac{64\pi}{5}$$

(4)

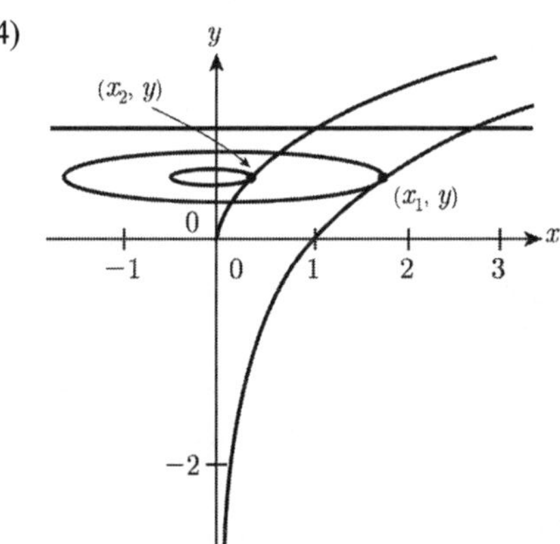

$$y=\ln x_1 \Rightarrow x_1=e^y$$
$$y=\sqrt{x_2} \Rightarrow x_2=y^2$$

$$V=\pi\int_0^1({x_1}^2-{x_2}^2)dy$$
$$=\pi\int_0^1(e^{2y}-y^4)\,dy$$
$$=\pi\left[\frac{1}{2}e^{2y}-\frac{1}{5}y^5\right]_0^1$$
$$=\pi\left[\left(\frac{1}{2}e^2-\frac{1}{5}\right)-\frac{1}{2}\right]$$
$$=\pi\left(\frac{e^2}{2}-\frac{7}{10}\right)$$

EX8

(1) $r(x)=0$ and $R(x)=1-x^2$

$$\Rightarrow V=\int_{-1}^1\pi([R(x)]^2-[r(x)]^2)dx$$
$$=\pi\int_{-1}^1(1-x^2)^2dx=\pi\int_{-1}^1(1-2x^2+x^4)dx$$
$$=\pi\left[x-\frac{2x^3}{3}+\frac{x^5}{5}\right]_{-1}^1=2\pi\left(1-\frac{2}{3}+\frac{1}{5}\right)$$
$$=2\pi\left(\frac{15-10+3}{15}\right)=\frac{16\pi}{15}$$

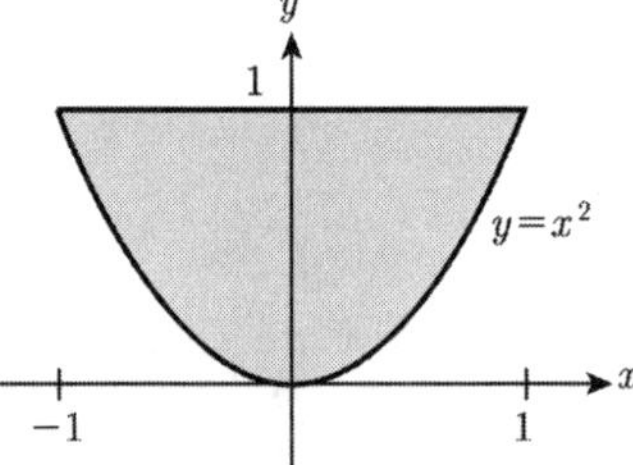

(2) $r(x)=1$ and $R(x)=2-x^2 \Rightarrow$

$$V=\int_{-1}^1\pi([R(x)]^2-[r(x)]^2)dx$$
$$=\pi\int_{-1}^1[(2-x^2)^2-1]dx$$

$$= \pi \int_{-1}^{1} (4-4x^2+x^4-1)dx$$

$$= \pi \int_{-1}^{1} (3-4x^2+x^4)dx$$

$$= \pi \left[3x - \frac{4}{3}x^3 + \frac{x^5}{5} \right]_{-1}^{1} = 2\pi \left(3 - \frac{4}{3} + \frac{1}{5} \right)$$

$$= \frac{2\pi}{15}(45-20+3) = \frac{56\pi}{15}$$

EX 9

(1) $r(y)=0$ and $R(y)=1-\frac{y}{2}$

$$\Rightarrow V = \int_0^2 \pi([R(y)]^2-[r(y)]^2)\,dy$$

$$= \pi \int_0^2 \left(1-\frac{y}{2}\right)^2 dy = \pi \int_0^2 \left(1-y+\frac{y^2}{4}\right)dy$$

$$= \pi \left[y - \frac{y^2}{2} + \frac{y^3}{12} \right]_0^2 = \pi \left(2 - \frac{4}{2} + \frac{8}{12} \right) = \frac{2\pi}{3}$$

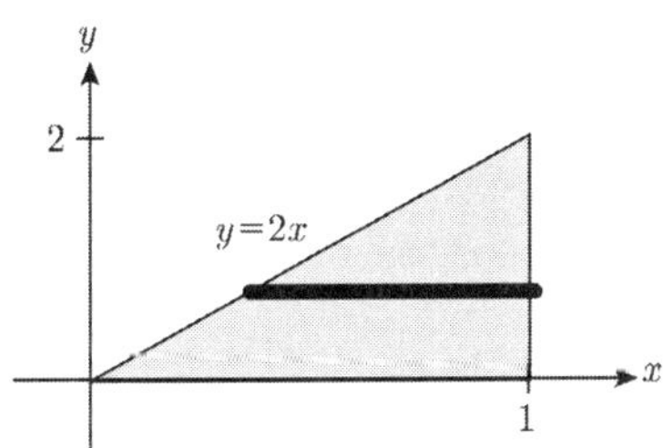

(2) $r(y)=1$ and $R(y)=2-\frac{y}{2}$

$$\Rightarrow V = \int_0^2 \pi([R(y)]^2-[r(y)]^2)dy$$

$$= \pi \int_0^2 \left[\left(2-\frac{y}{2}\right)^2 - 1 \right] dy$$

$$= \pi \int_0^2 \left(4-2y+\frac{y^2}{4}-1 \right) dy$$

$$= \pi \int_0^2 \left(3-2y+\frac{y^2}{4} \right) dy$$

$$= \pi \left[3y - y^2 + \frac{y^3}{12} \right]_0^2$$

$$= \pi \left(6-4+\frac{8}{12} \right) = \pi \left(2+\frac{2}{3} \right) = \frac{8\pi}{3}$$

EX 10

$e^{2x-x^2}=2$ when $x=0.446057,\ 1.553943$

Let $P=0.446057$ and $Q=1.553943$

(1) Area of $R = \int_P^Q (e^{2x-x^2}-2)dx = 0.514$

(2) $e^{2x-x^2}=1$ when $x=0,\ 2$

Area of $S = \int_0^2 (e^{2x-x^2}-1)dx$ − Area of R

$= 2.06016$ − Area of R $= 1.546$

OR

$$\int_0^P (e^{2x-x^2}-1)dx + (Q-P) \cdot 1 + \int_Q^2 (e^{2x-x^2}-1)dx$$

$$= 0.219064 + 1.107886 + 0.219064 = 1.546$$

(3) Volume $= \pi \int_P^Q ((e^{2x-x^2}-1)^2-(2-1)^2)dx$

EX 11

(1) Area

$$= \int_0^9 (6-2\sqrt{x})dx = \left(6x - \frac{4}{3}x^{3/2} \right) \Big|_{x=0}^{x=9} = 18$$

(2) Volume $= \pi \int_0^9 ((7-2\sqrt{x})^2-(7-6)^2)dx$

(3) Solving $y=2\sqrt{x}$ for x yields $x=\frac{y^2}{4}$

Each rectangular cross section has area

$$\left(3\frac{y^2}{4}\right)\left(\frac{y^2}{4}\right) = \frac{3}{16}y^4$$

$$\text{Volume} = \int_0^6 \frac{3}{16}y^4 dy$$

EX 12

(1) $5\pi/6$

(2)

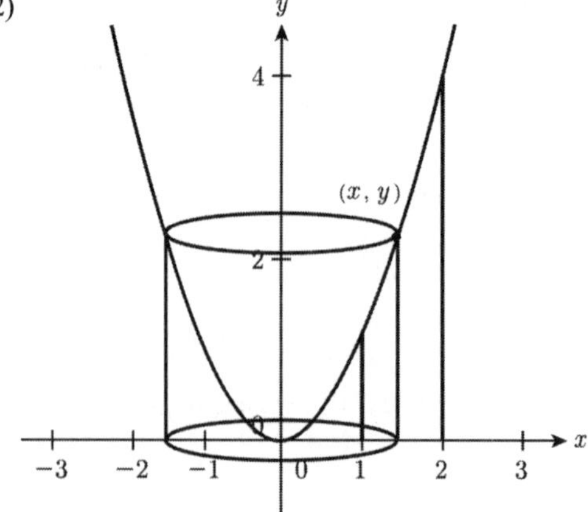

$$
\begin{aligned}
V &= 2\pi \int_1^2 xy\,dx \\
&= 2\pi \int_1^2 x \cdot x^2 dx \\
&= 2\pi \int_1^2 x^3 dx \\
&= 2\pi \left(\frac{1}{4}x^4\right)_1^2 \\
&= 2\pi\left(4 - \frac{1}{4}\right) = \frac{15\pi}{2}
\end{aligned}
$$

(3) $7\pi/15$

(4)

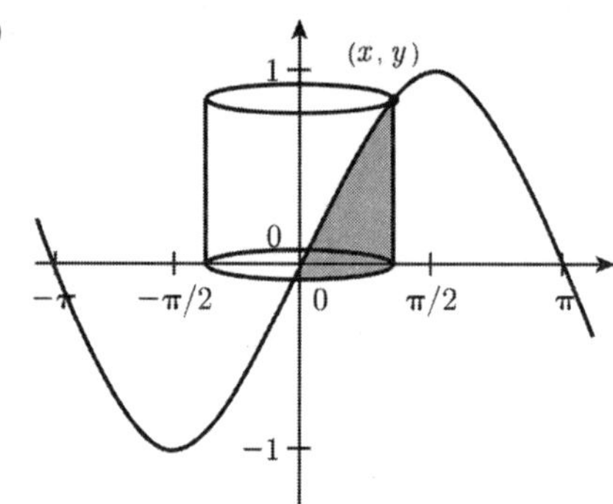

$$
\begin{aligned}
V &= 2\pi \int_0^\pi xy\,dx \\
&= 2\pi \int_0^\pi x \sin x\,dx
\end{aligned}
$$

(Use repeated integration by parts)

$$
\begin{aligned}
&= 2\pi[x(-\cos x) - 1(-\sin x)]_0^\pi \\
&= 2\pi[-x\cos x + \sin x]_0^\pi \\
&= 2\pi[\pi + 0 - (0+0)] \\
&= 2\pi^2
\end{aligned}
$$

(5) $8\pi/3$

(6)

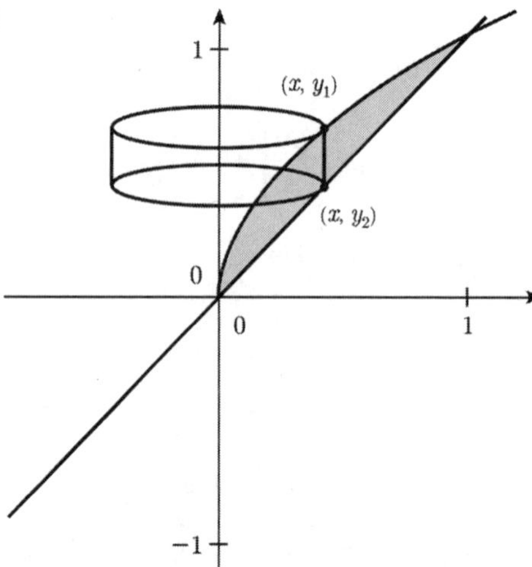

$\sqrt{x} = x \Rightarrow x = 0, 1$

$$
\begin{aligned}
V &= 2\pi \int_0^1 x(y_1 - y_2)dx \\
&= 2\pi \int_0^1 x(\sqrt{x} - x)dx \\
&= 2\pi \int_0^1 \left(x^{\frac{3}{2}} - x^2\right)dx \\
&= 2\pi \left[\frac{2}{5}x^{\frac{5}{2}} - \frac{1}{3}x^3\right]_0^1 \\
&= 2\pi\left(\frac{2}{5} - \frac{1}{3}\right) = \frac{2\pi}{15}
\end{aligned}
$$

EX 13

(1) $5\pi/6$

(2) $c = 0, d = 1$

$$
\begin{aligned}
V &= \int_c^d 2\pi \begin{pmatrix} shell \\ radius \end{pmatrix} \begin{pmatrix} shell \\ height \end{pmatrix} dy \\
&= \int_0^1 2\pi\, y(2y - y^2 - y)dy \\
&= 2\pi \int_0^1 y(y - y^2)dy = 2\pi \int_0^1 (y^2 - y^3)dy \\
&= 2\pi \left[\frac{y^3}{3} - \frac{y^4}{4}\right]_0^1 = 2\pi\left(\frac{1}{3} - \frac{1}{4}\right) = \frac{\pi}{6}
\end{aligned}
$$

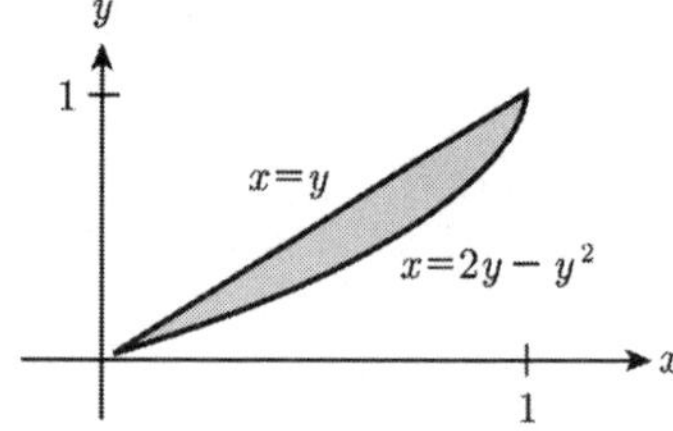

(3) $8\pi/3$

(4) $c=0, d=2;$

$$V=\int_c^d 2\pi\begin{pmatrix}shell\\radius\end{pmatrix}\begin{pmatrix}shell\\height\end{pmatrix}dy$$
$$=\int_0^2 2\pi y(2y-y^2)dy$$
$$=2\pi\int_0^2(2y^2-y^3)dy=2\pi\left[\frac{2y^3}{3}-\frac{y^4}{4}\right]_0^2$$
$$=2\pi\left(\frac{16}{3}-\frac{16}{4}\right)$$
$$=32\pi\left(\frac{1}{3}-\frac{1}{4}\right)=\frac{32\pi}{12}=\frac{8\pi}{3}$$

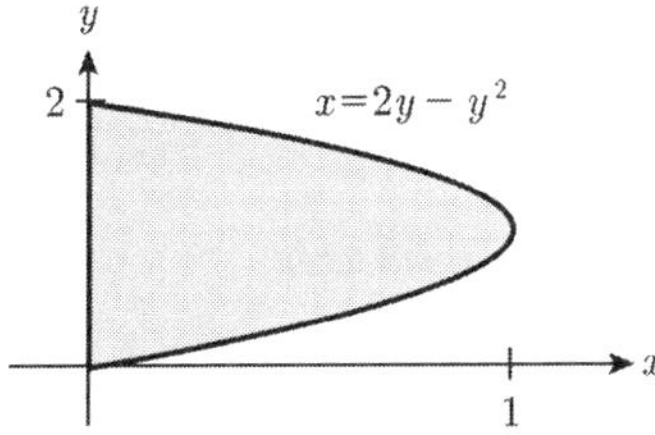

EX14

(1) $V=\int_a^b 2\pi\begin{pmatrix}shell\\radius\end{pmatrix}\begin{pmatrix}shell\\height\end{pmatrix}dx$
$$=\int_0^2 2\pi(3-x)(8-x^3)dx$$
$$=2\pi\int_0^2(24-8x-3x^3+x^4)dx$$
$$=2\pi\left[24x-4x^2-\frac{3}{4}x^4+\frac{1}{5}x^5\right]_0^2$$
$$=2\pi\left(48-16-12+\frac{32}{5}\right)=\frac{264\pi}{5}$$

(2) $V=\int_a^b 2\pi\begin{pmatrix}shell\\radius\end{pmatrix}\begin{pmatrix}shell\\height\end{pmatrix}dx$
$$=\int_0^2 2\pi(x+2)(8-x^3)dx$$
$$=2\pi\int_0^2(16+8x-2x^3-x^4)dx$$
$$=2\pi\left[16x+4x^2-\frac{1}{2}x^4-\frac{1}{5}x^5\right]_0^2$$
$$=2\pi\left(32+16-8-\frac{32}{5}\right)=\frac{336\pi}{5}$$

(3) $V=\int_c^d 2\pi\begin{pmatrix}shell\\radius\end{pmatrix}\begin{pmatrix}shell\\height\end{pmatrix}dy$
$$=\int_0^8 2\pi(8-y)y^{1/3}dy=2\pi\int_0^8(8y^{1/3}-y^{4/3})dy$$
$$=2\pi\left[6y^{4/3}-\frac{3}{7}y^{7/3}\right]_0^8=2\pi\left(96-\frac{384}{7}\right)=\frac{576\pi}{7}$$

(4) $V=\int_c^d 2\pi\begin{pmatrix}shell\\radius\end{pmatrix}\begin{pmatrix}shell\\height\end{pmatrix}dy$
$$=\int_0^8 2\pi(y+1)y^{1/3}dy=2\pi\int_0^8(y^{4/3}+y^{1/3})dy$$
$$=2\pi\left[\frac{3}{7}y^{7/3}+\frac{3}{4}y^{4/3}\right]_0^8=2\pi\left(\frac{384}{7}+12\right)=\frac{936\pi}{7}$$

EX15

$$V=\int_c^d 2\pi\begin{pmatrix}shell\\radius\end{pmatrix}\begin{pmatrix}shell\\height\end{pmatrix}dy$$
$$=\int_0^2 2\pi(5-y)\left[\frac{y^2}{2}-\left(\frac{y^4}{4}-\frac{y^2}{2}\right)\right]dy$$
$$=\int_0^2 2\pi(5-y)\left(y^2-\frac{y^4}{4}\right)dy$$
$$=2\pi\int_0^2\left(5y^2-\frac{5}{4}y^4-y^3+\frac{y^5}{4}\right)dy$$
$$=2\pi\left[\frac{5y^3}{3}-\frac{5y^5}{20}-\frac{y^4}{4}+\frac{y^6}{24}\right]_0^2$$
$$=2\pi\left(\frac{40}{3}-\frac{160}{20}-\frac{16}{4}+\frac{64}{24}\right)=8\pi$$

6.5 Definite Integral as Net Change

EX1

(1) $\int_0^h(10+40t)dt=100$. This occurs at $h=2$ hours after 8 A.M. $\therefore$ 10 A.M.

(2) Total gallons

$$=\lim_{n\to\infty}\sum_{k=1}^n 1000e^{-0.3t_k}\Delta t=\int_0^8 1000e^{-0.3t}dt\approx 3031.$$

(3) The total amount dumped after 7 weeks is

$$\int_0^{49}\frac{\sqrt{t}}{180}dt\simeq 1.27 \text{ tons}.$$

(4) $\int_0^T te^{-0.3t}dt=9$. The answer is about 10.2 weeks.

(5) The amount by which your investment of $1000 today will increase in 6 yr is $\int_0^6 80e^{0.08t}dt$ or $616.07, so it will be worth $(1000+616.07).

Note: $1000e^{0.08(6)} = 1616.07$

(6) $\int_0^3 (-0.1t^2 - 0.3t + 2)dt \simeq 3.75$ gal.

EX2

(1) $s(t) = s(0) + \int_0^t v(t)dt$
$= 2 + \int_0^t (3t^2 + 2t)dt$
$= 2 + [t^3 + t^2]_0^t$
$= t^3 + t^2 + 2$

(2) $s(t) = s(\pi) + \int_\pi^t 4\cos 2t\,dt$
$= -3 + [2\sin 2t]_\pi^t$
$= -3 + 2\sin 2t$

(3) $s(t) = s(3) + \int_3^t v(t)dt$
$= 1 + \int_3^t \frac{1}{\sqrt{1+t}}dt$
$= 1 + [2\sqrt{1+t}\,]_3^t$
$= 1 + 2\sqrt{1+t} - 4$
$= 2\sqrt{1+t} - 3$

(4) $s(t) = s(1) + \int_1^t v(t)dt$
$= 2 + \int_1^t \frac{\ln t}{t}dt$
$= 2 + \left[\frac{1}{2}(\ln t)^2\right]_1^t$
$= 2 + \frac{1}{2}(\ln t)^2$

(5) $v(t) = v(1) + \int_1^t (6y - 3)dt$
$= 2 + [3t^2 - 3t]_1^t$
$= 3t^2 - 3t + 2$

$s(t) = s(0) + \int_0^t (3t^2 - 3t + 2)dt$
$= 2 + \left[t^3 - \frac{3}{2}t^2 + 2t\right]_0^t$
$= t^3 - \frac{3}{2}t^2 + 2t + 2$

(6) $v(t) = v(0) + \int_0^t 2\sin 2t\,dt$
$= 3 + [-\cos 2t\,]_0^t$
$= 3 - \cos 2t - (-1)$
$= 4 - \cos 2t$

$s(t) = s(0) + \int_0^t (4 - \cos 2t)\,dt$
$= -2 + \left[4t - \frac{1}{2}\sin 2t\right]_0^t$
$= -2 + 4t - \frac{1}{2}\sin 2t$

EX3

(1) $f(3) = f(1) + \int_1^3 f'(x)\,dx$
$= 3 + \int_1^3 3\sin(x^2)\,dx = 4.390$

(2) $f(2) = f(-1) + \int_{-1}^2 f'(x)\,dx$
$= 2 + \int_{-1}^2 e^{-x^2}dx = 3.629$

(3) $s(4) = s(0) + \int_0^4 v(t)dt$
$= 2 + \int_0^4 (e^{\sin t} - 1)dt = 4.796$

(4) $v(5) = v(1) + \int_1^3 a(t)dt$
$= 3 + \int_1^3 2\cos(t + \ln t)dt = 0.673$

(5) Net change
$= \int_{10}^{30} 500e^{-0.02t}dt = 6747.977 \approx 6748$

EX4

(1) $\int_0^{12} P(t)dt \approx 46 \cdot 4 + 57 \cdot 4 + 62 \cdot 4$
$= 660\,ft^3$

(2) $\int_0^{12} R(t)dt = 225.594\,ft^3$

(3) $1000 + \int_0^{12} P(t)dt - \int_0^{12} R(t)dt = 1434.406$

At time $t = 12$ hours, the volome of water in the pool is approximately 1434 ft^3

(4) $V'(t)=P(t)-R(t)$

$V'(8)=P(8)-R(8)=60-25e^{-0.4}$

$=43.241$ or $43.242\,ft^3/hr$

$V=\pi(12)^2h$

$\dfrac{dV}{dt}=144\pi\dfrac{dh}{dt}$

$\left.\dfrac{dh}{dt}\right|_{t=8}=\dfrac{1}{144\pi}\cdot\left.\dfrac{dV}{dt}\right|_{t=8}$

$=0.095$ or $0.096\,ft/hr$

EX5

(1) $H'(3.5)\approx\dfrac{H(5)-H(2)}{5-2}$

$=\dfrac{52-60}{3}=-2.666$ or -2.667 degrees Celsius per minute

(2) $\dfrac{1}{10}\displaystyle\int_0^{10}H(t)\,dt$ is the average temperature of the tea, in degrees Celsus, over the 10 minutes.

$$\frac{1}{10}\int_0^{10}H(t)dt\approx\frac{1}{10}\left(2\cdot\frac{66+60}{2}+3\cdot\frac{60+52}{2}+4\cdot\frac{52+44}{2}+1\cdot\frac{44+43}{2}\right)$$

$=52.95$

(3) $\displaystyle\int_0^{10}H'(t)dt=H(10)-H(0)$

$=43-66=-23$

The temperature of the tea drops 23 degrees Celsius from time $t=0$ to time $t=10$ minutes.

(4) $B(10)=100+\displaystyle\int_0^{10}B'(t)dt=34.18275;$

$H(10)-B(10)=8.817$

The biscuits are 8.817 degrees Celsius cooler than the tea.

6.6 Motion Along a Straight Line(1-Dimension)

EX1

(1) Displacement $=\dfrac{5\pi}{6}-\dfrac{1}{2}$

Distance$=\dfrac{5\pi}{6}-\dfrac{1}{2}$

(2)

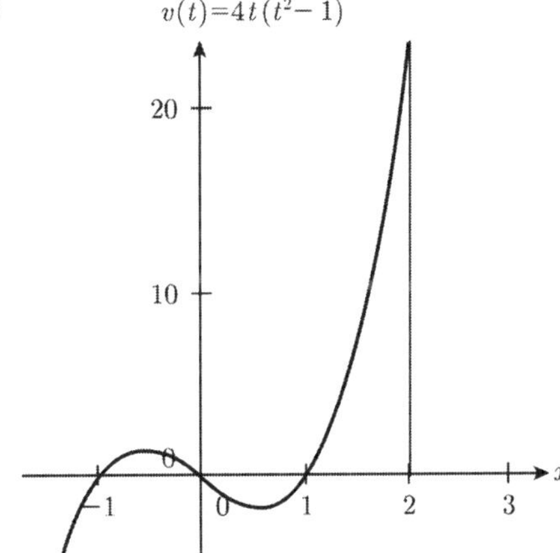

Displacement $=\displaystyle\int_0^2 v(t)dt=\int_0^2(4t^3-4t)dt$

$=[t^4-2t^2]_0^2=16-8=8$

Distance

$=\displaystyle\int_0^2|v(t)|dt$

$=-\displaystyle\int_0^1(4t^3-4t)\,dt+\int_1^2(4t^3-4t)dt$

$=-\left[t^4-2t^2\right]_0^1+\left[t^4-2t^2\right]_1^2$

$=-[1-2]+(16-8)-(1-2)$

$=1+9=10$

(3) Displacement $=0$

Distance$=8$

(4)

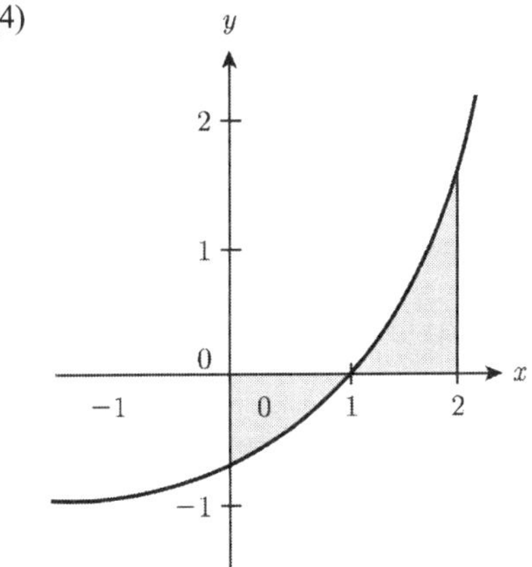

$v(t)=v(1)+\displaystyle\int_1^t e^{t-1}dt$

$=0+[e^{t-1}]_1^t$

$=e^{t-1}-1$

Displacement

$= \int_0^2 v(t)dt = \int_0^2 (e^{t-1}-1)dt = [e^{t-1}-t]_0^2$

$= (e^1-2)-(e^{-1}) = e-2-\frac{1}{e}$

Distance

$= \int_0^2 |v(t)|dt$

$= -\int_0^1 (e^{t-1}-1)\,dt + \int_1^2 (e^{t-1}-1)dt$

$= -[e^{t-1}-t]_0^1 + [e^{t-1}-t]_1^2$

$= -[(1-1)-e^{-1}] + [e-2-(1-1)]$

$= e^{-1}+e-2 = \frac{1}{e}+e-2$

EX2

(1) The squirrel changes direction whenever its velocity changes sign.

This occurs at $t=9$ and $t=15$.

(2) Velocity is 0 at $t=0$, $t=9$, and $t=15$.

t	position at time t
0	0
9	$\frac{9+5}{2} \cdot 20 = 140$
15	$140 - \frac{6+4}{2} \cdot 10 = 90$
18	$90 + \frac{3+2}{2} \cdot 10 = 115$

The squirrel is farthest from building A at time $t=9$;

its greatest distance from the building is 140

(3) The total distance traveled is

$\int_0^{18} |v(t)|dt = 140+50+25 = 215$

(4) For $7<t<10$, $a(t) = \frac{20-(-10)}{7-10} = -10$

$v(t) = 20-10(t-7) = -10t+90$

$x(7) = \frac{7+5}{2} \cdot 20 = 120$

$x(t) = x(7) + \int_7^t (-10u+90)du$

$= 120 + (-5u^2+90u)\big|_{u=7}^{u=t}$

$= -5t^2+90t-265$

EX3

(1) $a(7.5) = v'(7.5) = \frac{v(8)-v(7)}{8-7} = -0.1$ miles/minute^2

(2) $\int_0^{12} |v(t)|dt$ is the distance, in miles, that Caren rode during the 12 minutes from $t=0$ to $t=12$.

$\int_0^{12} |v(t)|dt$

$= \int_0^2 v(t)\,dt - \int_2^4 v(t)dt + \int_4^{12} v(t)dt$

$= 0.2+0.2+1.4 = 1.8$ miles

(3) Caren turns around to go back home at time $t=2$ minutes. This is the time at which her velocity changes from positive to negative.

(4) $\int_0^{12} w(t)dt = 1.6$; Larry lives 1.6 miles from school.

$\int_0^{12} v(t)\,dt = 1.4$; Caren lives 1.4 miles from school.

Therefore, Caren lives closer to school.

EX4

(1) $a(5) \approx \frac{v(10)-v(0)}{10-0} = \frac{0.3}{10} = 0.03$ meters/sec^2

(2) $\int_0^{60} |v(t)|dt$ is the total distance, in meters, Ben rides over the 60-second interval $t=0$ to $t=60$.

$\int_0^{60} |v(t)|dt \approx 2.0 \cdot 10$

$+2.3(40-10)+2.5(60-40) = 139$

meters.

(3) Because $\frac{B(60)-B(40)}{60-40} = \frac{49-9}{20} = 2$,

The Mean Value Theorem implies there is a time t, $40<t<60$, such that $v(t)=2$

(4) $2L(t)L'(t) = 2B(t)B'(t)$

$L'(40) = \frac{B(40)B'(40)}{L(40)} = \frac{9 \cdot 2.5}{\sqrt{144+81}} = \frac{3}{2}$

meters/sec

EX5

(1) $\int_0^6 f(t)dt = 142.274$ or 142.275 cubic feet

(2) Rate of change is $f(8)-g(8) = -59.582$ or -59.583 cubic feet per hour.

(3) $h(0)=0$

For

$0<t\le 6,\ h(t)=h(0)+\int_0^t g(s)ds$

$=0+\int_0^t 0\,ds=0.$

For

$6<t\le 7,\ h(t)=h(6)+\int_6^t g(s)ds$

$=0+\int_6^t 125\,ds = 125(t-6).$

For

$7<t\le 9,\ h(t)=h(7)+\int_7^t g(s)ds$

$=125+\int_7^t 108ds = 125+108(t-7).$

Thus, $h(t)=\begin{cases}0\\125(t-6)\\125+108(t-7)\end{cases}$

$\text{for } 0\le t\le 6$
$\text{for } 6<t\le 7$
$\text{for } 7<t\le 9$

(4) Amount of snow is

$\int_0^9 f(t)dt-h(9)=26.334$ or 26.335

cubic feet.

6.7 Motion Along a Plane Curve

EX1

(1) $v=(-2\sin 2t+c_1)i+(2\cos t+c_2)j$, and since $v=2j$ when $t=0$, it follows that $c_1=c_2=0$.

So $v=-2\sin 2ti+2\cos tj$. Also

$R=(\cos 2t+c_3)i+(2\sin t+c_4)j$; and since

$R=i$ when $t=0$, we see that $c_3=c_4=0$.

Finally, then,

$R=\cos 2ti+2\sin tj.$

(2) From (1) the parametric equations of motion are

$x=\cos 2t, \quad y=2\sin t.$

By a trigonometric identity,

$x=1-2\sin^2 t=1-\frac{y^2}{2}.$

P travels in a counterclockwise direction along part of a parabola that has its ver-tex at $(1,0)$ and opens to the left. The Path of the particle is sketched in Figure N8-1;

note that $-1\le x\le 1,\ -2\le y\le 2.$

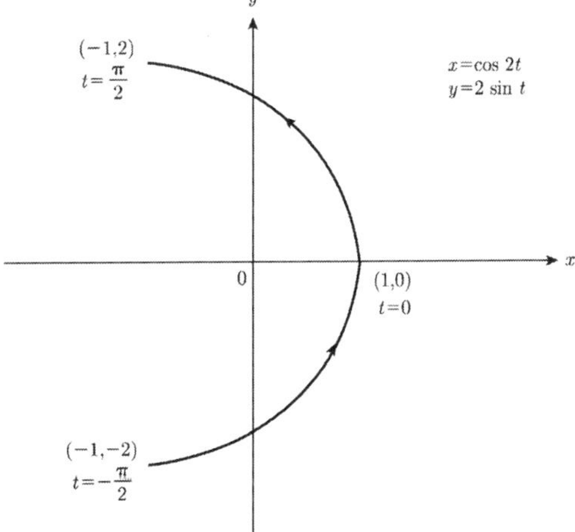

EX2

(1) $\frac{dy}{dt}=0$ only when $t=0.36735$. Let $b=0.36735$.

The maximum vertical distance from the water surface to the diver's shoulders is

$y(b)=11.4+\int_0^b \frac{dy}{dt}dt=12.061$ meters.

Alternatively, $y(t)=11.4+3.6t-4.9t^2$,

so $y(b)=12.061$ meters.

(2) $y(A)=11.4+\int_0^A \frac{dy}{dt}dt=11.4+3.6A-4.9A^2=0$

when

$A=1.936$ seconds.

(3) $\int_0^A \sqrt{\left(\frac{dx}{dt}\right)^2+\left(\frac{dy}{dt}\right)^2}\,dt=12.946$ meters

(4) At time A, $\frac{dy}{dx}=\frac{dy/dt}{dx/dt}\bigg|_{t=A}=-19.21913.$

The angle between the path of the diver and the water is $\tan^{-1}(19.21913)=1.518$ or 1.519.

EX3

(1) Speed$=\sqrt{(x'(3))^2+(y'(3))^2}=2.828$ meters per second

(2) $x'(t)=2t-4$

Distance

$=\int_0^4 \sqrt{(2t-4)^2+(te^{t-3}-1)^2}\,dt$ meters

$=11.587$ or 11.588

(3) $\frac{dy}{dx}=\frac{dy/dt}{dx/dt}=0$ when $te^{t-3}-1=0$ and

$2t-4\neq 0$

This occurs at $t=2.20794$.

Since $x'(2.20794)>0$, the particle is moving toward the right at time $t=2.207$ or 2.208.

(4) $x(t)=5$ at $t=1$ and $t=3$

At time $t=1$, the slope is

$\frac{dy}{dx}\bigg|_{t=1}=\frac{dy/dt}{dx/dt}\bigg|_{t=1}=0.432.$

At time $t=3$, the slope is

$\frac{dy}{dx}\bigg|_{t=3}=\frac{dy/dt}{dx/dt}\bigg|_{t=3}=1.$

$y(1)=y(3)=3+\frac{1}{e}+\int_2^3 \frac{dy}{dt}dt=4$

EX4

(1) Speed

$=\sqrt{(x'(3))^2+(y'(3))^2}=13.006$ or 13.007

Acceleration

$=<x''(3), y''(3)>$

$=<4, -5.566>$ or $<4, -5.467>$

(2) Slope $=\frac{y'(3)}{x'(3)}=0.031$ or 0.032

(3) $x(3)=0+\int_0^3 \frac{dx}{dt}dt=21$

$y(3)=-4+\int_0^3 \frac{dy}{dt}dt=-3.226$

At time $t=3$, the particle is at position $(21, -3.226)$.

(4) Distance $=\int_0^3 \sqrt{\left(\frac{dx}{dt}\right)^2+\left(\frac{dy}{dt}\right)^2}\,dt=21.091$

6.8 Arc Length

EX1

Since $x-2=2y^{3/2}$ and $\frac{dx}{dy}=3y^{1/2}$,

Equation above yields $s=\int_0^1 \sqrt{1+9y}\,dy\approx 2.268$, again using the calculator to evaluate the definite integral.

EX2

$$\int_0^4 \sqrt{1+\frac{9}{4}x}\,dx=\frac{8}{27}(10^{\frac{3}{2}}-1)$$

EX3

Here $\frac{dy}{dx}=\frac{\sec x\tan x}{\sec x}$,

$$s=\int_0^{\pi/3}\sqrt{1+\tan^2 x}\,dx=\int_0^{\pi/3}\sec x\,dx$$

$$=\ln(\sec x+\tan x)\,|_0^{\pi/3}=\ln(2+\sqrt{3}).$$

EX4

$$s=\int_1^2 \sqrt{(t^2+1)^2}\,dt=\int_1^2 (t^2+1)dt$$

$$=\frac{t^3}{3}+t\bigg|_1^2=\frac{10}{3}.$$

EX5

$$\int_0^{2\pi}\sqrt{(1-\cos t)^2+(\sin t)^2}\,dt=\int_0^{2\pi}\sqrt{2-2\cos t}\,dt.$$

$$=\int_0^{2\pi}\sqrt{2-2(1-2\sin^2\frac{t}{2})}\,dt=\int_0^{2\pi}2\sin\frac{t}{2}dt=8$$

EX6

$$\int_2^3 \sqrt{(-e^t \sin t + e^t \cos t)^2 + (e^t \cos t + e^t \sin t)^2}\, dt$$

$$= \sqrt{2} e^t \Big|_2^3 = \sqrt{2}(e^3 - e^2)$$

EX7

(1) $f'(x) = 2e^{2x}$

Perimeter $= 1 + k + e^{2k} + \int_0^k \sqrt{1 + (2e^{2x})^2}\, dx$

(2) Volume

$$= \pi \int_0^k (e^{2x})^2 dx = \pi \int_0^k e^{4x} dx = \frac{\pi}{4} e^{4x} \Big|_{x=0}^{x=k}$$

$$= \frac{\pi}{4} e^{4k} - \frac{\pi}{4}$$

(3) $\frac{dV}{dt} = \pi e^{4k} \frac{dk}{dt}$

When $k = \frac{1}{2}$, $\frac{dV}{dt} = \pi e^2 \cdot \frac{1}{3}$

EX8

(1) $Area = 30 \cdot 20 - \int_0^{30} f(x) dx = 218.028 \text{ cm}^2$

(2) volume $= \int_0^{30} \frac{\pi}{2} \left(\frac{f(x)}{2} \right)^2 dx = 2356.194 \text{ cm}^3$

Therefore, the baker needs

$2356.194 \times 0.05 = 117.809$ or 117.810

grams of chocolate.

(3) Perimeter

$$= 30 + \int_0^{30} \sqrt{1 + (f'(x))^2}\, dx$$

$= 81.803$ or 81.804

6.9 Improper Integral

EX1

(1) $\int_1^\infty \frac{dx}{\sqrt{x}} = \lim_{b\to\infty} \int_1^b x^{-1/2} dx$

$$= \lim_{b\to\infty} 2\sqrt{x} \Big|_1^b = \lim_{b\to\infty} 2(\sqrt{b} - 1) = +\infty$$

Then $\int_1^\infty \frac{dx}{\sqrt{x}}$ diverges.

(2) $\int_0^\infty \frac{dx}{x^2+9} = \lim_{b\to\infty} \int_0^b \frac{dx}{x^2+9}$

$$= \lim_{b\to\infty} \frac{1}{3} \tan^{-1} \frac{x}{3} \Big|_0^b = \lim_{b\to\infty} \frac{1}{3} \cdot \frac{\pi}{2} = \frac{\pi}{6}.$$

(3) $\int_0^\infty \frac{dy}{e^y} = \lim_{b\to\infty} \int_0^b e^{-y} dy = \lim_{b\to\infty} -(e^{-b} - 1) = 1.$

(4) $\int_{-\infty}^0 \frac{dz}{(z-1)^2} - \lim_{b\to-\infty} \int_b^0 (z-1)^{-2} dz$

$$= \lim_{b\to-\infty} -\frac{1}{z-1} \Big|_b^0$$

$$= \lim_{b\to-\infty} -\left(-1 - \frac{1}{b-1}\right) = 1.$$

(5) $\int_0^\infty \cos x\, dx = \lim_{b\to\infty} \sin x \Big|_b^0 = \lim_{b\to\infty} \sin b.$

Since this limit does not exist ($\sin b$ takes on values between -1 and 1 as $b\to\infty$), if follows that the given integral diverges. Note, however, that it does not become infinite; rather, it diverges by oscillation.

(6) $\int_0^1 \frac{dx}{x^3} - \lim_{k\to0^+} \int_k^1 x^{-3} dx$

$$= \lim_{k\to0^+} -\frac{1}{2x^2} \Big|_k^1 = \lim_{k\to0^+} -\frac{1}{2}\left(1 - \frac{1}{k^2}\right) = \infty.$$

Therefore, this integral diverges.

(7) $\int_0^2 \frac{dy}{\sqrt{4-y^2}} = \lim_{k\to2^-} \int_0^k \frac{dy}{\sqrt{4-y^2}} = \lim_{k\to2^-} \sin^{-1} \frac{y}{2} \Big|_0^k$

$$= \sin^{-1} 1 - \sin^{-1} 0 = \frac{\pi}{2}.$$

(8) $\int_0^2 \frac{dx}{(x-1)^{2/3}}$

$$= \lim_{k\to1^-} \int_0^k (x-1)^{-2/3} dx + \lim_{m\to1^+} \int_m^2 (x-1)^{-2/3} dx$$

$$= \lim_{k\to1^-} 3(x-1)^{1/3} \Big|_0^k + \lim_{m\to1^+} 3(x-1)^{1/3} \Big|_m^2$$

$$= 3(0+1) + 3(1-0) = 6.$$

(9) $\int_{-2}^2 \frac{dx}{x^2} = \lim_{k\to0^-} \int_{-2}^k x^{-2} dx + \lim_{m\to0^+} \int_m^2 x^{-2} dx$

$$= \lim_{k\to0^-} -\frac{1}{x} \Big|_{-2}^k + \lim_{m\to0^+} -\frac{1}{x} \Big|_m^2$$

Neither limit exists; the integral diverges.

(10) $\displaystyle\int_0^{\pi/2}\frac{\sin x}{\sqrt{1-\cos x}}dx=\lim_{k\to(\pi/2)^-}2\sqrt{1-\cos x}\,\Big|_0^k=2$

EX2

(1)

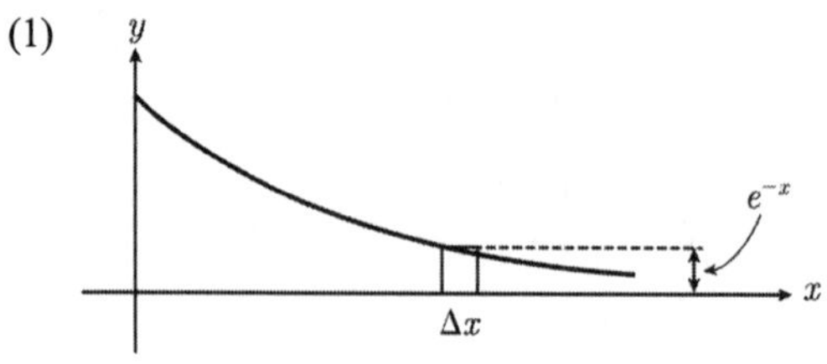

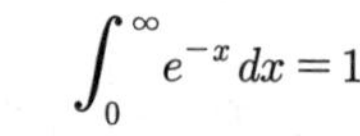

$$\int_0^{\infty}e^{-x}\,dx=1$$

(2)

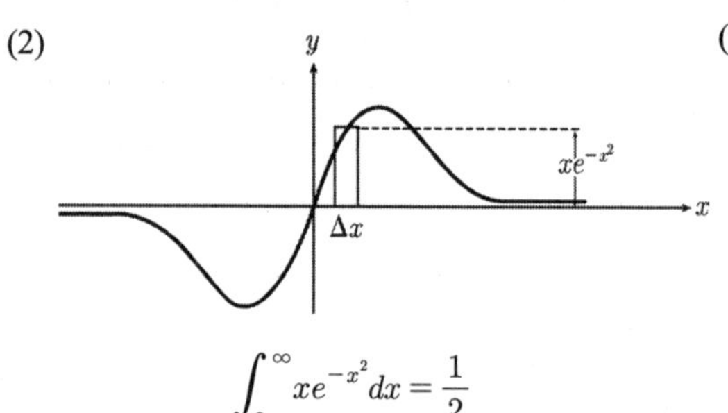

$$\int_0^{\infty}xe^{-x^2}dx=\frac{1}{2}$$

(3)

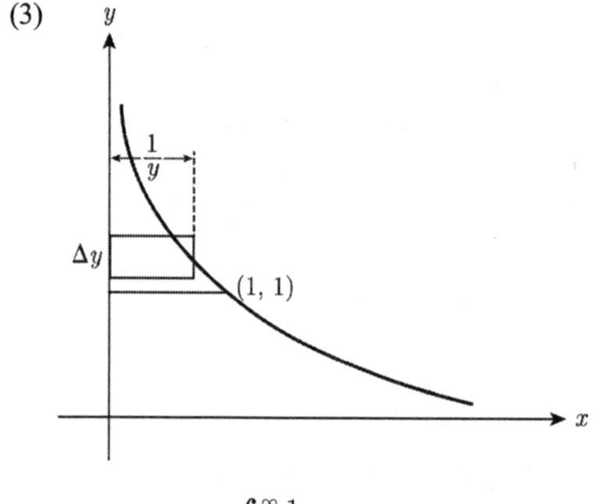

$$\int_1^{\infty}\frac{1}{y}dx=\infty$$

(4)

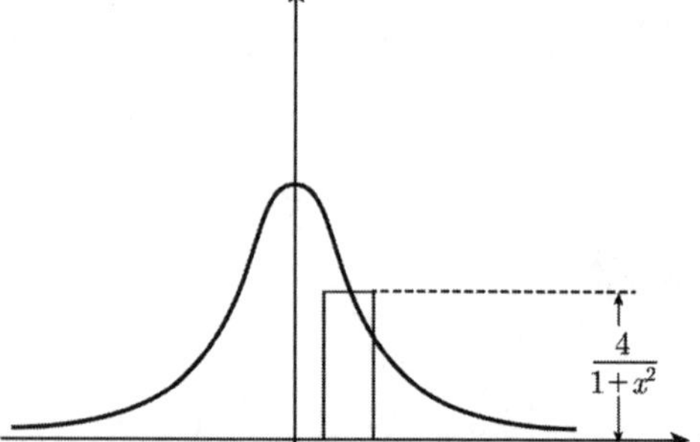

$$\int_{-\infty}^{\infty}\frac{4}{1+x^2}dx=\lim_{b\to\infty}4\tan^{-1}x\Big|_{-b}^{b}=4\pi$$

(5)

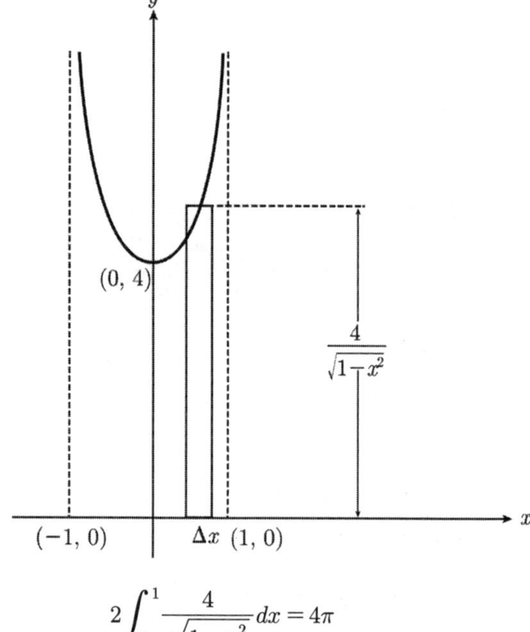

$$2\int_0^{1}\frac{4}{\sqrt{1-x^2}}dx=4\pi$$

EX3

(1)

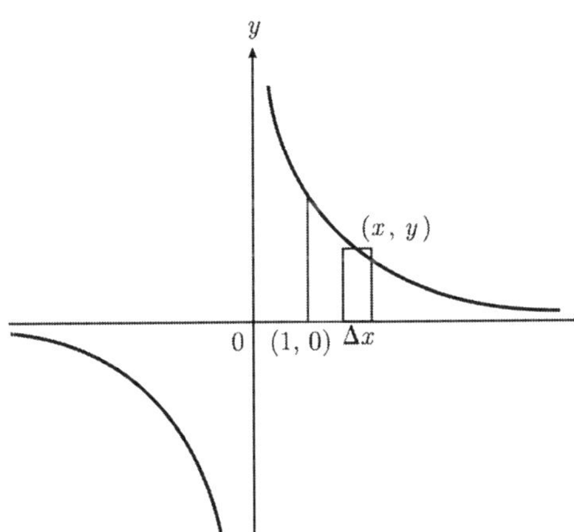

About the x − axis, disk

$\triangle V = \pi y^2 \Delta x = \dfrac{\pi}{x^2}\Delta x, \int_1^{\infty} \dfrac{\pi}{x^2} dx = \pi$

(2)

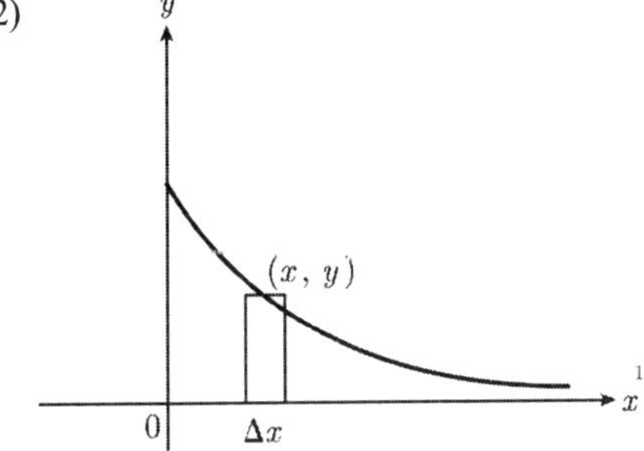

About the x − axis, disk

$\triangle V = \pi y^2 \Delta x = \pi e^{-2x} \Delta x, \pi \int_1^{\infty} e^{-2x} dx = \dfrac{\pi}{2}$

EX4

(1) $g'(x) = \dfrac{4(1+4x^2) - 4x(8x)}{(1+4x^2)^2} = \dfrac{4(1-4x^2)}{(1+4x^2)^2}$

For $x > 0$, $g'(x) = 0$ for $x = \dfrac{1}{2}$.

$g'(x) > 0$ for $0 < x < \dfrac{1}{2}$

$g'(x) < 0$ for $x > \dfrac{1}{2}$

$g\left(\dfrac{1}{2}\right) = 1$

Therefore g has a maximum value of 1 at $x = \dfrac{1}{2}$, and g has no minimum value on the open interval $(0, \infty)$

(2) $\displaystyle\int_1^{\infty} (f(x) - g(x))dx = \lim_{b\to\infty} \int_1^{b} f(x) - g(x))\, dx$

$= \displaystyle\lim_{b\to\infty} \left(\ln(x) - \frac{1}{2}\ln(1+4x^2) \right) \Bigg|_{x=1}^{x=b}$

$= \displaystyle\lim_{b\to\infty} \ln\left(\frac{b\sqrt{5}}{\sqrt{1+4b^2}} \right)$

$= \displaystyle\lim_{b\to\infty} \ln\left(\frac{\sqrt{5b^2}}{\sqrt{1+4b^2}} \right)$

$= \displaystyle\frac{1}{2}\lim_{b\to\infty} \ln\left(\frac{5b^2}{1+4b^2} \right)$

$= \dfrac{1}{2}\ln\dfrac{5}{4}$

Multiple Choice

01 D 02 B 03 B 04 A 05 D 06 A 07 D 08 E 09 E 10 E 11 B 12 D 13 B 14 D 15 B 16 C 17 C 18 D 19 C 20 D 21 A 22 E 23 B 24 B

07 Differential Equations

7.1 Slope Fields(Direction Fields)

EX1

(1)

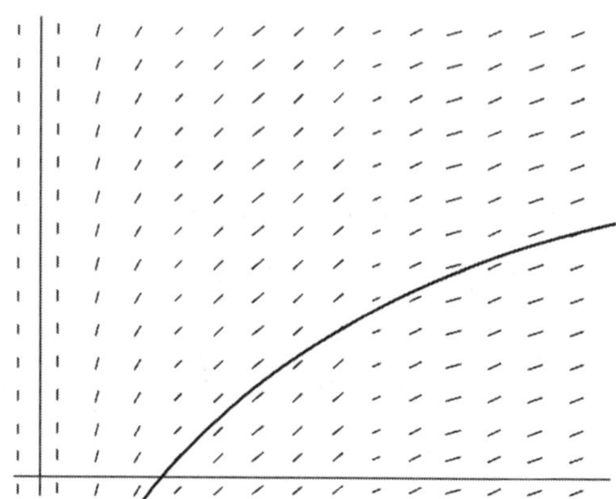

(2) Since we already know that, if

$$\frac{dy}{dx} = \frac{1}{x}, \text{ then } y = \int \frac{1}{x} dx = \ln x + C$$

EX2

(1)

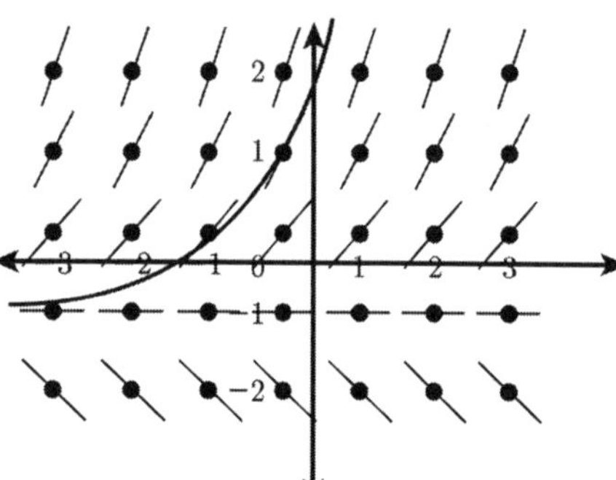

(2)

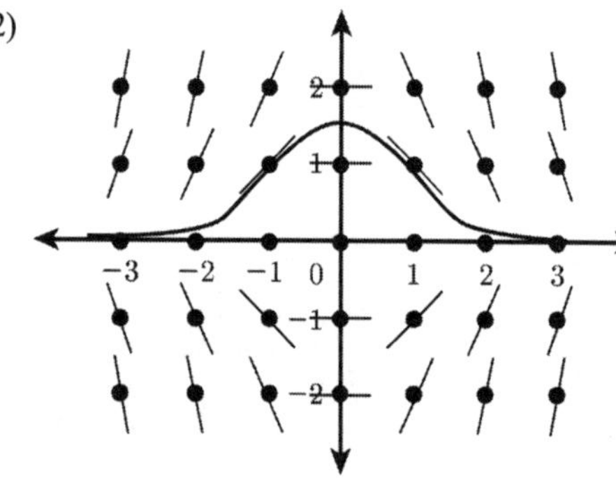

(3)

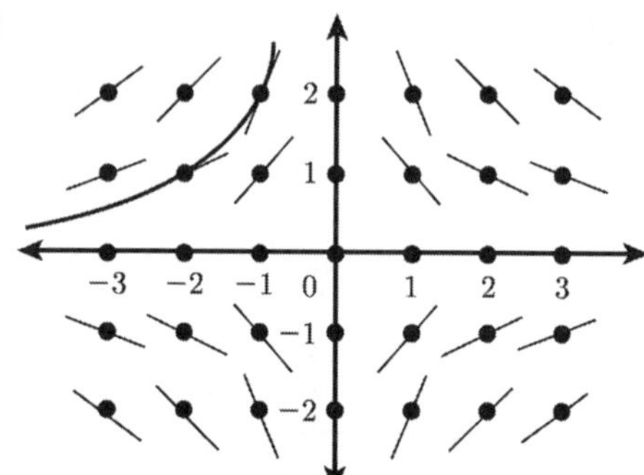

(4)

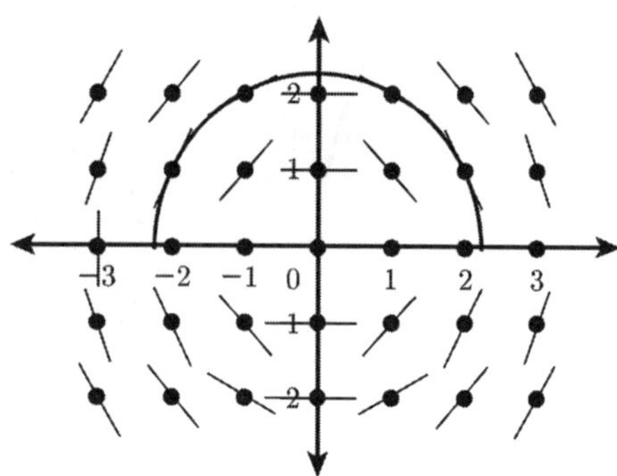

(5)

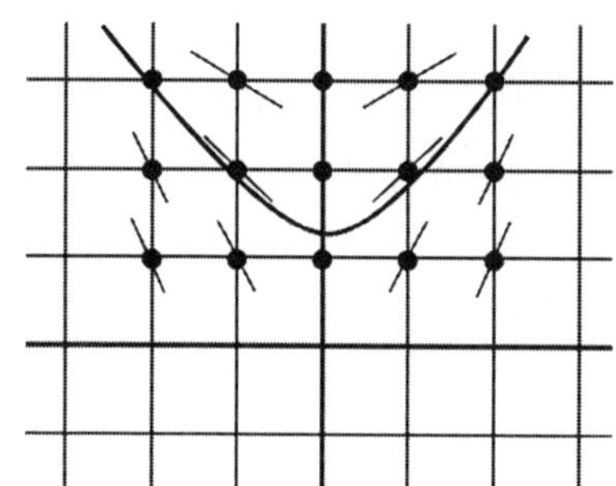

(6)

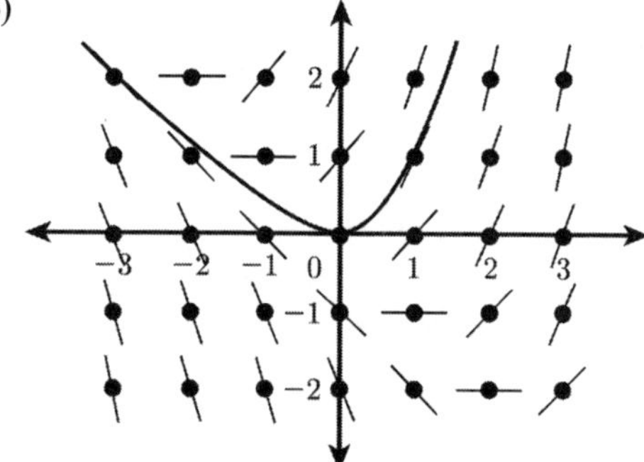

EX3

(1) (b)

(2) (c)

(3) (d)

(4) (a)

EX4

(1) (e)

(2) (c)

(3) (b)

(4) (a)

(5) (d)

(6) (f)

EX5

(1) (c)

(2) (a)

(3) (b)

(4) (d)

7.2 First Order Separable Differential Equations

EX1

(1) $y=\left(\frac{1}{3}x^{3/2}+1\right)^2$

(2) $\frac{dy}{y^3}=-2x\,dx$

$$\int\frac{dy}{y^3}dy=\int-2x\,dx$$

$$-\frac{1}{2y^2}=-x^2+C$$

$$f(0)=1:\quad -\frac{1}{2}=0+C,\quad C=-\frac{1}{2}$$

$$-\frac{1}{2y^2}=-x^2-\frac{1}{2}$$

$$\frac{1}{y^2}=2x^2+1$$

$$y^2=\frac{1}{2x^2+1}\Rightarrow y=\pm\sqrt{\frac{1}{2x^2+1}}$$

Since the curve passes through (0,1),

$$y=\sqrt{\frac{1}{2x^2+1}}$$

(3) $y=x$

(4) $2x\frac{dy}{dx}=3y$

$$\int\frac{2dy}{y}=\int\frac{3dx}{x}$$

$$2\ln|y|=3\ln|x|+C$$

$$\ln|y|=\frac{3}{2}\ln|x|+\frac{1}{2}C$$

$$|y|=e^{\frac{3}{2}\ln|x|+\frac{1}{2}C}$$

$$y=Ae^{\frac{3}{2}\ln|x|}=A|x|^{\frac{3}{2}\ln e}=A|x|^{\frac{3}{2}}$$

$$y(1)=100:\quad 100=A\cdot 1$$

$$A=100$$

$$\therefore y=100|x|^{3/2}$$

EX2

(1) $y=e^{\sqrt{x}-2}$

(2) $\frac{dy}{dx}=y^2\cos 2x$

$$\int\frac{dy}{y^2}=\int\cos 2x\,dx$$

$$-\frac{1}{y}=\frac{1}{2}\sin 2x+C$$

$$f\left(\frac{\pi}{4}\right)=-1:\ 1=\frac{1}{2}\sin\left(\frac{\pi}{2}\right)+C,\ C=\frac{1}{2}$$

$$-\frac{1}{y}=\frac{1}{2}\sin 2x+\frac{1}{2}$$

$$\frac{1}{y}=-\frac{1}{2}\sin 2x-\frac{1}{2}$$

$$y=\frac{1}{-\frac{1}{2}\sin 2x-\frac{1}{2}}\qquad\therefore y=\frac{-2}{\sin 2x+1}$$

(3) $\frac{1}{2}\ln^2 y=\ln|x|+\frac{1}{2}$

(4) $\frac{dy}{dx}=\frac{y+2}{x}$

$$\int\frac{dy}{y+2}=\int\frac{dx}{x}$$

$$\ln|y+2|=\ln|x|+C$$

$$|y+2|=e^{\ln|x|+C}$$

$$y+2=Ae^{\ln|x|}$$

$$y+2=A|x|$$

$$y(-1)=3:\ 5=A\cdot 1$$

$$A=5$$

$$y+2=5|x|$$

$$\therefore y=5|x|-2$$

(5) $y=-\ln(-x+2)$

(6) $\int \frac{dy}{\sqrt{y}} = \int \frac{2}{x^2+1}dx$

$2\sqrt{y} = 2\tan^{-1}x + C$

$y(0) = 4 : \ 4 = 2\tan^{-1}(0) + C, \ C = 4$

$2\sqrt{y} = 2\tan^{-1}x + 4$

$\sqrt{y} = \tan^{-1}x + 2$

$y = (\tan^{-1}x + 2)^2$

(7) $e^y dy = \frac{2dx}{\cos^2 x}$

$\int e^y dy = \int 2\sec^2 x\, dx$

$e^y = 2\tan x + C$

$f\left(-\frac{\pi}{4}\right) = 0 : \ e^0 = 2\tan\left(-\frac{\pi}{4}\right) + C$

$1 = -2 + C, \quad C = 3$

$e^y = 2\tan x + 3$

$\therefore y = \ln(2\tan x + 3)$

(8) $x\frac{dy}{dx} = y^2 + 4$

$\int \frac{dy}{y^2+4} = \int \frac{dx}{x}$

$\frac{1}{2}\tan^{-1}\left(\frac{y}{2}\right) = \ln|x| + C$

$f(1) = 0 : \ \frac{1}{2}\tan^{-1}(0) = \ln 1 + C, \ C = 0$

$\tan^{-1}\left(\frac{y}{2}\right) = 2\ln|x|$

$\frac{y}{2} = \tan(2\ln|x|)$

$\therefore y = 2\tan(2\ln|x|)$

EX3

(1) $f(x) = \pm\sqrt{x^2+1}$

(2) $\frac{dy}{dx} = y^2 + 1$

$\int \frac{dy}{y^2+1} = \int dx$

$\tan^{-1}y = x + C$

$f(0) = \sqrt{3} : \ \tan^{-1}(\sqrt{3}) = 0 + C$

$\therefore C = \frac{\pi}{3}$

$\tan^{-1}(y) = x + \frac{\pi}{3}$

$\therefore y = \tan\left(x + \frac{\pi}{3}\right)$

$f\left(\frac{\pi}{2}\right) = \tan\left(\frac{\pi}{2} + \frac{\pi}{3}\right)$

$= \tan\left(\frac{5\pi}{6}\right)$

$= -\frac{1}{\sqrt{3}}$

7.3 Exponential Growth and Decay

EX1

(1) 9

(2) 7.6 hr

EX2

$\frac{dy}{dt} = ky$

$\int \frac{dy}{y} = \int kdt$

$\ln|y| = kt + C$

$\therefore y = Ae^{kt}$

Let the initial amount be y_0.

Then $y = y_0$ when $t = 0$.

$\therefore y = y_0 e^{kt}$

When $t = 45, y = 0.8y_0$.

$0.8y_0 = y_0 e^{45k}$

$0.8 = e^{45k}, \ 45k = \ln(0.8)$

$k = \frac{\ln(0.8)}{45}$

$\therefore y = y_0 e^{\frac{t}{45}\ln(0.8)}$

Find t when $y = 0.5y_0$.

$0.5y_0 = y_0 e^{\frac{t}{45}\ln(0.8)}$

$0.5 = e^{\frac{t}{45}\ln(0.8)}$

$\ln(0.5) = \frac{t}{45}\ln(0.8)$

$t = \frac{45\ln(0.5)}{\ln(0.8)}$

$= 139.783 \ hours$

EX3

$17.33\,yr$

EX4

$$\frac{dy}{dt}=ky$$
$$\int\frac{dy}{y}=\int kdt$$
$$\ln|y|=kt+C$$
$$\therefore y=Ae^{kt}$$

When $t=0,\ y=4.$

$$4=Ae^{0}$$
$$A=4$$
$$\therefore y=4e^{kt}$$

When $t=2, y=10$

$$10=4e^{2k}$$
$$\frac{5}{2}=e^{2k},\ 2k=\ln\left(\frac{5}{2}\right)$$
$$k=\frac{1}{2}\ln\left(\frac{5}{2}\right)$$
$$\therefore y=4e^{\frac{1}{2}\ln\left(\frac{5}{2}\right)t}$$

Find y when $t=5.$

$$y=4e^{\frac{5}{2}\ln\left(\frac{5}{2}\right)}$$
$$=4\left(\frac{5}{2}\right)^{\frac{5}{2}}$$
$$=39.528$$

EX5

This is an example of Case (the temperature is decreasing toward the limiting temperature 20°F).

(1) If R(t) is the temperature of the roast at time t, then

$\dfrac{dR(t)}{dt}=-k[R(t)-20]$ and $R(t)=20+ce^{-kt}$

Since $R(0)=68$°F, we have

$68=20+c,$

$c=48,$

$R(t)=20+48e^{-kt},$

Also, $R(2)=40$°F,

so $40=20+48e^{-k\cdot 2}$

and $e^{-k}=0.65,$

yielding $R(t)=20+48(0.65)^{t}$ (*)

and, finally, $R(5)=20+48(0.65)^{5}=26$°F

(2) Equation (*) in part (1) gives the roast's temperature at time t. We must find t when R=21;

$$21=20+48(0.65)^{t},$$
$$\frac{1}{48}=(0.65)^{t},$$
$$-\ln 48=t\ln(0.65),$$
$$t=9\,hr.$$

EX6

$$\frac{dT}{dt}=-k(T-41)$$
$$\int\frac{dT}{T-41}=\int -kdt$$
$$\ln|T-41|=-kt+C$$
$$\therefore T-41=Ae^{-kt}$$

When $t=0,\ T=77$

$$77-41=Ae^{0}$$
$$A=36$$
$$\therefore T-41=36e^{-kt}$$

When $t=10,\ T=68$

$$27=36e^{-10k}$$
$$\frac{3}{4}=e^{-10k},\ -10k=\ln\left(\frac{3}{4}\right)$$
$$k=-\frac{1}{10}\ln\left(\frac{3}{4}\right)$$
$$\therefore T-41=36e^{\frac{t}{10}\ln\left(\frac{3}{4}\right)}$$

Find t when $T=45$

$$4=36e^{\frac{t}{10}\ln\left(\frac{3}{4}\right)}$$
$$\frac{1}{9}=e^{\frac{t}{10}\ln\left(\frac{3}{4}\right)}$$
$$\ln\left(\frac{1}{9}\right)=\frac{t}{10}\ln\left(\frac{3}{4}\right)$$
$$t=\frac{10\ln(1/9)}{\ln(3/4)}$$
$$=76.337\text{ min}$$

EX7

Let N(t) be the nimber of people aware of the product at time t.

$N'(t)=k[15{,}000-N(t)]$

The solution of this d.e is

$N(t)=15{,}000-ce^{-kt}$

Since $N(0)=0, c=15{,}000$ and

$$N(t)=15{,}000(1-e^{-kt}).$$

when $t=6,\ N=1500$

$$1500=15{,}000(1-e^{-6k}),$$
$$e^{-6k}=0.9,$$
$$k=\frac{\ln 0.9}{-6}=0.018.$$

Find when $N=2700$

$$2700=15{,}000(1-e^{-0.0018t})$$
$$0.18=1-e^{-0.018t},$$
$$e^{-0.018t}=0.82,$$
$$t\simeq 11\,days.$$

EX8

$$\frac{dT}{dt}=k(250-T)$$
$$\frac{dT}{dt}=-k(T-250)$$
$$\int\frac{dt}{T-250}=\int -kdt$$
$$\ln|T-250| = -kt+C$$
$$\therefore T-250=Ae^{-kt}$$

When $t=0, T=-5$.

$$-5-250=Ae^{0}$$
$$A=-255$$
$$\therefore T-250=-255e^{-kt}$$

When $t=7, T=100$

$$-150=-255e^{-7k}$$
$$\frac{10}{17}=e^{-7k},\ -7k=\ln\left(\frac{10}{17}\right)$$
$$k=-\frac{1}{7}\ln\left(\frac{10}{17}\right)$$
$$\therefore\ T-250=-255e^{\frac{t}{7}\ln\left(\frac{10}{17}\right)}$$

Find t when $T=200$.

$$-50=-255e^{\frac{t}{7}\ln\left(\frac{10}{17}\right)}$$
$$\frac{10}{51}=e^{\frac{t}{7}\ln\left(\frac{10}{17}\right)}$$
$$\ln\left(\frac{10}{51}\right)=\frac{t}{7}\ln\left(\frac{10}{17}\right)$$
$$t=\frac{7\ln(10/51)}{\ln(10/17)}$$
$$=21.493\text{ min}$$

EX9

$$\frac{dT}{dt}=-k(T-20)$$
$$\int\frac{dT}{T-20}=\int -kdt$$
$$\ln|T-20|=-kt+C$$
$$\therefore T-20=Ae^{-kt}$$

When $t=0,\ T=25$.

$$25-20=Ae^{0}$$
$$A=5$$
$$\therefore T-20=5e^{-kt}$$

When $t=2,\ T=23$.

$$3=5e^{-2k}$$
$$\frac{3}{5}=e^{-2k},\quad -2k=\ln\left(\frac{3}{5}\right)$$
$$k=-\frac{1}{2}\ln\left(\frac{3}{5}\right)$$
$$\therefore T-20=5e^{\frac{t}{2}\ln\left(\frac{3}{5}\right)}$$

Find t when $T=37$.

$$17=5e^{\frac{t}{2}\ln\left(\frac{3}{5}\right)}$$
$$\frac{17}{5}=e^{\frac{t}{2}\ln\left(\frac{3}{5}\right)}$$
$$\ln\left(\frac{17}{5}\right)=\frac{t}{2}\ln\left(\frac{3}{5}\right)$$
$$t=\frac{2\ln(17/5)}{\ln(3/5)}$$
$$=-4.791\ hours$$
$$=4hrs\ 47\text{min}\ ago$$

$\therefore$ The murder occurred at 7:13 PM.

7.4 Euler's Method

EX1

(1) $y_1=y_0+\left(1-\dfrac{y_0}{x_0}\right)dx$

$$=-1+\left(1-\frac{-1}{2}\right)(.5)=-0.25,$$

$$y_2=y_1+\left(1-\frac{y_1}{x_1}\right)dx$$
$$=-0.25+\left(1-\frac{-0.25}{2.5}\right)(.5)=0.3,$$

$$y_3 = y_2 + \left(1 - \frac{y_2}{x_2}\right)dx$$
$$= 0.3 + \left(1 - \frac{0.3}{3}\right)(.5) = 0.75;$$

(2) $y_1 = y_0 + x_0(1 - y_0)dx$
$= 0 + 1(1-0)(.2) = .2,$
$y_2 = y_1 + x_1(1 - y_1)dx$
$= .2 + 1.2(1 - .2)(.2) = .392,$
$y_3 = y_2 + x_2(1 - y_2)dx$
$= .392 + 1.4(1 - .392)(.2) = .5622;$

(3) $y_1 = y_0 + (2x_0y_0 + 2y_0)dx$
$= 3 + [2(0)(3) + 2(3)](.2) = 4.2,$
$y_2 = y_1 + (2x_1y_1 + 2y_1)dx$
$= 4.2 + [2(.2)(4.2) + 2(4.2)](.2) = 6.216,$
$y_3 = y_2 + (2x_2y_2 + 2y_2)dx$
$= 6.216 + [2(.4)(6.216) + 2(6.216)](.2)$
$= 9.6969;$

(4) $y_1 = y_0 + y_0^2(1 + 2x_0)dx$
$= 1 + 1^2[1 + 2(-1)](.5) = .5,$
$y_2 = y_1 + y_1^2(1 + 2x_1)dx$
$= .5 + (.5)^2[1 + 2(-.5)](.5) = .5,$
$y_3 = y_2 + y_2^2(1 + 2x_2)dx$
$= .5 + (.5)^2[1 + 2(0)](.5) = .625;$

EX2

(1) $\lim_{x\to 0}(f(x) + 1) = -1 + 1$
$= 0$ and $\lim_{x\to 0} \sin x = 0$

Using L' Hospital's Rule,

$$\lim_{x\to 0}\frac{f(x)+1}{\sin x} = \lim_{x\to 0}\frac{f'(x)}{\cos x} = \frac{f'(0)}{\cos 0}$$
$$= \frac{(-1)^2 \cdot 2}{1} = 2$$

(2) $f\left(\frac{1}{4}\right) \approx f(0) + f'(0)\left(\frac{1}{4}\right)$
$= -1 + (2)\left(\frac{1}{4}\right) = -\frac{1}{2}$

$f\left(\frac{1}{2}\right) \approx f\left(\frac{1}{4}\right) + f'\left(\frac{1}{4}\right)\left(\frac{1}{4}\right)$
$= -\frac{1}{2} + \left(-\frac{1}{2}\right)^2\left(2 \cdot \frac{1}{4} + 2\right)\left(\frac{1}{2}\right)$
$= -\frac{11}{32}$

(3) $\frac{dy}{dx} = y^2(2x+2)$

$\frac{dy}{y^2} = (2x+2)dx$

$\int \frac{dy}{y^2} = \int (2x+2)dx$

$-\frac{1}{y} = x^2 + 2x + C$

$-\frac{1}{-1} = 0^2 + 2 \cdot 0 + C \Rightarrow C = 1$

$-\frac{1}{y} = x^2 + 2x + 1$

$y = -\frac{1}{x^2+2x+1} = -\frac{1}{(x+1)^2}$

EX3

(1) $f\left(\frac{1}{2}\right) = f(1) + \left(\frac{dy}{dx}\Big|_{(1,0)}\right) \cdot \triangle x$
$= 0 + 1 \cdot \left(-\frac{1}{2}\right) = -\frac{1}{2}$

$f(0) = f\left(\frac{1}{2}\right) + \left(\frac{dy}{dx}\Big|_{\left(\frac{1}{2}, -\frac{1}{2}\right)}\right) \cdot \triangle x$
$= -\frac{1}{2} + \frac{3}{2} \cdot \left(-\frac{1}{2}\right) = -\frac{5}{4}$

(2) Since f is differentiable at $x = 1$, f is continuous at $x = 1$, So,

$\lim_{x\to 1} f(x) = 0 = \lim_{x\to 1}(x^3 - 1)$ and we may apply L' Hospital's Rule.

$$\lim_{x\to 1}\frac{f(x)}{x^3-1} = \lim_{x\to 1}\frac{f'(x)}{3x^2} = \frac{\lim_{x\to 1} f'(x)}{\lim_{x\to 1} 3x^2} = \frac{1}{3}$$

(3) $\frac{dy}{dx} = 1 - y$

$\int \frac{1}{1-y}dy = \int 1dx$

$-\ln|1-y| = x + C$

$-\ln 1 = 1 + C \Rightarrow -1$

$\ln|1-y| = 1 - x$

$|1-y| = e^{1-x}$

$f(x) = 1 - e^{1-x}$

EX4

(1) $\frac{d^2y}{dx^2}=3+2\frac{dy}{dx}=3+2(3x+2y+1)$
$=6x+4y+5$

(2) if $y=mx+b+e^{rx}$ is a solution, then

$m+re^{rx}=3x+2(mx+b+e^{rx})+1.$

if $r\neq 0:\ m=2b+1,\ r=2,\ 0=3+2m,$

$so\ m=-\frac{3}{2},\ r=2,$ and $b=-\frac{5}{4}$

OR

if $r=0:\ m=2b+3,\ r=0,\ 0=3+2m,$

$so\ m=-\frac{3}{2},\ r=0,\ b=-\frac{9}{4}.$

(3) $f\left(\frac{1}{2}\right)\approx f(0)+f'(0)\cdot\frac{1}{2}$

$=-2+(-3)\cdot\frac{1}{2}=-\frac{7}{2}$

$f'\left(\frac{1}{2}\right)\approx 3\left(\frac{1}{2}\right)+2\left(-\frac{7}{2}\right)+1=-\frac{9}{2}$

$f(1)\approx f\left(\frac{1}{2}\right)+f'\left(\frac{1}{2}\right)\cdot\frac{1}{2}$

$=-\frac{7}{2}+\left(-\frac{9}{2}\right)\cdot\frac{1}{2}=-\frac{23}{4}$

(4) $g'(0)=3\cdot 0+2\cdot k+1=2k+1$

$g(1)\approx g(0)+g'(0)\cdot 1$

$=k+(2k+1)=3k+1=0$

$k=-\frac{1}{3}$

7.5 Logistic Growth

EX1

(1) $k=0.0015$, and $A=150$, and

$y=\frac{150}{1+ce^{-0.225t}}$

initial condition :

$y(0)=6,\ \Rightarrow 6=\frac{150}{1+c}\Rightarrow 1+c=25\Rightarrow c=24$

Formula : $y=\frac{150}{1+24e^{-0.225t}}$

(2) $100=\frac{150}{1+24e^{-0.225t}}\Rightarrow 1+24e^{-0.225t}$

$=\frac{3}{2}\Rightarrow 24e^{-0.225t}=\frac{1}{2}\Rightarrow e^{-0.225t}$

$=\frac{1}{48}\Rightarrow -0.225t=-\ln 48$

$\Rightarrow t=\frac{\ln 48}{0.225}\approx 17.21\,weeks$

$125=\frac{150}{1+24e^{-0.225t}}\Rightarrow 1+24e^{-0.225t}$

$=\frac{6}{5}\Rightarrow 24e^{-0.225t}=\frac{1}{5}\Rightarrow e^{-0.225t}$

$=\frac{1}{120}\Rightarrow -0.225t=-\ln 120$

$\Rightarrow t=\frac{\ln 120}{0.225}\approx 21.28$

It will take about 17.21 weeks to reach 100 guppies, and about 21.28 weeks to reach 125 guppies.

EX2

(1)

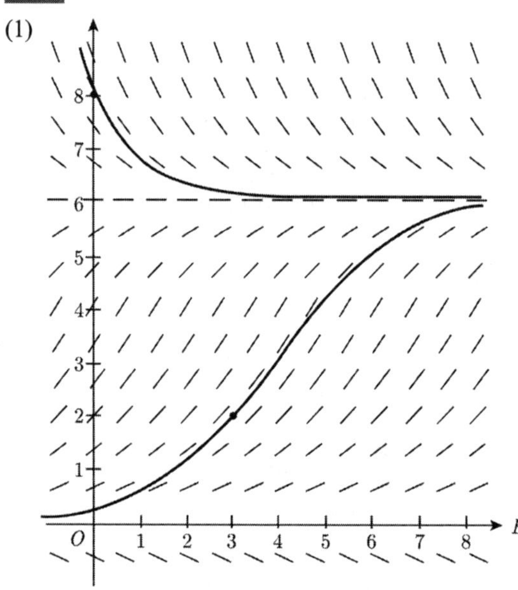

(2) $f\left(\frac{1}{2}\right)\approx 8+(-2)\left(\frac{1}{2}\right)=7$

$f(1)\approx 7+\left(-\frac{7}{8}\right)\left(\frac{1}{2}\right)=\frac{105}{16}$

(3) $\frac{d^2y}{dt^2} = \frac{1}{8}\frac{dy}{dt}(6-y) + \frac{y}{8}\left(-\frac{dy}{dt}\right)$

$\left. f(0) = 8;\ f'(0) = \frac{dy}{dt} \right\}$

The second-degree Taylor polynomial for f about

$t = 0$ is $P_2(t) = 8 - 2t + \frac{5}{4}t^2$.

$$f(1) \approx P_2(1) = \frac{29}{4}$$

(4) The range of f for $t \geq 0$ is $6 < y \leq 8$.

EX3

(1) We are told that $N'(t) = k \cdot N \cdot (50{,}000 - N)$, that $N(0) = 100$, and that $N(1) = 130$.

The d.e. describing logistic growth leads to

$$N(t) = \frac{50{,}000}{1 + ce^{-50{,}000kt}}.$$

From $N(0) = 100$, *we get*

$$100 = \frac{50{,}000}{1+c},$$

which yields $c = 499$, From $N(1) = 130$, we get

$$130 = \frac{50{,}000}{1 + 499e^{-50{,}000k}},$$

$$130(1 + 499\,e^{-50{,}000k}) = 50{,}000,$$

$$e^{-50{,}000k} = 0.77.$$

Then $$N(t) = \frac{50{,}000}{1 + 499(0.77)^t}.$$

(2) We must find $N(8)$, Since $t = 0$ represents yesterday:

$$N(8) = \frac{50{,}000}{1 + 499(0.77)^8} = 798\,people.$$

Multiple Choice

01 C 02 C 03 C 04 C 05 D 06 C 07 E 08 A 09 C 10 C 11 E 12 A 13 C 14 B 15 B
16 C 17 D 18 D 19 C 20 C 21 E 22 D 23 B 24 A 25 C

Index

저자 소개

Mansik Min
Department of AI and Big Data, Suwon University
Ph.D. in Statistics, Korea University

Sangmin Hwang
Department of AI and Big Data, Suwon University
Ph.D. Coursework Compleleted in mathematics, Sungkyunkwan University

Jungbin Yoon
St.Johnsbury Academy Jeju

Sangwoo Jeon
Department of AI and Big Data, Suwon University
Ph.D. in Mathematics, Korea University

Single Variable Calculus

저　　자 Mansik Min · Sangmin Hwang · Jungbin Yoon · Sangwoo Jeon
발 행 인 김지영
발 행 처 자유아카데미
주　　소 경기도 파주시 회동길 37-42
　　　　 파주출판도시
전　　화 031-955-1321
팩　　스 031-955-1322
홈페이지 www.freeaca.com
전자우편 main@freeaca.com(대표)
　　　　 editor@freeaca.com(편집)
　　　　 crm@freeaca.com(영업)
등　　록 제406-2003-017호, 1980. 7. 12
제1판1쇄 2026년 1월 25일 발행

저자와의 협의하에 인지생략

ISBN 979-11-5808-803-3 93410